AF880657

The Asymptotic Distribution of Eigenvalues of Partial Differential Operators

TRANSLATIONS OF
MATHEMATICAL
MONOGRAPHS
VOLUME 155

Yu. Safarov
D. Vassiliev

The Asymptotic Distribution of Eigenvalues of Partial Differential Operators

American Mathematical Society

Д. ВАСИЛЬЕВ И Ю. САФАРОВ

АСИМПТОТИЧЕСКОЕ РАСПРЕДЕЛЕНИЕ
СОБСТВЕННЫХ ЗНАЧЕНИЙ
ДИФФЕРЕНЦИАЛЬНЫХ ОПЕРАТОРОВ
В ЧАСТНЫХ ПРОИЗВОДНЫХ

Translated by the authors from an original Russian manuscript

EDITORIAL COMMITTEE

AMS Subcommittee
Robert D. MacPherson
Grigorii A. Margulis
James D. Stasheff (Chair)
ASL Subcommittee Steffen Lempp (Chair)
IMS Subcommittee Mark I. Freidlin (Chair)

1991 *Mathematics Subject Classification.* Primary 35P20, 58G25;
Secondary 35J55, 73J06, 73L20.

ABSTRACT. The book studies the eigenvalues of elliptic linear boundary value problems. The principal result of the book is a collection of asymptotic formulae which describe the distribution of eigenvalues with high sequential numbers. The use of these asymptotic formulae is illustrated on standard eigenvalue problems of mechanics and mathematical physics. The book is intended for pure as well as applied mathematicians specialising in partial differential equations. The book is mostly self-contained and provides a basic introduction to all the necessary mathematical concepts and tools, such as microlocal analysis, billiards, symplectic geometry and Tauberian theorems.

Library of Congress Cataloging-in-Publication Data

Safarov, Yu.
The asymptotic distribution of eigenvalues of partial differential operators / Yu. Safarov, D. Vassiliev.
p. cm. — (Translations of mathematical monographs, ISSN 0065-9282 ; v. 155)
Translated by the authors from an unpublished Russian manuscript.
Includes bibliographical references (p. –) and indexes.
ISBN 0-8218-4577-2 (acid-free paper)
1. Partial differential operators. 2. Eigenvalues. 3. Asymptotic distribution (Probability theory) 4. Theory of distributions (Functional analysis) I. Vassiliev, D. II. Title. III. Series.
QA329.42.S24 1996
515′.7242—dc20

96-32942
CIP

Copying and reprinting. Individual readers of this publication, and nonprofit libraries acting for them, are permitted to make fair use of the material, such as to copy a chapter for use in teaching or research. Permission is granted to quote brief passages from this publication in reviews, provided the customary acknowledgment of the source is given.

Republication, systematic copying, or multiple reproduction of any material in this publication (including abstracts) is permitted only under license from the American Mathematical Society. Requests for such permission should be addressed to the Assistant to the Publisher, American Mathematical Society, P. O. Box 6248, Providence, Rhode Island 02940-6248. Requests can also be made by e-mail to reprint-permission@ams.org.

© 1997 by the American Mathematical Society. All rights reserved.
Reprint with corrections 1998.
Translation authorized by the
All-Union Agency for Author's Rights, Moscow.
The American Mathematical Society retains all rights
except those granted to the United States Government.
Printed in the United States of America.

∞ The paper used in this book is acid-free and falls within the guidelines established to ensure permanence and durability.

10 9 8 7 6 5 4 3 2   02 01 00 99 98

1029 6094
THE NOTTINGHAM TRENT
UNIVERSITY LIS
BOOK 517.927 SAF

DDC 10296094
UNIVERSI
515·7242 SAF

Contents

Preface xi

Chapter 1. Main Results 1
1.1. Statement of the spectral problem 1
1.2. One-term asymptotic formula for $N(\lambda)$ 9
1.3. Hamiltonian billiards I: basic definitions and results 11
1.4. Hamiltonian billiards II: reflection matrix 27
1.5. Hamiltonian billiards III: Maslov index 32
1.6. Classical two-term asymptotic formula for $N(\lambda)$ 37
1.7. Nonclassical two-term asymptotic formulae for $N(\lambda)$ 49
1.8. Two-term asymptotic formulae for the spectral function 56

Chapter 2. Oscillatory Integrals 63
2.1. Local oscillatory integrals and pseudodifferential operators 63
2.2. Global oscillatory integrals 70
2.3. Homogeneous canonical transformations 74
2.4. Phase functions associated with homogeneous canonical transformations 78
2.5. Restriction of phase functions to the boundary 89
2.6. Extension of phase functions from the boundary 91
2.7. Standard oscillatory integrals associated with homogeneous canonical transformations 100
2.8. Boundary layer oscillatory integrals associated with homogeneous canonical transformations 112
2.9. Boundary oscillatory integrals associated with homogeneous canonical transformations 115
2.10. Parameter-dependent oscillatory integrals 119

Chapter 3. Construction of the Wave Group 129
3.1. Characteristic properties of distributions associated with the wave group 130
3.2. Representation of the wave group by means of oscillatory integrals: sufficient conditions 140
3.3. Representation of the wave group by means of oscillatory integrals: effective construction for manifolds without boundary 142
3.4. Representation of the wave group by means of oscillatory integrals: effective construction for manifolds with boundary 148
3.5. Construction of the wave group when the source is close to the boundary 161

Chapter 4. Singularities of the Wave Group 167
4.1. Singularities of Lagrangian distributions 168
4.2. Singularity of the wave group at $t = 0$ 176
4.3. Singularities of the wave group at $t \neq 0$ for admissible pseudodifferential cut-offs 179
4.4. Singularities of the wave group at $t \neq 0$ for nonadmissible pseudodifferential cut-offs 183
4.5. Singularity of the wave group at $t = 0$ when the source is close to the boundary 190

Chapter 5. Proof of Main Results 193
5.1. Partition of the manifold M into three zones 193
5.2. Asymptotics of the trace of the spectral projection in the interior zone 198
5.3. Asymptotics of the trace of the spectral projection in the intermediate zone 201
5.4. Asymptotics of the trace of the spectral projection in the boundary zone 205
5.5. Asymptotics of the spectral function 224

Chapter 6. Mechanical Applications 229
6.1. Membranes and acoustic resonators 229
6.2. Elastic plates 230
6.3. Two- and three-dimensional elasticity 236
6.4. Elastic shells 239
6.5. Hydroelasticity 247

Appendix A. Spectral Problem on the Half-line by A. Holst 251
A.1. Basic facts 251
A.2. The reflection matrix 263
A.3. Trace formulae 276
A.4. Dependence on parameters 292

Appendix B. Fourier Tauberian Theorems by M. Levitin 297
B.1. Introductory remarks 297
B.2. Basic theorem 298
B.3. Rough estimate for the nonzero singularities 300
B.4. General refined theorem 301
B.5. Special version of the general refined theorem 304

Appendix C. Stationary Phase Formula 307

Appendix D. Hamiltonian Billiards: Proofs 313
D.1. Measure of "awkward" starting points 313
D.2. Dead-end trajectories 318
D.3. Convexity and concavity 321
D.4. Measurability of sets and functions 324
D.5. Lengths of loops and periodic trajectories 326
D.6. Maslov index 328

Appendix E. Factorization of Smooth Functions and Taylor-type Formulae 335

References 343

Principal Notation 349

Index 353

Preface

Spectral asymptotics for partial differential operators have been the subject of extensive research for over a century. It has attracted the attention of many outstanding mathematicians and physicists.

As a characteristic example let us consider the following spectral problem:

$$(0.0.1) \qquad -\Delta v = \lambda^2 v \quad \text{in} \quad M\,, \qquad v|_{\partial M} = 0\,,$$

where M is a bounded domain in $\mathbb{R}^3$, and Δ is the Laplace operator. The problem (0.0.1) has nontrivial solutions v only for a discrete set of $\lambda = \lambda_k$, which are called eigenvalues. Let us enumerate the eigenvalues in increasing order: $0 < \lambda_1 < \lambda_2 \leqslant \lambda_3 \leqslant \cdots$. In the general case the eigenvalues λ_k cannot be evaluated explicitly. Moreover, for large k it is difficult to evaluate them numerically. So it is natural to look for asymptotic formulae for λ_k as $k \to \infty$.

However, for a number of reasons it is traditional in such problems to deal with the matter the other way round, i.e., to study the sequential number k as a function of λ. Namely, let us introduce the counting function $N(\lambda)$ defined as the number of eigenvalues λ_k less than a given λ. Then our asymptotic problem is reformulated as the study of the asymptotic behaviour of $N(\lambda)$ as $\lambda \to +\infty$. The derivation of asymptotic formulae for $N(\lambda)$ is the main subject of this book.

It is well known that for the problem (0.0.1)

$$(0.0.2) \qquad N(\lambda) = \frac{V}{6\pi^2}\,\lambda^3 + o(\lambda^3)\,, \qquad \lambda \to +\infty\,,$$

where V is the volume of M. The asymptotic formula (0.0.2) has been known for a long time; it appeared already in the works of Rayleigh. Written in a slightly different form it is known in theoretical physics as the Rayleigh–Jeans law.

Rayleigh [**Ra**] arrived at (0.0.2) by considering the case when the domain M is a cube of side a. Then, solving the problem (0.0.1) by separation of variables one obtains

$$N(\lambda) = \#\{\vec{q} \in \mathbb{N}^3 : |\vec{q}| < R\}\,,$$

where $R = a\lambda/\pi$. In other words, $N(\lambda)$ is the number of integer lattice points in an octant of a ball of radius R. Clearly, for large R we have

$$N(\lambda) \approx \frac{1}{8}\,(4\pi R^3/3) = \frac{a^3}{6\pi^2}\,\lambda^3 = \frac{V}{6\pi^2}\,\lambda^3\,.$$

Now physical arguments suggest that the same formula should hold for a domain of arbitrary shape.

Formula (0.0.2) is remarkable not only for its role in the development of theoretical physics, but also for the fact that Rayleigh made a mistake by writing it without the coefficient $1/8$. This mistake was corrected by J. H. Jeans. As pointed out in [**Ja**], Jeans's only contribution to the "Rayleigh–Jeans" law was the statement: "It seems to me that Lord Rayleigh has introduced an unnecessary factor 8 by counting negative as well as positive values of his integers," [**Je,** p. 98].

The first rigorous proof of (0.0.2) was given by H. Weyl [**We1**]. Later, R. Courant and D. Hilbert included a proof of (0.0.2) in their classical textbook [**CouHilb**], which stimulated the study of asymptotic formulae of this type. The list of mathematicians who have contributed to this field includes S. Agmon, V. M. Babich, P. H. Bérard, M. S. Birman, T. Carleman, Y. Colin de Verdière, J. Duistermaat, B. V. Fedosov, L. Gårding, V. W. Guillemin, L. Hörmander, V. Ya. Ivrii, M. Kac, B. M. Levitan, R. B. Melrose, G. Métivier, Å. Pleijel, R. T. Seeley, M. A. Shubin, M. Z. Solomyak, A. Weinstein, and many others. An extensive bibliographical review can be found in [**RoSoSh**]. Physicists also worked on spectral asymptotics and have made essential contributions. Being less familiar with the physical literature we shall only mention the names of M. V. Berry, P. Debye, and L. Onsager; see also [**BaHilf**] for further references.

The asymptotic formula (0.0.2) is remarkably simple: the asymptotic coefficient is determined only by the volume of the domain and is independent of its shape. Moreover, a similar one-term asymptotic formula has been established in a very general setting, namely, for an elliptic self-adjoint partial differential operator with variable coefficients acting on a manifold subject to reasonably good boundary conditions.

However, this simplicity and high degree of generality indicate the weaknesses of (0.0.2) and its analogues. First, such formulae involve only the most basic geometric characteristics of M: say, the eigenvalues of the problem (0.0.1) for a cube and a long narrow parallelepiped of the same volume are obviously quite different, but (0.0.2) does not feel this difference. Second, one-term asymptotic formulae do not depend on the boundary conditions: say, if we replace in (0.0.1) the Dirichlet boundary condition by the Neumann one the eigenvalues will change substantially, which cannot be noticed from (0.0.2). These deficiencies motivated the search for sharper results.

In 1913 H. Weyl put forward [**We2**] a conjecture concerning the existence of a second asymptotic term. Namely, he predicted that for the problem (0.0.1)

$$N(\lambda) = \frac{V}{6\pi^2}\lambda^3 - \frac{S}{16\pi}\lambda^2 + o(\lambda^2), \qquad \lambda \to +\infty, \tag{0.0.3}$$

where S is the surface area of ∂M. Formula (0.0.3) became known as *Weyl's conjecture*. It was finally justified, under a certain condition on periodic billiard trajectories, by Ivrii [**Iv1**] and Melrose [**Me**] only in 1980. This revived interest to such problems. In particular, in subsequent years Ivrii extended his result on two-term asymptotics to much more general classes of boundary value problems. As our book does not aim to provide a full bibliographic review and reflects the research interests of its authors, we refer only to Ivrii's publications [**Iv2**]–[**Iv4**] where the reader can find further references.

Our contribution to the problem concerns the following aspects.

First, we are interested in deriving two-term asymptotic formulae for higher order differential operators.

Second, we study the case when the condition on periodic billiard trajectories, which guarantees the existence of a classical second term in Weyl's formula, fails.

In this case the second asymptotic term may contain an oscillating function, which depends on the structure of the set of periodic billiard trajectories.

Third, we obtain two-term asymptotic formulae for the spectral function. In this case one has to deal with loops instead of periodic billiard trajectories.

The basic idea which we use for the derivation of spectral asymptotics is due to Levitan [**Ltan**]. It involves the study of the singularities of the corresponding evolutionary problem (say, in the case of (0.0.1) this would be the wave equation), and the subsequent application of Fourier Tauberian theorems. This approach produces the sharpest possible results. Levitan's method was developed by Hörmander, Duistermaat, Guillemin, and Melrose (see [**Hö1**], [**DuiGui**], [**DuiGuiHö**], [**Me**]). The most advanced version of this method is due to Ivrii [**Iv1**]–[**Iv4**]. Our approach, however, is somewhat different from that of Ivrii, even in the case of the Laplace operator.

We tried to make the book self-contained and all our constructions explicit. The main results are collected in Chapter 1. Chapter 2 introduces the reader to the main technical tools; it can be regarded as a brief introduction to microlocal analysis. Chapters 3–5 are devoted to the proofs of our main results. Chapter 6 lists the basic mechanical applications; it is intended mostly for applied mathematicians and does not require a sophisticated mathematical background. The book also has a number of appendices. Some of them can be read separately from the main text, others contain cumbersome proofs. Appendix A was written by A. Holst, and Appendix B by M. Levitin.

We do not aim at achieving the highest possible degree of generality in our book. In particular, we do not discuss

1. Systems; see [**Sa4**], [**Sa5**], [**SaVa1**], [**Va4**], [**Va6**].
2. Piecewise smooth boundaries; see [**Va6**].
3. Highly nonsmooth (fractal) boundaries; see [**FlLtinVa1**], [**FlLtinVa2**], [**FlVa1**], [**FlVa2**], [**LtinVa1**], [**LtinVa2**], [**Va10**].

This book was preceded by survey papers [**GolVa**], [**Sa7**], [**SaVa2**] describing our main results.

We take this opportunity to express our gratitude to our teachers, V. B. Lidskiĭ and M. Z. Solomyak, for guiding us through our first steps in modern analysis and introducing us to the spectral theory of partial differential operators. We would also like to thank our colleagues S. Agmon, E. B. Davies, A. Holst, M. Levitin, Yu. Netrusov, L. Parnovski, A. V. Sobolev, and T. Weidl for their help and useful comments during the preparation of this manuscript. We thank our graduate students W. Nicoll and A. Roth for providing technical support. Last, but not least, we thank S. I. Gelfand for his patience and understanding in waiting all these years for our manuscript.

The first author was supported by the Royal Society and the Engineering and Physical Sciences Research Council (grant B/93/AF/1559), and the second author by the Nuffield Foundation.

CHAPTER 1

Main Results

1.1. Statement of the spectral problem

The principal objective of this book is the study of the spectral problem

$$Av = \lambda^{2m} v, \tag{1.1.1}$$

$$\left(B^{(j)} v\right)\Big|_{\partial M} = 0, \qquad j = 1, 2, \dots, m, \tag{1.1.2}$$

where $\lambda > 0$ is the spectral parameter and A is a positive definite self-adjoint elliptic linear differential operator of order $2m$ ($m \in \mathbb{N}$) acting on a compact n-dimensional ($n \geqslant 2$) manifold M with boundary ∂M or without boundary ($\partial M = \varnothing$). The $B^{(j)}$ are "boundary" linear differential operators describing boundary conditions in the case $\partial M \neq \varnothing$.

In this section we discuss the problem (1.1.1), (1.1.2) and introduce basic notation.

1. Spectral parameter. First of all we must explain the appearance of the $2m$th power on the right-hand side of (1.1.1). It seems more natural to write

$$Av = \nu v \tag{1.1.1$'$}$$

with ν as spectral parameter, but this leads to fractional powers of ν in the resulting spectral asymptotics. So for the sake of convenience from the very beginning we have made a change of spectral parameter

$$\nu = \lambda^{2m}. \tag{1.1.3}$$

There is a discrete set of $\lambda = \lambda_k$, $0 < \lambda_1 \leqslant \lambda_2 \leqslant \lambda_3 \leqslant \cdots \leqslant \lambda_k \leqslant \cdots$, $\lambda_k \to \infty$, called *eigenvalues*, for which the problem (1.1.1), (1.1.2) has nontrivial solutions v. The number of linearly independent solutions v corresponding to a given λ_k is called the *multiplicity* of λ_k. We always enumerate our eigenvalues taking into account their multiplicities.

2. Coordinates. The compact manifold M and its $(n-1)$-dimensional boundary ∂M are assumed to be infinitely smooth. Without loss of generality M is assumed to be connected (otherwise the problem (1.1.1), (1.1.2) decomposes into separate subproblems). By $\overset{\circ}{M} := M \setminus \partial M$ we denote the interior of M.

By T^*M, $T^*\overset{\circ}{M}$, $T^*\partial M$ we denote the respective cotangent bundles, and by $T'M$, $T'\overset{\circ}{M}$, $T'\partial M$ the cotangent bundles T^*M, $T^*\overset{\circ}{M}$, $T^*\partial M$ with the zero section ($\xi = 0$ or $\xi' = 0$, see below) excluded.

Local coordinates (and points) on M, ∂M are denoted by $x = (x_1, x_2, \dots, x_n)$, $x' = (x_1, x_2, \dots, x_{n-1})$ respectively, and their dual coordinates on the fibres $T^*_x M$,

$T_x^*\partial M$ by $\xi = (\xi_1, \xi_2, \dots, \xi_n)$, $\xi' = (\xi_1, \xi_2, \dots, \xi_{n-1})$. Thus, (x, ξ), (x', ξ') are local coordinates on (or points of) the cotangent bundles T^*M, $T^*\partial M$.

Near ∂M we always use special local coordinates $x = (x', x_n)$ such that $\partial M = \{x_n = 0\}$, and $x_n > 0$ for points in $\overset{\circ}{M}$; consequently $\xi = (\xi', \xi_n)$. We call x' the *boundary coordinates* (because when $x_n = 0$ the x' are indeed local coordinates on ∂M), ξ' the *coboundary coordinates*, x_n the *"normal" coordinate*, and ξ_n the *conormal coordinate* (note that the covector $\xi = (0, \xi_n)$ is conormal to ∂M).

Moreover, near ∂M we use only coordinates for which x_n is the same on the intersection of different coordinate patches. We fix once and for all the choice of the coordinate x_n near ∂M. One of the possible ways of defining invariantly such a coordinate x_n is with the help of Hamiltonian trajectories introduced in Section 1.3: trajectories originating from points (y, η) with $y \in \partial M$, $\eta' = 0$, can be used for this purpose, with the value of x_n being the time it takes for a trajectory to reach the point $x \in M$. Here $y' \in \partial M$ has to be chosen in such a way that the ray goes through x; this condition defines y' uniquely for x sufficiently close to ∂M and sufficiently small t.

In spectral problems of the type (1.1.1), (1.1.2) arising from mechanical applications there is always an obvious choice of the coordinate x_n near ∂M, which is invariant and convenient. Indeed, such problems are always associated with some natural Riemannian metric, not necessarily directly related to the differential operator A. This metric is either the Euclidean $\mathbb{R}^n$-metric (when $M \subset \mathbb{R}^n$), or an n-dimensional Riemannian metric generated by the Euclidean metric of the embedding in $(n+1)$-dimensional space (when M is an n-dimensional surface in $\mathbb{R}^{n+1}$). Then the standard definition of the "normal" coordinate is $x_n := \operatorname{dist}(x, \partial M)$, where $\operatorname{dist}(\cdot, \cdot)$ is the geodesic distance.

If the Hamiltonian is associated with the given Riemannian metric (see formula (1.1.14) and Example 1.2.4 below), then the two definitions of the coordinate x_n given in the two previous paragraphs coincide.

Throughout the book we shall often speak of open sets in $\overset{\circ}{M}$ and in M (in fact we have already used the notion of an open set in M describing local coordinates in a neighbourhood of ∂M). This needs some explanation as the notions of an open set in $\overset{\circ}{M}$ and in M are different. Local coordinates establish a local diffeomorphism between points of $\overset{\circ}{M}$, M and of $\mathbb{R}^n$, $\mathbb{R}^n_+$, respectively. Consequently, the Euclidean notion of convergence in $\mathbb{R}^n$, $\mathbb{R}^n_+$ (considered as metric spaces) generates the notion of convergence in $\overset{\circ}{M}$, M with the subsequent standard definition of an open set. Note that under this definition a set which is open in M may contain points from ∂M.

An important example of an open set in M is $M_\delta := \{x \in M : x_n < \delta\}$, with $\delta > 0$ sufficiently small. The set M_δ is called the *tubular neighbourhood* of ∂M. This name is due to the fact that M_δ is diffeomorphic to the cylinder $\partial M \times [0, \delta)$. Note, however, that we do not fix this diffeomorphism, because near ∂M we allow changes of coordinates of the type $\widetilde{x}' = \widetilde{x}'(x)$, $\widetilde{x}_n = x_n$, and consequently in the general case $\widetilde{x}'$ depends on x_n.

Sometimes it will be technically convenient for us to go outside the original manifold M. So by $\widehat{M}$ we shall denote some compact manifold without boundary which is a smooth extension of M, i.e., $M \subset \widehat{M}$; the existence of such an extension is established by a standard doubling argument (make two copies of M and smoothly connect their boundaries by cylinders). By $\widehat{A}$ we shall denote a differential operator on $\widehat{M}$ which is a smooth extension of A. We shall always assume that $\widehat{A}$ is chosen in such a way that it is elliptic and self-adjoint. The coordinates related to $\widehat{M}$ will also be marked by a "wide hat".

Running out of symbols later on we shall use (y,η) instead of (x,ξ) as coordinates on T^*M, and (y',η') instead of (x',ξ') as coordinates on $T^*\partial M$.

3. Symbols of differential operators. Throughout the book A and $B^{(j)}$ are linear differential operators of orders $2m$ and m_j, respectively ($m \geqslant 1$, $m_j \geqslant 0$ are integers), the coefficients of which are complex-valued infinitely differentiable functions of x, x'. (The set of orders m_j may be different on different connected components of ∂M.) By $A_{2m}(x,\xi)$, $B^{(j)}_{m_j}(x',\xi)$ we denote the principal symbols of the operators A, $(B^{(j)}\,\cdot)\big|_{\partial M}$, i.e., homogeneous polynomials in ξ of degrees $2m$, m_j obtained by leaving only the leading (of orders $2m$, m_j) derivatives in A, $(B^{(j)}\,\cdot)\big|_{\partial M}$, and replacing each $D_{x_k} = -i\partial/\partial x_k$ by ξ_k, $k = 1,2,\dots,n$.

It is well known that under changes of coordinates x the principal symbol $A_{2m}(x,\xi)$ behaves as a function on $T'M$. This means that in new coordinates $\widetilde{x}$ it takes the form

$$\widetilde{A}_{2m}(\widetilde{x},\widetilde{\xi}) = A_{2m}\left(x(\widetilde{x}),\,\sum_{k=1}^{n}\widetilde{\xi}_k\,\frac{\partial\widetilde{x}_k}{\partial x}\bigg|_{x=x(\widetilde{x})}\right).$$

Analogously, the principal symbols $B^{(j)}_{m_j}(x',\xi)$ behave as functions on $T'M|_{\partial M}$:

$$\widetilde{B}^{(j)}_{m_j}(\widetilde{x}',\widetilde{\xi}) = B^{(j)}_{m_j}\left(x'(\widetilde{x}',0),\,\sum_{k=1}^{n}\widetilde{\xi}_k\,\frac{\partial\widetilde{x}_k}{\partial x}\bigg|_{x=(x'(\widetilde{x}',0),0)}\right).$$

Recall that near ∂M the coordinate x_n is assumed to be fixed once and for all (see subsection 2 above), so $\partial\widetilde{x}_n/\partial x_n \equiv 1$.

Later on we will have to deal with the subprincipal symbol $A_{\mathrm{sub}}(x,\xi)$ of the operator A, which is also a function on $T'M$. The subprincipal symbol is a polynomial in ξ of degree $2m-1$, and it plays the role of the "second" symbol of A. The rigorous definition of $A_{\mathrm{sub}}(x,\xi)$ will be given in 2.1.3.

4. Ellipticity. The problem (1.1.1), (1.1.2) is assumed to be regular elliptic [**LioMag**, Chap. 2, Sect. 1.4], [**RoShSo**, Sect. 2.4]. This means that the following four conditions are fulfilled.

CONDITION 1.1.1. The operator A is elliptic, i.e.

$$A_{2m}(x,\xi) \neq 0, \qquad \forall (x,\xi) \in T'M. \tag{1.1.4}$$

CONDITION 1.1.2. The orders of the operators $B^{(j)}$ are different and lower than the order of the operator A, i.e.

$$0 \leqslant m_1 < m_2 < \ldots < m_m \leqslant 2m-1. \tag{1.1.5}$$

CONDITION 1.1.3. The operators $B^{(j)}$ can be resolved with respect to their leading conormal derivatives, i.e.

$$B^{(j)}_{m_j}(x',0,\xi_n) \neq 0, \qquad j=1,2,\dots,m, \quad \forall x' \in \partial M, \quad \forall \xi_n \neq 0. \tag{1.1.6}$$

CONDITION 1.1.4 (the Shapiro–Lopatinskiĭ condition). For all $(x',\xi')\in T'\partial M$ the equation $A_{2m}(x',0,\xi',\xi_n)=0$ has exactly m ξ_n-roots with $\operatorname{Im}\xi_n>0$ (for $n\geqslant 3$ this automatically follows from Condition 1.1.1) and $\nu=0$ is not an eigenvalue (see Definition 1.1.5 below) of the auxiliary one-dimensional spectral problem

$$A_{2m}(x',0,\xi',D_{x_n})v = \nu v\,, \tag{1.1.7}$$

$$\left(B_{m_j}^{(j)}(x',\xi',D_{x_n})v\right)\Big|_{x_n=0} = 0\,, \qquad j=1,2,\dots,m, \tag{1.1.8}$$

on the half-line $x_n\in\mathbb{R}_+=[0,+\infty)$.

Note that (1.1.7) is an ordinary differential equation with constant coefficients and (x',ξ') comes into the spectral problem (1.1.7), (1.1.8) simply as a parameter.

DEFINITION 1.1.5. We call the number ν an eigenvalue of the problem (1.1.7), (1.1.8) if for this value of ν it has a solution $v(x_n)\not\equiv 0$ which vanishes as $x_n\to+\infty$.

5. Self-adjointness. The problem (1.1.1), (1.1.2) is assumed to be formally self-adjoint with respect to some inner product $(\cdot\,,\cdot)$. Formal self-adjointness means that $(Av,w)=(v,Aw)$ for any $v,w\in C^\infty(M)$ satisfying the boundary conditions (1.1.2).

Usually it is assumed that v, w are functions and the inner product is defined by the formula

$$(v,w) = \int_M v(x)\overline{w(x)}\mu(x)\,dx \tag{1.1.9}$$

where $\mu\in C^\infty(M)$ is some positive density. We say that v is a *function* if it does not depend on the choice of local coordinates, i.e., $v(x)=\widetilde v(\widetilde x(x))$ where $\widetilde x=\widetilde x(x)$ are new local coordinates and $\widetilde v$ is the representation of v in the coordinates $\widetilde x$. We say that μ is a *density* if $\mu(x)=J(x)\widetilde\mu(\widetilde x(x))$ where $\widetilde\mu$ is the representation of μ in the coordinates $\widetilde x$ and $J=|\det\partial\widetilde x/\partial x|$. It is easy to see that the integrand in (1.1.9) is independent of the choice of local coordinates and so the inner product is well defined.

However, under such an approach the definition of the inner product and, consequently, the notion of self-adjointness depend on the choice of the density μ. This is somewhat inconvenient because it subsequently leads to a parasitic dependence of some quantities on μ (see, for example, the definition of the subprincipal symbol in 2.1.3). The following technical device allows us to avoid these inconveniences. Instead of functions and operators acting in spaces of functions one can consider half-densities and operators acting in spaces of half-densities.

We say that v is a *half-density* if $v(x)=J^{1/2}(x)\widetilde v(\widetilde x(x))$ where $\widetilde v$ is the representation of v in the coordinates $\widetilde x$. For half-densities the inner product is defined by the formula

$$(v,w) = \int_M v(x)\overline{w(x)}\,dx \tag{1.1.10}$$

(cf. (1.1.9)). Thus, the Hilbert space $L_2(M)$ and, consequently, the concept of self-adjointness are invariantly defined for half-densities without any auxiliary constructions.

The elementary substitution $v \to \mu^{1/2}v$, $A \to \mu^{1/2}A\mu^{-1/2}$, $B^{(j)} \to B^{(j)}\mu^{-1/2}$ transforms the spectral problem (1.1.1), (1.1.2) in functions into a spectral problem in half-densities. So, without loss of generality, we shall assume from now on that v, w are half-densities, A acts in the space of half-densities, $(B^{(j)}\,\cdot\,)\big|_{\partial M}$ are operators acting on half-densities, and the problem (1.1.1), (1.1.2) is formally self-adjoint with respect to the inner product (1.1.10).

Let v be a half-density on M. According to our convention Av is also a half-density on M. A separate question is an invariant interpretation of the expressions $(B^{(j)}v)\big|_{\partial M}$. An invariant interpretation of the expressions $(B^{(j)}v)\big|_{\partial M}$ is not really essential for the formulation of the eigenvalue problem (1.1.1), (1.1.2) because the boundary conditions are homogeneous. It is sufficient to have the boundary conditions (1.1.2) written in local coordinates on a finite number of coordinate patches, with them being equivalent on the intersections of different patches. (Here equivalence means that on intersections of coordinate patches the conditions (1.1.2) in different coordinates define the same linear relationships between $(\partial^k v/\partial x_n^k)\big|_{x_n=0}$, $k = 0, 1, 2, \ldots, 2m-1$, subject to the standard rules of transformation of the half-density v and of its partial derivatives under changes of coordinates.)

However, at some stage (in the course of the effective construction of oscillatory integrals related to our eigenvalue problem) we will be forced to consider various half-densities on M which, taken separately, do not necessarily satisfy the boundary conditions (1.1.2). So it is convenient to assign an invariant meaning to the expressions $(B^{(j)}v)\big|_{\partial M}$. Further on we shall assume that they are half-densities on ∂M. This can always be achieved by an adequate renormalization of the operators $B^{(j)}$ in some local coordinates, with subsequent transformation of these operators under changes of coordinates x' in the appropriate way. The identity operator is an example of an operator $B^{(j)}$ satisfying the required invariance condition: the expression $v|_{\partial M}$ is a half-density on ∂M because our "normal" coordinate x_n is specified once and for all (see subsection 2).

In mechanical applications assigning an invariant interpretation to the expressions $(B^{(j)}v)\big|_{\partial M}$ is usually easy because the boundary conditions have a clear mechanical meaning: they state that some mechanical quantity (say, normal displacement, angle of rotation, normal stress, flexural moment, etc.) is zero. These quantities are either functions or densities on ∂M, so one has only to make an elementary substitution of the form $B^{(j)} \to (\mu')^{\pm 1/2}B^{(j)}$, where $\mu'(x')$ is some smooth positive density on ∂M.

The principal symbol of the formal adjoint of a differential operator is the complex conjugate of the initial principal symbol. Therefore the principal symbol $A_{2m}(x,\xi)$ of the formally self-adjoint operator A is real.

6. Positiveness. The problem (1.1.1), (1.1.2) is assumed to be positive definite.

DEFINITION 1.1.6. We call the formally self-adjoint problem (1.1.1), (1.1.2) semi-bounded from below if there exists a real constant c such that

$$(Av, v) \geqslant c\,(v, v) \tag{1.1.11}$$

for all half-densities $v \in C^\infty(M)$ satisfying the boundary conditions (1.1.2). If (1.1.11) holds with $c > 0$ we call the problem (1.1.1), (1.1.2) positive definite.

Though it is not always easy to check whether a spectral problem is positive definite, one can usually check effectively whether it is semi-bounded from below (i.e., establish a somewhat weaker property). Namely, a regular elliptic formally self-adjoint spectral problem (1.1.1), (1.1.2) is semi-bounded from below if and only if the following two conditions are fulfilled.

CONDITION 1.1.1$'$. The principal symbol of the operator A is positive, i.e.,

$$A_{2m}(x,\xi) > 0\,, \qquad \forall (x,\xi) \in T'M \tag{1.1.12}$$

(cf. (1.1.4)).

CONDITION 1.1.4$'$. The auxiliary one-dimensional spectral problem (1.1.7), (1.1.8) on the half-line $0 \leqslant x_n < +\infty$ is positive definite, i.e., for all $(x',\xi') \in T'\partial M$ it does not have eigenvalues $\nu \leqslant 0$.

Indeed, the necessity of Conditions 1.1.1$'$, 1.1.4$'$ follows from the fact that if at least one of them does not hold, then we can effectively construct (in the form of a linear combination of up to m oscillating or decaying exponential functions modulated by smooth amplitudes) a sequence of half-densities $w_k(x) \in C^\infty(M)$, $k = 1,2,\ldots$, satisfying boundary conditions (1.1.2) and such that $(Aw_k,w_k)/(w_k,w_k) \to -\infty$. Sufficiency is proved, for example, in [**AgrVi**].

It is easy to see that Conditions 1.1.1, 1.1.4 follow from Conditions 1.1.1$'$, 1.1.4$'$. This means that formulating the notion of a semi-bounded from below self-adjoint elliptic boundary value problem we can replace Conditions 1.1.1, 1.1.4 by Conditions 1.1.1$'$, 1.1.4$'$.

REMARK 1.1.7. In practice it is quite enough to establish whether the spectral problem under consideration is semi-bounded from below. Indeed, any problem semi-bounded from below can be turned into a positive definite one by the elementary substitution

$$\widetilde{A} = A - cI\,, \qquad \widetilde{\nu} = \nu - c = \widetilde{\lambda}^{2m} \tag{1.1.13}$$

(cf. (1.1.11)); here I is the identity operator. Moreover, writing the resulting spectral asymptotics (see Sections 1.2, 1.4, 1.6, 1.7, 1.8) in terms of the spectral parameter $\widetilde{\nu} \to +\infty$ one may notice that all these asymptotics admit a formal interchange $\widetilde{\nu} \leftrightarrow \nu$. This is due to the fact that replacing $\widetilde{\nu}$ by ν we introduce an additional relative error of $o(\widetilde{\nu}^{-1})$, whereas the inherent relative error of these asymptotics is at least $o(\widetilde{\lambda}^{-1}) \equiv o(\widetilde{\nu}^{-1/2m})$ which is obviously greater than $o(\widetilde{\nu}^{-1})$.

7. Statement of our problem in the framework of operator theory. Let $H^{2m}(M)$ denote the Sobolev space of half-densities which belong to $L_2(M)$ together with all their partial derivatives of order $\leqslant 2m$. Under the conditions described above the differential operator A initially defined on

$$\{v \in C^\infty(M) : \left.\left(B^{(j)}v\right)\right|_{\partial M} = 0\,, \quad j = 1,2,\ldots,m\}$$

is essentially self-adjoint and admits a self-adjoint closure $\mathcal{A}$ in $L_2(M)$ with domain of definition

$$D(\mathcal{A}) = \{v \in H^{2m}(M) : \left.\left(B^{(j)}v\right)\right|_{\partial M} = 0\,, \quad j = 1,2,\ldots,m\}\,.$$

It is well known [**RoShSo,** Sect. 5.1] that the operator $\mathcal{A}$ has a positive discrete spectrum $0 < \nu_1 \leqslant \nu_2 \leqslant \dots$ accumulating to $+\infty$ (we enumerate the eigenvalues taking into account their multiplicities). The numbers $\lambda_k = \nu_k^{1/2m}$, $k = 1, 2, \dots$ (see (1.1.1′), (1.1.3)) may be interpreted as the eigenvalues of the operator $\mathcal{A}^{1/(2m)}$. It is also well known that the respective eigenfunctions (more precisely, half-densities) v_k are infinitely smooth on M, satisfy (1.1.1), (1.1.2) and form an orthonormal basis in $L_2(M)$.

REMARK 1.1.8. Throughout the book we denote by A the differential expression on the left-hand side of (1.1.1), i.e., we do not normally assign to A any particular domain of definition in any particular space. However, for the sake of simplicity we refer to A as to an operator. This should not cause confusion because the real operator $\mathcal{A}$ is distinguished by different script.

EXAMPLE 1.1.9. Let A be a formally self-adjoint second order differential operator (more precisely, differential expression) with $A_2(x, \xi) > 0$. Then the Dirichlet and Neumann boundary value problems for A satisfy Conditions 1.1.1′, 1.1.2, 1.1.3, 1.1.4′ and generate self-adjoint operators $\mathcal{A}$.

EXAMPLE 1.1.10. Let A be a formally self-adjoint differential operator of order $2m$ with $A_{2m}(x, \xi) > 0$. Then the Dirichlet boundary value problem ($B^{(j)} = \partial^{j-1}/\partial x_n^{j-1}$, $j = 1, 2, \dots, m$) for A satisfies Conditions 1.1.1′, 1.1.2, 1.1.3, 1.1.4′ and generates a self-adjoint operator $\mathcal{A}$.

8. Pseudodifferential case. For manifolds without boundary ($\partial M = \varnothing$) one can also consider a somewhat more general case when A is a positive definite self-adjoint elliptic pseudodifferential operator (see Section 2.1) of positive order $2m$; here m is an arbitrary positive real number. In this case the spectrum of the problem (1.1.1) remains discrete, positive and accumulating to $+\infty$. Subsequent asymptotic analysis differs insignificantly from the case of a differential operator; only the reasoning in Remark 1.1.7 suffers for $2m \leqslant 1$. Note that spectral problems for pseudodifferential operators are a normal occurrence in applications; see Sections 6.4, 6.5.

9. Mechanical interpretation. In mechanical and physical applications the spectral problem (1.1.1) (if $\partial M = \varnothing$) or (1.1.1), (1.1.2) (if $\partial M \neq \varnothing$) usually describes free harmonic oscillations of some system (elastic body, resonator, etc.) with ν being the frequency parameter proportional to some integer power of the natural frequency. Self-adjointness of the spectral problem means conservation of the full energy in the oscillating system, and positiveness means stability (i.e., absence of movements with amplitude exponentially growing in time). Some examples arising from applications will be considered in Chapter 6.

10. Canonical differential forms and measures. In order to formulate our results we will have to deal with some invariant differential forms and measures (see [**Ar2**], [**Tr,** Vol. 2], [**Hö3,** Vol. 3] for details).

The differential forms on T^*M, $T^*\partial M$ defined in local coordinates by

$$\langle \xi, dx \rangle = \xi_1 dx_1 + \xi_2 dx_2 + \dots + \xi_n dx_n ,$$
$$\langle \xi', dx' \rangle = \xi_1 dx_1 + \xi_2 dx_2 + \dots + \xi_{n-1} dx_{n-1}$$

respectively, are called symplectic 1-forms. One can easily check that these forms are independent of the choice of local coordinates. Their differentials are the canonical 2-forms

$$\begin{aligned} d\xi \wedge dx &= d\xi_1 \wedge dx_1 + d\xi_2 \wedge dx_2 + \cdots + d\xi_n \wedge dx_n \,, \\ d\xi' \wedge dx' &= d\xi_1 \wedge dx_1 + d\xi_2 \wedge dx_2 + \cdots + d\xi_{n-1} \wedge dx_{n-1} \,. \end{aligned}$$

As usual we denote

$$\begin{aligned} dx &= dx_1\, dx_2 \cdots dx_n \,, & d\xi &= d\xi_1\, d\xi_2 \cdots d\xi_n \,, \\ dx' &= dx_1\, dx_2 \cdots dx_{n-1} \,, & d\xi' &= d\xi_1\, d\xi_2 \cdots d\xi_{n-1} \,. \end{aligned}$$

Then the elements of symplectic volumes $\mathrm{vol}(\cdot)$, $\mathrm{vol}'(\cdot)$ on the cotangent bundles T^*M, $T^*\partial M$ are defined in local coordinates as $dx\, d\xi$, $dx'\, d\xi'$. Obviously, these expressions are also independent of the choice of coordinates.

We shall also use the notation

$$\begin{aligned} {d\!\!\!{}^{-}}\xi &= (2\pi)^{-n}\, d\xi, & dx\, {d\!\!\!{}^{-}}\xi &= (2\pi)^{-n}\, dx\, d\xi, \\ {d\!\!\!{}^{-}}\xi' &= (2\pi)^{1-n}\, d\xi', & dx'\, {d\!\!\!{}^{-}}\xi' &= (2\pi)^{1-n}\, dx'\, d\xi' \,. \end{aligned}$$

Throughout the book we shall extensively use the function

$$h(x,\xi) = (A_{2m}(x,\xi))^{1/(2m)} > 0 \tag{1.1.14}$$

which we shall call the *Hamiltonian*. The set

$$S^*M = \{(x,\xi) \in T'M \,:\, h(x,\xi) = 1\}$$

is said to be the *unit cosphere bundle*. Having fixed $x \in M$, we shall also deal with the *unit cosphere*

$$S^*_x M = \{\xi \in T'_y M \,:\, h(x,\xi) = 1\}$$

(the fibre of S^*M over the point x).

There exists a natural measure meas_x on the unit cosphere $S^*_x M$ with element $\widetilde{d\xi}$ defined by the equality $d\xi = \widetilde{d\xi}\, dh$. Note that the Euclidean measure dS_x on $S^*_x M$ is defined by $d\xi = dS_x\, d\rho$, where $\rho(\xi)$ is the Euclidean distance from ξ to $S^*_x M$, so $\widetilde{d\xi} = |\nabla_\xi h|^{-1}\, dS_x$. Equivalently, one can define $\widetilde{d\xi}$ by the condition that for any function $f(\xi)$ positively homogeneous in ξ of degree $d > -n$,

$$(n+d) \int_{h(x,\xi)\leqslant 1} f\, d\xi = \int_{h(x,\xi)=1} f\, \widetilde{d\xi} \,. \tag{1.1.15}$$

Clearly, for all measurable sets $\Omega \subset S^*_x M$ and functions f defined on $S^*_x M$, the integrals $\int_\Omega f\, \widetilde{d\xi}$ (in particular, $\mathrm{meas}_x\, \Omega_x = \int_\Omega \widetilde{d\xi}$) depend on the choice of the coordinates x and behave as densities in x under changes of coordinates. This means that the measure meas_x takes its values in the space of densities on M.

The natural measure meas on the unit cosphere bundle S^*M is defined as $dx\, \widetilde{d\xi}$. We shall denote

$$\widetilde{{d\!\!\!{}^{-}}\xi} = (2\pi)^{-n}\, \widetilde{d\xi} \,, \qquad dx\, \widetilde{{d\!\!\!{}^{-}}\xi} = (2\pi)^{-n}\, dx\, \widetilde{d\xi} \,.$$

The element $d\widetilde{\xi}$ of the measure meas_x can be written in local coordinates $\widetilde{\xi}$ on S_x^*M. In particular, on a conic subset of $T_x'M$ on which $h_{\xi_n} \neq 0$ one can choose as local coordinates $(\widetilde{\xi}, h)$, where

$$\widetilde{\xi} = (\widetilde{\xi}_1, \widetilde{\xi}_2, \ldots, \widetilde{\xi}_{n-1}) = (\xi_1/h, \xi_2/h, \ldots, \xi_{n-1}/h).$$

Then $\widetilde{\xi}$ are local coordinates on S_x^*M and

$$d\widetilde{\xi} = (h_{\xi_n})^{-1} d\widetilde{\xi}_1 d\widetilde{\xi}_2 \cdots d\widetilde{\xi}_{n-1}.$$

1.2. One-term asymptotic formula for $N(\lambda)$

1. Statement of the result. Let us introduce the *counting function* $N(\lambda)$ which is defined as the number of eigenvalues λ_k of the problem (1.1.1), (1.1.2) less than a given λ:

$$N(\lambda) = \#\{k : \lambda_k < \lambda\}.$$

Our final aim is to describe the asymptotic behaviour of $N(\lambda)$ as $\lambda \to +\infty$. Of course one can afterwards invert such formulae (see Remarks 1.2.2 and 1.6.2) and obtain asymptotic formulae for λ_k as $k \to +\infty$.

The following theorem gives a one-term asymptotic formula for $N(\lambda)$. Under additional restrictions it was established in [**Hö1**] ($\partial M = \varnothing$) and [**Se1**], [**Se2**] ($m = 1$). The final result appeared in [**Va3**], [**Va4**], [**Va7**].

THEOREM 1.2.1.

$$N(\lambda) = c_0 \lambda^n + O(\lambda^{n-1}), \qquad \lambda \to +\infty, \tag{1.2.1}$$

where

$$c_0 = (2\pi)^{-n} \operatorname{vol}\{(x,\xi) : A_{2m}(x,\xi) \leqslant 1\} = \int_{A_{2m} \leqslant 1} dx\, d\!\!{}^-\xi. \tag{1.2.2}$$

By (1.1.15) the Weyl constant (1.2.2) can also be written as

$$c_0 = \frac{1}{n\,(2\pi)^n} \operatorname{meas} S^*M = \frac{1}{n} \int_{S^*M} dy\, d\!\!{}^-\widetilde{\eta}. \tag{1.2.2'}$$

REMARK 1.2.2. Formula (1.2.1) can be written down in the equivalent form

$$\lambda_k = (c_0)^{-1/n} k^{1/n} + O(1), \qquad k \to \infty. \tag{1.2.1'}$$

EXAMPLE 1.2.3. Let M be a region in the Euclidean space $\mathbb{R}^n$ and $A = (-\Delta)^m$, where

$$\Delta = \partial^2/\partial x_1^2 + \partial^2/\partial x_2^2 + \cdots + \partial^2/\partial x_n^2$$

is the Laplacian in Cartesian coordinates. In this case $A_{2m} = |\xi|^{2m}$ and

$$c_0 = (2\pi)^{-n} \omega_n \operatorname{Vol} M, \tag{1.2.3}$$

where $\operatorname{Vol} M$ is the n-dimensional volume of M and ω_n is the volume of the unit ball in $\mathbb{R}^n$. In particular, for $n = 2$ we have $c_0 = S/4\pi$ where S is the surface area of M, and for $n = 3$ we have $c_0 = V/6\pi^2$ where V is the three-dimensional volume of M.

EXAMPLE 1.2.4 (generalization of Example 1.2.3). Let M be a Riemannian manifold with metric tensor $\{g_{ij}\}$, $1 \leqslant i,j \leqslant n$, and contravariant metric tensor $\{g^{ij}\} = \{g_{ij}\}^{-1}$. Let $A = (-\Delta)^m$ where

$$\Delta = g^{-1/4}\left(\sum_i \frac{\partial}{\partial x_i}\left(\sum_j g^{ij}\sqrt{g}\,\frac{\partial}{\partial x_j}\right)\right) g^{-1/4}$$

is the Laplacian on half-densities; here $g := \det\{g_{ij}\}$, and $\sqrt{g}$ is the standard Riemannian density. In this case $A_{2m}(x,\xi) = |\xi|_x^{2m}$ where $|\xi|_x = \left(\sum_{i,j} g^{ij}\xi_i\xi_j\right)^{1/2}$, and formula (1.2.3) holds with $\operatorname{Vol} M = \int_M \sqrt{g}\,dx$.

2. Discussion of the result. Theorem 1.2.1 is remarkable in two ways. First of all, the one-term asymptotic formula (1.2.1) does not depend on the boundary conditions (1.1.2). This fact suggests that in practice the asymptotic formula (1.2.1) is not very accurate and that it is natural to search for a refined two-term asymptotic formula with the second asymptotic term describing boundary phenomena.

The second remarkable fact about Theorem 1.2.1 is that the remainder estimate in (1.2.1) is sharp, that is, in (1.2.1) one cannot replace $O(\lambda^{n-1})$ by $o(\lambda^{n-1})$. This will become evident after we single out the second asymptotic term in Sections 1.6, 1.7; as we shall see, the second term is of the order of λ^{n-1} and in the general situation is not identically zero. However, the simplest way of checking that the remainder estimate in (1.2.1) is sharp is by analyzing the following simple example.

EXAMPLE 1.2.5. Let M be a unit n-dimensional sphere,

$$M = \mathbb{S}^n = \{x \in \mathbb{R}^{n+1} : |x| = 1\},$$

and $A = (-\Delta + (n-1)^2/4)^m$, where Δ is the Laplacian (see Example 1.2.4). In this case the spectrum of the operator $\mathcal{A}^{1/2m}$ consists of the eigenvalues

$$\Lambda_j = j + (n-1)/2, \qquad j = 0,1,2,\ldots,$$

with multiplicities

$$\frac{(n+j-2)!\,(n+2j-1)}{(n-1)!\,j!}$$

(see, e.g., [**Sh,** Sect. 22]). Therefore

$$\lambda_k = j + (n-1)/2, \qquad k = 1,2,\ldots,$$

where $j = j(k)$ is the maximal integer satisfying the inequality

$$\sum_{i=0}^{j} \frac{(n+i-2)!\,(n+2i-1)}{(n-1)!\,i!} \leqslant k\,.$$

Elementary calculations show that

$$N(\Lambda_j) = \frac{2}{n!}\Lambda_j^n - \frac{n^2-3n+3}{(n-1)!}\Lambda_j^{n-1} + O(\Lambda_j^{n-2}), \tag{1.2.4}$$

$$N(\Lambda_j + 0) = \frac{2}{n!}\Lambda_j^n + \frac{n^2-3n+3}{(n-1)!}\Lambda_j^{n-1} + O(\Lambda_j^{n-2}), \tag{1.2.5}$$

(here $c_0 = 2/n!$). Formulae (1.2.4), (1.2.5) demonstrate that in the general case the order of the remainder estimate in (1.2.1) cannot be decreased.

3. Pseudodifferential case. Theorem 1.2.1 remains true for a pseudodifferential operator acting on a manifold without boundary [**Hö1**], [**DuiGuiHö**].

1.3. Hamiltonian billiards I: basic definitions and results

Singling out the second term in the asymptotic formula for $N(\lambda)$ requires the study of some global geometric characteristics of the problem. These characteristics are formulated in terms of billiard (or Hamiltonian) trajectories generated by the principal symbol of the differential operator A. In this section we introduce the concept of branching Hamiltonian billiards and describe their basic properties. More detailed discussions and proofs are contained in Appendix D, and, partially, in various sections of the main text.

The notions introduced in this section are necessary for the formulation of all our results on two-term spectral asymptotics (Sections 1.6–1.8).

The concept of branching Hamiltonian billiards is a generalization of the classical concepts of Hamiltonian flows and geodesic billiards. Therefore we shall first consider Hamiltonian flows (subsection 1) and geodesic billiards (subsection 2); branching Hamiltonian billiards as such will appear in subsection 3.

1. Hamiltonian flow. Let us consider the case when M is a manifold without boundary.

For $t \in \mathbb{R}$ we shall denote by

$$(x^*(t;y,\eta),\xi^*(t;y,\eta)) \tag{1.3.1}$$

the integral curve of the Hamiltonian vector field generated by the Hamiltonian (1.1.14). The curve (1.3.1) lies in $T'M$ and (y,η) denotes its starting point, i.e.,

$$(x^*(0;y,\eta),\xi^*(0;y,\eta)) = (y,\eta) \in T'M\,. \tag{1.3.2}$$

In other words, (1.3.1) is the solution of the Hamiltonian system of equations

$$\dot{x}^* = h_\xi(x^*,\xi^*), \qquad \dot{\xi}^* = -h_x(x^*,\xi^*) \tag{1.3.3}$$

with initial condition (1.3.2) at $t=0$. Here and further on the dot stands for the time derivative, and subscripts are used for denoting respective partial derivatives.

In our notation for Hamiltonian trajectories we shall sometimes omit (for the sake of brevity) the variables (y,η), leaving only the variable t; i.e., we shall sometimes write $(x^*(t),\xi^*(t))$ instead of (1.3.1).

Denote by Φ^t the group of shifts along the trajectories (1.3.1). This group is called the *Hamiltonian flow*. The group Φ^t preserves the symplectic differential 1-form and, consequently, the symplectic differential 2-form and the symplectic volume vol (see Section 2.3).

We obviously have

$$\frac{d}{dt}h(x^*,\xi^*) = \langle h_x,\dot{x}\rangle + \langle h_\xi,\dot{\xi}\rangle = 0\,,$$

so

$$h(x^*(t;y,\eta),\xi^*(t;y,\eta)) = h(y,\eta)\,,$$

i.e., the Hamiltonian h is preserved by the Hamiltonian flow Φ^t. Since the Hamiltonian h is positively homogeneous in ξ of degree 1, the Hamiltonian flow is also homogeneous:

$$(x^*(t;y,\mu\eta),\xi^*(t;y,\mu\eta)) = (x^*(t;y,\eta),\mu\xi^*(t;y,\eta))\,, \qquad \forall\mu>0\,. \tag{1.3.4}$$

Therefore it is convenient to consider the restriction of the Hamiltonian flow Φ^t to the unit cosphere bundle S^*M. We retain for this restriction the original notation Φ^t. Since $dy\,d\eta = dy\,d\tilde{\eta}\,dh$, the flow Φ^t preserves the canonical measure meas on S^*M.

We shall call the curves $x^*(t)$ *rays* (by analogy with geometric optics). Note that $\dot{x}^* \neq 0$, i.e., the ray cannot stop. This fact follows from Euler's formula $\langle \xi, h_\xi \rangle = h \neq 0$ and the conservation of the Hamiltonian.

In the special case when M is a Riemannian manifold and $h(x,\xi) = |\xi|_x$ (see Example 1.2.4) the Hamiltonian flow is called *geodesic* flow. In this situation the rays x^* are geodesics and $\xi_i^* = \sum_j g_{ij}(x^*)\,\dot{x}_j^*$.

Having defined the Hamiltonian flow Φ^t we can now introduce the notions of periodicity and absolute periodicity which are decisive in obtaining refined spectral asymptotics.

DEFINITION 1.3.1. The point $(y,\eta) \in T'M$ and the trajectory (1.3.1) originating from this point are called *T-periodic*, $T > 0$, if

$$(x^*(T;y,\eta), \xi^*(T;y,\eta)) = (y,\eta). \tag{1.3.5}$$

The point (y,η) and the trajectory (1.3.1) are called *periodic* if they are T-periodic for some $T > 0$.

DEFINITION 1.3.2. The point $(y_0,\eta_0) \in T'M$ and the trajectory (1.3.1) originating from this point are called *absolutely T-periodic*, $T > 0$, if the function

$$|x^*(T;y,\eta) - y|^2 + |\xi^*(T;y,\eta) - \eta|^2 \tag{1.3.6}$$

of the variables (y,η) has an infinite order zero at (y_0,η_0) (clearly, this property does not depend on the choice of coordinates). The point (y_0,η_0) and the trajectory (1.3.1) originating from this point are called *absolutely periodic* if they are absolutely T-periodic for some $T > 0$.

REMARK 1.3.3. In the analytic case (when M is a real-analytic manifold and h is a real-analytic function) the function (1.3.6) is analytic as well. Therefore in the analytic case the existence of one absolutely T-periodic trajectory implies that all the trajectories are absolutely T-periodic with the same T. (Recall that the manifold M is assumed to be connected.)

It is important to know how rich are the sets of periodic and absolutely periodic trajectories, because this determines the structure of the two-term spectral asymptotics.

Denote by Π_T, Π_T^a, Π, and Π^a the sets of T-periodic, absolutely T-periodic, periodic, and absolutely periodic points in S^*M, respectively. Of course,

$$\Pi = \bigcup_{T>0} \Pi_T, \qquad \Pi^a = \bigcup_{T>0} \Pi_T^a, \qquad \Pi_T^a \subset \Pi_T, \qquad \Pi^a \subset \Pi.$$

LEMMA 1.3.4. *The set $\Pi \subset S^*M$ is measurable. Moreover, for all $T_+ > 0$ the sets $\bigcup_{0<T\leqslant T_+} \Pi_T$ are also measurable.*

As a rule, the set Π has a very complicated structure, whereas the set Π_a is organized essentially more simply (see Remark 1.3.3). Moreover, in realistic situations the set of absolutely periodic points is usually empty. Nevertheless, the following lemma shows that the set Π is insignificantly richer than Π^a.

LEMMA 1.3.5.

$$\operatorname{meas}\left(\bigcup_{T>0}(\Pi_T \setminus \Pi_T^a)\right) = 0. \tag{1.3.7}$$

From (1.3.7) we immediately obtain

COROLLARY 1.3.6. *Almost all periodic points are absolutely periodic, i.e.,*

$$\operatorname{meas}(\Pi \setminus \Pi^a) = 0. \tag{1.3.8}$$

Now we are prepared to introduce the notion of nonperiodicity which will play a key role in the study of refined spectral asymptotics.

DEFINITION 1.3.7. We will say that the *nonperiodicity condition* is fulfilled if

$$\operatorname{meas}\Pi = 0. \tag{1.3.9}$$

It follows from Corollary 1.3.6 that (1.3.9) is equivalent to

$$\operatorname{meas}\Pi^a = 0. \tag{1.3.10}$$

Therefore, defining the concept of nonperiodicity we can choose between (1.3.9) and (1.3.10) depending on the circumstances. This observation often allows us to simplify matters. For example, when proving spectral results it is more convenient to deal with the seemingly more restrictive condition (1.3.9). On the other hand, applying spectral results to concrete problems it is much easier to check the condition (1.3.10). In particular, from Remark 1.3.3 and Corollary 1.3.6 we obtain

LEMMA 1.3.8. *In the analytic case there are only two possibilities: either the nonperiodicity condition is fulfilled or all the trajectories are periodic with the same period* $T>0$.

It is clear that failures of the nonperiodicity condition are very rare (at least in the analytic situation). For instance, in the analytic case it is sufficient to find one nonperiodic trajectory or two periodic trajectories with incommensurable periods to make sure that the nonperiodicity condition is fulfilled.

Failure of the nonperiodicity condition is usually due to some strong symmetries. The most obvious example is the geodesic flow on the sphere $\mathbb{S}^n$. However, there exist nontrivial examples.

EXAMPLE 1.3.9. Let M be a closed (i.e., without boundary) connected compact real-analytic two-dimensional surface of revolution in $\mathbb{R}^3$ without self-intersections. We shall call a surface M of this class a *Zoll surface* if there exists a $T>0$ such that all the geodesics on M are T-periodic (i.e., if the nonperiodicity condition fails). Following [**Be**] we explicitly describe below all Zoll surfaces.

Let us introduce cylindrical coordinates (φ, r, z) according to the formulae

$$y_1 = r\cos\varphi, \qquad y_2 = r\sin\varphi, \qquad y_3 = z,$$

where (y_1, y_2, y_3) are the Cartesian coordinates in $\mathbb{R}^3$ and the y_3-axis is the axis of revolution. Let us consider the meridian curve of M in the half-plane

$$\varphi = \text{const}, \qquad 0 \leqslant r < +\infty, \qquad -\infty < z < +\infty.$$

It turns out that the surface M is a Zoll surface if and only if its meridian curve can be parametrized as

$$(1.3.11)\quad r = R\cos\theta\,, \quad z = R\int_0^\theta \sqrt{(1+f(\psi))^2 - \sin^2\psi}\,d\psi\,, \quad -\frac{\pi}{2} \leqslant \theta \leqslant \frac{\pi}{2}\,,$$

with some constant $R > 0$ and some real-valued function f satisfying the following conditions:

1. f is odd.
2. $f(\psi) \geqslant |\sin\psi| - 1\,, \ \forall\psi \in [-\pi/2, \pi/2]\,.$
3. f is real-analytic on $\mathbb{R}$.
4. f is even with respect to the point $\pi/2$.
5. The Taylor expansions of the function $(1+f(\psi))^2 - \sin^2\psi$ at the points $\pm\pi/2$ start with terms of order $4q_\pm - 2$, where $q_\pm \in \mathbb{N}$.
6. The function z defined by (1.3.11) satisfies the inequality $z(\theta) > z(-\theta)$, $\forall\theta \in (0, \pi/2]$.

We choose the sign of the square root in (1.3.11) in such a way that the function $F(\psi) = \sqrt{(1+f(\psi))^2 - \sin^2\psi}$ is real-analytic and $F(0) > 0$.

Conditions (1), (2) are the main ones which single out the class of Zoll surfaces. The other ones play only an auxiliary role, ensuring the analyticity of the surface M and excluding self-intersections.

A trivial example of a function f satisfying all the above conditions is $f \equiv 0$. In this case (1.3.11) parametrizes a semicircle which means that M is a sphere. A nontrivial example is $f(\theta) = (1/3)\sin\theta\cos^2\theta$; the Zoll surface for this f is shown in Figure 1. It gives a side view of the Zoll surface. The external egg-shaped curve is the meridian of the Zoll surface, and the internal boomerang-shaped curve is the projection of a geodesic (namely, the geodesic intersecting the equator at the angle $\pi/4$).

The reader can find other examples of manifolds with periodic geodesic flow and interesting discussions of this subject in [**Be**], [**Gui**].

Now we proceed to manifolds with boundary. For the sake of simplicity we shall consider only trajectories $(x^*(t;y,\eta), \xi^*(t;y,\eta))$ originating from points $(y,\eta) \in T'\overset{\circ}{M}$ (not $T'M$), i.e., strictly from the interior. This will be sufficient for our aims because, as we shall see later, only sets of nonzero measure influence the second asymptotic term of $N(\lambda)$, whereas

$$\operatorname{vol}\left(T'M\big|_{\partial M}\right) = \operatorname{meas}\left(S^*M\big|_{\partial M}\right) = 0\,.$$

2. Geodesic billiards. Let M be a Riemannian manifold with boundary, and let $h(x,\xi) = |\xi|_x$ (see Example 1.2.4). In this case the Hamiltonian billiards described below are called *geodesic billiards*.

The trajectory $(x^*(t), \xi^*(t))$ may hit the boundary at some time $t = \tau \neq 0$ (i.e., $x^*(\tau) \in \partial M$) and the question arises how to extend it. Geometric optics gives a natural reflection law: the angle of incidence of a ray equals the angle of reflection. Certainly, the notion of angle is understood in the sense of Riemannian geometry, i.e., at the point x_0 angles are measured in local coordinates x in which $|\xi|_{x_0} = |\xi| = \sqrt{\xi_1^2 + \xi_2^2 + \cdots + \xi_n^2}\,.$

We shall now reformulate the *reflection law* in a more explicit manner, in terms of Hamiltonian trajectories. Consider the points $(x^*(\tau - 0), \xi^*(\tau - 0))$ and

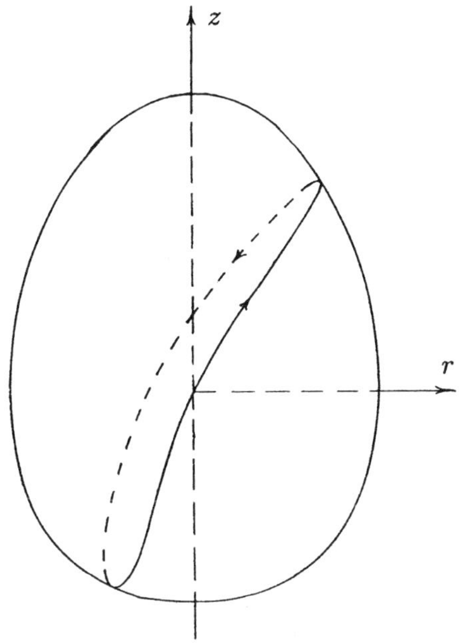

FIGURE 1. Example of a Zoll surface. Side view.

$(x^*(\tau+0), \xi^*(\tau+0))$ which we shall call the *point of incidence* and *point of reflection*, respectively. As $\xi^*(t)$ determines the direction of the ray $x^*(t)$, the ξ-components of these two points are, generally speaking, different, and our aim is to determine the unknown covector ξ, which is $\xi^*(\tau+0)$ or $\xi^*(\tau-0)$ depending on whether $\tau > 0$ or $\tau < 0$.

Let us impose the natural condition that the Hamiltonian h is preserved under reflection, i.e.,

$$h(x^*(\tau-0), \xi^*(\tau-0)) = h(x^*(\tau+0), \xi^*(\tau+0)). \tag{1.3.12}$$

From this equality and from the geometric reflection law it follows that the points of incidence and reflection differ only in their ξ_n-components. Therefore we can denote these points as

$$(x^*(\tau-0), \xi^*(\tau-0)) = (x^{*\prime}, 0, \xi^{*\prime}, \xi_n^-)$$

and

$$(x^*(\tau+0), \xi^*(\tau+0)) = (x^{*\prime}, 0, \xi^{*\prime}, \xi_n^+),$$

respectively (recall that near ∂M we use special local coordinates x, see 1.1.2). Now we can rewrite (1.3.12) as

$$h(x^{*\prime}, 0, \xi^{*\prime}, \xi_n^-) = h(x^{*\prime}, 0, \xi^{*\prime}, \xi_n^+). \tag{1.3.13}$$

This is the equation which describes the reflection law in terms of Hamiltonian trajectories. Note that both the equation (1.3.13) and the condition that the points of incidence and reflection differ only in their conormal components are invariant under changes of coordinates x.

Having fixed $(x^{*\prime}, \xi^{*\prime})$, let us consider $h(\xi_n) := h(x^{*\prime}, 0, \xi^{*\prime}, \xi_n)$ as a function of the variable $\xi_n \in \mathbb{R}$. Since h is the square root of a quadratic polynomial it

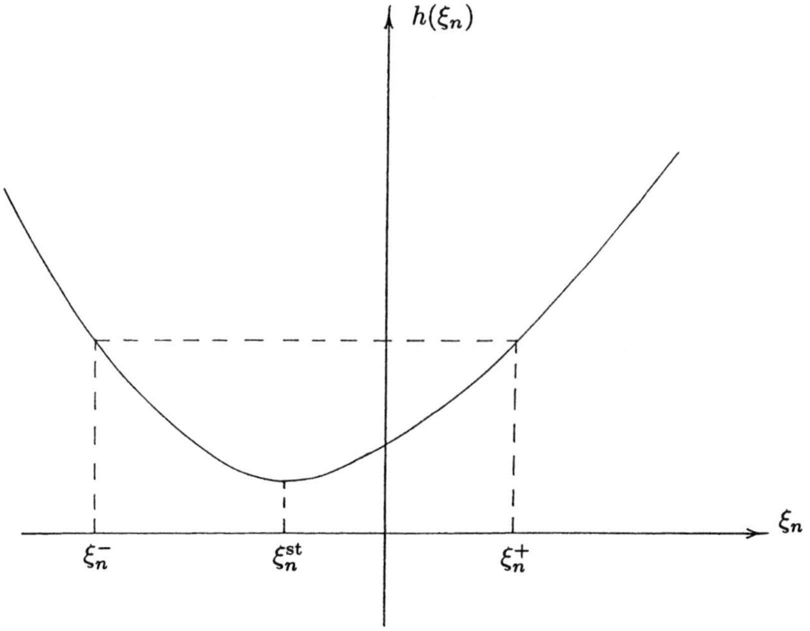

FIGURE 2. Graph of the function $h(\xi_n)$ in the case of a geodesic billiard system.

has a unique local (and global) minimum at some point $\xi_n = \xi_n^{\mathrm{st}}$. As x_n^* increases from ∂M into the interior of M and $\dot{x}_n^* = h_{\xi_n}$, we have

$$\xi_n^- \leqslant \xi_n^{\mathrm{st}}, \qquad \xi_n^+ \geqslant \xi_n^{\mathrm{st}}, \tag{1.3.14}$$

$$h_{\xi_n}(\xi_n^-) \leqslant 0, \qquad h_{\xi_n}(\xi_n^+) \geqslant 0. \tag{1.3.15}$$

The equalities $\xi_n^- = \xi_n^{\mathrm{st}}$ $(\tau > 0)$ or $\xi_n^+ = \xi_n^{\mathrm{st}}$ $(\tau < 0)$ describe the situation when the arriving (from $t = 0$) ray $x^*(t)$ is tangent to ∂M at the point $x^*(\tau \mp 0)$. Such trajectories $(x^*(t), \xi^*(t))$ are called *grazing*. We shall exclude this case from consideration because the set of grazing trajectories is sparse (see Lemma 1.3.11 below), and for our aims there is no need to deal with them.

Having excluded grazing trajectories, we obtain instead of (1.3.14), (1.3.15) the strict inequalities

$$\xi_n^- < \xi_n^{\mathrm{st}}, \qquad \xi_n^+ > \xi_n^{\mathrm{st}}, \tag{1.3.16}$$

$$h_{\xi_n}(\xi_n^-) < 0, \qquad h_{\xi_n}(\xi_n^+) > 0. \tag{1.3.17}$$

Formulae (1.3.13), (1.3.16), and (1.3.17) are illustrated in Figure 2.

It is clear from Figure 2 that, given ξ_n^-, one can uniquely determine ξ_n^+. This means that for $\tau > 0$ we uniquely determine $(x^*(\tau+0), \xi^*(\tau+0))$ which serves as the originating point of the reflected Hamiltonian trajectory. Conversely, if $\tau < 0$, then given ξ_n^+ one can uniquely determine $(x^*(\tau - 0), \xi^*(\tau - 0))$ which is the starting point of the extended Hamiltonian trajectory. In the second case we have deliberately avoided the word "reflected" because according to our terminology the

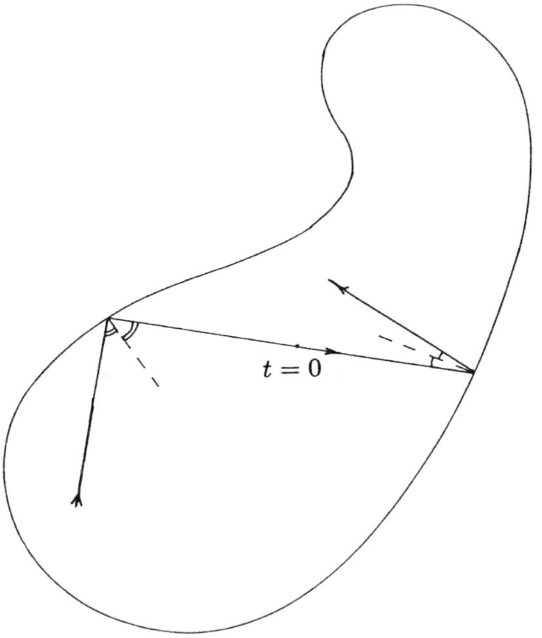

FIGURE 3. Transversal reflections.

point of reflection is $(x^*(\tau+0), \xi^*(\tau+0))$, whereas we are extending the trajectory back in time from the point $(x^*(\tau-0), \xi^*(\tau-0))$.

Exclusion of grazing trajectories means that we consider only *transversal* reflections, i.e., reflections at nonzero angles. Figure 3 illustrates transversal reflections of a ray.

DEFINITION 1.3.10. A trajectory obtained by consecutive transversal reflections is called a *billiard* trajectory.

We shall retain for billiard trajectories the same notation $(x^*(t;y,\eta), \xi^*(t;y,\eta))$ as for Hamiltonian trajectories.

As the Hamiltonian h is preserved along the billiard trajectories and as these trajectories are homogeneous in η (see (1.3.4)), it is natural to consider only trajectories lying in S^*M.

It is easy to see that only two factors may prevent us from extending a billiard trajectory $(x^*(t;y,\eta), \xi^*(t;y,\eta))$ to all $t \in \mathbb{R}$.

First of all, it may occur that after a finite number of transversal reflections we obtain a grazing Hamiltonian trajectory. We shall denote the set of starting points $(y,\widetilde{\eta}) \in S^*\overset{\circ}{M}$ of such billiard trajectories by $\mathcal{P}^{\mathrm{g}}$.

Secondly, it may occur that a billiard trajectory experiences an infinite number of transversal reflections in a finite (positive or negative) time; see Figure 4.

We shall call such a billiard trajectory a *dead-end* trajectory. By $\mathcal{P}^{\mathrm{d}}$ we shall denote the set of starting points $(y,\widetilde{\eta}) \in S^*\overset{\circ}{M}$ of dead-end billiard trajectories.

LEMMA 1.3.11 ([**CorFomSin,** Sect. 6.1]). *For a geodesic billiard system,*

$$\operatorname{meas} \mathcal{P}^{\mathrm{g}} = 0 \tag{1.3.18}$$

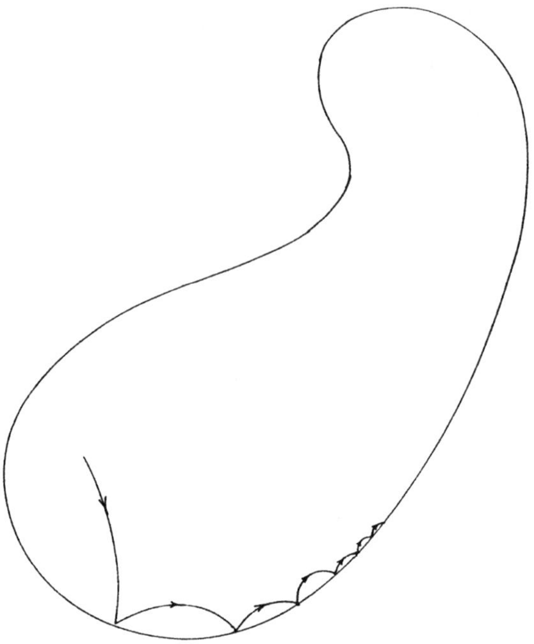

FIGURE 4. A dead-end trajectory.

and

$$\text{meas}\,\mathcal{P}^{\mathrm{d}} = 0\,. \tag{1.3.19}$$

Lemma 1.3.11 means that almost all billiard trajectories can be extended to all $t \in \mathbb{R}$. The group Φ^t of shifts along the billiard trajectories is called the *billiard flow*. The billiard flow preserves the symplectic differential 1-form and, consequently, the symplectic differential 2-form, the symplectic volume and the canonical measure on S^*M.

Adding the phrase "after a finite number of transversal reflections" in Definitions 1.3.1, 1.3.2 we obtain definitions of T-periodic, absolutely T-periodic, periodic and absolutely periodic billiard trajectories and points. As before, we shall denote the sets of such points in $S^*\overset{\circ}{M}$ by Π_T, Π_T^a, Π, and Π^a, respectively. Lemmas 1.3.4, 1.3.5 and Corollary 1.3.6 remain valid in this case. We shall retain for geodesic billiards Definition 1.3.7 describing the nonperiodicity condition.

Let us now introduce the important notions of convexity and concavity.

DEFINITION 1.3.12. Denote

$$\mathbf{k}(x',\xi') := \{h_{\xi_n}, h\}|_{x=(x',0),\xi=(\xi',\xi_n^{\mathrm{st}}(x',\xi'))}\,, \tag{1.3.20}$$

$(x',\xi') \in T'\partial M$. Here $\{\cdot,\cdot\}$ are the Poisson brackets (see Principal Notation for the proper sign), and $\xi_n^{\mathrm{st}}(x',\xi')$ is the real ξ_n-root of the equation

$$h_{\xi_n}(x',0,\xi',\xi_n) = 0\,; \tag{1.3.21}$$

see also Figure 2. We shall call $\mathbf{k}$ the *Hamiltonian curvature* of ∂M.

Note that formulae (1.3.20), (1.3.21) have a simple interpretation: $\mathbf{k} = -\ddot{x}^*_n|_{t=0}$, where $x^*(t) \subset \widehat{M}$ is a ray emitted from the point $x^*(0) = (x', 0) \in \partial M$ in the direction $\dot{x}^*(0) = h_\xi(x', 0, \xi', \xi_n^{\mathrm{st}}(x', \xi'))$ tangent to ∂M. Here we deliberately consider the ray on the extended manifold $\widehat{M}$ (see 1.1.2) because this ray may leave M at arbitrarily small $t \neq 0$. Obviously, under the assumption that the coordinate ξ_n is fixed (see 1.1.2) $\mathbf{k}$ is a well-defined function on $T'\partial M$.

DEFINITION 1.3.13. We shall say that the manifold M is *convex* if $\mathbf{k}(x', \xi') \geqslant 0$, $\forall (x', \xi') \in T'\partial M$, and *strongly convex* if $\mathbf{k}(x', \xi') > 0$, $\forall (x', \xi') \in T'\partial M$.

DEFINITION 1.3.14. We shall say that the manifold M is *concave* if $\mathbf{k}(x', \xi') \leqslant 0$, $\forall (x', \xi') \in T'\partial M$, and *strongly concave* if $\mathbf{k}(x', \xi') < 0$, $\forall (x', \xi') \in T'\partial M$.

The notion of convexity is illustrated by the following

EXAMPLE 1.3.15 (Euclidean case). Let $M \subset \mathbb{R}^n$ and $h = |\eta|$; in this example we denote by y Cartesian coordinates in $\mathbb{R}^n$, and by η the corresponding duals. Definition 1.3.13 in this case is equivalent to the traditional definition of convexity: the set M is called convex if for any two points $y, z \in M$ the entire segment of the straight line connecting y and z belongs to M. Moreover, the quantity $\mathbf{k}$ in this case has a clear geometric meaning. Let us introduce (curvilinear) local coordinates x such that x_n is the Euclidean distance from the point to ∂M, with positive sign when the point is in $\overset{\circ}{M}$ and negative when the point is in $\mathbb{R}^n \setminus M$. Then $\mathbf{k}(x', \xi')$ is the curvature (in the traditional sense) of the normal section of the surface ∂M at the point $y = y(x', 0)$ in the direction $\eta = \eta(x', 0, \xi', \xi_n^{\mathrm{st}}(x', \xi'))$.

The Euclidean case is ill-suited for illustrating the notion of concavity because concave manifolds in this case are noncompact. Therefore we shall consider the following less trivial

EXAMPLE 1.3.16. Let M be a spherical cap, that is, part of a unit n-dimensional sphere $\mathbb{S}^n \subset \mathbb{R}^{n+1}$ cut off by a hyperplane in $\mathbb{R}^{n+1}$:

$$M = \{y \in \mathbb{R}^{n+1} : |y| = 1, \ y_{n+1} \geqslant c\},$$

where c is some constant, $|c| < 1$, and y are Cartesian coordinates in $\mathbb{R}^{n+1}$. The metric on M is assumed to be the standard one induced by the Euclidean metric in $\mathbb{R}^{n+1}$. In this case the manifold M is convex if it is less than or equal to a hemisphere ($c \geqslant 0$), strongly convex if it is strictly less than a hemisphere ($c > 0$), concave if it is greater than or equal to a hemisphere ($c \leqslant 0$), and strongly concave if it is strictly greater than a hemisphere ($c < 0$); see Figure 5.

LEMMA 1.3.17. *If the manifold M is strongly convex, then there are no grazing billiard trajectories (i.e., $\mathcal{P}^{\mathrm{g}} = \varnothing$). If the manifold M is strongly concave or strongly convex, then there are no dead-end billiard trajectories (i.e., $\mathcal{P}^{\mathrm{d}} = \varnothing$).*

It is easy to see that a somewhat weaker version of Remark 1.3.3 holds for geodesic billiards.

REMARK 1.3.18. In the analytic case, if a trajectory is absolutely T-periodic, then all sufficiently close billiard trajectories are T-periodic with the same period T.

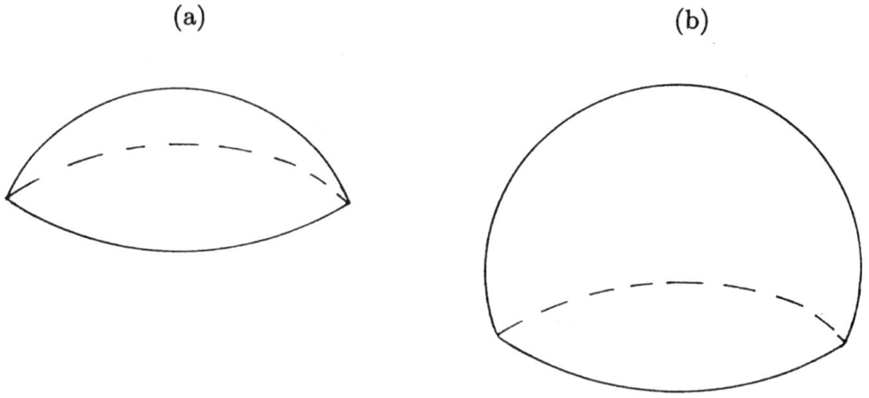

FIGURE 5. Spherical caps: (a) convex, (b) concave.

Unfortunately, in the general case it is impossible to extend analytically the property of absolute periodicity to all points from $S^*\overset{\circ}{M}$, because analyticity may be lost on $\mathcal{P}^{\mathrm{g}}$. A lucky exception is the case of a convex manifold, see Lemma 1.3.17. For a convex manifold, Remark 1.3.3 and Lemma 1.3.8 still hold. Moreover, we have the following important

LEMMA 1.3.19. *Let us consider the analytic case and let the manifold M be convex, $\partial M \neq \varnothing$, and $\mathbf{k}(x',\xi') \not\equiv 0$. Then the nonperiodicity condition is fulfilled.*

Thus, Lemma 1.3.19 gives effective sufficient conditions guaranteeing nonperiodicity.

In particular, applying Lemma 1.3.19 to the Euclidean case (Example 1.3.15) we conclude that if M is convex (in the traditional sense) and ∂M is analytic, then the nonperiodicity condition is fulfilled.

Applying Lemma 1.3.19 to Example 1.3.16 we conclude that if M is strictly less than a hemisphere ($c > 0$), then the nonperiodicity condition is fulfilled. On the other hand, it is easy to see directly that if $c \leqslant 0$, then the nonperiodicity condition is not fulfilled: when M is a hemisphere ($c = 0$) all geodesics are periodic after two reflections, and when M is strictly greater than a hemisphere ($c < 0$) the great circles which do not intersect ∂M form a rich set of periodic geodesics. So in this particular example strong convexity is necessary and sufficient for nonperiodicity.

It is worth noting that Lemma 1.3.19 does not have an analogue for manifolds without boundary. This leads us to the paradoxical conclusion that some manifolds with boundary are simpler objects than manifolds without boundary, at least as far as nonperiodicity is concerned.

Concluding this subsection, let us make the obvious observation that given a second order elliptic self-adjoint differential operator A acting on M one can always introduce a Riemannian metric based on the principal symbol of A and consider the corresponding geodesic billiard system.

3. Branching Hamiltonian billiards. Let us proceed at last to the most difficult case when M is a manifold with boundary and A is an operator of order $2m$, $m > 1$.

As in the preceding subsections, we start by introducing the trajectories

$$(x^*(t;y,\eta),\xi^*(t;y,\eta))\,,\qquad (y,\eta)\in T'\overset{\circ}{M}\,,$$

of the Hamiltonian system (1.3.3). The trajectory may hit the boundary at some time $t=\tau\neq 0$ and our task is to define the reflection law. Unfortunately, in the general case we cannot formulate the reflection law in simple geometric terms because the Hamiltonian $h(x,\xi)$ does not induce a Riemannian metric on M (with the exception of the special case $A_{2m}(x,x)=\big(A_2(x,\xi)\big)^m$, where $A_2(x,\xi)$ is the principal symbol of some second order differential operator). We can, however, formulate the reflection law on the basis of the Hamiltonian formalism developed in subsection 2. Let us recall that this formalism requires $x^*(t)$ and the first $n-1$ components of $\xi^*(t)$ to be continuous at $t=\tau$ and the nth component of $\xi^*(t)$ to satisfy the equality (1.3.13), $\xi_n^\pm=\xi_n^*(\tau\pm 0)$. For $\tau>0$, ξ_n^- is given and ξ_n^+ must be determined from (1.3.13); for $\tau<0$, ξ_n^+ is given and ξ_n^- must be determined from (1.3.13). As in subsection 2, we also require $\xi_n^\pm$ to satisfy the inequalities (1.3.17). In a weaker form (1.3.15) these inequalities are natural because the ray $x^*(t)$ has to stay on the manifold M. As in subsection 2, the exclusion of cases $h_{\xi_n}(\xi_n^-)=0$, $h_{\xi_n}(\xi_n^+)=0$ simplifies matters and (as we shall see later) leads to the loss of a set of initial points (y,η) of zero measure.

Thus, in the most general case we define the reflection law by the equality (1.3.13) with the additional conditions (1.3.17), plus the requirement that $x^*(t)$ and $\xi^{*\prime}(t)$ are continuous.

The application of this reflection law leads us to the discovery of a completely new phenomenon compared with the case of a second-order differential operator: for $m\geqslant 2$ the equation (1.3.13) may have several solutions ξ_n^+ ($\tau>0$) or ξ_n^- ($\tau<0$). This becomes clearer if we transform (1.3.13) to an equivalent form by raising both parts to the $2m$th power:

$$A_{2m}(x',0,\xi',\xi_n^-)\;=\;A_{2m}(x',0,\xi',\xi_n^+)\,. \tag{1.3.22}$$

Multiplying (1.3.17) by $2mh^{2m-1}$, we obtain

$$A'_{2m}(x',0,\xi',\xi_n^-)<0\,,\qquad A'_{2m}(x',0,\xi',\xi_n^+)>0\,, \tag{1.3.23}$$

where the prime at A_{2m} denotes differentiation with respect to ξ_n. At this stage we assume, of course, that

$$\operatorname{Im}\xi_n^\pm\;=\;0\,; \tag{1.3.24}$$

nonreal roots will be dealt with later, in 2.6.4.

With respect to the unknown quantity ξ_n^+ (or ξ_n^-), (1.3.22) is an algebraic equation of degree $2m$ with real coefficients. Obviously, this equation has at most m ξ_n^+-roots (or ξ_n^--roots) satisfying the conditions (1.3.23), (1.3.24) if ξ_n^- (or ξ_n^+) is given. Moreover, one can easily construct an example when the equation (1.3.22) has exactly m ξ_n^+-roots and m ξ_n^--roots satisfying the conditions (1.3.23), (1.3.24). Thus, the case $m=1$ considered in subsection 2 was a very special one: there we had a unique ξ_n^+ for each given ξ_n^- and vise versa. Figure 6 illustrates the nonuniqueness of the solution ξ_n^+ (here ξ_n^- is given).

There is no reason to exclude any of the possible ways of continuation of the trajectory $(x^*(t),\xi^*(t))$ at each reflection. We shall retain the notation $(x^*(t),\xi^*(t))$ for each of the possible continuations (in positive or negative time) of the initial

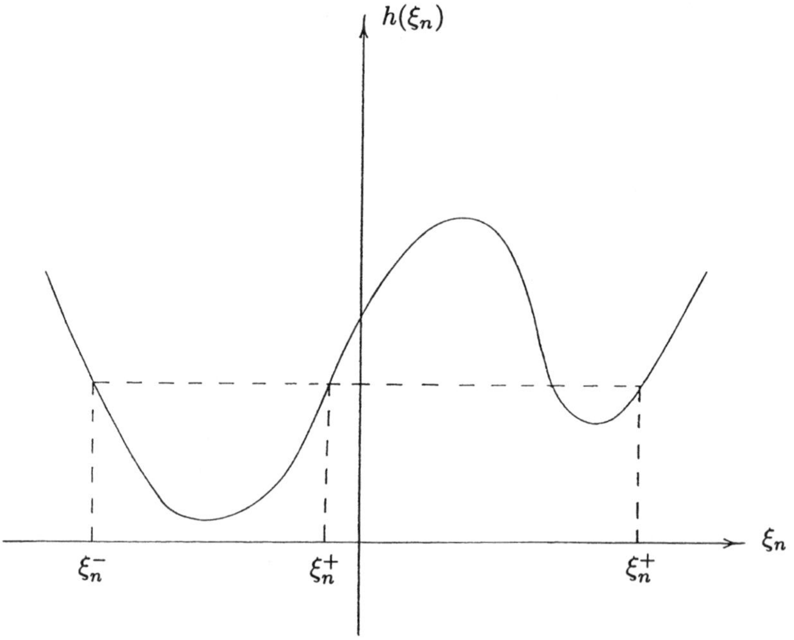

FIGURE 6. Graph of the function $h(\xi_n)$ in the case $m \geqslant 2$.

trajectory obtained by consecutive transversal reflections, and in line with Definition 1.3.10 we shall call $(x^*(t), \xi^*(t))$ a billiard trajectory. Here transversality means that both of the inequalities (1.3.17) must be fulfilled; in the case $m \geqslant 2$ these inequalities may not follow from one another as in the case $m = 1$. Considering billiard trajectories of finite length and with finitely many legs we shall always assume that $x^*(t; y, \eta)$ and $\xi^*(t; y, \eta)$ depend smoothly on $(t; y, \eta)$ at the points $(t; y, \eta)$ such that $x^*(t; y, \eta) \notin \partial M$ (see also the end of this subsection, where we introduce the notion of type of a billiard trajectory).

If for some incident trajectory there exist several reflected ones, then we shall say that the trajectory *branches*, see Figure 7. Note that if we have branching, the set of trajectories originating from a fixed point (y, η) may be uncountable.

REMARK 1.3.20. Branching of trajectories occurs also in transmission problems, in which several manifolds are connected by common stretches of boundary with appropriate boundary conditions; see [**Sa1**], [**Sa3**].

We shall call a billiard system with a Hamiltonian of general form (that is, not necessarily the square root of a quadratic polynomial) a *branching Hamiltonian billiard system*.

Let us now examine the factors that may prevent the extension of the billiard trajectory $(x^*(t; y, \eta), \xi^*(t; y, \eta))$ to all $t \in \mathbb{R}$.

Let $\mathcal{P}^{\mathrm{g}}$ be the set of points $(y, \widetilde{\eta}) \in S^*\overset{\circ}{M}$ such that at least one trajectory originating from $(y, \widetilde{\eta})$, after a finite number of transversal reflections (in positive or negative time), hits a point $(x', 0, \xi', \xi_n) \in T'M|_{\partial M}$ at which the equation $h(x', 0, \xi', \zeta) = 1$ has a multiple real ζ-root. (It may well be that $\zeta = \xi_n$ is itself a multiple root.)

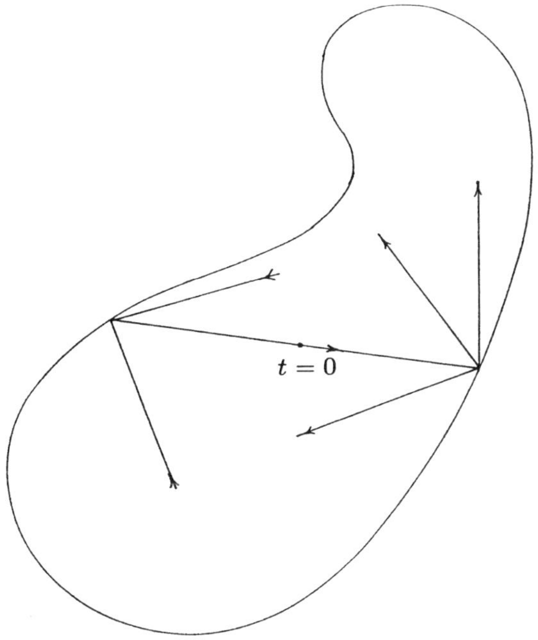

FIGURE 7. Branching of trajectories.

Let $\mathcal{P}^{\mathrm{d}}$ be the set of points $(y, \widetilde{\eta}) \in S^*\overset{\circ}{M}$ such that at least one trajectory originating from $(y, \widetilde{\eta})$ experiences an infinite number of transversal reflections in a finite (positive or negative) time. As before we call such trajectories dead-end.

If $(y, \widetilde{\eta}) \notin \mathcal{P}^{\mathrm{g}} \cup \mathcal{P}^{\mathrm{d}}$, then all trajectories originating from $(y, \widetilde{\eta})$ can be extended to all $t \in \mathbb{R}$.

LEMMA 1.3.21. *For a branching Hamiltonian billiard system,*

$$\operatorname{meas} \mathcal{P}^{\mathrm{g}} = 0. \tag{1.3.25}$$

In the general case we cannot guarantee the fulfilment of equality (1.3.19). Moreover, there exists an effective (even analytic) example [**SaVa1**] of a branching Hamiltonian billiard system with $\operatorname{meas} \mathcal{P}^{\mathrm{d}} \neq 0$. Therefore we will have to impose (1.3.19) as a condition.

DEFINITION 1.3.22. We shall say that the *nonblocking condition* is fulfilled if $\operatorname{meas} \mathcal{P}^{\mathrm{d}} = 0$.

The word *nonblocking* is used here to exhibit the fact that under this condition almost all (in terms of the measure of their starting points) trajectories can be extended to all $t \in \mathbb{R}$.

When the nonblocking condition is fulfilled, the set $\mathcal{P}^{\mathrm{g}} \cup \mathcal{P}^{\mathrm{d}}$ not only has zero measure, but it is also a set of first category in the sense of Baire.

Let us define T-periodic, absolutely T-periodic, periodic and absolutely periodic billiard trajectories in the same manner as in subsections 1 and 2 (Definitions 1.3.1, 1.3.2 plus the phrase "after a finite number of transversal reflections"). As before, we shall denote the sets of starting points of such trajectories in $S^*\overset{\circ}{M}$ by Π_T, Π_T^a, Π, and Π^a, respectively. However, we shall refrain from calling the points themselves periodic because some trajectories originating from these points may not be periodic.

The following results are analogues of Lemmas 1.3.4, 1.3.5 and Corollary 1.3.6 for branching Hamiltonian billiards.

LEMMA 1.3.23. *If the nonblocking condition is fulfilled, then the sets* Π *and* $\bigcup_{0<T\leqslant T_+}\Pi_T$, $\forall T_+>0$, *are measurable.*

LEMMA 1.3.24. *If the nonblocking condition is fulfilled, then* (1.3.7) *and* (1.3.8) *hold.*

In accordance with Definition 1.3.7, we say that the nonperiodicity condition is fulfilled if

$$\operatorname{meas}\Pi = 0\,. \tag{1.3.26}$$

It follows from Lemma 1.3.24 that if the nonblocking condition is fulfilled, then (1.3.26) is equivalent to

$$\operatorname{meas}\Pi^a = 0\,, \tag{1.3.27}$$

which is very useful in obtaining effective sufficient conditions for nonperiodicity.

Further on it will be convenient for us to avoid considering reflections at negative times. Since our Hamiltonian (1.1.14) is an even function of ξ we have

$$(x^*(-t;y,\eta),\xi^*(-t;y,\eta)) = (x^*(t;y,-\eta),-\xi^*(t;y,-\eta))\,, \tag{1.3.28}$$

so all the results for negative t can be deduced from those for positive t. We shall use this observation throughout the book. Note that if A is a pseudodifferential operator, formula (1.3.28) does not necessarily hold, so in this case one has to consider negative as well as positive t.

In the process of the effective construction of the wave group (Chapter 3) we will often be forced to restrict ourselves to some finite interval on the time axis and to exclude some awkward reflections. We introduce the following

DEFINITION 1.3.25. Let T_+ be a finite positive number. We shall say that the trajectory

$$(x^*(t;y,\eta),\xi^*(t;y,\eta))\,, \qquad 0\leqslant t\leqslant T_+\,, \quad (y,\eta)\in T'\overset{\circ}{M}\,, \tag{1.3.29}$$

is *admissible* if it satisfies the following conditions.

(1) It experiences a finite number of reflections.

(2) At each moment of reflection $t=\tau\in(0,T_+]$ all the ζ-roots of the algebraic equation

$$A_{2m}(x^{*\prime}(\tau;y,\eta),0,\xi^{*\prime}(\tau;y,\eta),\zeta) = A_{2m}(y,\eta) \tag{1.3.30}$$

are simple.

(3) At each moment of reflection $t=\tau\in(0,T_+]$ the number $\nu=A_{2m}(y,\eta)$ is not an eigenvalue of the auxiliary one-dimensional spectral problem (1.1.7), (1.1.8) with $(x',\xi')=(x^{*\prime}(\tau;y,\eta),\xi^{*\prime}(\tau;y,\eta))\in T^*\partial M$.

The second condition of Definition 1.3.25 implies, in particular, that the reflections are transversal. However this condition has wider implications. When we say "the ζ-roots of the algebraic equation are simple" we mean *all the roots in the complex plane*, not only the real ones. The role of the nonreal roots will become clear in Sections 2.6 and 2.8 when we introduce the concept of a boundary layer oscillatory integral. In a sense, the nonreal ζ-roots correspond to Hamiltonian trajectories which leave the real space after reflection and become complex (in the analytic situation this statement has a precise meaning). Obviously, the equation (1.3.30) may have nonreal ζ-roots only if $m \geqslant 2$.

Note also that for a second order operator the third condition of Definition 1.3.25 is always fulfilled.

DEFINITION 1.3.26. We shall say that $(y,\eta) \in T'\overset{\circ}{M}$ is a T_+-admissible point if all the trajectories (1.3.29) originating from this point are T_+-admissible. We shall say that $(y,\eta) \in T'\overset{\circ}{M}$ is an admissible point if it is T_+-admissible for all $T_+ > 0$. A subset of $T'\overset{\circ}{M}$ is said to be T_+-admissible (admissible) if all its points are T_+-admissible (admissible).

Denote by O_{T_+} and O_∞ the sets of all T_+-admissible and admissible points in $T'\overset{\circ}{M}$, respectively.

LEMMA 1.3.27. *For any finite positive T_+ the set O_{T_+} is open.*

LEMMA 1.3.28. *If the nonblocking condition is fulfilled, then* $\operatorname{vol}(T'\overset{\circ}{M} \setminus O_\infty) = \operatorname{meas}(S^*\overset{\circ}{M} \setminus O_\infty) = 0$.

Further on we prefer to deal with T_+-admissible billiard trajectories. Most of our results and definitions can be extended to all well-defined billiard trajectories, so it is a rather technical assumption. In view of Lemma 1.3.28 this will be sufficient for our purposes. Often we shall not distinguish between Π_{T_+} and $\Pi_{T_+} \cap O_{T_+}$, $\Pi^a_{T_+}$ and $\Pi^a_{T_+} \cap O_{T_+}$, Π and $\Pi \cap O_\infty$, Π^a and $\Pi^a \cap O_\infty$.

Consider a T_+-admissible billiard trajectory (1.3.29) experiencing $\mathbf{r}$ reflections on the time interval $(0, T_+)$. We can associate with this trajectory its *type*, which is the multiindex $\mathfrak{m} = \mathfrak{m}_1|\mathfrak{m}_2|\dots|\mathfrak{m}_i|\dots|\mathfrak{m}_\mathbf{r}$ defined as follows. If $\tau_i \in (0, T_+)$ is the moment of the ith reflection, then $\xi^*_n(\tau_i + 0; y, \eta)$ is the $2\mathfrak{m}_i$th (in increasing order) real ζ-root of the equation (1.3.30). The notion of a billiard trajectory being T_+-admissible ensures that all the real ζ-roots of the equation (1.3.30) are simple; consequently, for any real root with even sequential number we have $h_{\xi_n}(x^{*\prime}(\tau_i; y, \eta), 0, \xi^{*\prime}(\tau_i; y, \eta), \zeta) > 0$, i.e., it really corresponds to a reflected trajectory.

LEMMA 1.3.29. *The number of types of billiard trajectories originating from a T_+-admissible point is finite.*

Let $(x^*(t; y_0, \eta_0), \xi^*(t; y_0, \eta_0))$ be a T_+-admissible billiard trajectory of type $\mathfrak{m}$ such that T_+ is not a moment of reflection. Clearly, if we fix the type $\mathfrak{m}$, then $(x^*(t; y, \eta), \xi^*(t; y, \eta))$ smoothly depends on $(t; y, \eta)$ provided (y, η) is sufficiently close to (y_0, η_0) and t is not a moment of reflection.

The branching of trajectories does not allow us to introduce the group of shifts Φ^t along billiard trajectories (cf. subsection 2). However, shifts along admissible billiard trajectories of fixed type preserve the symplectic differential 1-form and 2-form, as well as the symplectic volume (see Section 2.3).

4. Simple reflection. An important class of Hamiltonian billiard systems is the class of systems satisfying the simple reflection condition.

DEFINITION 1.3.30. We shall say that the *simple reflection condition* is fulfilled if for any $(x', \xi') \in T'\partial M$ the Hamiltonian $h(x', 0, \xi', \xi_n)$ has only one local (and hence global) minimum as a function of $\xi_n \in \mathbb{R}$.

In other words, the simple reflection condition is fulfilled if and only if there is no branching.

LEMMA 1.3.31 (generalization of the second part of Lemma 1.3.11). *If the Hamiltonian billiard system satisfies the simple reflection condition, then the non-blocking condition is fulfilled, i.e.,* $\operatorname{meas} \mathcal{P}^{\mathrm{d}} = 0$.

For a Hamiltonian billiard system satisfying the simple reflection condition we can introduce the group of shifts Φ^t along billiard trajectories. This billiard flow has properties similar to those of the geodesic billiard flow.

DEFINITION 1.3.32. We shall say that the *strong simple reflection condition* is fulfilled if for any $(x', \xi') \in T'\partial M$ the equation $h_{\xi_n}(x', 0, \xi', \xi_n) = 0$ has only one real ξ_n-root and this root is simple ($h_{\xi_n \xi_n} \neq 0$).

REMARK 1.3.33. Geodesic billiards obviously satisfy the strong simple reflection condition.

Hamiltonian billiards satisfying the strong simple reflection condition are a subclass of Hamiltonian billiards satisfying the simple reflection condition. Billiards from this subclass are of special interest because **all** the facts described in subsection 2 (devoted to geodesic billiards) remain true for them as well. In particular, Lemma 1.3.19 remains true; due to its importance we shall state it again.

LEMMA 1.3.34 (generalization of Lemma 1.3.19). *Let us consider the analytic case and let the strong simple reflection condition be satisfied, the manifold M be convex (see Definitions* 1.3.12, 1.3.13), *$\partial M \neq \varnothing$, and $\mathbf{k}(x', \xi') \not\equiv 0$. Then the nonperiodicity condition is fulfilled.*

Note that under the assumptions of Lemma 1.3.34 the nonblocking condition is fulfilled as well, by virtue of Lemma 1.3.31.

In the case of a manifold without boundary there are no reflections, but in order to simplify our subsequent statements we shall assume throughout the book that a billiard system without reflections satisfies the strong simple reflection condition.

Concluding this subsection, let us make a remark on the relationship between the concepts of nonperiodicity and ergodicity. As pointed out earlier, a Hamiltonian billiard system satisfying the simple reflection condition generates the group of shifts Φ^t which preserves the symplectic volume; such groups are called *flows*. A flow is called *ergodic* if any set invariant with respect to this flow has zero or full measure [**CorFomSin**]. If the billiard flow is ergodic, then the nonperiodicity condition is fulfilled. The converse is not true in general. An example is the case of Euclidean billiards in a closed domain $M \subset \mathbb{R}^2$ bounded by an ellipse. The fact that such a billiard system is not ergodic is well known [**CorFomSin,** Sect. 6.3], whereas its nonperiodicity follows from Lemma 1.3.19 or Lemma 1.3.34. Thus, nonperiodicity is a weaker (in fact, considerably weaker) requirement than ergodicity. This explains why constructing effective sufficient conditions for nonperiodicity is easier and more rewarding than constructing effective sufficient conditions for ergodicity.

5. Euclidean billiards. Let us finally look at the most basic billiards, that is, Euclidean billiards. Euclidean billiards are a special case of geodesic billiards.

In this case the manifold M is a region in $\mathbb{R}^n$, the metric is Euclidean, the rays are straight lines, and reflections from the boundary satisfy the usual "angle of incidence equals angle of reflection" law.

For Euclidean billiards the nonblocking condition is always fulfilled due to simple reflection. But nonperiodicity has to be checked.

We state the following

CONJECTURE 1.3.35. *In the case of a Euclidean billiard system there are no absolutely periodic billiard trajectories.*

Note that this conjecture, if proven, would imply the nonperiodicity of Euclidean billiard systems (see Lemma 1.3.6).

The arguments in favour of this conjecture are the following.

If one assumes the existence of an absolutely periodic billiard trajectory, this leads to certain algebraic equations involving the Taylor coefficients which describe the shape of the boundary near the points of reflection. There are infinitely many algebraic equations and they involve infinitely many Taylor coefficients, which makes analysis difficult. However, one can truncate this infinite system of equations and look at the resulting finite subsystem of l equations. One can check that if l is sufficiently large, then the number of unknowns (Taylor coefficients) is less than l. Therefore it is highly unlikely that this system has a solution.

There is also a physical argument in favour of Conjecture 1.3.35. If the conjecture were not true, one could construct an ideal optical system with a finite number of mirrors. Here "ideal" means focusing up to infinite order, i.e., without any aberrations.

Conjecture 1.3.35 is purely geometrical, and is remarkably simple and natural. It contains no direct reference to partial differential equations, spectral theory, symplectic geometry, measure theory, etc. The authors, however, are unaware of a mathematical proof of this conjecture.

We know Conjecture 1.3.35 to be true only for special classes of shapes of M. In particular, we know it to be true when M is convex (in the usual sense) and ∂M analytic; see proof of Lemma 1.3.34 given in Appendix D. If one allows piecewise smooth boundaries [**Va6**], we can state two other sets of conditions under which the conjecture is true.

1. Each smooth component of ∂M has nonpositive normal curvature (say, a polyhedron satisfies this condition). In this case a divergent beam of rays becomes more and more divergent after each reflection and cannot focus. This is a standard argument from ergodic theory.
2. A very special set of conditions [**Wo**].

1.4. Hamiltonian billiards II: reflection matrix

As we shall discover later (Sections 3.3, 3.4), the physical meaning of a branching Hamiltonian billiard system is that it describes the propagation of waves governed by the equation $D_t^{2m}u = Au$ (analogue of the wave equation). Billiard trajectories trace the movement of these waves, and our billiard trajectories play the role of a "skeleton" upon which we will consequently build the wave group $\exp(-it\mathcal{A}^{1/(2m)})$.

Tracing the path of a wave is only the first (and simplest) step. The next step is to describe the partition of energy between different branches of reflected waves. The notion of the reflection matrix introduced in this section serves this purpose.

The reflection matrix is not used directly in the formulation of the main result of Section 1.6. However, it is used later on in Section 1.6 when we perform the effective calculation of the second asymptotic coefficient.

Under the simple reflection condition energy is not redistributed at reflections (there is no branching), and the reflection matrix is just a complex number of the form $e^{i\mu}$, $\mu \in \mathbb{R}$. In this case it is natural to study the *phase shift* which the wave gains in the process of its propagation. The phase shift comes from three sources:

1. Phase shift induced by reflections from the boundary (the sum of arguments μ of the corresponding reflection matrices).
2. Phase shift induced by the passage of the trajectory through caustics.
3. Phase shift induced by the subprincipal symbol of the differential operator A.

This section deals with the first source and Section 1.5 deals with the second one. Dealing with the third source will not require a special section because the corresponding phase shift will be given by a simple integral over the billiard trajectory (see 1.7.2). The notion of the phase shift is necessary for the formulation of the results of Sections 1.7 and 1.8.

Let us fix an arbitrary point $(x', \xi') \in T^*\partial M$ and consider the one-dimensional spectral problem (1.1.7), (1.1.8) on the half-line $x_n \in \mathbb{R}_+$. For the time being let us omit the parameters (x', ξ') and rewrite (1.1.7), (1.1.8) as

$$A(D_{x_n})v = \nu v\,, \tag{1.4.1}$$

$$\left(B^{(j)}(D_{x_n})v\right)\Big|_{x_n=0} = 0\,, \qquad j = 1, 2, \dots, m\,, \tag{1.4.2}$$

where $A(D_{x_n}) = A_{2m}(x', 0, \xi', D_{x_n})$, $B^{(j)}(D_{x_n}) = B^{(j)}_{m_j}(x', \xi', D_{x_n})$. Obviously, the spectral problem (1.4.1), (1.4.2) is formally self-adjoint with respect to the inner product

$$(v, w)_+ = \int_0^{+\infty} v(x_n)\,\overline{w(x_n)}\,dx_n\,, \tag{1.4.3}$$

and one can associate with (1.4.1), (1.4.2) a self-adjoint operator $\mathbf{A}^+$ in $L_2(\mathbb{R}_+)$.

The equation (1.4.1) is an ordinary differential equation of order $2m$ with constant coefficients and it can be explicitly solved in exponential functions. This simple observation opens the way to the effective study of the spectral problem (1.4.1), (1.4.2). In fact, all the constructions described below (as well as in 1.6.3 and Appendix A) are of purely algebraic nature.

In this section we give a list of basic properties of the spectral problem (1.4.1), (1.4.2), referring the reader to Appendix A for details.

The problem (1.4.1), (1.4.2) has a finite number of eigenvalues. Moreover, this number is uniformly bounded over all problems of this type with fixed order $2m$; see Proposition A.1.13.

The problem (1.4.1), (1.4.2) has no singular continuous spectrum (see Remark A.2.8). As the number of eigenvalues is finite, the continuous spectrum coincides with the essential one. Thus, the spectrum of (1.4.1), (1.4.2) is the union of the (absolutely) continuous spectrum and the set of eigenvalues. Note that some eigenvalues may be embedded in the continuous spectrum.

The continuous spectrum of the problem (1.4.1), (1.4.2) is the interval $[\nu_1^{\mathrm{st}}, +\infty)$, where $\nu_1^{\mathrm{st}} = \min_{\xi_n \in \mathbb{R}} A(\xi_n)$ (see A.1.4 and A.1.5).

We shall call the number ν^{st} a *threshold* (or a *stationary value* of the symbol) if the equation $A(\xi_n) = \nu^{\mathrm{st}}$ has a multiple real ξ_n-root. Let us enumerate the thresholds: $\nu_1^{\mathrm{st}} < \nu_2^{\mathrm{st}} < \cdots < \nu_s^{\mathrm{st}}$. Clearly, $1 \leqslant s \leqslant 2m-1$. After the removal of thresholds the continuous spectrum separates into zones

$$(\nu_1^{\mathrm{st}}, \nu_2^{\mathrm{st}}),\ (\nu_2^{\mathrm{st}}, \nu_3^{\mathrm{st}}), \ldots,\ (\nu_{s-1}^{\mathrm{st}}, \nu_s^{\mathrm{st}}),\ (\nu_s^{\mathrm{st}}, +\infty). \tag{1.4.4}$$

Further on we use the (scalar) variable ζ instead of ξ_n in order to simplify notation.

Assume that ν belongs to one of the zones (1.4.4), and consider the algebraic equation

$$A(\zeta) = \nu. \tag{1.4.5}$$

By $2q$ we shall denote the number of real ζ-roots of this equation. The number q is called the *multiplicity of the continuous spectrum* at the point ν. Obviously, this number is the same for all ν from a given zone (1.4.4), and $1 \leqslant q \leqslant m$.

It will be convenient for us to impose temporarily the following technical conditions (which are related to the admissibility condition).

CONDITION 1.4.1. Our ν is such that the complex ζ-roots of (1.4.5) are simple.

CONDITION 1.4.2. Our ν is not an eigenvalue of (1.4.1), (1.4.2).

Let us denote the real ζ-roots of (1.4.5) by $\zeta^{\mp}(\nu)$, where the superscript $\mp$ indicates the sign of the derivative $A'(\zeta^{\mp}(\nu))$, $A' \equiv dA/d\zeta$; note that this derivative is nonzero because our ν is not a threshold. Let us enumerate the real roots $\zeta^{\mp}(\nu)$ in increasing order:

$$\zeta_1^-(\nu) < \zeta_1^+(\nu) < \zeta_2^-(\nu) < \zeta_2^+(\nu) < \cdots < \zeta_q^-(\nu) < \zeta_q^+(\nu). \tag{1.4.6}$$

Formula (1.4.6) is illustrated by Figure 8 (here $q = 2$). Note that apart from different notation Figure 8 is essentially the same as Figure 6 from 1.3.3.

Let us denote by $\zeta_l^{\mp}(\nu)$, $l = q+1,\ q+2, \ldots,\ m$, the complex roots of (1.4.5) with negative and positive imaginary part, respectively. Note that under Condition 1.4.1 we have $A'(\zeta^{\mp}(\nu)) \neq 0$, $l = q+1,\ q+2, \ldots,\ m$.

Let us call nontrivial bounded solutions of (1.4.1), (1.4.2) *generalized eigenfunctions* (or *eigenfunctions of the continuous spectrum*) corresponding to our ν. As we have excluded the eigenvalues (Condition 1.4.2), the generalized eigenfunctions defined in this way are not in $L_2(\mathbb{R}_+)$. With account of Condition 1.4.1 we shall search for generalized eigenfunctions in the form

$$v(x_n) = \sum_{l=1}^{q} \frac{a_l^- \, e^{ix_n\zeta_l^-(\nu)}}{\sqrt{-2\pi A'(\zeta_l^-(\nu))}} + \sum_{l=1}^{m} \frac{a_l^+ \, e^{ix_n\zeta_l^+(\nu)}}{\sqrt{2\pi A'(\zeta_l^+(\nu))}}, \tag{1.4.7}$$

where the a_l^-, $l = 1, 2, \ldots, q$, and a_l^+, $l = 1, 2, \ldots, m$, are some complex constants. On the right-hand side of (1.4.7) all the square roots in the first sum and the first q square roots in the second sum are chosen to be positive, whereas the last $m-q$ square roots in the second sum can be chosen arbitrarily. The normalizing factors

$$\frac{1}{\sqrt{\mp 2\pi A'(\zeta_l^{\mp}(\nu))}}$$

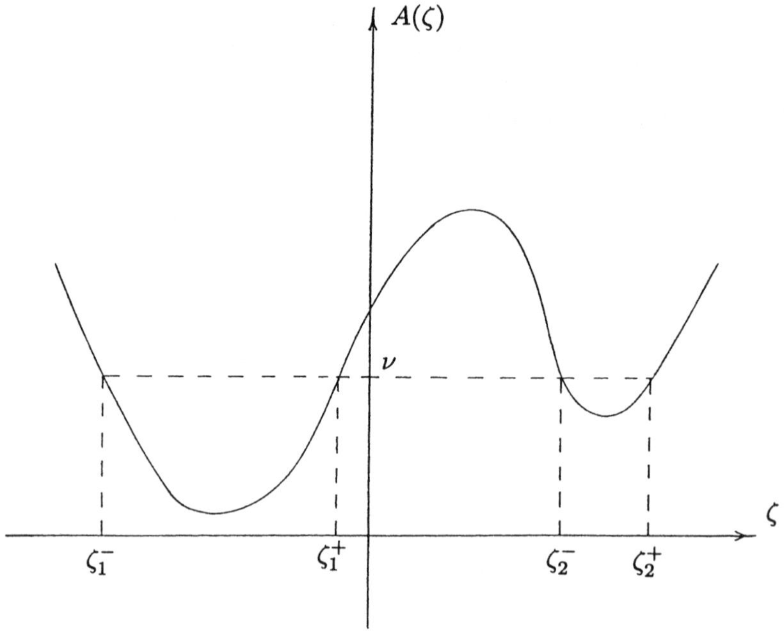

FIGURE 8. Real roots of the equation (1.4.5).

in (1.4.7) are introduced for the sake of convenience, in order to make the reflection matrix defined below a unitary one. Another reason for introducing these normalizing factors is that they will naturally appear when we will start dealing with our global invariant oscillatory integrals, see Lemmas 2.9.10, 2.9.11.

Proposition A.2.1 establishes that the linear space of generalized eigenfunctions is q-dimensional (this justifies our notion of multiplicity defined above) and that each generalized eigenfunction (1.4.7) is uniquely defined by the set of coefficients a_l^-, $l=1,2,\ldots,q$, or by the set of coefficients a_l^+, $l=1,2,\ldots,q$. Denote by a^- and a^+ the columns of coefficients a_l^-, $l=1,2,\ldots,q$, and a_l^+, $l=1,2,\ldots,q$, respectively. As a generalized eigenfunction is uniquely determined by either of the two columns a^- or a^+, we have a linear relation

$$a^+ = R(\nu)\,a^-, \tag{1.4.8}$$

where $R(\nu)$ is an invertible $q\times q$ matrix. The matrix $R(\nu)$ can be effectively constructed by substituting (1.4.7) into the boundary conditions (1.4.2) and solving the resulting system of linear algebraic equations with respect to the a_l^+.

The matrix $R(\nu)$ is called the *reflection matrix*. By Proposition A.2.1 the reflection matrix is unitary.

Obviously, the matrix $R(\nu)$ is real analytic as a function of ν in each of the zones (1.4.4) apart, maybe, from a finite number of points at which either of Conditions 1.4.1 or 1.4.2 fails. By Proposition A.2.1 $R(\nu)$ admits an analytic extension to such points. Thus $R(\nu)$ is well defined and real-analytic in each of the zones (1.4.4).

The thresholds are, generally speaking, branching points for the elements of $R(\nu)$. In addition, passage through a threshold normally results in a change of the size of $R(\nu)$; this corresponds to the change of the multiplicity of the continuous spectrum.

Note, however, that the elements of the reflection matrix always have one-sided limits at thresholds. This follows from the fact that these elements are bounded (recall that the matrix is unitary!) and that the branching is of root type.

DEFINITION 1.4.3. We shall say that the one-dimensional problem (1.4.1), (1.4.2) satisfies the *simple reflection condition* if the continuous spectrum has multiplicity one in all the zones (1.4.4).

Of course, in the case of simple reflection the reflection matrix $R(\nu)$ is of size 1×1 (complex number) in all the zones (1.4.4), and in view of (1.4.8) $R(\nu) = a_1^+/a_1^-$. Naturally, $|R(\nu)| = 1$. The quantity $\arg R(\nu)$ is called the *phase shift generated by the reflection*. This phase shift is defined modulo 2π.

DEFINITION 1.4.4. We call the threshold ν^{st} *normal* if the algebraic equation $A(\zeta) = \nu^{\mathrm{st}}$ has only one multiple real root $\zeta = \zeta^{\mathrm{st}}$ and $A''(\zeta^{\mathrm{st}}) \neq 0$.

Note that the situation of a normal threshold can be viewed as a generic one, in the sense that all the thresholds can be made normal by a small self-adjoint perturbation of the coefficients of our spectral problem (1.4.1), (1.4.2).

DEFINITION 1.4.5. We shall say that the one-dimensional problem (1.4.1), (1.4.2) satisfies the *strong simple reflection condition* if the problem has only one threshold and this threshold is normal.

Definitions 1.4.3, 1.4.5 given above match well with Definitions 1.3.30, 1.3.32. Namely, the one-dimensional problem (1.4.1), (1.4.2) satisfies the simple reflection condition (strong simple reflection condition) for all $(x',\xi') \in T'\partial M$ if and only if the Hamiltonian billiard system satisfies the simple reflection condition (strong simple reflection condition).

From now on it will be convenient for us to indicate the dependence of all our quantities on the parameters $(x',\xi') \in T^*\partial M$.

Let us examine how the reflection matrix $R(\nu; x', \xi')$ behaves under changes of local coordinates. Let $x = x(\widetilde{x})$ be a change of local coordinates on M such that $x_n \equiv \widetilde{x}_n$ (see 1.1.2), and let

$$f'(\widetilde{x}') := \left. \frac{\partial x'}{\partial \widetilde{x}_n} \right|_{\widetilde{x}_n = 0}.$$

Our original change of local coordinates on M generates a change of local coordinates $x' = x'(\widetilde{x}') := x'(\widetilde{x}', 0)$ on ∂M, which leads to a change of dual coordinates

$$\xi' = \xi'(\widetilde{x}', \widetilde{\xi}') := \sum_{k=1}^{n-1} \widetilde{\xi}_k \left. \frac{\partial \widetilde{x}_k}{\partial x'} \right|_{x' = x'(\widetilde{x}')}$$

on the fibres of $T'\partial M$.

Suppose now that $v(x_n; x', \xi')$ is a solution of the problem (1.1.7), (1.1.8). Set

$$\widetilde{v}(\widetilde{x}_n; \widetilde{x}', \widetilde{\xi}') := e^{i\widetilde{x}_n \langle f'(\widetilde{x}'), \xi'(\widetilde{x}', \widetilde{\xi}')\rangle} v(\widetilde{x}_n; x'(\widetilde{x}'), \xi'(\widetilde{x}', \widetilde{\xi}')). \tag{1.4.9}$$

Direct substitution demonstrates that this function is a solution of the problem

$$A_{2m}(\widetilde{x}', 0, \widetilde{\xi}', D_{\widetilde{x}_n})\widetilde{v} = \nu \widetilde{v},$$

$$\left(B^{(j)}_{m_j}(\widetilde{x}', \widetilde{\xi}', D_{\widetilde{x}_n})\widetilde{v}\right)\Big|_{\widetilde{x}_n=0} = 0\,, \qquad j = 1, 2, \ldots, m\,.$$

Formula (1.4.9) implies that under changes of local coordinates x the ordering of the real roots (1.4.6) is preserved and, moreover, the reflection matrix $R(\nu; x', \xi')$ behaves as a function on $T'\partial M$. In the case of simple reflection the same applies to the phase shift.

It remains only to relate the quantities defined above to billiard trajectories. Let $(x^*(t; y, \eta), \xi^*(t; y, \eta))$ be a trajectory which experiences a reflection at $t = t^*(y, \eta)$. Then

$$\nu = A_{2m}(y, \eta)\,, \qquad (x', \xi') = (x^{*\prime}(t^*(y, \eta); y, \eta), \xi^{*\prime}(t^*(y, \eta); y, \eta)) \in T^*\partial M$$

are the values of parameters corresponding to this particular reflection, and these are the values to be substituted into all the functions defined in this section. In particular, we get

$$\xi^*_n(t^*(y, \eta) - 0; y, \eta) = \zeta^-_k(A_{2m}(y, \eta); x^{*\prime}(t^*(y, \eta); y, \eta), \xi^{*\prime}(t^*(y, \eta); y, \eta))\,,$$
$$\xi^*_n(t^*(y, \eta) + 0; y, \eta) = \zeta^+_l(A_{2m}(y, \eta); x^{*\prime}(t^*(y, \eta); y, \eta), \xi^{*\prime}(t^*(y, \eta); y, \eta))$$

for some k and l. Consequently

$$R_{lk}(A_{2m}(y, \eta); x^{*\prime}(t^*(y, \eta); y, \eta), \xi^{*\prime}(t^*(y, \eta); y, \eta))$$

is the element of the reflection matrix corresponding to our particular reflection; here l is the number of the row, and k is the number of the column.

In the case of simple reflection,

$$\arg R(A_{2m}(y, \eta); x^{*\prime}(t^*(y, \eta); y, \eta), \xi^{*\prime}(t^*(y, \eta); y, \eta))$$

is the phase shift generated by our particular reflection. Recall that in the case of simple reflection,

$$|R(A_{2m}(y, \eta); x^{*\prime}(t^*(y, \eta); y, \eta), \xi^{*\prime}(t^*(y, \eta); y, \eta))| = 1\,,$$

and consequently the 1×1 reflection matrix is uniquely determined by the phase shift. The phase shift itself is, of course, defined modulo 2π; we shall always choose it to be locally continuous in (y, η).

1.5. Hamiltonian billiards III: Maslov index

In this section we introduce an important geometric characteristic of a billiard trajectory which is called the *Maslov index*. We also state some simple results which justify the definitions and can be used for the calculation of the Maslov index. These results will be proved in Appendix D.

In geometrical optics the Maslov index, multiplied by $-\pi/2$, is interpreted as the phase shift generated by the passage of the trajectory through caustics. It should be mentioned that the words "Maslov index" are often used for absolutely different objects and, probably, in our case it would be more appropriate to call it the Morse index rather than the Maslov index. However, we will follow the tradition.

The notion of the Maslov index introduced in this section is necessary for the formulation of the results of Sections 1.7, 1.8, but it is not needed in Section 1.6.

Consider a T-admissible billiard trajectory

$$(1.5.1)_0 \qquad \Gamma = \left(x^*(t; y_0, \eta_0)\,, \xi^*(t; y_0, \eta_0)\right), \qquad 0 \leqslant t \leqslant T\,,$$

such that $x^*(T;y_0,\eta_0) \notin \partial M$. Let $O \in T'M$ be a sufficiently small conic neighbourhood of the point (y_0,η_0) and

$$(1.5.1) \qquad (x^*(t;y,\eta)\,,\,\xi^*(t;y,\eta))\,, \qquad 0 \leqslant t \leqslant T\,, \quad (y,\eta) \in O\,,$$

be the family of T-admissible billiard trajectories with the same type (see the end of 1.3.3). We denote by $0 < t_1^*(y,\eta) < \cdots < t_{\mathbf{r}}^*(y,\eta) < T$ the moments of reflection.

Let us introduce in local coordinates the matrices x_η^* and ξ_η^* with elements $(x_j^*)_{\eta_i}$ and $(\xi_j^*)_{\eta_i}$, respectively (j being the number of the row and i that of the column). Certainly, at the moments of reflection these matrices are not smooth in t, and at these points we have to consider the matrices $x_\eta^*(t_k^*(y,\eta)-0;y,\eta)$, $\xi_\eta^*(t_k^*(y,\eta)-0;y,\eta)$ and $x_\eta^*(t_k^*(y,\eta)+0;y,\eta)$, $\xi_\eta^*(t_k^*(y,\eta)+0;y,\eta)$.

Throughout this section we assume that

$$(1.5.2) \qquad x_\eta^*(T;y_0,\eta_0) \;=\; 0\,.$$

Under a change of coordinates, x_η^* behaves as a vector in x and as a vector in y, i.e., in new coordinates $\widetilde{x}_{\widetilde{\eta}}^* = (\partial\widetilde{x}/\partial x)\cdot x_\eta^*\cdot(\partial\widetilde{y}/\partial y)^T$ where $\partial\widetilde{x}/\partial x$ and $\partial\widetilde{y}/\partial y$ are the Jacobi matrices. Therefore the condition (1.5.2) is invariant under changes of coordinates.

The matrix ξ_η^* does not behave as a tensor. We will discuss its properties in Section 2.3. In particular, we will prove (Lemma 2.3.2) that for any fixed point $(t_0;y_0,\eta_0)$ and for any coordinates y one can choose coordinates x in a neighbourhood of $x^*(t_0;y_0,\eta_0)$ such that

$$(1.5.3) \qquad \det \xi_\eta^*(t;y,\eta) \;\neq\; 0$$

for $(t;y,\eta)$ close to $(t_0;y_0,\eta_0)$.

1. First definition. We start with a definition which is closer to the standard definition of the Morse index.

Let us choose functions $t_j(y,\eta) \in C^\infty(O)$, $j = 0,\dots,N$, and coordinate patches $\Omega_j \subset M_x$, $j = 0,\dots,N-1$, in such a way that

1. $0 \equiv t_0(y,\eta) < t_1(y,\eta) < \cdots < t_{N-1}(y,\eta) < t_N(y,\eta) \equiv T$.
2. For each $k = 1,2,\dots,\mathbf{r}$ there exists a j_k such that $t_{j_k}(y,\eta) \equiv t_k^*(y,\eta)$.
3. $x^*(t;y,\eta) \in \Omega_j$ for all $(t;y,\eta) \in \check{\mathfrak{D}}_j$, $j = 0,1,\dots,N-1$.
4. The local coordinates defined on Ω_j satisfy (1.5.3) for all $(t;y,\eta) \in \check{\mathfrak{D}}_j$, $j = 0,\dots,N-1$ (here we may need to use coordinates in which $\partial M \neq \{x_n = 0\}$, see Remark 2.3.3).

We use the notation

$$(1.5.4) \qquad \check{\mathfrak{D}}_j \;:=\; \{(t;y,\eta)\;:\;(y,\eta) \in O\,,\; t \in [t_j(y,\eta), t_{j+1}(y,\eta)]\}\,,$$

$j = 0,1,\dots,N-1$. Clearly, conditions (1), (2) imply that $N \geqslant \mathbf{r}+1$.

Let us introduce the matrix functions $C_j = C_j(t;y,\eta) := (\xi_\eta^*)^T\cdot x_\eta^*$ defined on the sets (1.5.4). If $t = t_{j_k}(y,\eta) = t_k^*(y,\eta)$ is a moment of reflection, then we set $C_{j_k-1}(t;y,\eta) := C_{j_k-1}(t-0;y,\eta)$ and $C_{j_k}(t;y,\eta) := C_{j_k}(t+0;y,\eta)$. Since the shift along billiard trajectories preserves the canonical 2-form $dx\wedge d\xi$, the matrices C_j are symmetric, i.e., $C_j = (x_\eta^*)^T\cdot\xi_\eta^*$ (see Section 2.3). Since the functions x_j^* are homogeneous in η of degree 0, by the Euler identity we have $x_\eta^*\eta \equiv 0$. Therefore $\operatorname{rank} x_\eta^*(t;y,\eta) \leqslant n-1$ and, consequently, $\operatorname{rank} C_j(t;y,\eta) \leqslant n-1$ for all j and $(t;y,\eta) \in \check{\mathfrak{D}}_j$.

Let $r_j^-(t;y,\eta)$ be the number of strictly negative eigenvalues of the matrix $C_j(t;y,\eta)$. By $r_j^-(t\pm 0;y,\eta)$ we shall denote the one-sided limits of the functions r_j^- (whenever these limits exist).

DEFINITION 1.5.1. The integer

$$\alpha_\Gamma \;=\; -\sum_{j=0}^{N-1}\left(r_j^-(t_{j+1}(y_0,\eta_0);y_0,\eta_0) \;-\; r_j^-(t_j(y_0,\eta_0);y_0,\eta_0)\right) \tag{1.5.5}$$

is called the Maslov index of the trajectory Γ.

PROPOSITION 1.5.2. *Let $\widetilde{x}$ be another coordinate system on Ω_j satisfying (1.5.3) and $\widetilde{r}_j^-(t;y,\eta)$ be the number of negative eigenvalues of the corresponding matrix $\widetilde{C}_j(t;y,\eta)$. Then*

$$r_j^-(t;y,\eta) \;-\; \widetilde{r}_j^-(t;y,\eta) \;=\; \text{const}\,,$$

i.e., this difference is independent of $(t;y,\eta)\in\check{\mathfrak{D}}_j$.

By Proposition 1.5.2 the difference

$$r_j^-(t_{j+1}(y_0,\eta_0);y_0,\eta_0) \;-\; r_j^-(t_j(y_0,\eta_0);y_0,\eta_0)$$

does not depend on the choice of the coordinates x. This implies that the definition of the Maslov index is independent of the choice of the t_j, the covering $\{\Omega_j\}$ and the coordinate systems on Ω_j. Besides, Proposition 1.5.2 shows that the jumps $r_j^-(t+0;y,\eta)-r_j^-(t;y,\eta)$ and $r_j^-(t;y,\eta)-r_j^-(t-0;y,\eta)$ are invariant objects. Up to the factor -1, the Maslov index α_Γ is equal to the sum of all these jumps along the billiard trajectory Γ provided the number of jumps is finite.

LEMMA 1.5.3. *For all $k=1,\dots,\mathbf{r}$*

$$\operatorname{rank} x_\eta^*(t_k^*(y,\eta)-0;y,\eta) \;=\; \operatorname{rank} x_\eta^*(t_k^*(y,\eta)+0;y,\eta)$$

(*here we first take the limit in t and then evaluate the rank*).

Let $\mathcal{R} = \mathcal{R}(t;y,\eta) := \operatorname{rank} x_\eta^*(t;y,\eta)$. By Lemma 1.5.3 $\mathcal{R}(t;y,\eta)$ is well defined at the points of reflection, and we can consider it as a function on $[0,T]\times O$.

If $\mathcal{R}(t;y,\eta)$ is constant on some subset of (1.5.4), then $r_j^-(t;y,\eta)$ is also constant on this subset. This means that the parts of the trajectory where $\mathcal{R}$ does not change do not contribute to the Maslov index. Thus, the Maslov index α_Γ depends only on the behaviour of the trajectory Γ in the neighbourhoods of the points at which $\mathcal{R}$ has jumps.

If $\mathcal{R}(t;y,\eta)<n-1$, then the point $x^*(t;y,\eta)$ is said to be the *conjugate point* of the ray $x^*(t;y,\eta)$, and the number $n-1-\mathcal{R}(t;y,\eta)$ is called its multiplicity. In geometrical optics the set of conjugate points of the rays starting at a fixed point y is often called the *caustic set* (or just *caustic*). It is known that when a ray $x^*(t;y,\eta)$ passes through the caustic set the wave amplitude is multiplied by

$$\exp\left(i\pi\,(r_j^-(t+0;y,\eta)-r_j^-(t-0;y,\eta))/2\right),$$

or, in other words, the wave acquires the phase shift

$$\pi\left(r_j^-(t+0;y,\eta) \;-\; r_j^-(t-0;y,\eta)\right)/2\,.$$

Thus, the quantity $\mathfrak{f}_c(t; y_0, \eta_0) = \alpha_\Gamma \pi/2$ is the total phase shift generated by the passage of the ray $x^*(t; y_0, \eta_0)$ through caustics.

REMARK 1.5.4. Since $C_0(0; y, \eta) \equiv 0$ and $C_{N-1}(T; y_0, \eta_0) = 0$,

$$\alpha_\Gamma = -\sum_{j=1}^{N-1} \left(r_{j-1}^-(t_j(y_0, \eta_0); y_0, \eta_0) - r_j^-(t_j(y_0, \eta_0); y_0, \eta_0) \right). \tag{1.5.6}$$

Let $\partial M = \varnothing$. Then by Proposition 1.5.2 for any $j = 1, \dots, N-1$ the expression

$$\sigma_{j-1,j} := r_{j-1}^-(t; y, \eta) - r_j^-(t; y, \eta), \qquad \forall (t; y, \eta) \in \check{\mathfrak{O}}_{j-1} \cap \check{\mathfrak{O}}_j,$$

is a constant integer (note that our $\sigma_{j-1,j}$ differ by sign from Hörmander's [**Hö2,** p. 92], [**DuiGuiHö,** p. 36(92)]). Let us assume that all the trajectories (1.5.1) are T-periodic and $\Omega_N = \Omega_1$. We identify the points $(0; y, \eta)$ and $(T; y, \eta)$, and then the set of intersections $\check{\mathfrak{O}}_{j-1} \cap \check{\mathfrak{O}}_j$ and the constant functions $\sigma_{j-1,j}$ defined on them form an integer cocycle on $\mathbb{S}^1 \times O$. From the representation (1.5.6) it is clear that $-\alpha_\Gamma$ coincides with the value of this cocycle on the closed curve $(t; y_0, \eta_0)$, $0 \leqslant t \leqslant T$, so α_Γ is determined by the corresponding Čech cohomology class.

2. Some explicit formulae. Let $r_+(x, \xi)$ and $r_-(x, \xi)$ be the numbers of the positive and negative eigenvalues of the matrix $\partial_{\xi\xi} h(x, \xi)$. Changing coordinates $x \to \widetilde{x}$ we obtain $h_{\widetilde{\xi}\widetilde{\xi}} = (\partial \widetilde{x}/\partial x) \cdot h_{\xi\xi} \cdot (\partial \widetilde{x}/\partial x)^T$. Therefore the definition of the functions $r_+(x, \xi)$ and $r_-(x, \xi)$ is independent of the choice of coordinates. Since $h_\xi(x, \xi)$ is positively homogeneous in ξ of degree zero, Euler's identity implies $\operatorname{rank} \partial_{\xi\xi} h(x, \xi) \leqslant n-1$ for all $(x, \xi) \in T'M$.

We have seen that in "good" cases the Maslov index α_Γ is the sum of the jumps of $r_j^-(t; y_0, \eta_0)$. It turns out that under some assumptions these jumps can be expressed in terms of $r_+(x^*, \xi^*)$, $r_-(x^*, \xi^*)$ and $\mathcal{R}(t; y, \eta)$. Here and below $(x^*, \xi^*) = (x^*(t; y, \eta), \xi^*(t; y, \eta))$.

LEMMA 1.5.5. *Let* $x_\eta^*(t; y, \eta) = 0$ *and* $\operatorname{rank} \partial_{\xi\xi} h(x^*, \xi^*) = n-1$ *for some* $(t; y, \eta) \in \check{\mathfrak{O}}_j$. *If* $s \neq 0$ *is sufficiently small and* $t + s \in [t_j(y, \eta), t_{j+1}(y, \eta)]$, *then*

$$\operatorname{rank} C_j(t+s; y, \eta) = n-1 \tag{1.5.7}$$

and

$$r_j^-(t+s; y, \eta) = \begin{cases} r_+(x^*, \xi^*), & s < 0, \\ r_-(x^*, \xi^*), & s > 0. \end{cases}$$

LEMMA 1.5.6. *Let* $r_+(x^*, \xi^*) = n-1$ *for some* $(t; y, \eta) \in \check{\mathfrak{O}}_j$. *If* s *is sufficiently small and* $t + s \in [t_j(y, \eta), t_{j+1}(y, \eta)]$, *then*

$$r_j^-(t+s; y, \eta) = \begin{cases} r_j^-(t; y, \eta) - \mathcal{R}(t; y, \eta) + n - 1, & s < 0, \\ r_j^-(t; y, \eta), & s \geqslant 0, \end{cases}$$

and (1.5.7) *hold.*

REMARK 1.5.7. In the general case, when the conditions of Lemma 1.5.5 or Lemma 1.5.6 are not fulfilled, the jumps of r_j^- depend not only on the matrix $\partial_{\xi\xi} h$ but also on the higher order derivatives of h.

EXAMPLE 1.5.8. Let M be a Riemannian manifold and $h(x,\xi) = |\xi|_x$ (see Example 1.2.4). Then $r_+(x,\xi) = n-1$ for all $(x,\xi) \in T'M$. Therefore by Lemma 1.5.6 and Definition 1.5.1 the Maslov index is the number of conjugate points counted with their multiplicities. Note that in this case the number of conjugate points on the trajectory Γ is finite; this fact follows from (1.5.7) and the obvious formula $\mathcal{R}(t;y,\eta) = \operatorname{rank} C_j(t;y,\eta)$.

3. Another definition. In this subsection we give another definition of the Maslov index which is more traditional. In fact, it is a modification of the classical definition suggested by V. Arnol'd [**Ar1**].

In Section 2.4 we will associate with the kth leg of our family of billiard trajectories (1.5.1) a class $\mathfrak{F}_k$ of smooth complex-valued nondegenerate phase functions φ_k which are defined in a neighbourhood of the set

$$\{\,(t,x;y,\eta)\,:\, t\in[t_k^*(y,\eta),t_{k+1}^*(y,\eta)]\,,\; x=x^*(t;y,\eta)\,\}\,,\qquad k=0,\dots,\mathbf{r},$$

where $t_0^*(y,\eta) := 0$ and $t_{\mathbf{r}+1}^*(y,\eta) := T$. The fact that $\varphi_k(t,x;y,\eta)$ is a phase function means that $\operatorname{Im}\varphi_k \geqslant 0$ and φ_k is positively homogeneous in η of degree 1. A phase function φ_k belongs to $\mathfrak{F}_k$ if

$$(1.5.8)\qquad (\varphi_k)_\eta(t,x;y,\eta)\big|_{x=x^*} = 0\,,\quad (\varphi_k)_x(t,x;y,\eta)\big|_{x=x^*} = \xi^*(t;y,\eta)\,,$$

and

$$\det(\varphi_k)_{x\eta}(t,x;y,\eta)\big|_{x=x^*} \neq 0$$

for all $t\in[t_k^*(y,\eta),t_{k+1}^*(y,\eta)]$ (see Section 2.4 for more precise definitions).

Let us choose a matching sequence of phase functions $\varphi_k\in\mathfrak{F}_k$, $k=0,\dots,\mathbf{r}$, for which

$$(1.5.9)\qquad \varphi_k(t,x;y,\eta)\big|_{x\in\partial M} = \varphi_{k+1}(t,x;y,\eta)\big|_{x\in\partial M}$$

(by Lemma 2.6.3 such a sequence always exists). The matrices $(\varphi_k)_{x\eta}$ behave as tensors under changes of coordinates, i.e., they are multiplied by the Jacoby matrices (see Section 2.2). So under changes of coordinates the expressions $(\det(\varphi_k)_{x\eta})^2$ are multiplied by positive numbers, which does not change their arguments. Moreover, the condition (1.5.9) implies that

$$\begin{aligned}\left(\det{}^2(\varphi_k)_{x\eta}\,/\,|\det{}^2(\varphi_k)_{x\eta}|\right)\Big|_{x=x^*(t_{k+1}^*;y,\eta)}\\ = \left(\det{}^2(\varphi_{k+1})_{x\eta}\,/\,|\det{}^2(\varphi_{k+1})_{x\eta}|\right)\Big|_{x=x^*(t_{k+1}^*;y,\eta)}\end{aligned}$$

(see formula (2.6.14)). Therefore we can introduce on $[0,T]\times O$ the continuous function f defined by the equalities

$$(1.5.10)\qquad f(t;y,\eta) := \left(\det{}^2(\varphi_k)_{x\eta}\,/\,|\det{}^2(\varphi_k)_{x\eta}|\right)\Big|_{x=x^*(t;y,\eta)}\,,$$

where $t\in[t_k^*(y,\eta),t_{k+1}^*(y,\eta)]$, $k=0,\dots,\mathbf{r}$.

Obviously, $|f|\equiv 1$. Since $x^*|_{t=0}\equiv y$, the conditions (1.5.8) imply that $(\varphi_0)_{x\eta}\equiv I$ for $t=0$, $x=y$ (see 2.4.1) and, consequently, $f(0;y,\eta)\equiv 1$. Therefore there exists a unique continuous branch $\arg_0 f(t;y,\eta)$ of the argument $\arg f(t;y,\eta)$ such that $\arg_0 f(0;y,\eta)\equiv 0$.

Under condition (1.5.2) the matrix $(\varphi_{\mathbf{r}})_{x\eta}$ is real for $t=T$, $x=x^*(T;y_0,\eta_0)$. Consequently, $\arg_0 f(T;y_0,\eta_0)$ is a multiple of 2π.

DEFINITION 1.5.9. The integer $-(2\pi)^{-1}\arg_0 f(T;y_0,\eta_0)$ is called the Maslov index of the trajectory Γ.

In other words, the Maslov index is equal to the sum over $k=0,\dots,\mathbf{r}$ of the variations of $-(2\pi)^{-1}\arg(\det(\varphi_k)_{x\eta})^2\big|_{x=x^*(t;y_0,\eta_0)}$ as t goes from $t_k^*(y_0,\eta_0)$ to $t_{k+1}^*(y_0,\eta_0)$.

The equivalence of Definitions 1.5.1 and 1.5.9 will be proved in Appendix D.

REMARK 1.5.10. When $\partial M=\varnothing$ we have only one phase function $\varphi=\varphi_0$ (instead of the sequence $\{\varphi_k\}$). Let us introduce the differential 1-form

$$-(2\pi)^{-1}\,d\left(\arg(\det\varphi_{x\eta})^2\big|_{x=x^*}\right).$$

Obviously, the Maslov index α_Γ coincides with the integral of this 1-form over the curve $(t;y_0,\eta_0)$, $0\leqslant t\leqslant T$. Under the assumptions of Remark 1.5.4 this integral is determined by the corresponding de Rham cohomology class, which is the image of the Čech cohomology class from Remark 1.5.4 provided by the standard isomorphism of the Čech and de Rham cohomology groups (see [**LapSaVa**] for details).

1.6. Classical two-term asymptotic formula for $N(\lambda)$

1. Statement of the result. Having introduced the geometrical concepts of nonperiodicity (see Definition 1.3.7 and its subsequent generalizations in 1.3.2, 1.3.3) and nonblocking (Definition 1.3.22), we can now formulate the theorem which gives the classical two-term asymptotic formula for $N(\lambda)$. For the case $\partial M=\varnothing$ it was established in [**DuiGui**], and for the case $m=1$ in [**Iv1**], [**Me**]. The general result appeared in [**Va3**], [**Va4**], [**Va7**].

THEOREM 1.6.1. *If the nonperiodicity and nonblocking conditions are fulfilled, then*

$$N(\lambda)=c_0\lambda^n+c_1\lambda^{n-1}+o(\lambda^{n-1}),\qquad \lambda\to+\infty. \tag{1.6.1}$$

Here the coefficient c_0 *is the same as in Theorem* 1.2.1 *and*

$$c_1=\int_{T^*\partial M}\operatorname{shift}^+(1;x',\xi')\,dx'\,đ\xi', \tag{1.6.2}$$

where $\operatorname{shift}^+(\nu;x',\xi')$ *for fixed* $(x',\xi')\in T^*\partial M$ *is the spectral shift* (*see subsection* 3) *of the auxiliary one-dimensional problem* (1.1.7), (1.1.8) *on the half-line. For manifolds without boundary* $c_1=0$.

REMARK 1.6.2. Formula (1.6.1) can be written down in an equivalent form

$$\lambda_k=c_0^{-1/n}k^{1/n}-c_1(nc_0)^{-1}+o(1),\qquad k\to+\infty. \tag{1.6.1'}$$

The remainder of this section is split into five subsections. In subsection 2 we briefly discuss the conditions appearing in Theorem 1.6.1. In subsection 3 we define the spectral shift and produce formulae for its effective evaluation. In subsection 4 we discuss the formula for the second asymptotic coefficient and give two elementary examples of its calculation. In subsection 5 we formulate a modified version of Theorem 1.6.1 suited for the case of a pseudodifferential operator acting on a manifold without boundary.

2. Discussion of the result. Let us turn once again to Example 1.2.5 (Laplacian on a sphere). We have already noted that in this example the second asymptotic term of $N(\lambda)$ is of the order of $\lambda^{(n-1)/(2m)}$. However, it is not of the form $c_1\lambda^{(n-1)/(2m)}$, i.e., the classical two-term Weyl formula (1.6.1) does not hold. Indeed, it follows from (1.2.4), (1.2.5) that there is a sequence $\Lambda_j \to +\infty$ such that $N(\Lambda_j+0)-N(\Lambda_j) \geqslant \text{const}\,\Lambda_j^{n-1}$, const >0, whereas (1.6.1) would imply $N(\lambda+0)-N(\lambda)=o(\lambda^{n-1})$ as $\lambda\to+\infty$. In other words, looking at the fine structure of the spectrum we see that in Example 1.2.5 the eigenvalues are unevenly distributed along the spectrum: they cluster into groups of growing multiplicity.

The reason for the uneven distribution of eigenvalues in Example 1.2.5 is clear: in this case the spectral problem (1.1.1), (1.1.2) has a very rich group of symmetries, and consequently, the eigenvalues have very high multiplicities. Such "pathologically" (in the terminology of [**DuiGui**]) symmetric cases have to be excluded if we want to obtain a classical (polynomial) two-term asymptotic formula for $N(\lambda)$, and this is why we need the nonperiodicity condition. It was first introduced in [**DuiGui**] for manifolds without boundary, and it appears in all the subsequent works on classical two-term spectral asymptotics.

The fact that the (asymptotic) symmetries of the spectral problem (1.1.1), (1.1.2) can be described in terms of billiard trajectories is a nontrivial one. The precise role of billiard trajectories will be revealed in further chapters in the course of our proof. However, the underlying idea is worth mentioning here. Let us introduce the time variable t by a change $\lambda\to D_t$ $(=-i\partial/\partial t)$, i.e., let us consider the nonstationary equation

$$Av = D_t^{2m}v \tag{1.6.3}$$

with boundary conditions (1.1.2). One can single out (see Section 3.1) solutions of (1.6.3), (1.1.2) of the form $v(x,t)=\exp(-itA^{1/(2m)})v_0(x)$. Clearly, such solutions contain full information about the spectral problem (1.1.1), (1.1.2). Analysis of the nonstationary problem (1.6.3), (1.1.2) shows that the singularities of such solutions propagate along the billiard trajectories defined in Section 1.3, so it is natural to use them for describing global geometrical characteristics.

Another way of explaining the appearance of billiard trajectories is to try to solve the spectral problem (1.1.1), (1.1.2) asymptotically. Let us search for the eigenfunction v in the form

$$v = a(x,\lambda)\exp(i\lambda\varphi(x)), \tag{1.6.4}$$

where $a(x,\lambda)=a_0(x)+\lambda^{-1}a_{-1}(x)+\lambda^{-2}a_{-2}(x)+\cdots$, $\lambda\to+\infty$. Substituting (1.6.4) into (1.1.1) and leaving only the terms with the leading power of λ, we obtain the eikonal equation $A_{2m}(x,\varphi_x)=1$ the solution of which can be expressed in terms of Hamiltonian trajectories (1.3.1); see, e.g., [**Sh**]. A similar construction with m compensating exponents added to (1.6.4) allows one to satisfy asymptotically the boundary conditions (1.1.2), with the phase functions $\varphi^{(l)}$ of the compensating exponents being expressed in terms of reflected billiard trajectories. Unfortunately, the exponential representation technique works only locally, in the neighbourhood of a fixed point $(x_0,\varphi_x|_{x=x_0})\in T'M$. In the general case one cannot solve the equations (1.1.1) and (1.1.2) on the whole of M and ∂M using this technique, but it gives an idea why billiard trajectories appear in spectral asymptotics.

The nonblocking condition is more technical and its necessity in Theorem 1.6.1 is less obvious. Spectral problems which do not satisfy the nonblocking condition have not yet been studied; we can only say that for such problems the geometric picture may be very complicated and modern microlocal techniques fail to describe it adequately.

The nonblocking condition was introduced by the authors in two independent works [**Va4**] and [**Sa1**].

3. Spectral shift of the auxiliary one-dimensional problem. Let us consider the one-dimensional spectral problem (1.1.7), (1.1.8). In this subsection we define for (1.1.7), (1.1.8) the notion of the spectral shift, and give (without proof) a convenient trace formula for its evaluation. A more detailed analysis with the proofs is carried out in Appendix A.

As in Section 1.4, let us omit for the time being the parameter (x', ξ'), and rewrite (1.1.7), (1.1.8) as (1.4.1), (1.4.2). Recall that by $\mathbf{A}^+$ we denote the self-adjoint operator in $L_2(\mathbb{R}_+)$ associated with (1.4.1), (1.4.2).

Let us also consider the spectral problem (1.4.1) on the whole line $\mathbb{R}$ without boundary conditions, and let us denote by $\mathbf{A}$ the corresponding self-adjoint operator in $L_2(\mathbb{R})$.

Denote by $\mathbf{E}_\nu^+$, $\mathbf{E}_\nu$ the spectral projections of the operators $\mathbf{A}^+$, $\mathbf{A}$, respectively. For definiteness we choose $\mathbf{E}_\nu^+$, $\mathbf{E}_\nu$ to be left-continuous in ν. For each fixed ν our spectral projections are integral operators

$$\mathbf{E}_\nu^+ = \int_0^{+\infty} \mathbf{e}^+(\nu, x_n, y_n)\,(\,\cdot\,)\,dy_n\,, \qquad \mathbf{E}_\nu = \int_{-\infty}^{+\infty} \mathbf{e}(\nu, x_n, y_n)\,(\,\cdot\,)\,dy_n$$

with continuous kernels $\mathbf{e}^+$, $\mathbf{e}$.

Denote by $\theta : L_2(\mathbb{R}) \to L_2(\mathbb{R}_+)$ the restriction operator; its adjoint θ^*: $L_2(\mathbb{R}_+) \to L_2(\mathbb{R})$ is the extension operator defined by

$$(\theta^* v)(x_n) = \begin{cases} 0, & x_n < 0, \\ v(x_n), & x_n \geqslant 0. \end{cases}$$

By θ_X we shall denote the characteristic function of the interval $[0, X]$, and by Tr we shall denote the trace of an operator.

DEFINITION 1.6.3. The *spectral shift* of the problem (1.4.1), (1.4.2) is the function on $\mathbb{R}$ defined as follows:

1. For ν which are not thresholds (see Section 1.4)

$$\begin{aligned} \operatorname{shift}^+(\nu) &= \int_0^{+\infty} (\mathbf{e}^+(\nu, x_n, x_n) - \mathbf{e}(\nu, x_n, x_n))\,dx_n \\ &\equiv \lim_{X\to+\infty} \operatorname{Tr}(\theta_X(\mathbf{E}_\nu^+ - \theta\mathbf{E}_\nu\theta^*)\theta_X)\,. \end{aligned} \tag{1.6.5}$$

2. At thresholds $\operatorname{shift}^+(\nu)$ is defined by left-continuity.

Convergence of the integral (1.6.5) follows from the asymptotic formula

$$\begin{aligned} &\mathbf{e}^+(\nu, x_n, x_n) \;-\; \mathbf{e}(\nu, x_n, x_n) \\ &= \sum_{k,l=1}^{q} \frac{\sqrt{-\,A'(\zeta_l^+(\nu))\,A'(\zeta_k^-(\nu))}}{\pi\, x_n\,\big(A'(\zeta_l^+(\nu)) - A'(\zeta_k^-(\nu))\big)}\; \mathrm{Im}\big(R_{lk}(\nu)\, e^{i x_n (\zeta_l^+(\nu) - \zeta_k^-(\nu))}\big) \\ &\quad +\; O(x_n^{-2}) \end{aligned} \tag{1.6.6}$$

as $x_n \to +\infty$; here the $R_{lk}(\nu)$ are the elements of the reflection matrix and the $\zeta_l^{\pm}(\nu)$ are the real ζ-roots of the algebraic equation $A(\zeta) = \nu$; see Section 1.4 for details. Formula (1.6.6) implies that normally the integral (1.6.5) is not absolutely convergent. The absence of absolute convergence in (1.6.5) implies, in turn, that the operator $\mathbf{E}_\nu^+ - \theta \mathbf{E}_\nu \theta^*$ is not of trace class.

For a positive self-adjoint operator with a purely discrete spectrum the trace of the spectral projection is the counting function (i.e., the number of eigenvalues below a given ν). Hence the notion of the spectral shift is very similar to that of the counting function. The difference is that the spectral shift may be defined for self-adjoint operators the spectra of which are not purely discrete; in this situation the spectral projection is not of trace class and one has to perform a regularization procedure. In our particular case (1.6.5) regularization is performed in two stages: firstly, by subtracting the spectral projection of a basic (reference) operator, and secondly, by introducing the regularizing spatial cut-off θ_X.

Denote by $\sigma(\mathbf{A}^+)$, $\sigma(\mathbf{A})$ the spectra of the operators $\mathbf{A}^+$, $\mathbf{A}$. We have

$$[\nu_1^{\mathrm{st}}, +\infty) = \sigma(\mathbf{A}) \subset \sigma(\mathbf{A}^+) \subset [0, +\infty),$$

where ν_1^{st} is as defined in Section 1.4, and the last inclusion is a consequence of Conditions 1.1.1′, 1.1.4′. Thus, for any $\mu \in \mathbb{C} \setminus \sigma(\mathbf{A}^+)$ we can define the resolvents $\mathbf{R}_\mu^+ = (\mathbf{A}^+ - \mu I)^{-1}$ and $\mathbf{R}_\mu = (\mathbf{A} - \mu I)^{-1}$. For each fixed μ these resolvents are integral operators

$$\mathbf{R}_\mu^+ = \int_0^{+\infty} \mathbf{r}^+(\mu, x_n, y_n)\,(\,\cdot\,)\, dy_n\,, \qquad \mathbf{R}_\mu = \int_{-\infty}^{+\infty} \mathbf{r}(\mu, x_n, y_n)\,(\,\cdot\,)\, dy_n$$

with continuous kernels $\mathbf{r}^+$, $\mathbf{r}$.

For any $\mu \in \mathbb{C} \setminus \sigma(\mathbf{A}^+)$ the operator $\mathbf{R}_\mu^+ - \theta \mathbf{R}_\mu \theta^*$ is of trace class (moreover, it is of finite rank), and $\mathbf{r}^+(\mu, x_n, x_n) - \mathbf{r}(\mu, x_n, x_n)$ tends to zero exponentially as x_n tends to $+\infty$. Thus, the expression

$$\mathbf{f}^+(\mu) := \mathrm{Tr}(\mathbf{R}_\mu^+ - \theta \mathbf{R}_\mu \theta^*) \equiv \int_0^{+\infty} (\mathbf{r}^+(\mu, x_n, x_n) - \mathbf{r}(\mu, x_n, x_n))\, dx_n \tag{1.6.7}$$

is well defined. The function $\mathbf{f}^+$ is analytic in $\mathbb{C} \setminus \sigma(\mathbf{A}^+)$, and for any given $\nu \in \sigma(\mathbf{A}^+)$ which is not a threshold and not an eigenvalue of the problem (1.4.1), (1.4.2), $\mathbf{f}^+(\mu)$ is bounded as $\mathbb{C} \setminus \sigma(\mathbf{A}^+) \ni \mu \to \nu$.

Definition 1.6.3 is equivalent to

DEFINITION 1.6.4. The *spectral shift* of the problem (1.4.1), (1.4.2) is the function on $\mathbb{R}$ defined as follows.

1. $\mathrm{shift}^+(\nu) = 0$ for $\nu \leqslant 0$.

2. For positive ν which are not thresholds and are not eigenvalues,

$$\text{shift}^+(\nu) = (-2\pi i)^{-1} \int_{L(\nu)} \mathbf{f}^+(\mu)\, d\mu \,, \tag{1.6.8}$$

where $L(\nu)$ is the oriented arc of the circumference $|\mu| = \nu$ in the complex μ-plane going counterclockwise from $\nu + i0$ to $\nu - i0$.

3. At thresholds and eigenvalues $\text{shift}^+(\nu)$ is defined by left-continuity.

Let us now list the basic properties of our spectral shift.

LEMMA 1.6.5. *The function* shift^+ *is bounded:*

$$|\,\text{shift}^+(\nu)| \leqslant m, \qquad \forall \nu \in \mathbb{R}.$$

Recall that $2m$ is the order of the differential equation (1.4.1).

Let us denote by $\mathbf{N}^+(\nu)$ the counting function of the spectral problem (1.4.1), (1.4.2), that is the number of eigenvalues below a given ν. Here we count both the eigenvalues outside the continuous spectrum and those embedded in the continuous spectrum.

LEMMA 1.6.6. *The function* $\text{shift}^+(\nu) - \mathbf{N}^+(\nu)$ *is continuous in each of the zones* (1.4.4) *of the continuous spectrum.*

Lemma 1.6.6 implies that the function shift^+ may be discontinuous only at thresholds and eigenvalues. With account of Proposition A.1.13 this means that the number of discontinuities of the function shift^+ is finite, and, moreover, uniformly bounded over all problems of this type with fixed order $2m$.

LEMMA 1.6.7. *For any* $\mu \in \mathbb{C} \setminus [\min \sigma(\mathbf{A}^+), +\infty)$,

$$\mathbf{f}^+(\mu) = \int_0^{+\infty} \frac{\text{shift}^+(\nu)}{(\nu - \mu)^2}\, d\nu \,.$$

A consequence of Lemma 1.6.7 is the following result which can be used for the evaluation of the jumps of the function shift^+.

COROLLARY 1.6.8. *For any* $\nu \in \mathbb{R}$,

$$\text{shift}^+(\nu + 0) - \text{shift}^+(\nu) = \lim_{\substack{\mu \to \nu \\ \mu \in \mathbb{C} \setminus \sigma(\mathbf{A}^+)}} (\nu - \mu)\, \mathbf{f}^+(\mu) \,.$$

If ν is an eigenvalue which is not a threshold, then it it is not necessary to use Corollary 1.6.8 because by Lemma 1.6.6 $\text{shift}^+(\nu + 0) - \text{shift}^+(\nu)$ is just the multiplicity of the eigenvalue ν. Corollary 1.6.8 is really needed for computing the jumps of the function shift^+ at thresholds. Corollary 1.6.8 is, however, not very convenient. Below we consider the important special case of a normal threshold (see Definition 1.4.4). Then the formula for the jump of the function shift^+ becomes completely explicit.

DEFINITION 1.6.9. We call the normal threshold ν^{st} *soft* if for $\nu = \nu^{\text{st}}$ the problem (1.4.1), (1.4.2) has a solution of the form

$$v(x_n) = e^{i x_n \zeta^{\text{st}}} + w(x_n)\,, \quad \text{where} \quad w(x_n) \to 0 \quad \text{as} \quad x_n \to +\infty\,,$$

and *rigid* if it has no solution of this form. Here ζ^{st} is as in Definition 1.4.4.

LEMMA 1.6.10. *If the threshold* ν^{st} *is normal, then*

$$\mathrm{shift}^+(\nu^{\mathrm{st}}+0) - \mathrm{shift}^+(\nu^{\mathrm{st}}) = \mathbf{N}^+(\nu^{\mathrm{st}}+0) - \mathbf{N}^+(\nu^{\mathrm{st}}) \pm \frac{1}{4},$$

where the plus or minus sign is chosen according to whether the threshold ν^{st} *is soft or rigid, respectively.*

It follows from Lemma 1.6.10 that a normal threshold produces a jump in the spectral shift which in absolute value is one quarter of the jump produced by an eigenvalue.

Lemmas 1.6.6, 1.6.10 and Corollary 1.6.8 provide us with an effective description of the jumps of the spectral shift, but we are still left with the problem of determining its continuous part. Using Definitions 1.6.3 or 1.6.4 for this purpose is inconvenient. Indeed, Definition 1.6.3 involves the spectral function $\mathbf{e}^+(\nu, x_n, y_n)$ of the problem (1.4.1), (1.4.2), the effective construction of which is not completely trivial, and, moreover, it involves integration in x_n. Definition 1.6.4 is better in the sense that it does not use the spectral function, but on the other hand it contains a double integral (see (1.6.7), (1.6.8)). Fortunately, this inconvenience can be overcome. The following lemma gives a formula for the continuous part of the spectral shift which does not contain the spectral function or any integrations. Everything is expressed through the reflection matrix $R(\nu)$ introduced in Section 1.4.

LEMMA 1.6.11. *For each zone* (1.4.4) *of the continuous spectrum there exists a continuous branch* $\arg_0 \det\big(i\,R(\nu)\big)$ *of the argument of* $\det\big(i\,R(\nu)\big)$ *such that*

$$\mathrm{shift}^+(\nu) = \mathbf{N}^+(\nu) + \frac{\arg_0 \det\big(i\,R(\nu)\big)}{2\pi} \tag{1.6.9}$$

in this zone. For ν *lying below the continuous spectrum* $\mathrm{shift}^+(\nu) = \mathbf{N}^+(\nu)$.

REMARK 1.6.12. Formula (1.6.9) is a special case of the standard trace formula from scattering theory

$$\mathrm{shift}(\nu) = \mathbf{N}(\nu) + \frac{\arg_0 \det S(\nu)}{2\pi}, \tag{1.6.10}$$

where $\mathbf{N}(\nu)$ is the counting function of the perturbed problem and $S(\nu)$ is the scattering matrix; see, e.g., [**Ya**]. There is, however, a technical difficulty in that we compare two spectral problems in different Hilbert spaces, namely, $L_2(\mathbb{R}_+)$ and $L_2(\mathbb{R})$. The natural way of overcoming this difficulty is to consider instead of the problem on the half-line $\mathbb{R}_+$ a problem on the perforated line $\mathbb{R}\setminus\{0\}$ with boundary conditions

$$\left(B^{(j)}(D_{x_n})v\right)\Big|_{x_n=+0} = 0, \qquad j=1,2,\dots,m, \tag{1.6.11$^+$}$$

$$\left(\overline{B^{(j)}}(D_{x_n})v\right)\Big|_{x_n=-0} = 0, \qquad j=1,2,\dots,m \tag{1.6.11$^-$}$$

(cf. (1.4.2)), where $\overline{B^{(j)}}(D_{x_n}) \equiv \left(B^{(j)}(D_{x_n})\right)^*$. Then the problem (1.4.1), (1.6.11$^+$), (1.6.11$^-$) on the perforated line $\mathbb{R}_+\setminus\{0\}$ is a spectral problem in the Hilbert space $L_2(\mathbb{R})$, and one can view it as a perturbation of the reference problem (1.4.1) on the full line $\mathbb{R}$ in the same Hilbert space $L_2(\mathbb{R})$. In this case one can develop a consistent scattering theory (see Appendix A) and show that (1.6.10) holds. Moreover, the scattering matrix in this case is given by the formula

$$S(\nu) = \begin{pmatrix} 0 & R(\nu) \\ R^T(\nu) & 0 \end{pmatrix}, \tag{1.6.12}$$

where $R(\nu)$ is the reflection matrix defined in Section 1.4. Formulae (1.6.10), (1.6.12) imply

$$\operatorname{shift}(\nu) = \mathbf{N}(\nu) + \frac{\arg_0 \det^2(i\,R(\nu))}{2\pi}. \tag{1.6.13}$$

But the problem (1.4.1), ($1.6.11^+$), ($1.6.11^-$) on the perforated line $\mathbb{R}_+ \setminus \{0\}$ is invariant under the change $v(x_n) \to v(-x_n)$, and consequently

$$\operatorname{shift}(\nu) = 2\operatorname{shift}^+(\nu), \qquad \mathbf{N}(\nu) = 2\,\mathbf{N}^+(\nu). \tag{1.6.14}$$

From (1.6.13), (1.6.14) we obtain

$$\operatorname{shift}^+(\nu) = \mathbf{N}^+(\nu) + \frac{\arg_0\bigl(\pm\det(i\,R(\nu))\bigr)}{2\pi}$$

with either a plus or a minus. A more detailed analysis shows that the sign in the latter formula is a plus, and we arrive at (1.6.9).

Further on we use the notation $\arg_0\det\bigl(i\,R(\nu)\bigr)$ for the particular branch of the argument $\arg\det\bigl(i\,R(\nu)\bigr)$ specified by Lemma 1.6.11. It will be convenient for us to set $\arg_0\det\bigl(i\,R(\nu)\bigr) \equiv 0$ for ν lying below the continuous spectrum, and to define $\arg_0\det\bigl(i\,R(\nu)\bigr)$ at thresholds by left-continuity. Under such a convention the trace formula (1.6.9) holds for all $\nu \in \mathbb{R}$.

Lemmas 1.6.10, 1.6.11 immediately imply

COROLLARY 1.6.13. *If the threshold ν^{st} is normal, then*

$$\bigl|\arg_0\det\bigl(i\,R(\nu^{\mathrm{st}}+0)\bigr) - \arg_0\det\bigl(i\,R(\nu^{\mathrm{st}})\bigr)\bigr| = \frac{\pi}{2}.$$

If ν^{st} is a normal threshold and if we know the choice of the branch $\arg_0\det\bigl(i\,R(\nu)\bigr)$ of the argument $\arg\det\bigl(i\,R(\nu)\bigr)$ for $\nu \leqslant \nu^{\mathrm{st}}$, then Corollary 1.6.13 allows us to choose uniquely the branch $\arg_0\det\bigl(i\,R(\nu)\bigr)$ of the argument $\arg\det\bigl(i\,R(\nu)\bigr)$ for $\nu > \nu^{\mathrm{st}}$. This is possible because according to Corollary 1.6.13 $\arg_0\det\bigl(i\,R(\nu^{\mathrm{st}}+0)\bigr)$ belongs to the interval

$$\bigl[\arg_0\det\bigl(i\,R(\nu^{\mathrm{st}})\bigr) - \pi/2\,,\ \arg_0\det\bigl(i\,R(\nu^{\mathrm{st}})\bigr) + \pi/2\bigr]$$

the length of which is less than 2π. One does not even have to check whether the threshold ν^{st} is soft or rigid. Thus, if all the thresholds are normal, we can determine consecutively the branches $\arg_0\det\bigl(i\,R(\nu)\bigr)$ on all the intervals (1.4.4) of the continuous spectrum.

Matters are facilitated even further if our one-dimensional problem (1.4.1), (1.4.2) satisfies the strong simple reflection condition (see Definition 1.4.5). In this case

$$\arg_0\det\bigl(i\,R(\nu)\bigr) = \begin{cases} 0\,, & \text{if } \nu \leqslant \nu_1^{\mathrm{st}}\,, \\ \arg_0\left(\dfrac{i\,a_1^+(\nu)}{a_1^-(\nu)}\right), & \text{if } \nu > \nu_1^{\mathrm{st}}\,, \end{cases} \tag{1.6.15}$$

where the $a_1^\mp(\nu)$ are the coefficients from (1.4.7) and the branch of the argument is uniquely specified by the condition

$$\left|\arg_0\left(\frac{i\,a_1^+(\nu_1^{\mathrm{st}}+0)}{a_1^-(\nu_1^{\mathrm{st}}+0)}\right)\right| = \frac{\pi}{2}. \tag{1.6.16}$$

Further on we indicate the dependence of all our quantities on the parameter $(x',\xi')\in T'\partial M$. The reasoning from the end of Section 1.4 (see formula (1.4.9)) shows that under changes of local coordinates, $\mathrm{shift}^+(\nu;x',\xi')$, $\mathbf{N}^+(\nu;x',\xi')$, and $\arg_0\det\big(i\,R(\nu;x',\xi')\big)$ behave as functions on $T'\partial M$.

We shall often be using the following two properties of the spectral shift:

$$\mathrm{shift}^+(\lambda^{2m}\nu;x',\lambda\xi') = \mathrm{shift}^+(\nu;x',\xi'), \qquad \forall\lambda>0; \tag{1.6.17}$$

for any given local coordinate system x there exists a positive constant such that

$$\mathrm{shift}^+(\nu;x',\xi') = 0 \qquad \text{for} \qquad |\xi'|^{2m} > \mathrm{const}\,\nu. \tag{1.6.18}$$

The rescaling property (1.6.17) follows from the homogeneity of $A(x,\xi)$, $B_{m_j}^{(j)}(x',\xi)$ in ξ. The property (1.6.18) follows from (1.6.17) and the fact that for any given $\xi'\neq 0$ the spectra of the operators $\mathbf{A}^+$, $\mathbf{A}$ are strictly positive.

4. The second asymptotic coefficient: discussion and examples. First, let us show that the integral (1.6.2) exists in the usual Riemann sense. In view of (1.6.18) the domain of integration in (1.6.2) is, in fact, bounded. Denote by Σ' the set of points $(x',\xi')\in T^*\partial M$ such that 1 is an eigenvalue or a threshold of the one-dimensional problem (1.1.7), (1.1.8); obviously, the set Σ' is closed and bounded. By Lemma A.4.1 the function $\mathrm{shift}(1;x',\xi')$ is continuous on $T^*\partial M\setminus\Sigma'$, and by Lemma 1.6.5 it is uniformly bounded on $T^*\partial M$. So in order to prove Riemann integrability it is sufficient to show that the $(2n-2)$-dimensional Jordan measure of Σ' is zero. As the set Σ' is closed and bounded, this is equivalent to the $(2n-2)$-dimensional Lebesgue measure of Σ' being zero. In view of Tonelli's theorem, in order to prove the latter it is sufficient to show that for any $(x_0',\xi_0')\in T'\partial M$ the ray $\{(x_0',\lambda\xi_0'),\ \lambda>0\}$ intersects Σ' at a finite number of points. By rescaling, a point $(x_0',\lambda\xi_0')$ belongs to Σ' if and only if $\nu=\lambda^{-2m}$ is an eigenvalue or a threshold of the one-dimensional problem (1.1.7), (1.1.8) with $(x',\xi')\equiv(x_0',\xi_0')$, and we already know that the number of such ν is finite. This completes the proof of Riemann integrability.

Substituting (1.6.14) into (1.6.2) we obtain

$$c_1 = \int_{T^*\partial M}\left(\mathbf{N}^+(1;x',\xi') + \frac{\arg_0\det\big(i\,R(1;x',\xi')\big)}{2\pi}\right)dx'\,đ\xi'. \tag{1.6.19}$$

Formula (1.6.19) proves to be very convenient for the practical evaluation of the coefficient c_1.

Let us illustrate the algorithm of computing the coefficient c_1 by a few elementary examples.

EXAMPLE 1.6.14. Let us consider the spectral problem

$$-\Delta v = \lambda^2 v, \tag{1.6.20}$$

$$v|_{\partial M} = 0 \tag{1.6.21}$$

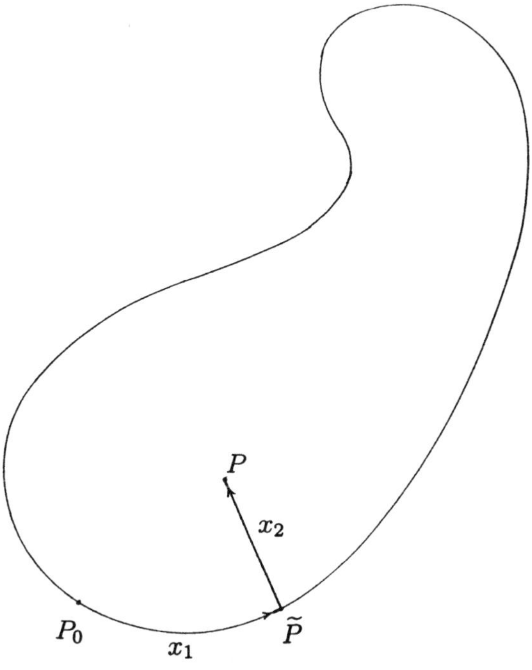

FIGURE 9. Local coordinates x.

in a region $M \subset \mathbb{R}^2$. Here $\Delta = \partial^2/\partial y_1^2 + \partial^2/\partial y_2^2$ is the Laplacian and $y = (y_1, y_2)$ are Cartesian coordinates in $\mathbb{R}^2$. We already know (see Example 1.2.3) that for this problem $c_0 = S/(4\pi)$, where S is the surface area of M. For computing the coefficient c_1 we introduce special local coordinates $x = (x_1, x_2)$ in the small neighbourhood of ∂M (see Figure 9): we associate with a point $P \in M$ the nearest point $\widetilde{P} \in \partial M$ and then take x_2 to be the length of the (straight) line segment $P\widetilde{P}$, and x_1 the length of the (curvilinear) arc $P_0\widetilde{P} \subset \partial M$, where $P_0 \in \partial M$ is some fixed point.

The auxiliary one-dimensional spectral problem associated with (1.6.20), (1.6.21) is

$$-d^2v/dx_2^2 + \xi_1^2 v = \nu v, \tag{1.6.22}$$

$$v|_{x_2=0} = 0, \tag{1.6.23}$$

where $v \equiv v(x_2)$, $x_2 \in \mathbb{R}_+$. In the subsequent analysis we assume that $\xi_1 \neq 0$; this can be done because the integral (1.6.19) does not depend on the value of the integrand at a particular ξ_1.

The problem (1.6.22), (1.6.23) has no eigenvalues, so

$$\mathbf{N}^+(\nu; x_1, \xi_1) \equiv 0. \tag{1.6.24}$$

The problem (1.6.22), (1.6.23) has only one threshold $\nu_1^{\mathrm{st}} = \xi_1^2$, and the continuous spectrum is the semi-infinite interval $[\xi_1^2, +\infty)$. The points $\nu > \xi_1^2$ of the continuous spectrum have multiplicity one, and the corresponding generalized eigenfunctions have the form

$$v(x_2) = \sin\left(x_2\sqrt{\nu-\xi_1^2}\right) = \frac{a_1^-\, e^{ix_2\zeta_1^-(\nu)}}{\sqrt{-2\pi A'(\zeta_1^-(\nu))}} + \frac{a_1^+\, e^{ix_2\zeta_1^+(\nu)}}{\sqrt{2\pi A'(\zeta_1^+(\nu))}}$$

(cf. (1.4.7)), where $\zeta_1^\pm(\nu) = \pm\sqrt{\nu-\xi_1^2}$ and

$$a_1^\pm = \mp i\sqrt{\pi\sqrt{\nu-\xi_1^2}}\,. \tag{1.6.25}$$

As the strong simple reflection condition is satisfied, we can use formulae (1.6.15), (1.6.16). Substituting (1.6.25) into (1.6.15) we obtain for $\nu > \xi_1^2$

$$\arg_0 \det\bigl(i\, R(\nu; x_1, \xi_1)\bigr) = -\pi/2 + 2\pi k \tag{1.6.26}$$

with an unknown integer k. Substituting (1.6.26) into (1.6.16) we establish that $k = 0$. Thus,

$$\arg_0 \det\bigl(i\, R(\nu; x_1, \xi_1)\bigr) = \begin{cases} 0\,, & \text{if} \quad \nu \leqslant \xi_1^2\,, \\ -\pi/2\,, & \text{if} \quad \nu > \xi_1^2\,. \end{cases} \tag{1.6.27}$$

Finally, substituting (1.6.24) and (1.6.27) into (1.6.19) we obtain

$$\begin{aligned} c_1 &= \int_0^L \int_{-\infty}^{+\infty} \frac{\arg_0 \det\bigl(i\, R(1; x_1, \xi_1)\bigr)}{2\pi}\, đ\xi_1\, dx_1 \\ &= -\frac{1}{4}\int_0^L \int_{-1}^{1} đ\xi_1\, dx_1 = -\frac{L}{4\pi}\,, \end{aligned} \tag{1.6.28}$$

where L is the length of ∂M.

It is interesting to determine whether in our case the normal threshold ν_1^{st} is soft or rigid, and check whether Lemma 1.6.10 gives the proper sign for the jump. The function $v(x_n)$ from Definition 1.6.9 in this case has the form

$$v(x_2) \equiv 1\,. \tag{1.6.29}$$

Obviously, the function (1.6.29) does not satisfy the boundary condition (1.6.23), so the threshold is rigid. Accordingly, the sign of $\arg_0 \det\bigl(i\, R(\nu_1^{\mathrm{st}} + 0; x_1, \xi_1)\bigr)$ is negative.

EXAMPLE 1.6.15. Consider the same spectral problem as in Example 1.6.14, but with the Neumann boundary condition

$$\partial v/\partial x_2|_{\partial M} = 0 \tag{1.6.30}$$

instead of (1.6.21). Of course, the coefficient c_0 is the same as in Example 1.6.14. In order to compute the coefficient c_1 we have to consider the auxiliary one-dimensional spectral problem associated with (1.6.20), (1.6.30). This one-dimensional problem is described by the equation (1.6.22) with boundary condition

$$dv/dx_2|_{x_2=0} = 0\,. \tag{1.6.31}$$

The problem (1.6.22), (1.6.31) has no eigenvalues, so (1.6.24) remains true.

The threshold and the continuous spectrum are the same as in Example 1.6.14, but the generalized eigenfunctions now have the form

$$v(x_2) = \cos\left(x_2\sqrt{\nu-\xi_1^2}\right) = \frac{a_1^-\, e^{ix_2\zeta_1^-(\nu)}}{\sqrt{-2\pi A'(\zeta_1^-(\nu))}} + \frac{a_1^+\, e^{ix_2\zeta_1^+(\nu)}}{\sqrt{2\pi A'(\zeta_1^+(\nu))}},$$

where

$$a_1^+ = a_1^- = \sqrt{\pi\sqrt{\nu-\xi_1^2}} \tag{1.6.32}$$

(cf. (1.6.25)). Substituting (1.6.32) into (1.6.15) we obtain for $\nu > \xi_1^2$

$$\arg_0 \det\bigl(i\,R(\nu;x_1,\xi_1)\bigr) = \pi/2 + 2\pi k \tag{1.6.33}$$

(cf. (1.6.26)) with an unknown integer k. Substituting (1.6.33) into (1.6.16) we establish that $k=0$. Thus,

$$\arg_0 \det\bigl(i\,R(\nu;x_1,\xi_1)\bigr) = \begin{cases} 0, & \text{if } \nu \leqslant \xi_1^2, \\ \pi/2, & \text{if } \nu > \xi_1^2, \end{cases} \tag{1.6.34}$$

which differs from (1.6.27) only in sign.

Finally, substituting (1.6.24) and (1.6.34) into (1.6.19) we obtain

$$c_1 = \frac{L}{4\pi}, \tag{1.6.35}$$

which also differs from (1.6.28) only in sign.

It is easy to see that the function (1.6.29) satisfies the boundary condition (1.6.31), so the threshold in Example 1.6.15 is soft. Accordingly, the sign of $\arg_0 \det\bigl(i\,R(\nu_1^{\mathrm{st}}+0;x_1,\xi_1)\bigr)$ is positive.

We are now prepared to explain the origin of the names "rigid" and "soft" with respect to normal thresholds. The spectral problem considered in Examples 1.6.14, 1.6.15 can be interpreted as the problem of free harmonic vibrations of a membrane. The Dirichlet boundary condition (1.6.21) (Example 1.6.14) describes the situation when the edge of the membrane is fixed ("rigid" boundary condition), and, accordingly, the threshold is called "rigid". The Neumann boundary condition (1.6.30) (Example 1.6.15) describes the situation when the edge of the membrane is free in the direction normal to the surface ("soft" boundary condition), and, accordingly, the stationary point is called "soft".

EXAMPLE 1.6.16. Let M be an n-dimensional Riemannian manifold and $A = -\Delta$, where Δ is the Laplacian (see Example 1.2.4). We shall consider the Dirichlet $v|_{x_n} = 0$ or Neumann $\partial v/\partial x_n|_{x_n} = 0$ boundary conditions, where x_n is the geodesic distance to ∂M. Direct calculations along the lines of Examples 1.6.14, 1.6.15 show that

$$c_1 = \mp\frac{1}{4}\,(2\pi)^{1-n}\,\omega_{n-1}\,\mathrm{Meas}\,\partial M,$$

where the minus corresponds to the Dirichlet condition, the plus to the Neumann one, $\mathrm{Meas}\,\partial M$ is Riemannian $(n-1)$-dimensional volume of ∂M, and ω_{n-1} is the volume of the unit ball in $\mathbb{R}^{n-1}$. Note that in deriving the above formula it is convenient to fix a point $x_0 = (x_0', 0) \in \partial M$, choose normal geodesic coordinates with origin x_0, and first integrate in (1.6.19) with respect to $đ\xi'$ only. The result will be a density on ∂M, so the integral with respect to dx' will be well defined and lead to the appearance of the factor $\operatorname{Meas}\partial M$.

Examples with nontrivial functions $\mathbf{N}^+(\nu;x',\xi')$ and $\arg_0\det\big(i\,R(\nu;x',\xi')\big)$ will be considered in Section 6.2.

Let us now compare the structure of the terms $c_0\,\lambda^n$ and $c_1\,\lambda^{n-1}$ from the asymptotic expansion (1.6.1).

By (1.2.2) we have

$$c_0\,\lambda^n = \int_{T'M} \mathbf{N}_A(\nu;x,\xi)\,dx\,đ\xi\,, \tag{1.6.36}$$

where

$$\mathbf{N}_A(\nu;x,\xi) = \begin{cases} 0, & \text{if} \quad \nu \leqslant A_{2m}(x,\xi)\,, \\ 1, & \text{if} \quad \nu > A_{2m}(x,\xi) \end{cases} \tag{1.6.37}$$

(the subscript $_A$ in $\mathbf{N}_A$ indicates that this quantity depends only on the differential operator A, but not on the boundary conditions). For fixed $(x,\xi)\in T'M$ the function (1.6.37) is the counting function of the operator of multiplication by $A_{2m}(x,\xi)$ in $\mathbb{R}$. In other words, (1.6.37) is the counting function of the 1×1 matrix $A_{2m}(x,\xi)$. This trivial spectral problem

$$A_{2m}(x,\xi)v = \nu v \tag{1.6.38}$$

(where v is a real number, not a function) has only one eigenvalue $\nu = A_{2m}(x,\xi)$, and (1.6.37) is indeed the corresponding counting function.

Using (1.6.19) and the rescaling properties of the functions $\mathbf{N}^+(\nu;x',\xi')$ and $\arg_0\det\big(i\,R(\nu;x',\xi')\big)$ (cf. (1.6.17)) we have

$$c_1\,\lambda^{n-1} = \int_{T^*\partial M} \left(\mathbf{N}^+(\nu;x',\xi') + \frac{\arg_0\det\big(i\,R(\nu;x',\xi')\big)}{2\pi} \right) dx'\,đ\xi'\,. \tag{1.6.39}$$

Comparing formulae (1.6.36) and (1.6.39) we see that the terms $c_0\,\lambda^n$ and $c_1\,\lambda^{n-1}$ have roughly the same structure. In both cases integration is carried out over a cotangent bundle and the integrand contains the counting function ($\mathbf{N}_A$ or $\mathbf{N}^+$) of some elementary spectral problem. The analogy seems to be spoiled by the presence of the term $\arg_0\det\big(i\,R(\nu;x',\xi')\big)$ in (1.6.39), but this is in fact quite natural: it is known that such a quantity plays the role of the counting function for the continuous spectrum.

It is interesting to note that formula (1.6.19) for the coefficient c_1 has a mechanical interpretation.

If for some $(x',\xi')\in T'\partial M$ the auxiliary one-dimensional spectral problem (1.1.7), (1.1.8) has an eigenvalue, then the original spectral problem (1.1.1), (1.1.2) usually has a subsequence of eigenvalues corresponding to eigenfunctions which are localized in a small neighbourhood of ∂M ($x_n \lesssim \lambda^{-1}$, with exponential decay for $x_n \gg \lambda^{-1}$). Such eigenfunctions are well known in the theory of elasticity and are associated with the so-called Rayleigh surface waves. In scalar problems this boundary localization effect may also occur; see Section 6.2 in which we consider the two-dimensional biharmonic operator with free boundary conditions. The quantity

$$\int_{T^*\partial M} \mathbf{N}^+(\nu;x',\xi')\,dx'\,đ\xi'$$

is the contribution of the subsequence of eigenvalues described above to the total counting function $N(\lambda)$.

The mechanical interpretation of the quantity

$$\int_{T^*\partial M} \frac{\arg_0 \det\big(i\,R(\nu;x',\xi')\big)}{2\pi}\,dx'\,đ\xi' \tag{1.6.40}$$

is somewhat more complicated. It describes the influence of the boundary conditions on the "generic" eigenfunctions of the problem (1.1.1), (1.1.2), i.e., on those eigenfunctions which are not localized near the boundary. If the strong simple reflection condition is fulfilled, then the integrand in (1.6.40) is expressed (see (1.6.15), (1.6.16)) through the phase shift induced by the reflection. The acquisition of a phase shift is equivalent to a slight displacement of the boundary, and (1.6.40) can be viewed as the resulting correction to the first term of the asymptotic expansion of $N(\lambda)$.

5. Pseudodifferential case. In 1.1.8 we stated that apart from the main spectral problem (1.1.1), (1.1.2) in which A is the differential operator, we are also interested in the case when A is a pseudodifferential operator acting on a manifold without boundary. In order to honour this commitment we state below a modified version of Theorem 1.6.1, which is due to Duistermaat and Guillemin [**DuiGui**], [**DuiGuiHö**].

THEOREM 1.6.1$'$. *Let $\partial M=\varnothing$ and A be a pseudodifferential operator. If the nonperiodicity condition is fulfilled, then* (1.6.1) *holds. Here c_0 is the same as in Theorem* 1.2.1, *and*

$$\begin{aligned} c_1 &= -\frac{1}{2m}\int_{S^*M} A_{\text{sub}}(x,\widetilde{\xi})\,dx\,đ\widetilde{\xi} \\ &= -\left(1+\frac{n-1}{2m}\right)\int_{A_{2m}\leqslant 1} A_{\text{sub}}(x,\xi)\,dx\,đ\xi\,. \end{aligned} \tag{1.6.41}$$

Note that for differential operators of even order the subprincipal symbol is an odd function with respect to ξ, $A_{\text{sub}}(x,-\xi)=-A_{\text{sub}}(x,\xi)$, so for such operators the integral (1.6.41) is zero. This explains why formula (1.6.2) does not contain the integral (1.6.41) as an additional term.

1.7. Nonclassical two-term asymptotic formulae for $N(\lambda)$

1. Preliminary discussion. In this section we turn to the case when the measure of the set Π is nonzero. One can expect that in this situation the two-term asymptotic formula contains an extra term of order $O(\lambda^{n-1})$ which reflects the influence of the periodic trajectories. The most natural conjecture is that

$$N(\lambda) = c_0\,\lambda^n + c_1\,\lambda^{n-1} + \widetilde{\mathbf{Q}}(\lambda)\,\lambda^{n-1} + o(\lambda^{n-1}), \qquad \lambda\to+\infty, \tag{1.7.1}$$

where c_0 and c_1 are the same constants as in (1.4.1), and $\widetilde{\mathbf{Q}}(\lambda)$ is a bounded function.

Of course, a function $\widetilde{\mathbf{Q}}$ satisfying (1.7.1) always exists; for example, we can take any function $\widetilde{\mathbf{Q}}$ of the form

$$\widetilde{\mathbf{Q}}(\lambda) = \lambda^{1-n}\,N(\lambda) - c_0\,\lambda - c_1 + o(1), \qquad \lambda\to+\infty.$$

But this observation is of no use unless we can say something more definite about $\widetilde{\mathbf{Q}}$. In particular, we would like this function to depend only on the higher order terms of the differential operator A and the boundary operators $B^{(k)}$.

However, Example 1.7.1 (see below) shows that, generally speaking, (1.7.1) cannot hold with a $\widetilde{\mathbf{Q}}$ independent of the lower order terms. Since in the general case the lower order terms are not invariantly defined, it seems to be impossible to obtain any general results concerning $\widetilde{\mathbf{Q}}$. Therefore we adopt another approach. We shall prove that there exists a bounded function $\mathbf{Q}$ depending only on the principal and subprincipal symbols of A and on the principal symbols of $B^{(k)}$, such that

$$\mathbf{Q}(\lambda+\varepsilon) - \mathbf{Q}(\lambda) \geqslant -\varepsilon\, n\, c_0\,, \qquad \forall \lambda \in \mathbb{R}, \quad \forall \varepsilon \geqslant 0, \tag{1.7.2}$$

and

$$\begin{aligned} c_0\,\lambda^n + c_1\,\lambda^{n-1} + \mathbf{Q}\,(\lambda - o(1))\,\lambda^{n-1} - o(\lambda^{n-1}) &\leqslant N(\lambda) \\ \leqslant c_0\,\lambda^n + c_1\,\lambda^{n-1} + \mathbf{Q}\,(\lambda + o(1))\,\lambda^{n-1} + o(\lambda^{n-1})&, \qquad \lambda \to +\infty, \end{aligned} \tag{1.7.3}$$

where $o(1)$ is some positive function tending to zero as $\lambda \to +\infty$ and $o(\lambda^{n-1}) = \lambda^{n-1} o(1)$. In view of (1.7.2)

$$\begin{aligned} &c_0\,\lambda^n + c_1\,\lambda^{n-1} + \mathbf{Q}(\lambda)\,\lambda^{n-1} \\ &\quad\leqslant c_0\,(\lambda+\varepsilon)^n + c_1\,\lambda^{n-1} + \mathbf{Q}(\lambda+\varepsilon)\,\lambda^{n-1} + o(\lambda^{n-1}), \qquad \lambda \to +\infty, \end{aligned}$$

for any fixed $\varepsilon \geqslant 0$, so (1.7.3) makes sense.

The asymptotic estimate (1.7.3) implies that the "distance" between the graphs of the functions

$$(N(\lambda))^{1/n} \quad \text{and} \quad \left(c_0\lambda^n + c_1\lambda^{n-1} + \mathbf{Q}(\lambda)\lambda^{n-1}\right)^{1/n}$$

tends to zero as $\lambda \to +\infty$. The asymptotic formula (1.7.1) would mean that the distance between these graphs in the vertical direction tends to zero, whereas (1.7.3) allows also an error $o(1)$ in the horizontal direction.

In other words, we shall prove that for all $\varepsilon > 0$

$$\begin{aligned} c_0\,(\lambda-\varepsilon)^n + c_1\,\lambda^{n-1} + \mathbf{Q}(\lambda-\varepsilon)\,\lambda^{n-1} - o(\lambda^{n-1}) &\leqslant N(\lambda) \\ \leqslant c_0\,(\lambda+\varepsilon)^n + c_1\,\lambda^{n-1} + \mathbf{Q}(\lambda+\varepsilon)\,\lambda^{n-1} + o(\lambda^{n-1})&, \qquad \lambda \to +\infty, \end{aligned} \tag{1.7.4}$$

where, generally speaking, $o(\lambda^{n-1})$ depends on ε. Obviously, (1.7.4) is equivalent to (1.7.3).

If the function $\mathbf{Q}$ is uniformly continuous, then we obtain from (1.7.4)

$$N(\lambda) = c_0\,\lambda^n + c_1\,\lambda^{n-1} + \mathbf{Q}(\lambda)\,\lambda^{n-1} + o(\lambda^{n-1}), \qquad \lambda \to +\infty. \tag{1.7.5}$$

We shall call (1.7.5) with a uniformly continuous $\mathbf{Q}$ the *quasi-Weyl* asymptotic formula.

If

$$\mathbf{Q}(\mu_j + \varepsilon_j) - \mathbf{Q}(\mu_j - \varepsilon_j) \geqslant c, \qquad j = 1, 2, \ldots, \tag{1.7.6}$$

where $\mu_j \to +\infty$, $\varepsilon_j \to +0$ are some positive sequences and c is a positive constant, then (1.7.4) implies

$$N(\mu_j + \varepsilon_j) - N(\mu_j - \varepsilon_j) \geqslant c\,\mu_j^{n-1} + o(\mu_j^{n-1}). \tag{1.7.7}$$

This means that the spectrum of the operator $\mathcal{A}^{1/(2m)}$ contains contracting groups of eigenvalues lying in the ε_j-neighbourhoods of the points μ_j, whose total multiplicities are estimated from below by const μ_j^{n-1}. Such groups of eigenvalues are called *clusters*. The precise location of eigenvalues inside the clusters normally depends on the lower order terms of the operators A and $B^{(k)}$. Therefore in the case of cluster asymptotics one cannot expect to have (1.7.5) with a function $\mathbf{Q}$ depending only on the higher order terms.

If the quasi-Weyl formula (1.7.5) holds, then there are no clusters in the spectrum. Indeed, since $\mathbf{Q}$ is uniformly continuous,

$$N(\mu_j+\varepsilon_j) \;-\; N(\mu_j-\varepsilon_j) \;=\; o(\mu_j^{n-1})$$

for all sequences $\mu_j \to +\infty$ and $\varepsilon_j \to 0$.

EXAMPLE 1.7.1. Let

$$A \;=\; (-\Delta + (n-1)^2/4)^m \;+\; V\,, \tag{1.7.8}$$

where Δ is the Laplacian on the unit sphere and V is the operator of multiplication by a smooth nonnegative function. Recall that the spectrum of $(-\Delta+(n-1)^2/4)^{1/2}$ consists of the eigenvalues

$$\Lambda_j \;=\; j+(n-1)/2\,, \qquad j=0,1,2,\ldots,$$

with multiplicities

$$\frac{(n+j-2)!\,(n+2j-1)}{(n-1)!\,j!} \;\geqslant\; \text{const}\ \Lambda_j^{n-1}$$

(see Example 1.2.5).

We shall see later that for the operator (1.7.8) formula (1.7.4) holds with a periodic function $\mathbf{Q}$ which has jumps at the points $j+(n-1)/2$ (Example 1.7.11). Such a function satisfies (1.7.6) with $\mu_j=\Lambda_j$, so we have clusters around the points Λ_j. This result can be also obtained by means of perturbation theory: if we add the lower order operator V to the operator $(-\Delta+(n-1)^2/4)^m$, then the multiple eigenvalues Λ_j of $(-\Delta+(n-1)^2/4)^m$ turn into contracting groups of eigenvalues with the same total multiplicities.

For a general potential V the quasi-Weyl formula (1.7.5) is not true. However, the difference

$$N(\lambda) \;-\; c_0\,\lambda^n \;-\; c_1\,\lambda^{n-1} \;-\; \mathbf{Q}(\lambda)\,\lambda^{n-1}$$

is zero outside the ε_j-neighbourhoods of the points $j+(n-1)/2$, for some $\varepsilon_j\to 0$. Note that there is no function $\widetilde{\mathbf{Q}}$ depending only on the principal and subprincipal symbols which satisfies (1.7.1) for all potentials V. Indeed, if $V\equiv V_0=\text{const}>0$, then the spectrum of $\mathcal{A}^{1/(2m)}$ consists of the eigenvalues

$$\Big((j+(n-1)/2)^{2m}+V_0\Big)^{1/(2m)}, \qquad j=0,1,2,\ldots,$$

with the same multiplicities as Λ_j. Since these eigenvalues depend on the constant V_0, the function $\widetilde{\mathbf{Q}}$ in (1.7.1) must also depend on V_0.

2. The case of simple reflection. Assume that the Hamiltonian billiard system satisfies the simple reflection condition (see Definition 1.3.30).

DEFINITION 1.7.2. For a T-admissible trajectory

$$\Gamma = (x^*(t;y,\eta), \xi^*(t;y,\eta)), \qquad 0 \leqslant t \leqslant T,$$

let

- $\mathfrak{f}_r(T;y,\eta)$ be the total phase shift generated by the reflections of the trajectory Γ, that is, the sum of the phase shifts generated by the reflections of Γ (see the end of Section 1.4);
- $\mathfrak{f}_s(T;y,\eta) = -(2m)^{-1} h^{1-2m}(y,\eta) \int_0^T A_{\mathrm{sub}}(x^*(t;y,\eta),\xi^*(t;y,\eta))\,dt$ (this quantity is interpreted as the phase shift generated by the subprincipal symbol).

When $x_\eta(T;y,\eta) = 0$ we also define

- $\mathfrak{f}_c(T;y,\eta)$ to be the phase shift generated by the passage of the trajectory Γ through caustics, that is, $\mathfrak{f}_c(T;y,\eta) = -\alpha_\Gamma \pi/2$ where α_Γ is the Maslov index of Γ (see Section 1.5).

The quantity

$$\mathfrak{f}(T;y,\eta) = \mathfrak{f}_r(T;y,\eta) + \mathfrak{f}_s(T;y,\eta) + \mathfrak{f}_c(T;y,\eta)$$

is said to be the *total phase shift* along the trajectory Γ.

The total phase shift is an additive function in the sense that

$$\mathfrak{f}(T_1+T_2;y,\eta) = \mathfrak{f}(T_1;y,\eta) + \mathfrak{f}(T_2;x^*(T_1;y,\eta),\xi^*(T_1;y,\eta)),$$

and the same is valid for $\mathfrak{f}_r$, $\mathfrak{f}_s$ and $\mathfrak{f}_c$.

DEFINITION 1.7.3. For a periodic point (y,η) we denote by $\mathbf{T}(y,\eta)$ the corresponding minimal positive period. If, in addition, (y,η) is absolutely $\mathbf{T}(y,\eta)$-periodic and admissible, then we set $\mathbf{q}(y,\eta) = \mathfrak{f}(\mathbf{T}(y,\eta);y,\eta)$.

Obviously, $\mathbf{q}(y,\eta)$ is the total phase shift along the primitive closed trajectory originating from (y,η). In view of Lemmas 1.3.5 and 1.3.28 the function $\mathbf{q}$ is defined almost everywhere (a.e.) on Π^a, as well as on Π.

LEMMA 1.7.4. *There exists a constant $C>0$ such that $\mathbf{T}(y,\tilde\eta) \geqslant C$ for all $(y,\tilde\eta) \in \Pi$.*

LEMMA 1.7.5. *The functions $\mathbf{T}$ and $\mathbf{q}$ are measurable.*

Let us denote by $\{\tau\}_{2\pi}$ the residue of the real number τ modulo 2π, that is,

$$\{\tau\}_{2\pi} := \tau + 2\pi k \in [-\pi,\pi), \qquad k \in \mathbb{Z}.$$

Clearly, $\{\tau\}_{2\pi} = 2\pi\{\tau/(2\pi)+1/2\} - \pi$, where $\{\cdot\}$ is the fractional part.

THEOREM 1.7.6 ([**Sa4**], [**Sa5**], [**Sa7**]). *If the Hamiltonian billiard system satisfies the simple reflection condition, then* (1.7.4) (*and* (1.7.3)) *hold with the function*

$$\mathbf{Q}(\lambda) = \int_{\Pi^a} \frac{\{\pi - \mathbf{q} - \lambda\mathbf{T}\}_{2\pi}}{\mathbf{T}}\, dy\, d\tilde\eta. \tag{1.7.9}$$

Note that for each fixed $(y,\eta)\in T^*M$ we have

$$\{\pi - \mathbf{q} - \lambda\mathbf{T}\}_{2\pi} = 2\sum_{k=1}^{\infty} k^{-1} \sin k(\lambda\mathbf{T}+\mathbf{q}) \tag{1.7.10}$$

where the series converges in the sense of distributions. From (1.7.9) and (1.7.10) one can easily deduce the following elementary properties of the function $\mathbf{Q}$:

1. $\mathbf{Q}$ is uniformly bounded, $|\mathbf{Q}(\lambda)| \leqslant \pi \int_{\Pi^a} \mathbf{T}^{-1}\, dy\, d\!\!\!{}^{-}\eta$.
2. $\mathbf{Q}$ is an oscillating function, i.e., the integral $\int_0^\lambda \mathbf{Q}(\mu)\, d\mu$ is bounded uniformly with respect to $\lambda \in \mathbb{R}$.
3. $\mathbf{Q}$ is left-continuous.
4. For all $\varepsilon > 0$ we have $\mathbf{Q}(\lambda+\varepsilon) - \mathbf{Q}(\lambda) \geqslant -\varepsilon(2\pi)^{-n} \operatorname{meas} \Pi^a$.
5. $\mathbf{Q}(\lambda+0) - \mathbf{Q}(\lambda) = 2\pi \int_{\Pi^a(\lambda)} \mathbf{T}^{-1}\, dy\, d\!\!\!{}^{-}\eta$, where

$$\Pi^a(\lambda) \ := \ \{ (y,\eta) \in \Pi^a \ : \ \mathbf{q}(y,\eta) + \lambda \mathbf{T}(y,\eta) = 0 \pmod{2\pi} \}.$$

6. If $\mathbf{T} = T = \text{const}$ a.e. on Π^a, then $\mathbf{Q}$ is $2\pi T^{-1}$-periodic.

In view of (1.2.2′), (4) implies (1.7.2). Moreover, from Theorem 1.7.6 and (4) it follows that

$$N(\lambda+\varepsilon) \ - \ N(\lambda) \ \geqslant \ \varepsilon\, (2\pi)^{-n} \ \operatorname{meas}(S^*M \setminus \Pi^a)\, \lambda^{n-1} \ + \ o(\lambda^{n-1}).$$

Therefore we obtain the following

COROLLARY 1.7.7. *Let the conditions of Theorem* 1.7.6 *be fulfilled. Assume that there exists a sequence of points* $\mu_j \to +\infty$ *and an* $\varepsilon > 0$ *such that the number of eigenvalues lying in the* ε*-neighbourhood of* μ_j *is* $o(\mu_j^{n-1})$. *Then* $\operatorname{meas}\Pi^a = \operatorname{meas} S^*M$.

The condition $\mathbf{T} = T = \text{const}$ in (6) seems to be very restrictive. However, the following lemma shows that it is always satisfied when M and h are analytic and the set Π^a is connected.

LEMMA 1.7.8. *In the analytic case,* $\mathbf{T} \equiv \text{const}$ *almost everywhere on any open connected subset of* Π^a.

Assume that (6) is fulfilled. Then, since a periodic function is uniformly continuous if and only if it is continuous, Theorem 1.7.6 and (5) immediately imply

COROLLARY 1.7.9. *Let the conditions of Theorem* 1.7.6 *be fulfilled and* $\mathbf{T} = T = \text{const}$ *a.e. on* Π^a. *Then the quasi-Weyl formula* (1.7.5) *holds with the function* $\mathbf{Q}$ *given by* (1.7.9) *if and only if*

$$\operatorname{meas}\{ (y,\eta) \in \Pi^a \ : \ \mathbf{q}(y,\eta) = \mu \} \ = 0 \ , \qquad \forall \mu \in \mathbb{R}.$$

If

$$\operatorname{meas}\{ (y,\eta) \in \Pi^a \ : \ \mathbf{q}(y,\eta) = q \} \ = \ C_q \ > \ 0 \tag{1.7.11}$$

for some constants q and C_q, then, in view of (5) and (6),

$$\mathbf{Q}\left(T^{-1}(q+2\pi j)+0\right) - \mathbf{Q}\left(T^{-1}(q+2\pi j)\right) = (2\pi)^{1-n}\, T^{-1}\, C_q, \qquad j = 1, 2, \ldots .$$

This implies

COROLLARY 1.7.10. *Let the simple reflection condition be fulfilled,* $\mathbf{T} = T = \text{const}$ *a.e. on* Π^a, *and* (1.7.11) *hold for some* q. *Then we have* (1.7.7) *with* $c = (2\pi)^{1-n}T^{-1}C_q$, $\mu_j = T^{-1}(q+2\pi j)$ *and some* $\varepsilon_j \to 0$, *i.e., the spectrum contains clusters.*

Thus, if $\mathbf{T} = T = \text{const}$ a.e. on Π^a, then either the quasi-Weyl formula holds or the spectrum contains clusters. If, modulo a set of measure zero, Π^a consists of several open connected components, then $\mathbf{Q}$ is the sum of functions corresponding to these components. Each of these functions either generates clusters or gives a contribution to the quasi-Weyl part of the asymptotics.

EXAMPLE 1.7.11. For the operator (1.7.8) we have $\Pi^a = S^*M$, $\mathbf{T} \equiv 2\pi$, $\mathbf{q} = \mathbf{q}_c$, and $\mathbf{q}_c \equiv \pi(n-1) \pmod{2\pi}$. Therefore

$$\mathbf{Q}(\lambda) = (2\pi)^{-n} \operatorname{meas} S^*M \, \frac{\{\pi - \pi(n-1) - 2\pi\,\lambda\}_{2\pi}}{2\pi} = n\,c_0\,(\{n/2 - \lambda\} - 1/2)\,,$$

where $\{n/2 - \lambda\}$ is the fractional part of $n/2 - \lambda$. The constant c_0 is equal to $c_0 = 2/n!$ (see Example 1.2.5), and $c_1 = 0$ since $\partial M = \varnothing$.

EXAMPLE 1.7.12. Let M be an n-dimensional unit sphere and $A = -\Delta + B$ where B is a self-adjoint first order differential operator. Again, we have $\Pi^a = S^*M$, $\mathbf{T} \equiv 2\pi$, and $\mathbf{q}_c \equiv \pi(n-1) \pmod{2\pi}$. However, now $\mathbf{q} = \mathbf{q}_c + \mathbf{q}_s$, where

$$\mathbf{q}_s(y,\eta) \;=\; -\frac{1}{2h}\int B_1\,ds\,,$$

B_1 is the principal symbol of B, and the integral is taken over the primitive periodic trajectory originating from (y,η). In this case the type of asymptotics is determined by the principal symbol B_1. It can even happen that $\mathbf{Q} \equiv 0$, and then we have the classical Weyl formula for $N(\lambda)$ (with the same constants as in Example 1.7.11) though all the trajectories are periodic. For instance, $\mathbf{Q}$ is identically zero if M is a unit three-dimensional sphere embedded in $\mathbb{R}^4$ and $B = \mathcal{B} + \mathcal{B}^*$, where $\mathcal{B}$ is the first order differential operator generated by the vector field with components $(-x_2, x_1, -x_4, x_3)$ (this vector field is tangent to the sphere); see [**Sa2**], [**Sa8**].

EXAMPLE 1.7.13. Let M be a two-dimensional hemisphere of radius R, and A the biharmonic operator Δ^2 on M with Dirichlet boundary condition. Then

$$c_0 \;=\; R^2/2\,, \qquad c_1 \;=\; -\frac{R}{2}\left(1 \,+\, \pi^{-1/2}\,\frac{\Gamma(3/4)}{\Gamma(5/4)}\right),$$

$$\mathbf{Q}(\lambda) \;=\; -R\left(\{R\lambda\} \,+\, (\cos\pi\{R\lambda\})_+^{1/2} \,-\, \frac{\mathbf{K}(\sqrt{2}/2) + \sqrt{2}}{2\,\mathbf{K}(\sqrt{2}/2)}\right),$$

where $\Gamma(\cdot)$ is the gamma function, $\{R\lambda\}$ is the fractional part of $R\lambda$, and $\mathbf{K}(\cdot)$ is the complete elliptic integral of the first kind. Note that the coefficient c_1 coincides with that for a plate with the same length of ∂M; see Section 6.2. The function $\mathbf{Q}$ is R^{-1}-periodic and continuous, so the quasi-Weyl formula holds. The graph of $\mathbf{Q}$ is given in Figure 10.

Elementary analysis of the formula for $\mathbf{Q}$ shows that the number of eigenvalues λ_k lying in the intervals $[(j - 1/2 + \varepsilon)/R, (j - \varepsilon)/R]$, $0 < \varepsilon < 1/4$, $j = 1, 2, \ldots$, is $o(j)$ as $j \to \infty$. This means that asymptotically the spectrum contains gaps. Another elementary observation is that the left derivative of $\mathbf{Q}$ is infinite at the points $(j - 1/2)/R$, so one expects the density of eigenvalues λ_k around these points to be abnormally high.

3. The general case. A similar result is true in the general case of branching Hamiltonian billiards.

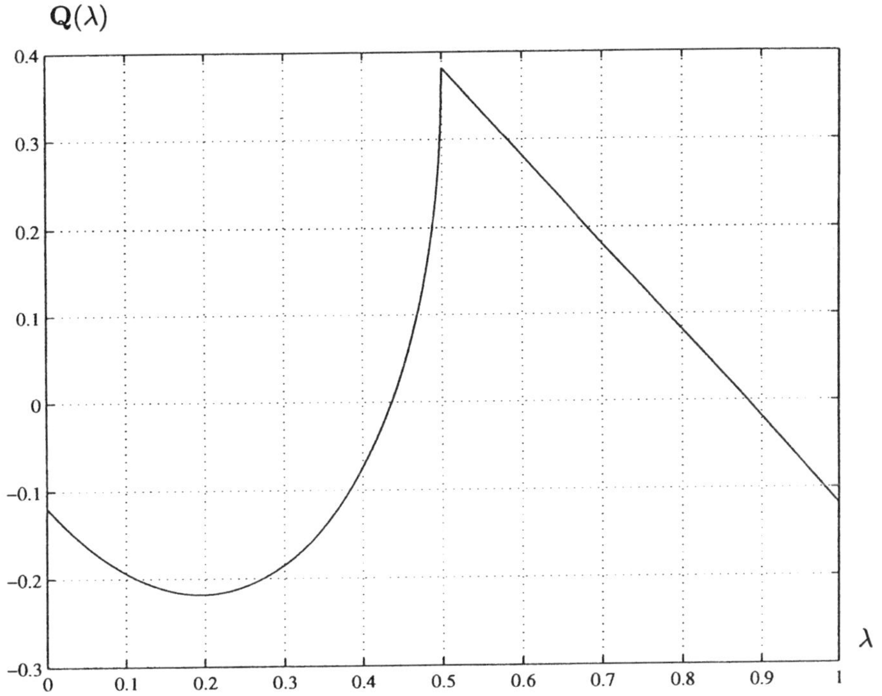

FIGURE 10. The function $\mathbf{Q}$ for the Dirichlet bi-Laplacian on a hemisphere.

THEOREM 1.7.14. *Let the nonblocking condition (Definition* 1.3.22) *be fulfilled. Assume, in addition, that there exist positive numbers* $T_1, T_2, \dots$ *tending to* $+\infty$ *such that* $\operatorname{meas}\left(\bigcup_{T\neq 0,\, T\neq T_j} \Pi_T^a\right) = 0$. *Then* (1.7.4) *holds with a bounded function* $\mathbf{Q}$ *depending only on the principal and subprincipal symbols of* A *and the principal symbols of* $B^{(k)}$, *and satisfying* (1.7.2). *The function* $\mathbf{Q}$ *is left-continuous and almost periodic with periods* $T_1, T_2, \dots$.

In the general case there is no simple formula for the function $\mathbf{Q}$. One can only prove that for positive λ this function is given by a trigonometric series

$$\mathbf{Q}(\lambda) = \sum_{j=1}^{\infty} C_j \sin \lambda T_j\,, \tag{1.7.12}$$

where C_j are constants depending on some characteristics of the corresponding periodic trajectories.

Theorem 1.7.14 is obtained in the same manner as Theorem 1.7.6, and therefore we shall not prove it in this book. A rigorous proof of (1.7.12) and formulae for C_j are given in [**Sa5**] where even a more general situation (with A being a system of partial differential operators) is examined.

4. Pseudodifferential case. Theorem 1.7.6 remains true for a pseudodifferential operator acting on a manifold without boundary [**Sa4**]. The only difference is that in this case the constant c_1 is given by formula (1.6.41).

1.8. Two-term asymptotic formulae for the spectral function

1. The spectral function. The *spectral function* of the operator $\mathcal{A}$ is defined by

$$e(\lambda,x,y)=\sum_{\lambda_k<\lambda} v_k(x)\,\overline{v_k(y)}\,,$$

where $\lambda_k, v_k(x)$, $k=1,2,\dots$, are the eigenvalues and the orthonormalized eigenfunctions of the problem (1.1.1), (1.1.2). Clearly, $e(\lambda,x,y)$ is a smooth half-density on $M\times M$ depending on the spectral parameter λ, and $e(\lambda,x,y)\equiv 0$ if $\lambda\leqslant 0$.

The operator

$$E_\lambda v\;=\;\sum_{\lambda_k<\lambda}(v,v_k)\,v_k\,,$$

where $(\cdot\,,\cdot)$ stands for the inner product (1.1.10), is called the *spectral projection* of the problem (1.1.1), (1.1.2) (or of the corresponding self-adjoint operator $\mathcal{A}^{1/(2m)}$). The spectral function is the integral kernel of the spectral projection:

$$E_\lambda v\;=\;\int_M e(\lambda,x,y)\,v(y)\,dy\,.$$

Clearly, $e(\lambda,y,y)$ is a density on M depending on λ. Integrating over M we obtain

$$N(\lambda)\;=\;\operatorname{Tr}E_\lambda\;=\;\int_M e(\lambda,y,y)\,dy\,. \tag{1.8.1}$$

Thus, $e(\lambda,y,y)$ contains full information about the spectrum.

In this section we shall discuss the asymptotic behaviour of $e(\lambda,y,y)$ as $\lambda\to+\infty$ at a fixed interior point y. More generally, we shall consider the functions

$$e_{P,Q}(\lambda,x,y)\;=\;\sum_{\lambda_k<\lambda} Q^*v_k(x)\,\overline{P^*v_k(y)}\,,$$

where P,Q are pseudodifferential operators of the class Ψ^l_0, $2l>1-n$. The principal and subprincipal symbols of P and Q are denoted by P_l,Q_l and P_{sub}, Q_{sub} respectively. (See Section 2.1 for definitions of pseudodifferential operators and their symbols.)

Obviously, $e_{P,Q}(\lambda,x,y)$ is the integral kernel of the operator $Q^*E_\lambda P$. We have

$$\begin{aligned}e_{P,Q}(\lambda,x,y)\;&=\frac14\,e_{P+Q,P+Q}(\lambda,x,y)\;-\;\frac14\,e_{P-Q,P-Q}(\lambda,x,y)\\&\quad+\;\frac i4\,e_{P+iQ,P+iQ}(\lambda,x,y)\;-\;\frac i4\,e_{P-iQ,P-iQ}(\lambda,x,y)\end{aligned}$$

(the *polarization* formula). Applying polarization, one can easily deduce the results for $e_{P,Q}(\lambda,x,y)$ from those for $e_{P,P}(\lambda,x,y)$.

2. Notation and definitions. Throughout this section we shall always assume that the Hamiltonian billiard system satisfies the simple reflection condition (see Definition 1.3.30).

Let us denote

$$\Pi_{y,T}\;=\;\{\tilde\eta\in S^*_yM\,:\,x^*(T;y,\tilde\eta)=y\}\,,\qquad \Pi_y\;=\;\bigcup_{T>0}\Pi_{y,T}\,.$$

Let $\Pi^a_{y,T} \subset \Pi_{y,T}$ be the set of $\widetilde{\eta}$ such that

$$\partial^\alpha_\eta \left(|y - x^*(T;y,\eta)|^2\right)\big|_{\eta=\widetilde{\eta}} = 0\,, \qquad \forall\alpha\,,$$

and $\Pi^a_y = \bigcup_{T>0} \Pi^a_{y,T}$.

DEFINITION 1.8.1. A point $y \in \overset{\circ}{M}$ is said to be *regular* if for almost all $\widetilde{\eta} \in S^*_y M$ the points $(y,\widetilde{\eta}) \in S^*M$ are admissible.

LEMMA 1.8.2. *If y is a regular point, then the sets Π_y and $\bigcup_{0<T\leqslant T_+} \Pi_{y,T}$, $\forall T_+ > 0$, are measurable.*

LEMMA 1.8.3. *If y is a regular point, then* $\operatorname{meas}_y(\Pi_y \setminus \Pi^a_y) = 0$. *Moreover,* $\operatorname{meas} \bigcup_{T>0} \left(\Pi_{y,T} \setminus \Pi^a_{y,T}\right) = 0$.

DEFINITION 1.8.4. A regular point $y \in \overset{\circ}{M}$ is said to be *focal* if $\operatorname{meas}_y \Pi^a_y > 0$.

3. Classical asymptotics. The following well-known result is an analogue of Theorem 1.2.1.

THEOREM 1.8.5. *If $y \in \overset{\circ}{M}$, then*

$$e_{P,Q}(\lambda, y, y) = c_{0;P,Q}(y)\,\lambda^{n+2l} + O(\lambda^{n+2l-1})\,, \qquad \lambda \to +\infty\,, \tag{1.8.2}$$

where

$$c_{0;P,Q}(y) = (n+2l)^{-1} \int_{S^*_y M} P_l(y,\widetilde{\eta})\,\overline{Q_l(y,\widetilde{\eta})}\,đ\widetilde{\eta}\,. \tag{1.8.3}$$

The asymptotic formula (1.8.2) *is uniform on compact subsets of* $\overset{\circ}{M}$.

Note that $e_{P,Q}(\lambda, y, y)$ is a density depending on λ, and the coefficient $c_{0;P,Q}(y)$ also behaves as a density on M. If P and Q are operators of multiplication by some functions $\widetilde{P}, \widetilde{Q} \in C^\infty_0(\overset{\circ}{M})$ and $\widetilde{P}(y) = \widetilde{Q}(y) = 1$, then $e_{P,Q}(\lambda, y, y) = e(\lambda, y, y)$ and (1.8.2) takes the form

$$e(\lambda, y, y) = c_0(y)\,\lambda^n + O(\lambda^{n-1})\,, \qquad \lambda \to +\infty\,, \tag{1.8.4}$$

where

$$c_0(y) = (2\pi)^{-n} \operatorname{vol}_y \{\eta : A_{2m}(y,\eta) \leqslant 1\} = \int_{A_{2m}(y,\eta)\leqslant 1} đ\eta\,.$$

If M is a manifold without boundary, then Theorem 1.2.1 is obtained by integrating (1.8.4) over M.

REMARK 1.8.6. By (1.8.2), if $\operatorname{ord} \widetilde{P} > \operatorname{ord} P$ and the principal symbols of $\widetilde{P}$ and P do not vanish at y, then

$$\frac{\sum_{\lambda_k<\lambda} |\widetilde{P} v_k(y)|^2}{\sum_{\lambda_k<\lambda} |P v_k(y)|^2} \to +\infty\,, \qquad \lambda \to +\infty\,.$$

This means that, in a sense, the eigenfunctions v_k become more and more oscillating as $k \to \infty$.

THEOREM 1.8.7 ([**Sa6**]). *Let* $y \in \overset{\circ}{M}$ *be a regular nonfocal point. Then*

$$e_{P,P}(\lambda, y, y) = c_{0;P}(y)\,\lambda^{n+2l} + c_{1;P}(y)\,\lambda^{n+2l-1} + o(\lambda^{n+2l-1}) \tag{1.8.5}$$

as $\lambda \to +\infty$, *where* $c_{0;P}(y) = c_{0;P,P}(y)$ *and*

$$\begin{aligned} c_{1;P}(y) &= (n+2l-1)^{-1} \int_{S_y^* M} \left(2\,\mathrm{Re}(P_l\,\overline{P_{\mathrm{sub}}}) + \frac{\mathrm{Im}\{\overline{P_l}, P_l\}}{2} \right) d\widetilde{\eta} \\ &+ \int_{S_y^* M} \left(\mathrm{Im}(P_l\{h, \overline{P_l}\}) - \frac{A_{\mathrm{sub}}\,|P_l|^2}{2m} \right) d\widetilde{\eta}\,. \end{aligned} \tag{1.8.6}$$

The asymptotic formula (1.8.5) *is uniform on compact subsets of* $\overset{\circ}{M}$ *which do not contain any focal or nonregular points.*

Assuming that P is the operator of multiplication by a C_0^∞-function which is equal to one in a neighbourhood of y, we obtain

$$e(\lambda, y, y) = c_0(y)\,\lambda^n + o(\lambda^{n-1}), \qquad \lambda \to +\infty\,.$$

Indeed, in this case $P_{\mathrm{sub}} = 0$ at the point y and the second term on the right-hand side of (1.8.6) disappears because A_{sub} is an odd function of $\widetilde{\eta}$. If M is a manifold without boundary and all the points $y \in M$ are nonfocal, then the last formula implies (1.6.1) (with $c_1 = 0$).

4. Nonclassical asymptotics. Given $\widetilde{\eta} \in \Pi_y$ we denote

$$\mathbf{T}_y(\widetilde{\eta}) = \min\{\, t > 0 \,:\, x^*(t; y, \widetilde{\eta}) = y \,\}\,.$$

If, in addition, $\widetilde{\eta} \in \Pi^a_{y,\mathbf{T}_y(\widetilde{\eta})}$ and $(y, \widetilde{\eta})$ is admissible, then we set

$$\mathbf{q}_y(\widetilde{\eta}) = \mathfrak{f}\,(\mathbf{T}_y(\widetilde{\eta}); y, \widetilde{\eta})\,,$$

where $\mathfrak{f}$ is the total phase shift from Definition 1.7.2. In the case of a geodesic flow $\mathbf{T}_y(\widetilde{\eta})$ is the length of the minimal geodesic loop originating from the point y and going in the direction $\widetilde{\eta}$, and $\mathbf{q}_y(\widetilde{\eta})$ is the total phase shift along this loop.

Further on in this section we assume that the point y is regular. Then Definition 1.8.1 and Lemma 1.8.3 imply that the function $\mathbf{q}_y$ is defined a.e. on Π^a_y, as well as on Π_y.

Let us define a map $\Phi_y : \Pi^a_y \to S^*_y M$ by

$$\Phi_y \widetilde{\eta} = \xi^*\,(\mathbf{T}_y(\widetilde{\eta}); y, \widetilde{\eta})\,,$$

and let

$$J_y(\widetilde{\eta}) = |\det \xi^*_\eta\,(\mathbf{T}_y(\widetilde{\eta}); y, \widetilde{\eta})\,|\,.$$

If $x^*_\eta(t; y, \eta) = 0$, then $|\det \xi^*_\eta(t; y, \eta)| \neq 0$ behaves under change of coordinates as a density in x and a density to the power -1 in y. Therefore the restriction of J_y to Π^a_y is independent of the choice of coordinates y, assuming that we take the same coordinates for x and for y. Moreover, since the shift along billiard trajectories is a nondegenerate map and $x^*_\eta = 0$ on Π^a_y, we have $J_y \geqslant \text{const} > 0$ uniformly on Π^a_y.

We shall need the following two lemmas.

LEMMA 1.8.8. *The functions* $\mathbf{T}_y$ *and* $\mathbf{q}_y$ *are measurable.*

LEMMA 1.8.9. *For any measurable set* $\Omega \subset \Pi_y^a$ *the set* $\Phi_y\Omega \subset S_y^*M$ *is also measurable and*

$$\operatorname{meas}_y(\Phi_y\Omega) = \int_\Omega J_y(\widetilde{\eta})\, d\widetilde{\eta}\,. \tag{1.8.7}$$

In other words, Lemma 1.8.9 states that J_y is the Radon–Nikodym derivative of the measure $\operatorname{meas}_y(\Phi_y\,\cdot\,)$ with respect to meas_y.

Let $U_y : L_2(S_y^*M) \to L_2(S_y^*M)$ and $U_{y,\lambda} : L_2(S_y^*M) \to L_2(S_y^*M)$ be the linear operators defined as follows:

$$(U_y f)(\widetilde{\eta}) = \begin{cases} 0\,, & \text{if } \widetilde{\eta} \notin \Pi_y^a\,,\\ e^{-i\mathbf{q}_y(\widetilde{\eta})}\sqrt{J_y(\widetilde{\eta})}\, f(\Phi_y\widetilde{\eta})\,, & \text{if } \widetilde{\eta} \in \Pi_y^a\,, \end{cases}$$

$$(U_{y,\lambda} f)(\widetilde{\eta}) = e^{-i\lambda\mathbf{T}_y(\widetilde{\eta})}(U_y f)(\widetilde{\eta})$$

(here λ is considered as a parameter). By Lemma 1.8.8 the operators U_y and $U_{y,\lambda}$ are partially isometric with kernel

$$\{f \in L_2(S_y^*M) : \operatorname{supp} f \cap \Phi_y\Pi_y^a = \varnothing\}$$

and range

$$U_y L_2(S_y^*M) = U_{y,\lambda}L_2(S_y^*M) = \{f \in L_2(S_y^*M) : \operatorname{supp} f \subset \Pi_y^a\}\,.$$

LEMMA 1.8.10. *For each compact set* $K \subset \overset{\circ}{M}$ *there exists a constant* $C_K > 0$ *such that* $\mathbf{T}_y(\widetilde{\eta}) \geqslant C_K$ *for all* $y \in K$, $\widetilde{\eta} \in \Pi_y$.

LEMMA 1.8.11. *For all* $f \in L_2(S_y^*M)$ *the series*

$$\|f\|^2_{L_2(S_y^*M)} + \sum_{k=1}^{\infty} 2\operatorname{Re}\left(f, U_{y,\lambda}^k f\right)_{L_2(S_y^*M)}$$

converges in the sense of distributions in λ. *Its sum is a positive Borel measure on* $\mathbb{R}$.

By Lemma 1.8.11 the series

$$\sum_{k=1}^{\infty}\left(U_{y,\lambda}^k + (U_{y,\lambda}^k)^*\right) \tag{1.8.8}$$

converges in the weak operator topology and defines a Borel measure with values in the space of bounded operators in $L_2(S_y^*M)$.

LEMMA 1.8.12. *The Fourier transform* $\mathcal{F}_{\lambda\to t}\left[\sum_{k=1}^{\infty}\left(U_{y,\lambda}^k + (U_{y,\lambda}^k)^*\right)\right]$ *vanishes in a neighbourhood of zero.*

Let $\mathbf{Q}(y,\lambda)$ be the operator-valued distribution function of the measure (1.8.8) such that $\mathcal{F}_{\lambda\to t}[\mathbf{Q}(y,\lambda)] = 0$ for sufficiently small t. Clearly,

$$\mathbf{Q}(y,\lambda) = \mathbf{Q}^*(y,\lambda), \qquad \forall\lambda \in \mathbb{R}\,. \tag{1.8.9}$$

LEMMA 1.8.13. *The function* $\mathbf{Q}(y,\lambda)$ *is uniformly bounded and oscillating in the sense that* $\int_0^\lambda \mathbf{Q}(y,\mu)\,d\mu$ *is uniformly bounded.*

Now we can state the following general result (it is an analogue of Theorem 1.7.6).

THEOREM 1.8.14 ([**Sa6**]). *Let* $y \in \overset{\circ}{M}$ *be a regular point. Then*
(1.8.10)
$$\begin{aligned} c_{0;P}(y)\,\lambda^{n+2l} &+ c_{1;P}(y)\,\lambda^{n+2l-1} + \mathbf{Q}_P(y,\lambda - o(\lambda))\,\lambda^{n+2l-1} - o(\lambda^{n+2l-1}) \\ &\leqslant e_{P,P}(\lambda,y,y) \\ &\leqslant c_{0;P}(y)\,\lambda^{n+2l} + c_{1;P}(y)\,\lambda^{n+2l-1} + \mathbf{Q}_P(y,\lambda + o(\lambda))\,\lambda^{n+2l-1} + o(\lambda^{n+2l-1}) \end{aligned}$$
as $\lambda \to +\infty$, *where* $c_{0;P}(y)$, $c_{1;P}(y)$ *are the same as in Theorem* 1.8.7 *and*
$$\mathbf{Q}_P(y,\lambda) = \left(P_1|_{S_y^*M}\,,\, \mathbf{Q}(y,\lambda)\; P_1|_{S_y^*M} \right)_{L_2(S_y^*M)}.$$

Note that, in view of Lemma 1.8.11,
$$\mathbf{Q}_P(y,\lambda+\varepsilon) \;-\; \mathbf{Q}_P(y,\lambda) \;\geqslant\; -\varepsilon\,(n+2l)\,c_{0;P}(y)\,, \qquad \forall\lambda\in\mathbb{R}, \quad \forall\varepsilon\geqslant 0\,,$$
so (1.8.10) makes sense.

5. The case $\mathbf{T}_y \equiv \mathrm{const}$. If the function $\mathbf{Q}_P(y,\lambda)$ is uniformly continuous in λ, then (1.8.10) implies the quasi-Weyl formula

$$\text{(1.8.11)} \qquad \begin{aligned} e_{P,P}(\lambda,y,y) \;&=\; c_{0;P}(y)\,\lambda^{n+2l} \;+\; c_{1;P}(y)\,\lambda^{n+2l-1} \\ &+\; \mathbf{Q}_P(y,\lambda)\,\lambda^{n+2l-1} \;+\; o(\lambda^{n+2l-1})\,. \end{aligned}$$

In the general case the definition of $U_{y,\lambda}$ involves two noncommuting operators (U_y and multiplications by $\mathbf{T}_y$) which makes it very difficult to give any simple sufficient conditions for the function $\mathbf{Q}_P(y,\lambda)$ to be uniformly continuous.

In this subsection we consider a special case assuming that there exists a constant T such that

$$\text{(1.8.12)} \qquad \mathbf{T}_y(\widetilde{\eta}) \;=\; T \quad \text{almost everywhere on } \Pi_y^a\,.$$

This assumption is motivated by the following

LEMMA 1.8.15. *In the analytic case* $\mathbf{T}_y \equiv \mathrm{const}$ *almost everywhere on any open connected subset of* Π_y^a.

Under the condition (1.8.12) $U_{y,\lambda} = e^{-i\lambda T}U_y$, so the operator-valued function $\mathbf{Q}(y,\lambda)$ is defined by the series
$$\sum_{k=1}^{\infty} \left((ikT)^{-1}\, e^{ik\lambda T}\, (U_y^k)^* \;-\; (ikT)^{-1}\, e^{-ik\lambda T}\, U_y^k \right).$$

In this case $\mathbf{Q}(y,\lambda)$ is $2\pi T^{-1}$-periodic in λ, so it is uniformly continuous if and only if it is continuous.

PROPOSITION 1.8.16. *Under the condition* (1.8.12)
$$\mathbf{Q}(y,\lambda) \;=\; \{\pi + \arg U_y - \lambda T\}_{2\pi}$$
where $\{\cdot\}_{2\pi}$ *denotes the residue modulo* 2π, *and the function* $\{\pi + \arg U_y - \lambda T\}_{2\pi}$ *of the contraction operator* U_y *is understood in the sense of the Sz.-Nagy–Foias calculus* [**Sz.-NaFoi**].

According to [**Sz.-NaFoi**], any contraction operator U in the Hilbert space H can be represented as $\mathbf{P}\widetilde{U}\big|_H$, where $\widetilde{U}$ is a unitary operator acting in a wider Hilbert space $\widetilde{H} \supset H$ and $\mathbf{P}$ is an orthogonal projection in $\widetilde{H}$ such that $\mathbf{P}\widetilde{H} = H$

and $U^j = \mathbf{P}\widetilde{U}^j\big|_H$, $j = 0, 1, 2, \dots$. The operator $\widetilde{U}$ is called a *unitary dilation* of U. For an arbitrary measurable function F on $\mathbb{S}$ one defines $F(U) = \mathbf{P}F(\widetilde{U})\big|_H$.

Proposition 1.8.16 implies

THEOREM 1.8.17 ([**Sa6**]). *Under the condition* (1.8.12) *the function* $\mathbf{Q}_P(y,\lambda)$ *is uniformly continuous with respect to* λ *if and only the function* $P_1|_{S_y^*M}$ *is orthogonal in* $L_2(S_y^*M)$ *to all the eigenfunctions of* U_y *corresponding to eigenvalues lying on* $\mathbb{S}$ *(i.e., with modulus* 1*).*

On the other hand, if $P_1|_{S_y^*M}$ is an eigenfunction of U_y corresponding to an eigenvalue e^{is_0}, $s_0 \in (-\pi, \pi]$, then

$$\mathbf{Q}_P(y,\lambda) \;=\; \{\pi + s_0 - \lambda T\}_{2\pi} \int_{S_y^*M} |P_1|^2\, d\tilde{\eta}$$

and $\mathbf{Q}_P(y,\lambda)$ has jumps at the points

$$\Lambda_j \;=\; T^{-1}\,(2\pi j + s_0)\,, \qquad j = 1, 2, \dots .$$

In this case, by Theorem 1.8.14,

$$e_{P,P}(y,y,\Lambda_{j+1}-\varepsilon) - e_{P,P}(y,y,\Lambda_j+\varepsilon) = o(\Lambda_j^{n+2l-1}), \quad \forall\varepsilon \in (0, \pi T^{-1}),\ \Lambda_j \to +\infty,$$

and there exist positive numbers $\varepsilon_j \to 0$ such that

$$e_{P,P}(y,y,\Lambda_j+\varepsilon_j) - e_{P,P}(y,y,\Lambda_j-\varepsilon_j) = 2\pi\,\Lambda_j^{n+2l-1} \int_{S_y^*M} |P_1|^2\, d\tilde{\eta} \;+\; o(\Lambda_j^{n+2l-1})\,.$$

In other words, $e_{P,P}(y,y,\lambda)$ has a purely cluster asymptotics with clusters around the points Λ_j. This implies, in particular, that there exist eigenvalues of $\mathcal{A}^{1/(2m)}$ lying in small neighbourhoods of the points Λ_j, i.e., the eigenvalue of U_y generates a series of eigenvalues of $\mathcal{A}^{1/(2m)}$.

EXAMPLE 1.8.18. Let M be a two-dimensional surface of revolution, and A the Laplace–Beltrami operator. Then $e(y,y,\lambda)$ has purely cluster asymptotics at the poles. The corresponding series of eigenvalues is $\lambda_j = \pi(j+1/2)/l$, where l is the length of the meridian.

EXAMPLE 1.8.19. Let M be a ball in $\mathbb{R}^n$, and A the Laplace operator subject to Dirichlet or Neumann boundary condition. Then (1.8.5) holds for all points y excluding the centre of the ball. At the centre, $e(y,y,\lambda)$ has a purely cluster asymptotics.

EXAMPLE 1.8.20. Let M be the flat domain bounded by an ellipse, and A the Laplace operator subject to Dirichlet or Neumann boundary condition. Then (1.8.5) (with $c_{0,P}^{(0)} = (4\pi)^{-1}$, $c_{1,P}^{(0)} = 0$) holds at all interior points excluding the foci of the ellipse. The foci are focal points. The corresponding operators U_y have purely continuous spectrum, so the quasi-Weyl formula (1.8.11) holds. In particular, if y is a focus and P is the operator of multiplication by a $C_0^\infty(\mathring{M})$-function equal to 1 at y, then

$$\mathbf{Q}_P(y,\lambda) \;=\; \frac{1}{8\pi^3 a}\int_{-\infty}^{+\infty} \left\{\mu\,\ln\frac{1+\varepsilon}{1-\varepsilon} - 4a\,\lambda\right\}_{2\pi} f^2(\mu)\,d\mu \;\equiv\; \sum_{k=1}^{\infty} b_k \sin(4ak\lambda)\,,$$

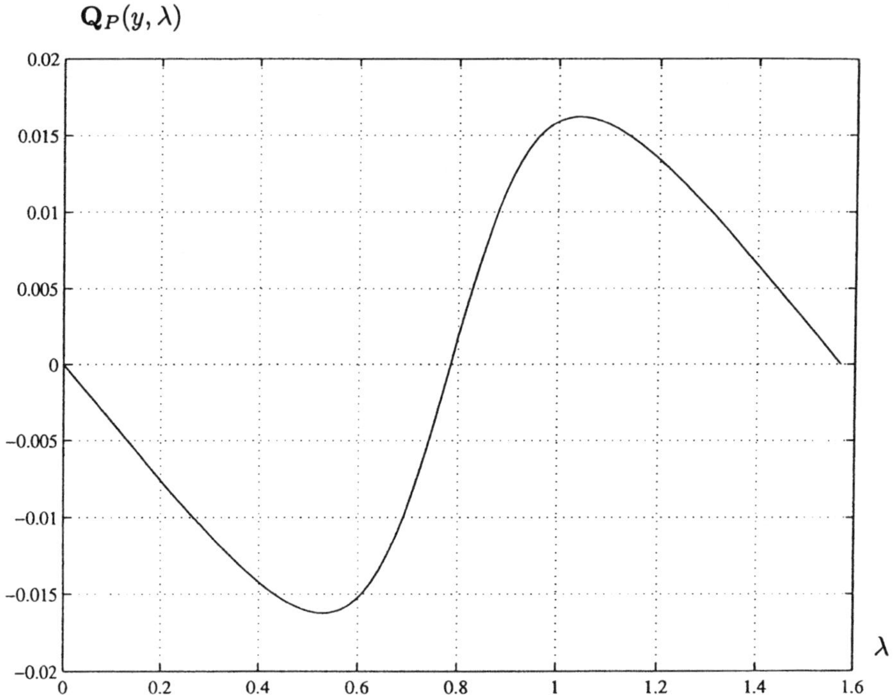

FIGURE 11. The function $\mathbf{Q}_P(y,\lambda)$ for $a=1$, $\varepsilon=1/2$.

where

$$f(\mu) = \int_{-\infty}^{+\infty} \frac{\cos \mu\tau}{\sqrt{\cosh 2\tau}}\, d\tau\,, \qquad b_k = \frac{(-1)^k}{4\pi^3 a k} \int_{-\infty}^{+\infty} \cos\left(k\mu \ln \frac{1+\varepsilon}{1-\varepsilon}\right) f^2(\mu)\, d\mu\,,$$

$2a$ is the length of the major axis and $0<\varepsilon<1$ is the eccentricity. The graph of this function for $\varepsilon = 1/2$ is given in Figure 11.

6. Off-diagonal asymptotics. We do not consider in this book the asymptotic behaviour of $e(\lambda,x,y)$ when $x \neq y$. This is a much more complicated problem, mostly because the function $e(\lambda,x,y)$ is not monotone in λ. It is known [**Hö1**], [**DuiGuiHö**] that at interior points

(1.8.13) $$e(\lambda,x,y) = O(\lambda^{n-1})\,, \qquad \lambda \to +\infty\,, \qquad x \neq y\,.$$

More advanced results can be found in [**Sa6**]. In particular, in [**Sa6**] it is shown that under some additional conditions,

$$e(\lambda,x,y) = o(\lambda^{n-1})\,, \qquad \lambda \to +\infty\,, \qquad x \neq y\,,$$

and that the estimate (1.8.13) cannot be improved in the general case.

7. Pseudodifferential case. All the main results of this section remain true for a pseudodifferential operator acting on a manifold without boundary [**Sa6**].

CHAPTER 2

Oscillatory Integrals

2.1. Local oscillatory integrals and pseudodifferential operators

1. Time-independent oscillatory integrals. Let Ω_x and Ω_y be sufficiently small open sets in $\overset{\circ}{M}$. By introducing local coordinates x and y we identify them with open sets in the Euclidean space $\mathbb{R}^n$.

Let us denote by S^l the class of complex-valued C^∞-functions $a(x,y,\theta)$ on $\Omega_x \times \Omega_y \times \mathbb{R}^n$ which admit an asymptotic expansion

$$a(x,y,\theta) \sim \sum_{k=0}^{\infty} a_{l-k}(x,y,\theta), \qquad |\theta| \to \infty, \tag{2.1.1}$$

with $a_{l-k}(x,y,\theta)$ positively homogeneous of degree $l-k$ in θ, i.e.,

$$a_{l-k}(x,y,\lambda\theta) \;=\; \lambda^{l-k} a_{l-k}(x,y,\theta), \qquad \forall \lambda > 0.$$

Using the sign $\sim$ we always mean that the expansion is uniform with respect to all the parameters running through a compact set, and that it can be differentiated arbitrarily many times; certainly, one of these parameters is $\theta/|\theta|$ (not θ itself).

We always assume that for $a \in S^l$

$$\operatorname{supp} a \;\subset\; K_x \times K_y \times \mathbb{R}^n,$$

where $K_x \subset \Omega_x$, $K_y \subset \Omega_y$ are some compacta. Thus, (2.1.1) is uniform with respect to $(x, y, \theta/|\theta|) \in K_x \times K_y \times \mathbb{S}^{n-1}$, and can be differentiated with respect to (x,y,θ). Note that the class S^l with $l \in \mathbb{N}$ contains all polynomials in θ of degree l with $C_0^\infty(\Omega_x \times \Omega_y)$-coefficients.

Obviously, $S^l \subset S^{l+1}$, because for $a \in S^l$ one can take $a_{l+1} \equiv 0$. It is also clear that for $a \in S^l$ we have $\partial_{x,y}^\alpha \partial_\theta^\beta a \;\in\; S^{l-|\beta|}$ for any multiindices α and β. The class $S^{-\infty} = \bigcap_{l \in \mathbb{R}} S^l$ consists of all C^∞-functions decreasing with all their derivatives faster than any power of $|\theta|$ as $|\theta| \to \infty$.

DEFINITION 2.1.1. For $a \in S^l$ we define

$$\operatorname{cone\,supp} a \;=\; \overline{\bigcup_{k \geqslant 0} \operatorname{supp} a_{l-k}} \;\subset\; \Omega_x \times \Omega_y \times (\mathbb{R}^n \setminus 0)$$

where a_{l-k} are the homogeneous functions from (2.1.1).

Further on we shall use (often implicitly) the following simple lemma (see, e.g., [**Sh,** Proposition 3.5]).

LEMMA 2.1.2. *For any sequence of functions*

$$a_{l-k}(x,y,\theta) \in S^{l-k}, \qquad k = 0,1,2,\dots$$

(or smooth for $\theta \neq 0$ homogeneous functions a_{l-k} of degree $l-k$), there exists a function $a(x,y,\theta)$ such that (2.1.1) holds. This function is determined uniquely modulo $S^{-\infty}$. Moreover, the function a can be chosen in such a way that $\operatorname{supp} a \subset \operatorname{cone supp} a$.

Let $\varphi(x,y,\theta)$ be a smooth complex-valued function on $\Omega_x \times \Omega_y \times (\mathbb{R}^n \setminus 0)$ positively homogeneous in θ of degree 1 and such that $\operatorname{Im}\varphi \geqslant 0$. Consider the formal integral

$$(2.1.2) \qquad \mathcal{I}_{\varphi,a}(x,y) = \int e^{i\varphi(x,y,\theta)} a(x,y,\theta)\, đ\theta, \qquad đ\theta = (2\pi)^{-n} d\theta_1 \cdots d\theta_n,$$

with a from S^l. Of course, when $l > -n$, the integral (2.1.2) does not converge in the usual sense. Nevertheless, under appropriate conditions on φ it can be considered as a distribution.

Let us introduce the set

$$C_\varphi = \{(x,y,\theta) : \varphi_\theta(x,y,\theta) = 0\}.$$

By Euler's identity for homogeneous functions $\varphi = \langle \theta, \varphi_\theta \rangle$. Therefore

$$(2.1.3) \qquad C_\varphi \subset \{\varphi = 0\}.$$

We shall always assume that

$$(2.1.4) \quad \varphi_x(x,y,\theta) \neq 0, \quad \varphi_y(x,y,\theta) \neq 0, \qquad \forall (x,y,\theta) \in C_\varphi \cap \operatorname{cone supp} a.$$

The inequalities (2.1.4) allow us to interpret (2.1.2) as a distribution with respect to one of the variables x or y smoothly dependent on the other. Indeed, for any test function $v(x) \in C^\infty(\Omega_x)$ we can define the value $\langle \mathcal{I}_{\varphi,a}, v\rangle$ of the distribution $I_{\varphi,a}$ on v in the following way.

Let $\mathcal{V}$ be a small conic neighbourhood of $\operatorname{cone supp} a$ such that $\overline{\mathcal{V}} \subset \Omega_x \times \Omega_y \times \mathbb{R}^n$ and the inequalities (2.1.4) still hold on $C_\varphi \cap \overline{\mathcal{V}}$. We split the amplitude a by smooth cut-off functions into a sum $a = \widetilde{a} + a_{-\infty}$ in such a way that $\operatorname{supp}\widetilde{a}$ lies in $\mathcal{V}$ and is separated from $\{\theta = 0\}$, and $a_{-\infty} \in S^{-\infty}$. Let $\omega \in C^\infty(\Omega_x \times \Omega_y \times (\mathbb{R}^n \setminus 0))$ be a nonnegative homogeneous cut-off function of degree 0 satisfying the following conditions:

1. $\omega = 0$ outside a small conic neighbourhood of C_φ.
2. $\omega = 1$ in a smaller conic neighbourhood of C_φ.
3. $|\varphi_x(x,y,\theta)| \neq 0$, $\forall (x,y,\theta) \in \operatorname{supp}\omega \cap \overline{\mathcal{V}}$.

Then we obviously have

$$|\varphi_x(x,y,\theta)| \geqslant C_1|\theta|, \qquad \forall (x,y,\theta) \in \operatorname{supp}\omega \cap \overline{\mathcal{V}} \cap (K_x \times K_y \times \mathbb{R}^n),$$
$$|\varphi_\theta(x,y,\theta)| \geqslant C_2, \qquad \forall (x,y,\theta) \in \operatorname{supp}(1-\omega) \cap (K_x \times K_y \times \mathbb{R}^n)$$

for some positive constants C_1 and C_2. Let us introduce the first order differential operators

$$\begin{aligned} L_1(x,y,\theta,D_\theta) &= |\varphi_\theta|^{-2} \sum_{j=1}^{n} \varphi_{\theta_j} D_{\theta_j}\,, \\ L_2(x,y,\theta,D_x) &= |\varphi_x|^{-2} \sum_{j=1}^{n} \varphi_{x_j} D_{x_j}\,, \end{aligned}$$

and let L_j^T be the transposed operators. Clearly, $L_j e^{i\varphi} = e^{i\varphi}$, $j=1,2$. We define

$$\begin{aligned} \langle \mathcal{I}_{\varphi,a}\,, v\rangle &= \int e^{i\varphi}\, a_{-\infty}\, v\, d\!\!{}^-\theta\, dx + \int e^{i\varphi}\, v\, (L_1^T)^k \left((1-\omega)\,\widetilde{a}\right) d\!\!{}^-\theta\, dx \\ &\quad + \int e^{i\varphi}\, (L_2^T)^k\, (\omega\, \widetilde{a}\, v)\, d\!\!{}^-\theta\, dx\,, \end{aligned} \tag{2.1.5}$$

where k is sufficiently large. (The first integral on the right-hand side of (2.1.5) is always absolutely convergent, whereas the second and the third are absolutely convergent for $k > n + l$.)

It can be shown that our definition is independent of the choice of $a_{-\infty}$, ω and k. So we can treat $\mathcal{I}_{\varphi,a}$ as an element from $C_0^\infty(\Omega_y; \mathcal{E}'(\Omega_x))$. By analogy we obtain also $\mathcal{I}_{\varphi,a} \in C_0^\infty(\Omega_x; \mathcal{E}'(\Omega_y))$. Therefore $\mathcal{I}_{\varphi,a}$ can be considered as the Schwartz kernel of some operator, mapping $\mathcal{D}'(\Omega_y)$ into $\mathcal{E}'(\Omega_x)$ and $C^\infty(\Omega_y)$ into $C_0^\infty(\Omega_x)$.

Here and further on we denote by $\mathcal{D}'$ the space of distributions (linear continuous functionals) acting on test functions from C_0^∞. By $\mathcal{E}'$ we denote the space of distributions with compact support acting on test functions from C^∞. (See [**Hö3**, Vol. 1] for precise definitions.)

DEFINITION 2.1.3. An integral of the type (2.1.2) is called *oscillatory integral*, φ is called *phase function*, and $a \in S^l$ is called *amplitude*. The number l is said to be the *order* of the oscillatory integral, and θ are said to be the *phase variables*.

REMARK 2.1.4. Considering the oscillatory integral (2.1.2) as a distribution, one can operate with it as with an absolutely convergent integral: formally integrate it by parts, differentiate under the integral sign, etc. A rigorous justification of all these operations is based on the formula (2.1.5).

Obviously, the smoothness of the distribution $\mathcal{I}_{\varphi,a}$ depends on the behaviour of the amplitude a as $|\theta| \to \infty$, i.e., on the order l. It becomes smoother and smoother as $l \to -\infty$; if $a \in S^{-\infty}$ (for example, if a has a compact support), then $\mathcal{I}_{\varphi,a} \in C^\infty$. Thus, the first integral in (2.1.5) gives a C^∞-contribution to (2.1.2).

As a rule, oscillatory integrals are used only for studying the singularities of distributions, and all calculations are carried out modulo $S^{-\infty}$. In particular, the amplitude a is usually given not by an exact formula but by an asymptotic expansion of the type (2.1.1), which is sufficient for the analysis of singularities in view of Lemma 2.1.1. Under such an approach we can assume a to be different from zero only in a small neighbourhood of $\operatorname{cone\,supp} a$ and outside a neighbourhood of $\{\theta = 0\}$.

Clearly, the distribution defined by the second integral in (2.1.5) is also a smooth function. Thus, all the singularities of the distribution (2.1.2) are contained in the third term of (2.1.5). Therefore $\mathcal{I}_{\varphi,a}$ is determined modulo C^∞ by the behaviour of a for large $|\theta|$ on a small conic neighbourhood of $C_\varphi \cap \operatorname{cone\,supp} a$. So further on, when studying singularities, we may always assume the amplitude a to be different from zero only for sufficiently large θ and near $C_\varphi \cap \operatorname{cone\,supp} a$.

DEFINITION 2.1.5. The minimal closed set containing all the singularities of the distribution u is called the singular support of $u \in \mathcal{E}'$ and is denoted by $\operatorname{sing\,supp} u$.

The above arguments demonstrate that $\operatorname{sing\,supp} \mathcal{I}_{\varphi,a}$ is contained in the projection of the set $C_\varphi \cap \operatorname{cone\,supp} a$ on $\Omega_x \times \Omega_y$.

2. Time-dependent oscillatory integrals. Without essential complications we can allow the phase function φ, the amplitude a with its homogeneous terms a_{l-k}, and consequently, the distribution (2.1.2) to depend on the time parameter $t \in (T_-, T_+) \subset \mathbb{R}$ (where $T_\mp$ may be finite or $\mp\infty$). In this case

$$\varphi \in C^\infty\big((T_-, T_+) \times \Omega_x \times \Omega_y \times (\mathbb{R}^n \setminus 0)\big),$$

$$a \in C^\infty\big((T_-, T_+) \times \Omega_x \times \Omega_y \times \mathbb{R}^n\big), \quad \operatorname{supp} a \subset (T_-, T_+) \times K_x \times K_y \times \mathbb{R}^n,$$
$$C_\varphi = \{(t,x,y,\theta) : \varphi_\theta(t,x,y,\theta) = 0\} \subset (T_-, T_+) \times \Omega_x \times \Omega_y \times \mathbb{R}^n,$$

and

$$\mathcal{I}_{\varphi,a} \in C^\infty((T_-, T_+) \times \Omega_y; \mathcal{E}'(\Omega_x)) \cap C^\infty((T_-, T_+) \times \Omega_x; \mathcal{E}'(\Omega_y)). \tag{2.1.6}$$

In addition to (2.1.4) we always suppose that

$$\varphi_t(t,x,y,\theta) \neq 0, \qquad \forall\, (t,x,y,\theta) \in C_\varphi \cap \operatorname{cone\,supp} a. \tag{2.1.7}$$

Then, assuming that $\varphi_t \neq 0$ on $\operatorname{supp}\omega \cap \overline{\mathcal{V}}$ and using in (2.1.5) the differential operator $(\varphi)_t^{-1} D_t$ instead of L_2, we obtain also

$$\mathcal{I}_{\varphi,a} \in C^\infty(\Omega_x \times \Omega_y; \mathcal{D}'(T_-, T_+)).$$

Here the space $\mathcal{D}'(T_-, T_+)$ appears instead of $\mathcal{E}'(T_-, T_+)$ because $\operatorname{supp} a$ is not necessarily detached from the endpoints of the interval (T_-, T_+), so we can act only on test functions from $C_0^\infty(T_-, T_+)$.

3. Pseudodifferential operators. Now let the local x- and y-coordinates be the same, $\Omega_x = \Omega_y = \Omega$, and $\varphi(x,y,\theta) = \langle x-y, \theta\rangle$. Then the oscillatory integral (2.1.2) with $a \in S^l$ takes the form

$$I_a(x,y) = \int e^{\langle x-y,\theta\rangle}\, a(x,y,\theta)\, đ\theta. \tag{2.1.8}$$

Obviously, for this phase function the inequalities (2.1.4) hold, and we can consider (2.1.8) as the Schwartz kernel of some operator P. An operator of such type is called *pseudodifferential operator* of *order* l.

The set C_φ in this case is given by the equation $x = y$. Consequently (see subsection 1), the kernel of a pseudodifferential operator can have singularities only on the diagonal $\{x = y\}$. Therefore a pseudodifferential operator possesses the property of *pseudolocality*:

$$\operatorname{sing\,supp} Pu \subset \operatorname{sing\,supp} u, \qquad \forall u \in \mathcal{E}'(\Omega).$$

Let us expand the amplitude $a(x,y,\theta)$ in (2.1.8) by Taylor's formula at the point $y=x$, replace $(y-x)^\alpha e^{i\langle x-y,\theta\rangle}$ by $(-D_\theta)^\alpha e^{i\langle x-y,\theta\rangle}$ and integrate by parts with respect to θ. Then the amplitude $a(x,y,\theta)$ will take the form

$$\sum_{|\alpha|\leqslant k} \frac{1}{\alpha!}\, D_\theta^\alpha \partial_y^\alpha a(x,y,\theta)\big|_{y=x} + \widetilde{a}^{(k)}(x,y,\theta)$$

where $\widetilde{a}^{(k)} \in S^{l-k-1}$, and k is an arbitrary natural number. By virtue of Lemma 2.1.2 there exists a function $\sigma_P(x,\theta)$ such that

$$\sigma_P(x,\theta) \sim \sum_\alpha \frac{1}{\alpha!}\, D_\theta^\alpha \partial_y^\alpha a(x,y,\theta)\big|_{y=x}$$

as $|\theta|\to\infty$, and then

$$\mathcal{I}_a(x,y) \;=\; \int e^{i\langle x-y,\theta\rangle} \sigma_P(x,\theta)\, đ\theta$$

modulo C^∞. The amplitude σ_P is called the *(full) symbol* of the pseudodifferential operator P. The full symbol is defined modulo $S^{-\infty}$ by the operator P, and, in turn, it defines the operator P modulo an operator with infinitely smooth kernel.

Analogously, one can write $I_a(x,y)$ as an oscillatory integral with an amplitude $\sigma'_P(y,\theta)$ independent of x. It is called the *dual symbol* of the pseudodifferential operator P. The symbols σ_P and σ'_P are related as follows :

$$\sigma_P(x,\theta) \sim \sum_\alpha D_\theta^\alpha \partial_x^\alpha \sigma'_P(x,\theta)/\alpha!\,, \qquad |\theta|\to\infty\,,$$

$$\sigma'_P(x,\theta) \sim \sum_\alpha D_\theta^\alpha (-\partial_x^\alpha) \sigma_P(x,\theta)/\alpha!\,, \qquad |\theta|\to\infty\,.$$

EXAMPLE 2.1.6. Let $a(x,y,\theta)$ be a polynomial in θ of degree l with coefficients smoothly dependent on x and y. Then P is a differential operator of order l. The polynomial $\sigma_P = \sum_{|\alpha|\leqslant l}(1/\alpha!) D_\theta^\alpha \partial_y^\alpha a\big|_{y=x}$ is its symbol; formally, $P=\sigma_P(x,D_x)$. The dual symbol σ'_P equals the symbol of the transposed operator $P^T(D_x,x)$. In particular, $P=I$ if and only if $\sigma_P\equiv\sigma'_P\equiv 1$.

Of course, the phase function $\langle x-y,\theta\rangle$ and, consequently, the full symbol σ_P depend on the choice of coordinates. However, the definition of pseudodifferential operators is invariant: if in some coordinate system the Schwartz kernel of the operator P can be represented in the form (2.1.8), then in any other coordinates it is also represented by an oscillatory integral of the form (2.1.8) with the phase function $\langle x-y,\theta\rangle$ and an amplitude from S^l. This fact allows us to define pseudodifferential operators on manifolds.

DEFINITION 2.1.7. The operator $P : C_0^\infty(\mathring{M}) \to C^\infty(\mathring{M})$ is called pseudodifferential of order l if its Schwartz kernel is infinitely smooth outside the diagonal $\{(x,y)\in \mathring{M}\times\mathring{M} \,:\, x=y\}$, and if for any point on the diagonal there exists a neighbourhood $\mathcal{W}\subset \mathring{M}\times\mathring{M}$ in which the Schwartz kernel of P can be written as an oscillatory integral of the form (2.1.8) with $a\in S^l$ and $\Omega\times\Omega\supset\mathcal{W}$.

Recall (see 1.1.5) that we always deal with operators acting in the space of half-densities, and this applies to our pseudodifferential operators.

The class of pseudodifferential operators of order l will be denoted by Ψ^l. By Ψ_0^l we will denote the subclass of Ψ^l consisting of operators the Schwartz kernels of which have compact support in $\overset{\circ}{M} \times \overset{\circ}{M}$.

Obviously, $P^{(0)} \in \Psi_0^l$ if and only if there exist a pseudodifferential operator $P \in \Psi^l$ and cut-off functions $\chi_1, \chi_2 \in C_0^\infty(\overset{\circ}{M})$ such that

$$P^{(0)} = \chi_1 P \chi_2 .$$

Then $P^{(0)} : C^\infty(\overset{\circ}{M}) \to C_0^\infty(\overset{\circ}{M})$. Further on we will deal with pseudodifferential operators only from Ψ_0^l, which will be sufficient for our purposes.

Let $P_l(x,\theta)$, $P_{l-1}(x,\theta), \dots$ be the positively homogeneous in θ functions which appear in the expansion of the full symbol σ_P as $|\theta| \to \infty$.

DEFINITION 2.1.8. The leading term P_l of the expansion of σ_P is called the *principal symbol* of the pseudodifferential operator P. The homogeneous function

$$P_{\mathrm{sub}}(x,\theta) = P_{l-1}(x,\theta) - (2i)^{-1} \sum_j \partial_{x_j} \partial_{\theta_j} P_l(x,\theta) \tag{2.1.9}$$

of degree $l-1$ is said to be the *subprincipal symbol*.

The principal and subprincipal symbols are transformed under a change of coordinates as functions on the cotangent bundle $T'\overset{\circ}{M}$, i.e., in new coordinates $\widetilde{x}$

$$\widetilde{P}_l(\widetilde{x},\widetilde{\xi}) = P_l(x(\widetilde{x}), ((\partial x/\partial\widetilde{x})^T)^{-1}\widetilde{\xi}),$$
$$\widetilde{P}_{\mathrm{sub}}(\widetilde{x},\widetilde{\xi}) = P_{\mathrm{sub}}(x(\widetilde{x}), ((\partial x/\partial\widetilde{x})^T)^{-1}\widetilde{\xi}).$$

They determine the operator P modulo a pseudodifferential operator of order $l-2$. The full symbol σ_P is transformed under a change of coordinates in a rather complicated way and, generally speaking, does not have an invariant meaning. Nevertheless, its conic support is well-defined as a subset of $T'\overset{\circ}{M}$.

DEFINITION 2.1.9. For a pseudodifferential operator $P \in \Psi_0^l$ let us denote

$$\operatorname{cone\,supp} P = \overline{\bigcup_{k \geqslant 0} \operatorname{supp} P_{l-k}} \subset T'\overset{\circ}{M},$$

where P_{l-k} are the homogeneous functions in the expansion of the full symbol σ_P (cf. Definition 2.1.1). The set $\operatorname{cone\,supp} P$ is called the *conic support* of the pseudodifferential operator P.

REMARK 2.1.10. Analogously, one can introduce the principal symbol $\sigma_{\widetilde{P}} \in C^\infty(T'\overset{\circ}{M})$ for a pseudodifferential operator $\widetilde{P}$ acting in the space of functions on M. However, in this case the subprincipal symbol given by (2.1.9) is not a function on $T'\overset{\circ}{M}$. Choosing a smooth positive density μ (and identifying then functions and half-densities, $v \leftrightarrow \mu^{1/2} v$) one can define the "second" symbol of $\widetilde{P}$ as

$$(\mu^{-1/2}\widetilde{P}\mu^{1/2})_{\mathrm{sub}} = \widetilde{P}_{l-1}(x,\theta) - (2i)^{-1} \sum_j \partial_{x_j}\partial_{\theta_j}\widetilde{P}_l(x,\theta) - (2i)^{-1} \sum_j \partial_{\theta_j}\widetilde{P}_l(x,\theta)\, \partial_{x_j}(\ln\mu).$$

This function is correctly defined on $T'\overset{\circ}{M}$ but it depends on the choice of μ.

Let $H^s(M)$ be the Sobolev space $W_2^s(M)$, and let $H^s_{\mathrm{comp}}(\overset{\circ}{M})$ be the subspace of $H^s(M)$ consisting of half-densities whose supports are compact subsets of $\overset{\circ}{M}$. By definition, $v_k \to v$ in $H^s_{\mathrm{comp}}(\overset{\circ}{M})$ if and only if $v_k \to v$ in $H^s(M)$ and $\operatorname{supp} v_k$ are contained in some fixed compact set $K \subset \overset{\circ}{M}$.

Now we list some basic facts from the classical theory of pseudodifferential operators [**Hö3,** Vol. 3], [**Sh**], [**Tr**].

PROPOSITION 2.1.11. *If $P \in \Psi_0^l$, then the formally adjoint operator P^* also belongs to Ψ_0^l, and*

$$\operatorname{cone\,supp} P^* = \operatorname{cone\,supp} P, \qquad (P^*)_l = \overline{P_l}, \qquad (P^*)_{\mathrm{sub}} = \overline{P_{\mathrm{sub}}}.$$

Proposition 2.1.11 allows as to extend P to an operator acting from $\mathcal{D}'(\overset{\circ}{M})$ to $\mathcal{E}'(\overset{\circ}{M})$: for $v \in \mathcal{D}'(\overset{\circ}{M})$ we define

$$\langle Pv, w\rangle = \langle v, \overline{P^*\overline{w}}\rangle, \qquad \forall w \in C^\infty(\overset{\circ}{M}).$$

PROPOSITION 2.1.12. *For any $s \in \mathbb{R}$ the pseudodifferential operator $P \in \Psi_0^l$ continuously maps $H^s(M)$ into $H^{s-l}_{\mathrm{comp}}(\overset{\circ}{M})$.*

PROPOSITION 2.1.13. *If $P^{(1)} \in \Psi_0^{l_1}$, $P^{(2)} \in \Psi_0^{l_2}$, then $P^{(1)}P^{(2)} \in \Psi_0^{l_1+l_2}$,*

$$\operatorname{cone\,supp} P^{(1)}P^{(2)} = \operatorname{cone\,supp} P^{(1)} \cap \operatorname{cone\,supp} P^{(2)},$$

and the following equalities hold for the principal and subprincipal symbols:

$$(P^{(1)}P^{(2)})_{l_1+l_2} = P^{(1)}_{l_1} P^{(2)}_{l_2},$$
$$(P^{(1)}P^{(2)})_{\mathrm{sub}} = P^{(1)}_{l_1} P^{(2)}_{\mathrm{sub}} + P^{(1)}_{\mathrm{sub}} P^{(2)}_{l_2} + (2i)^{-1}\{P^{(1)}_{l_1}, P^{(2)}_{l_2}\}.$$

REMARK 2.1.14. Proposition 2.1.11 immediately implies that the subprincipal symbol of a formally self-adjoint differential operator A of order $2m$ acting in the space of half-densities is equal to zero if the coefficients of A are real. Indeed, for a differential operator of even order with real coefficients the functions A_{2m-1}, $(2i)^{-1}A_{2m}$, and, consequently, A_{sub} are purely imaginary. But by Proposition 2.1.11 A_{sub} is real because of the formal self-adjointness. Therefore $A_{\mathrm{sub}} = 0$.

DEFINITION 2.1.15. Let $v \in \mathcal{E}'(\overset{\circ}{M})$. The *wave front set* $\mathrm{WF}\, v$ of this distribution is the conic subset of $T'\overset{\circ}{M}$ defined in the following way. A point (x_0, ξ_0) does not belong to $\mathrm{WF}\, v$ if there exist a coordinate chart and a cut-off function $\chi \in C_0^\infty(\overset{\circ}{M})$ with support lying in this coordinate chart such that $\chi(x_0) \neq 0$ and the Fourier transform

$$\mathcal{F}_{x\to\xi}[\chi(x)\, v(x)] = \int e^{-i\langle x,\xi\rangle}\chi(x)\, v(x)\, dx$$

vanishes faster than any negative power of $|\xi|$ as $|\xi| \to \infty$ uniformly over some conic neighbourhood of ξ_0.

It can be shown that if the rapid decrease property of the Fourier transform described in Definition 2.1.15 holds for some local coordinate system, then it holds for any other coordinates. Therefore $\mathrm{WF}\, v$ is well defined as a subset of $T'\overset{\circ}{M}$.

The projection of $\mathrm{WF}\,v$ on M coincides with $\operatorname{sing\,supp} v$. Roughly speaking, besides information on the singularities of v, $\mathrm{WF}\,v$ contains also information on the direction in which v is nondifferentiable.

PROPOSITION 2.1.16. *Let* $v \in \mathcal{E}'(\overset{\circ}{M})$ *and* $P \in \Psi_0^l$. *Then*

$$\mathrm{WF}(Pv) \subset \mathrm{WF}\,v \cap \operatorname{cone\,supp} P\,.$$

Propositions 2.1.13 and 2.1.16 show that in the study of wave front sets pseudodifferential operators can be used in the same way as cut-off functions are used in the study of local properties of distributions. For example, outside the boundary the identity operator can be represented as a sum of pseudodifferential operators with small conic supports, so the consideration can be restricted to a small conic subset of $T'\overset{\circ}{M}$. Such a procedure is called *microlocalization.*

4. Extension of oscillatory integrals up to the boundary. Now let Ω_x and Ω_y be sufficiently small open sets in M, not necessarily in $\overset{\circ}{M}$ (i.e., their intersections with ∂M may be nonempty; see 1.1.2). In local coordinates, Ω_x and Ω_y are identified with open sets in $\mathbb{R}^n_+ = \{x_n \geqslant 0\}$.

In this case it is often inconvenient to interpret our oscillatory integrals as distributions with respect to the spatial variables, so we shall mostly deal with time-dependent oscillatory integrals and view them as distributions in t. Then the fact that Ω_x or Ω_y may intersect ∂M does not lead to any complications. More precisely, under the condition (2.1.7) the oscillatory integral $\mathcal{I}_{\varphi,a}$ will be considered as a distribution from $C^\infty(\Omega_x \times \Omega_y;\, \mathcal{D}'(T_-\,,\,T_+))$. In particular, this implies that we have well-defined traces

$$\begin{aligned}(\partial^j_{x_n}\mathcal{I}_{\varphi,a})\big|_{\partial M_x\cap\Omega_x} &\in C^\infty((\partial M_x\cap\Omega_x)\times\Omega_y;\,\mathcal{D}'(T_-\,,\,T_+))\,,\\ (\partial^j_{y_n}\mathcal{I}_{\varphi,a})\big|_{\partial M_y\cap\Omega_y} &\in C^\infty((\partial M_y\cap\Omega_y)\times\Omega_x;\,\mathcal{D}'(T_-\,,\,T_+))\,.\end{aligned}$$

In view of (2.1.4) we also have

$$\mathcal{I}_{\varphi,a} \in C^\infty((T_-\,,\,T_+)\times\Omega_y;\,\mathcal{D}'(\Omega_x\cap\overset{\circ}{M}))\cap C^\infty((T_-\,,\,T_+)\times\Omega_x;\,\mathcal{D}'(\Omega_y\cap\overset{\circ}{M}))\,.$$

The disadvantage of this interpretation is that it is difficult to define traces of our distributions on ∂M.

2.2. Global oscillatory integrals

In this section we introduce global oscillatory integrals defined on the whole manifold. Usually global constructions are based on the partition of unity and on the subsequent consideration of finite sums of local oscillatory integrals. This leads, however, to a complicated procedure of matching local integrals. Our global construction allows us to avoid this inconvenience.

Further on we will deal with oscillatory integrals depending on the time parameter t from some interval $(T_-\,,\,T_+)$.

First, let us note that we can take the phase variables in an oscillatory integral not from an abstract Euclidean space $\mathbb{R}^n$ but from T'_yM or T'_xM. Indeed, choosing a coordinate system in a sufficiently small open set $\Omega \subset M$ we identify $T'\Omega$ with $\Omega\times(\mathbb{R}^n\setminus 0)$. Therefore we can assume the phase function and the amplitude in a local oscillatory integral to be defined on

$$(T_-\,,\,T_+)\times\Omega_x\times T'\Omega_y \qquad\text{or}\qquad (T_-\,,\,T_+)\times T'\Omega_x\times\Omega_y$$

instead of $(T_-, T_+) \times \Omega_x \times \Omega_y \times \mathbb{R}^n$. But then we do not really need the sets Ω_x and Ω_y to be small and can take x and y from the whole manifold M. This enables us to introduce oscillatory integrals with phase functions and amplitudes defined on $(T_-, T_+) \times M \times T'M$ or on conic subsets of $(T_-, T_+) \times M \times T'M$ (the latter usually makes sense when $\partial M \neq \varnothing$).

In accordance with our global ideology we shall assume our oscillatory integrals to be half-densities in x and y. We shall use the subscripts x and y to distinguish between the two copies of the manifold M.

For the sake of definiteness we will always use the phase variables $\eta \in T'_y M$. Let $\mathcal{O}$ be an open conic subset of $(T_-, T_+) \times M_x \times T'M_y$ (which may coincide with the latter), and $\varphi(t,x;y,\eta) \in C^\infty(\mathcal{O})$ a phase function, i.e., φ is positively homogeneous in η of degree 1 and $\operatorname{Im}\varphi \geqslant 0$. We define the class S^l of amplitude functions $a(t,x;y,\eta) \in C^\infty(\mathcal{O})$ such that

$$a(t,x;y,\eta) \sim \sum_{k=0}^{\infty} a_{l-k}(t,x;y,\eta), \qquad |\eta| \to \infty,$$

where $a_{l-k} \in C^\infty(\mathcal{O})$ are functions positively homogeneous in η of degree $l-k$. Let us emphasize that the open set $\mathcal{O}$ may contain interior points with $x \in \partial M$ or $y \in \partial M$ (see 1.1.2 where we explain the difference between an open set in M and in $\overset{\circ}{M}$), and at these points φ, a and a_{l-k} are supposed to be smooth up to $\{x \in \partial M\} \cup \{y \in \partial M\}$. Now $C_\varphi = \{\varphi_\eta = 0\}$, $\operatorname{supp} a$ and $\operatorname{cone\,supp} a$ are subsets of $\mathcal{O}$.

As in Section 2.1, we assume that

$$\varphi_t \neq 0, \quad \varphi_x \neq 0, \quad \varphi_y \neq 0 \tag{2.2.1}$$

on $C_\varphi \cap \operatorname{cone\,supp} a$.

DEFINITION 2.2.1. For a set $\mathcal{Q} \subset (T_-, T_+) \times M_x \times T'M_y$ we define its *conic hull* as

$$\operatorname{cone\,hull} \mathcal{Q} := \{ (t,x;y,\lambda\eta) \ : \ (t,x;y,\eta) \in \mathcal{Q},\ \lambda > 0 \}.$$

By $\overline{\operatorname{cone\,hull} \mathcal{Q}}$ we denote the closure of $\operatorname{cone\,hull}\mathcal{Q}$ in $(T_-, T_+) \times M_x \times T'M_y$.

Note that for an amplitude $a \in S^l$ we have $\operatorname{cone\,supp} a \subset \overline{\operatorname{cone\,hull}(\operatorname{supp} a)}$.

DEFINITION 2.2.2. The sets $\mathcal{Q}, \mathcal{R} \subset (T_-, T_+) \times M_x \times T'M_y$ are said to be *conically separated* if $\overline{\operatorname{cone\,hull}\mathcal{Q}} \cap \overline{\operatorname{cone\,hull}\mathcal{R}} = \varnothing$.

In subsequent sections we shall often be using the notions of conic hulls and of conic separation in various other situations, e.g., for subsets of $(T_-, T_+) \times T'M_y$, $T'M$, etc. These notions are defined similarly to Definitions 2.2.1, 2.2.2. In cases of possible confusion we shall indicate in which sense the closure of conic hulls is understood by saying that the sets in question are conically separated as subsets of some specified embedding set (say, as subsets of $(T_-, T_+) \times T'M_y$, $T'M$, etc.).

Denote by $\partial\mathcal{O}$ the boundary of $\mathcal{O}$ in $(T_-, T_+) \times M_x \times T'M_y$. We shall always assume that $C_\varphi \cap \operatorname{supp} a$ and $\partial\mathcal{O}$ are conically separated.

It seems natural to consider global oscillatory integrals of the form

$$\int e^{i\varphi(t,x;y,\eta)} a(t,x;y,\eta)\, \varsigma(t,x;y,\eta)\, đ\eta \tag{2.2.2}$$

with some smooth cut-off function ς the support of which lies in a small conic neighbourhood of C_φ (this cut-off is necessary in order for the support of the full amplitude $a\varsigma$ to be conically separated from $\partial\mathcal{O}$). However, such a definition of an oscillatory integral is inconvenient for our aims. We intend to deal with half-densities on M, whereas (2.2.2) behaves as a function in x and a density in y. Indeed, since $e^{i\varphi}$, a and ς are functions, they do not change under changes of coordinates x or y. But in new coordinates $\widetilde{y}$ we obtain $d\!\!\!{}^-\widetilde{\eta} = |\det(\partial y/\partial\widetilde{y})|\, d\!\!\!{}^-\eta$, and therefore the integral (2.2.2) is a density in y.

In order to turn (2.2.2) into a half-density in x and in y we introduce the following definitions.

DEFINITION 2.2.3. A phase function $\varphi \in C^\infty(\mathcal{O})$ is said to be *nondegenerate* if the matrix of second derivatives $\varphi_{x\eta}(t,x;y,\eta)$ is nonsingular for $(t,x;y,\eta) \in C_\varphi$.

DEFINITION 2.2.4. We call a nondegenerate phase function $\varphi \in C^\infty(\mathcal{O})$ *simple* if in a small conic neighbourhood of C_φ there exists a continuous (and, consequently, smooth) branch of the argument of $\det^2\varphi_{x\eta}$.

Both these definitions are invariant. Indeed, when we change the coordinates $x \to \widetilde{x}$ and $y \to \widetilde{y}$ we obtain

$$\varphi_{\widetilde{\eta}\widetilde{x}} = (\partial\widetilde{y}/\partial y)\cdot\varphi_{\eta x}\cdot(\partial\widetilde{x}/\partial x)^{-1} \tag{2.2.3}$$

where $\varphi_{\eta x} = (\varphi_{x\eta})^T$. This implies that $\det\varphi_{\widetilde{\eta}\widetilde{x}} \neq 0$ if and only if $\det\varphi_{x\eta} \neq 0$, and

$$\arg(\det{}^2\varphi_{x\eta}) = \arg(\det{}^2\varphi_{\widetilde{x}\widetilde{\eta}}).$$

Note that the last equality does not necessarily hold for arguments of $\det\varphi_{x\eta}$ and $\det\varphi_{\widetilde{x}\widetilde{\eta}}$ because $\det(\partial\widetilde{y}/\partial y)$ or $\det(\partial\widetilde{x}/\partial x)$ may be negative. That is why in Definition 2.2.2 we have used $\det^2\varphi_{x\eta}$ instead of $\det\varphi_{x\eta}$.

REMARK 2.2.5. Definition 2.2.4 seems to be very technical but it has a geometrical meaning. For example, a nondegenerate phase function associated with a canonical transformation is simple if and only if the Maslov cohomology class of the corresponding Lagrangian manifold is trivial [**LapSaVa**]; see also 2.6.3. We will not deliberate on this subject here, and note only that any real nondegenerate phase function is evidently simple.

Let the phase function φ be nondegenerate and simple, and let $\mathcal{O}_0 \subset \mathcal{O}$ be a small conic neighbourhood of C_φ such that $\det\varphi_{x\eta} \neq 0$ on $\mathcal{O}_0$ and there exists a smooth branch of $\arg(\det^2\varphi_{x\eta})$ on $\mathcal{O}_0$. Let us define on $\mathcal{O}_0$ the smooth "function"

$$d_\varphi(t,x;y,\eta) = (\det{}^2\varphi_{x\eta})^{1/4} = |\det\varphi_{x\eta}|^{1/2}\, e^{i\arg(\det^2\varphi_{x\eta})/4}. \tag{2.2.4}$$

We assume the branch of $\arg(\det^2\varphi_{x\eta})$ in (2.2.4) to be smooth and independent of the choice of coordinates x and y. Then by (2.2.3) d_φ behaves as a half-density in x and a half-density to the power -1 in y. Note that d_φ is uniquely defined on $\mathcal{O}_0$ up to the choice of a smooth branch of the argument (i.e., up to a factor $e^{i\varkappa\pi/2}$ where $\varkappa = 0, 1, 2,$ or 3). This branch is assumed to be fixed once and for all.

Further on in this book we shall consider only simple phase functions, and this (in view of Lemma 2.6.6) will prove to be sufficient for our purposes. Moreover, in order to simplify notation we shall establish the following convention throughout the book: whenever we consider a global oscillatory integral we assume that $\mathcal{O}_0 = \mathcal{O}$.

In other words, we shall assume the conic neighbourhood $\mathcal{O} \subset (T_-, T_+) \times M_x \times T'M_y$ of the set C_φ to be so small that $\det \varphi_{x\eta} \neq 0$ on $\mathcal{O}$ and there exists a smooth branch of $\arg(\det^2 \varphi_{x\eta})$ on $\mathcal{O}$. Note that for fixed C_φ a reduction of the neighbourhood $\mathcal{O}$ of the set C_φ does not spoil the condition that $C_\varphi \cap \operatorname{supp} a$ and $\partial\mathcal{O}$ are conically separated.

Let $\varsigma(t, x; y, \eta) \in C^\infty(\mathcal{O})$ be a cut-off function satisfying the following four conditions:

1. $\varsigma(t, x; y, \eta) = 0$ on the set $\{(t, x; y, \eta) : h(y, \eta) \leqslant 1/2\}$.
2. $\varsigma(t, x; y, \eta) = 1$ on the intersection of a small conic neighbourhood of C_φ with the set $\{(t, x; y, \eta) : h(y, \eta) \geqslant 1\}$.
3. $\varsigma(t, x; y, \lambda\eta) = \varsigma(t, x; y, \eta)$ for $h(y, \eta) \geqslant 1$, $\lambda \geqslant 1$.
4. The set $\operatorname{supp} \varsigma \cap \operatorname{supp} a$ is conically separated from $\partial\mathcal{O}$.

Here for h we can take an arbitrary positive smooth function on $T'M$ positively homogeneous in η of degree 1, for example, our Hamiltonian (1.1.14).

It is easy to choose a function ς satisfying conditions (1)–(3), whereas condition (4) can be satisfied because $C_\varphi \cap \operatorname{supp} a$ and $\partial\mathcal{O}$ are conically separated. Clearly, the cut-off function ς is independent of the amplitude a, as long as we take $\operatorname{supp} \varsigma$ to be small enough.

We now define the global oscillatory integral $\mathcal{I}_{\varphi,a}$ as

$$\mathcal{I}_{\varphi,a}(t, x, y) = \int e^{i\varphi(t,x;y,\eta)} a(t, x; y, \eta)\, \varsigma(t, x; y, \eta)\, d_\varphi(t, x; y, \eta)\, đ\eta . \tag{2.2.5}$$

Due to the weight factor d_φ, this integral is a half-density in x and in y. Using a partition of unity on M, we can represent (2.2.5) on any finite time interval (T'_-, T'_+), $T_- < T'_- < T'_+ < T_+$, as a finite sum of local oscillatory integrals introduced in the previous section. Therefore all the principal results from that section remain valid:

$$\mathcal{I}_{\varphi,a} \in C^\infty(M_x \times M_y; \mathcal{D}'(T_-, T_+)), \tag{2.2.6}$$

$\operatorname{sing\,supp} \mathcal{I}_{\varphi,a}$ is contained in the projection of $C_\varphi \cap \operatorname{cone\,supp} a$ on $(T_-, T_+) \times M_x \times M_y$, and we can operate with (2.2.5) as with an absolutely convergent integral. In particular, boundary traces of $\mathcal{I}_{\varphi,a}$ are well defined:

$$(B_x \mathcal{I}_{\varphi,a})|_{\partial M_x} \in C^\infty(\partial M_x \times M_y; \mathcal{D}'(T_-, T_+)), \tag{2.2.7}$$

where B_x is an arbitrary differential operator in the variable x.

Of course, we also have

$$\mathcal{I}_{\varphi,a} \in C^\infty((T_-, T_+) \times M_y; \mathcal{D}'(\mathring{M}_x)) \cap C^\infty((T_-, T_+) \times M_x; \mathcal{D}'(\mathring{M}_y)).$$

REMARK 2.2.6. For a given set C_φ, a phase φ and an amplitude a defined in some neighbourhood of C_φ, the choice of the particular neighbourhood $\mathcal{O}$ of the set C_φ and of the particular cut-off ς changes our oscillatory integral (2.2.5) only by a half-density from $C^\infty((T_-, T_+) \times M_x \times M_y)$. As all our subsequent constructions will be modulo C^∞, we will never need to specify the choice of $\mathcal{O}$ or ς.

REMARK 2.2.7. The main reason why we do not include the cut-off ς in the amplitude is that such an inclusion would prevent us (see Section 2.7) from reducing our oscillatory integrals to those with amplitudes independent of x. Indeed, if the amplitude $a \not\equiv 0$ is independent of x, its support cannot be conically separated from $\partial\mathcal{O}$ unless $\mathcal{O}$ is a direct product of M_x and some subset of $(T_-, T_+) \times T'M_y$.

REMARK 2.2.8. Later on we will deal only with simple phase functions. However, we could have introduced the concept of a global oscillatory integral for general nondegenerate phase functions as well. The most natural way of doing this is to replace in (2.2.6) d_φ by $|\det \varphi_{x\eta}|^{1/2}$ with no regard for $\arg\det\varphi_{x\eta}$ (see [**LapSaVa**]). Such a definition seems to be easier even for a simple phase function, but in that case we would lose the very useful Theorems 2.7.11, 2.8.9.

REMARK 2.2.9. When $\partial\mathcal{O}$ is good enough and $\varphi \in C^\infty(\overline{\mathcal{O}})$ (which is usually the case in all effective constructions) we can always smoothly extend φ from $\overline{\mathcal{O}}$ to the whole $(T_-, T_+) \times M_x \times T'M_y$ without changing the set $\overline{C_\varphi}$; for example, one can take $\operatorname{Im}\varphi > 0$ outside $\overline{\mathcal{O}}$. However, this does not contribute anything new to the understanding of our problems.

2.3. Homogeneous canonical transformations

1. Definitions and general properties. Let O be an open subset of $T'M$. The smooth nondegenerate transformation

$$O \ni (y,\eta) \;\to\; (x^*(y,\eta), \xi^*(y,\eta)) \in T'M \tag{2.3.1}$$

is said to be canonical if it preserves the symplectic 2-form $d\xi \wedge dx$. This means that in any coordinates for all k and l the following equalities hold:

$$\sum_j (x_j^*)_{y_k} (\xi_j^*)_{y_l} \;-\; \sum_j (x_j^*)_{y_l} (\xi_j^*)_{y_k} \;=\; 0, \tag{2.3.2}$$

$$\sum_j (x_j^*)_{\eta_k} (\xi_j^*)_{\eta_l} \;-\; \sum_j (x_j^*)_{\eta_l} (\xi_j^*)_{\eta_k} \;=\; 0, \tag{2.3.3}$$

$$\sum_j (x_j^*)_{y_k} (\xi_j^*)_{\eta_l} \;-\; \sum_j (x_j^*)_{\eta_l} (\xi_j^*)_{y_k} \;=\; \delta_l^k \tag{2.3.4}$$

(here δ_l^k is the Kronecker symbol). A canonical transformation automatically preserves the $2n$-form

$$\underbrace{(d\xi \wedge dx) \wedge \cdots \wedge (d\xi \wedge dx)}_{n \text{ times}} \;=\; n!\, d\xi_n \wedge d\xi_{n-1} \wedge \cdots \wedge d\xi_1 \wedge dx_1 \wedge dx_2 \wedge \cdots \wedge dx_n\,,$$

and, consequently, the symplectic volume $dx\, d\xi$.

The transformation $(y,\eta) \to (x^*, \xi^*)$ is called homogeneous if

$$(x^*(y,\lambda\eta), \xi^*(y,\lambda\eta)) \;=\; (x^*(y,\eta), \lambda\xi^*(y,\eta)), \qquad \forall\lambda > 0.$$

Later on we consider only homogeneous canonical transformations and always assume O to be a conic subset of $T'M$. By Euler's identity

$$\langle \eta, (x_j^*)_\eta \rangle \;=\; 0, \qquad \langle \eta, (\xi_j^*)_\eta \rangle \;=\; \xi_j^*, \qquad j = 1,2,\ldots,n.$$

These equalities together with (2.3.3), (2.3.4) imply that a homogeneous canonical transformation preserves also the symplectic 1-form $\langle \xi, dx \rangle$, i.e., in any local coordinates for all k we have

$$\langle \xi^*, x^*_{y_k} \rangle \;=\; \eta_k, \qquad \langle \xi^*, x^*_{\eta_k} \rangle \;=\; 0. \tag{2.3.5}$$

REMARK 2.3.1. Any transformation preserving the symplectic 1-form $\langle \xi, dx\rangle$ is canonical because $d\xi \wedge dx = d\langle \xi, dx\rangle$.

We denote by x^*_η and ξ^*_η matrices with elements $(x^*_j)_{\eta_k}$ and $(\xi^*_j)_{\eta_k}$, respectively (the index j is the number of the row, and k is that of the column). Then (2.3.3) is equivalent to

$$(\xi^*_\eta)^T \cdot x^*_\eta \;-\; (x^*_\eta)^T \cdot \xi^*_\eta \;=\; 0\,. \tag{2.3.6}$$

Changing coordinates $x \to \widetilde{x}$ and $y \to \widetilde{y}$ we obtain

$$\widetilde{x}^*_{\widetilde{\eta}} \;=\; (\partial\widetilde{x}/\partial x)\cdot x^*_\eta \cdot (\partial\widetilde{y}/\partial y)^T\,,$$

i.e., x^*_η behaves as a tensor. The matrix ξ^*_η also behaves as a tensor with respect to y. However, this is not true with respect to x. Indeed, passing from coordinates x to $\widetilde{x}$ we obtain

$$\widetilde{\xi}^*(y,\eta) \;=\; (\partial x/\partial\widetilde{x})^T\big|_{\widetilde{x}=\widetilde{x}^*}\, \xi^*(y,\eta)\,.$$

Differentiating this identity with respect to η we see that

$$(\partial x/\partial\widetilde{x})^T\big|_{\widetilde{x}=\widetilde{x}^*}\cdot \xi^*_\eta(y,\eta) \;=\; \widetilde{\xi}^*_\eta(y,\eta) \;-\; C(y,\eta)\cdot \widetilde{x}^*_\eta(y,\eta)$$

where $C=\{C_{ij}\}$ is the symmetric matrix function with elements

$$C_{ij}(y,\eta) \;=\; \sum_k \xi^*_k(y,\eta)\,(\partial^2 x_k/\partial\widetilde{x}_i\partial\widetilde{x}_j)\big|_{\widetilde{x}=\widetilde{x}^*(y,\eta)}\,. \tag{2.3.7}$$

Here $(\partial^2 x_k/\partial\widetilde{x}_i\partial\widetilde{x}_j)\big|_{\widetilde{x}=\widetilde{x}^*}$ are the second Taylor coefficients of $x_k(\widetilde{x})$ at the point $\widetilde{x}=\widetilde{x}^*$. Given coordinates $\widetilde{x}$, an arbitrary real symmetric matrix $C^{(0)}=\{C^{(0)}_{ij}\}$, and a fixed point (y,η), we can always find coordinates x such that in (2.3.7) $C_{ij}(y,\eta)=C^{(0)}_{ij}$.

This observation enables us to prove the following useful

LEMMA 2.3.2. *Let $(y_0,\eta_0)\in O$, and $x_0=x^*(y_0,\eta_0)$. Then in a neighbourhood of the point x_0 there exists a coordinate system x such that*

$$\det \xi^*_\eta(y_0,\eta_0) \neq 0\,. \tag{2.3.8}$$

PROOF. Let $\widetilde{x}$ be arbitrary coordinates in a neighbourhood of the point x_0. In view of (2.3.6)

$$\widetilde{x}^*_\eta(y_0,\eta_0) \;:\; \ker \widetilde{\xi}^*_\eta(y_0,\eta_0) \;\to\; \ker\big(\widetilde{\xi}^*_\eta(y_0,\eta_0)\big)^T\,,$$

and, since the transformation $(y,\eta)\to(x^*,\xi^*)$ is nondegenerate, the rank of this map is maximal. Let $C^{(0)}=\{C^{(0)}_{ij}\}$ be the orthogonal projection on the subspace $\ker\big(\widetilde{\xi}^*_\eta(y_0,\eta_0)\big)^T$. Then $C^{(0)}\cdot\widetilde{\xi}^*_\eta(y_0,\eta_0)=0$. Assume that for some vector $\vec{c}\in\mathbb{C}^n$ we have

$$\big(\widetilde{\xi}^*_\eta(y_0,\eta_0) \;-\; C^{(0)}\cdot\widetilde{x}^*_\eta(y_0,\eta_0)\big)\,\vec{c} \;=\; 0\,.$$

This implies

$$C^{(0)}\cdot\big(\widetilde{\xi}^*_\eta(y_0,\eta_0) \;-\; C^{(0)}\cdot\widetilde{x}^*_\eta(y_0,\eta_0)\big)\,\vec{c} \;=\; -\,C^{(0)}\cdot\widetilde{x}^*_\eta(y_0,\eta_0)\,\vec{c} \;=\; 0\,,$$

and, consequently, $\widetilde{\xi}^*_\eta(y_0,\eta_0)\,\vec{c}=0$. Therefore the kernel of the matrix

$$\widetilde{\xi}^*_\eta(y_0,\eta_0) \;-\; C^{(0)}\cdot\widetilde{x}^*_\eta(y_0,\eta_0)$$

contains only the zero vector. Choosing now coordinates x such that

$$C_{ij}^{(0)} = \sum_k \xi_k^*(y_0,\eta_0)\,(\partial^2 x_k/\partial\widetilde{x}_i\partial\widetilde{x}_j)\big|_{\widetilde{x}=\widetilde{x}^*(y_0,\eta_0)}, \tag{2.3.9}$$

we obtain (2.3.8). The proof is complete. □

REMARK 2.3.3. Lemma 2.3.2 may not be true if $x_0 \in \partial M$ and we restrict ourselves to coordinate systems in which $\partial M = \{x_n = 0\}$. Therefore at the boundary one should use this lemma with care.

REMARK 2.3.4. In the proof of Lemma 2.3.2 we used only the fact that the transformation is nondegenerate and canonical. So this result remains valid for nonhomogeneous transformations.

2. Transformations generated by Hamiltonian billiards. One of the most natural examples of homogeneous canonical transformations is the shift

$$O \ni (y,\eta) \;\to\; (x^*(t;y,\eta),\xi^*(t;y,\eta)) \in T'M \tag{2.3.10}$$

along the trajectories of the Hamiltonian system (1.3.3) or, more generally, along the corresponding billiard trajectories introduced in Section 1.3.

In (2.3.10) t is a parameter. If $\partial M = \varnothing$ and $(x^*(t;y,\eta),\xi^*(t;y,\eta))$ are Hamiltonian trajectories, then we can (and will) assume $O = T'M$ and $t \in \mathbb{R}$. But when $(x^*(t;y,\eta),\xi^*(t;y,\eta))$ are billiard trajectories, we need some restrictions on the domain of definition of the parameters t and (y,η).

Let us fix a $T_+ > 0$ and a T_+-admissible point $(y_0,\eta_0) \in T'\overset{\circ}{M}$. Without loss of generality we shall assume that $t = T_+$ is not a moment of reflection for any billiard trajectory originating from (y_0,η_0) (otherwise we slightly increase T_+).

We shall take

$$(y,\eta) \in O \tag{2.3.11}$$

where $O \subset T'\overset{\circ}{M}$ is a small conic T_+-admissible neighbourhood of (y_0,η_0). We shall assume that O is connected, simply connected, conically separated from $T'M|_{\partial M}$, and that $t = T_+$ is not a moment of reflection for any billiard trajectory originating from O. Such a neighbourhood exists in view of Lemmas 1.3.27, 1.3.29.

Let us fix the type $\mathfrak{m} = \mathfrak{m}_1|\mathfrak{m}_2|\dots|\mathfrak{m}_i|\dots|\mathfrak{m}_{\mathbf{r}}$ of our T_+-admissible billiard trajectories, and denote by $t_i^*(y,\eta)$, $i = 1,2,\dots,\mathbf{r}$, the moments of reflection. Clearly, $t_i^*(y,\eta) \in C^\infty(O)$ and $0 < t_1^* < t_2^* < \dots < t_{\mathbf{r}}^* < T_+$. Let us also choose a $T_- < 0$ such that Hamiltonian trajectories originating from O do not reach the boundary on the time interval $(T_-,0)$.

By $\mathfrak{T}_i(y,\eta)$ we shall denote the following subsets of the interval (T_-,T_+):

$$\mathfrak{T}_i(y,\eta) := \begin{cases} (T_-\,,\,T_+), & \text{if} \quad i = \mathbf{r} = 0, \\ (T_-\,,\,t_1^*(y,\eta)], & \text{if} \quad i = 0 < \mathbf{r}, \\ [t_i^*(y,\eta)\,,\,t_{i+1}^*(y,\eta)], & \text{if} \quad 0 < i < \mathbf{r}, \\ [t_i^*(y,\eta)\,,\,T_+), & \text{if} \quad 0 < i = \mathbf{r}. \end{cases} \tag{2.3.12}$$

Obviously, $(T_-\,,\,T_+) = \bigcup_{i=0}^{\mathbf{r}} \mathfrak{T}_i$.

Having fixed $i = 0,1,2,\dots,\mathbf{r}$ we shall take

$$t \in \mathfrak{T}_i(y,\eta). \tag{2.3.13}$$

Formulae (2.3.11), (2.3.13) describe the values of $(t; y, \eta)$ for which our canonical transformation, corresponding to the ith leg of the billiards, is defined.

We prove now that the shifts (2.3.10) along billiard trajectories are indeed canonical transformations, and consequently, all the results of the previous subsection are valid for each fixed t.

PROPOSITION 2.3.5. *The shifts* (2.3.10) *along billiard trajectories (with fixed type) preserve the canonical* 1*-form* $\langle \xi, dx \rangle$.

REMARK 2.3.6. Recall that the preservation of the canonical 1-form $\langle \xi, dx \rangle$ is equivalent to the fulfilment of conditions (2.3.5). So the precise formulation of Proposition 2.3.5 is that on each of the $\mathbf{r}$ legs of the billiards (i.e., for $(y, \eta) \in O$, $t \in \mathfrak{T}_i(y, \eta)$, $i = 0, 1, 2, \dots, \mathbf{r}$) we have (2.3.5). Note that the expressions on the left-hand sides of (2.3.5) are well defined for $t = t_i^*$, $i = 1, 2, \dots, \mathbf{r}$, if we understand them in the sense of lower and upper continuity in t; however, it is not *a priori* clear whether these limit values at $t = t_i^* - 0$ and $t = t_i^* + 0$ coincide.

PROOF OF PROPOSITION 2.3.5. Obviously, the equalities (2.3.5) hold for $t = 0 \in \mathfrak{T}_0$, so we must prove that

$$\frac{d}{dt} \langle \xi^*, x^*_{y_k} \rangle = 0, \qquad \frac{d}{dt} \langle \xi^*, x^*_{\eta_k} \rangle = 0 \tag{2.3.14}$$

for $t \in \mathfrak{T}_i$ $i = 0, 1, 2, \dots, \mathbf{r}$, and

$$\langle \xi^*, x^*_{y_k} \rangle \big|_{t=t_i^*-0} = \langle \xi^*, x^*_{y_k} \rangle \big|_{t=t_i^*+0}, \quad \langle \xi^*, x^*_{\eta_k} \rangle \big|_{t=t_i^*-0} = \langle \xi^*, x^*_{\eta_k} \rangle \big|_{t=t_i^*+0} \tag{2.3.15}$$

for $i = 1, 2, \dots, \mathbf{r}$.

Let us first prove (2.3.14). In view of Euler's identity and the preservation of the Hamiltonian h along Hamiltonian trajectories we have

$$\sum_j \xi_j^* \dot{x}_j^* = \sum_j \xi_j^* h_{\xi_j}(x^*, \xi^*) = h(x^*, \xi^*) = h(y, \eta). \tag{2.3.16}$$

Therefore

$$\begin{aligned} \frac{d}{dt} \langle \xi^*, x^*_{\eta_k} \rangle &= \sum_j \xi_j^* \frac{d}{d\eta_k} (h_{\xi_j}(x^*, \xi^*)) - \sum_j h_{x_j}(x^*, \xi^*) (x_j^*)_{\eta_k} \\ &= \sum_j \frac{d}{d\eta_k} (\xi_j^* h_{\xi_j}(x^*, \xi^*)) - \sum_j h_{x_j}(x^*, \xi^*) (x_j^*)_{\eta_k} - \sum_j h_{\xi_j}(x^*, \xi^*) (\xi_j^*)_{\eta_k} \\ &= \sum_j \frac{d}{d\eta_k} (\xi_j^* h_{\xi_j}(x^*, \xi^*)) - \sum_j \frac{d}{d\eta_k} h(x^*, \xi^*) = 0. \end{aligned}$$

The second formula (2.3.14) is proved analogously.

Let us now prove (2.3.15). According to the reflection law formulated in Section 1.3 we have

$$x^*(t_i^*(y, \eta) - 0; y, \eta) = x^*(t_i^*(y, \eta) + 0; y, \eta), \tag{2.3.17}$$

$$x_n^*(t_i^*(y, \eta) - 0; y, \eta) = x_n^*(t_i^*(y, \eta) + 0; y, \eta) \equiv 0, \tag{2.3.18}$$

$$\xi_j^*(t_i^*(y, \eta) - 0; y, \eta) = \xi_j^*(t_i^*(y, \eta) + 0; y, \eta) \tag{2.3.19}$$

for $j \neq n$, and

$$h(x^*, \xi^*)\big|_{t=t_i^*-0} = h(x^*, \xi^*)\big|_{t=t_i^*+0}. \tag{2.3.20}$$

Formulae (2.3.17)–(2.3.19) imply that

$$
\begin{aligned}
(t_i^*)_{\eta_k} \sum_j \left(\xi_j^*\, h_{\xi_j}(x^*,\xi^*)\right)\Big|_{t=t_i^*-0} &+ \sum_j \left(\xi_j^*\,(x_j^*)_{\eta_k}\right)\Big|_{t=t_i^*-0} \\
&= \sum_j \xi_j^*(t_i^*(y,\eta)-0;y,\eta)\,\frac{d}{d\eta_k} x_j^*(t_i^*(y,\eta)-0;y,\eta) \\
&= \sum_{j<n} \xi_j^*(t_i^*(y,\eta)-0;y,\eta)\,\frac{d}{d\eta_k} x_j^*(t_i^*(y,\eta)-0;y,\eta) \\
&= \sum_{j<n} \xi_j^*(t_i^*(y,\eta)+0;y,\eta)\,\frac{d}{d\eta_k} x_j^*(t_i^*(y,\eta)+0;y,\eta) \\
&= \sum_j \xi_j^*(t_i^*(y,\eta)+0;y,\eta)\,\frac{d}{d\eta_k} x_j^*(t_i^*(y,\eta)+0;y,\eta) \\
&= (t_i^*)_{\eta_k} \sum_j \left(\xi_j^*\, h_{\xi_j}(x^*,\xi^*)\right)\Big|_{t=t_i^*+0} + \sum_j \left(\xi_j^*\,(x_j^*)_{\eta_k}\right)\Big|_{t=t_i^*+0}.
\end{aligned}
$$

From this equality (see the first and the last lines), (2.3.16), and (2.3.20) it follows that

$$\sum_j \left(\xi_j^*\,(x_j^*)_{\eta_k}\right)\Big|_{t=t_i^*-0} = \sum_j \left(\xi_j^*\,(x_j^*)_{\eta_k}\right)\Big|_{t=t_i^*+0},$$

which coincides with the first formula (2.3.15). In the same way we obtain the second formula (2.3.15). □

2.4. Phase functions associated with homogeneous canonical transformations

Phase functions described in this section are a special (but most important for us) case of those introduced in Section 2.2.

1. Definition and basic properties. Let us consider the time-dependent homogeneous canonical transformation (2.3.10) generated by the shifts along billiard trajectories and defined on the set (2.3.11), (2.3.13), where i is the sequential number of the leg of our billiard trajectories, $0 \leqslant i \leqslant \mathbf{r}$. Let us denote

$$\mathfrak{D}_i := \{\, (t;y,\eta) \;:\; (y,\eta)\in O\,,\; t\in\mathfrak{T}_i(y,\eta)\,\}\,, \tag{2.4.1}$$

$$\mathfrak{C}_i := \{\, (t,x;y,\eta) \;:\; (t;y,\eta)\in\mathfrak{D}_i\,,\; x = x^*(t;y,\eta)\,\}\,. \tag{2.4.2}$$

DEFINITION 2.4.1. We say that the nondegenerate phase function φ is associated with the ith leg of the canonical transformation (2.3.10) if φ is defined on a conic neighbourhood $\mathcal{O}\subset(T_-\,,T_+)\times M_x\times O$ of the set $\mathfrak{C}_i$ and satisfies the following two conditions:

$$\varphi_\eta(t,x;y,\eta) = 0 \qquad \text{if and only if} \quad (t,x;y,\eta)\in\mathfrak{C}_i\,, \tag{2.4.3}$$

$$\varphi_x(t,x^*;y,\eta) = \xi^*, \qquad \forall\,(t;y,\eta)\in\mathfrak{D}_i\,. \tag{2.4.4}$$

By $\mathfrak{F}_i$ we denote the set of all nondegenerate phase functions associated with the ith leg of the canonical transformation (2.3.10), $0\leqslant i\leqslant\mathbf{r}$.

In this subsection it will often be convenient for us to treat ξ^* as a quantity defined on $\mathfrak{C}_i$. In this case ξ^* at a point $(t,x;y,\eta)\in\mathfrak{C}_i$ will be understood as

$\xi^*(t;y,\eta)$. Note that under this agreement formula (2.4.4) can be rewritten in the equivalent form

$$\varphi_x = \xi^* \quad \text{on } \mathfrak{C}_i\,. \tag{2.4.4$'$}$$

Certainly, (2.4.3) means that $C_\varphi = \mathfrak{C}_i$ (see the notation in Section 2.2). We introduced here new notation only to get rid of the subscript φ, because now we want to stress that the set $\mathfrak{C}_i$ does not depend on the choice of a particular phase function φ. The Hamiltonian billiard system and the sets $\mathfrak{C}_i$, $i = 0, 1, 2, \ldots, \mathbf{r}$, are primary objects, whereas the associated phase functions are secondary ones.

We write the subscript i in $\mathfrak{C}_i$ (and in $\mathfrak{O}_i$) to remind that we are considering the situation when billiard trajectories have already experienced i reflections, but have not yet experienced the $(i+1)$st. For different i these sets are, generally speaking, different. In this subsection the index i is fixed. The relation between the sets $\mathfrak{C}_i$ and $\mathfrak{C}_{i+1}$, as well as between the corresponding phase functions, will be discussed in subsection 2.

From Euler's identity and (2.4.3) it follows (see (2.1.3)) that

$$\varphi(t, x^*; y, \eta) = 0, \qquad \forall\,(t; y, \eta) \in \mathfrak{O}_i\,. \tag{2.4.5}$$

Differentiating this identity with respect to t and taking into account (2.4.4), (2.3.16) we obtain for any phase function $\varphi \in \mathfrak{F}_i$

$$\varphi_t(t, x^*; y, \eta) = -h(y, \eta)\,, \qquad \forall\,(t; y, \eta) \in \mathfrak{O}_i\,. \tag{2.4.6}$$

REMARK 2.4.2. Condition (2.4.3) in Definition 2.4.1 can be replaced by two weaker conditions: condition (2.4.5) and condition

$$\varphi_\eta \neq 0 \quad \text{on } \mathcal{O} \setminus \mathfrak{C}_i\,.$$

In this case we automatically get $\varphi_\eta = 0$ on $\mathfrak{C}_i$; this fact is established by differentiating the identity (2.4.5) with respect to η_k, and taking into account (2.4.4) as well as the second equality (2.3.5).

REMARK 2.4.3. By analogy (omitting the dependence on t and replacing $\mathfrak{O}_i$ by O) one can introduce the class of nondegenerate phase functions $\varphi(x; y, \eta)$ associated with a general time-independent transformation of the form (2.3.1). Practically all our further results are valid for this class of phase functions (with the exception of Lemma 2.6.6), and they are proved in the same way [**LapSaVa**]. However, later on we will not deal with general time-independent transformations, only with the identity transformation. It is obtained from (2.3.10) when we set $t = 0$, so in this case all the results follow from the corresponding results for (2.3.10) (see 2.7.5).

We define on the set $\mathfrak{O}_i$ the following matrix functions of second derivatives:

$$\begin{aligned} \Phi_{xx}(t; y, \eta) &:= \varphi_{xx}(t, x^*; y, \eta)\,, & \Phi_{\eta\eta}(t; y, \eta) &:= \varphi_{\eta\eta}(t, x^*; y, \eta)\,, \\ \Phi_{x\eta}(t; y, \eta) &:= \varphi_{x\eta}(t, x^*; y, \eta)\,, & \Phi_{\eta x}(t; y, \eta) &:= \varphi_{\eta x}(t, x^*; y, \eta) \end{aligned} \tag{2.4.7}$$

(obviously $\Phi_{\eta x} = (\Phi_{x\eta})^T$). Since $\operatorname{Im}\varphi \geqslant 0$ and $\operatorname{Im}\varphi|_{x=x^*} = 0$, the symmetric real matrix function $\operatorname{Im}\Phi_{xx}$ is nonnegative. Singling out the initial terms of Taylor's expansion of φ at the point $x = x^*$ and taking into account (2.4.5), (2.4.4), we obtain in an arbitrary coordinate system x

$$\begin{aligned}\varphi(t,x;y,\eta) = \langle\, x - x^*, \xi^*\,\rangle + \frac{1}{2}\langle\, \Phi_{xx}(x-x^*), x - x^*\,\rangle \\ + O(|x-x^*|^3)\,, \qquad \forall (t;y,\eta) \in \mathfrak{D}_i\,,\end{aligned} \tag{2.4.8}$$

as $x \to x^*$; in particular, the function $\operatorname{Im}\varphi$ has a second order zero at the point $x = x^*$. Differentiating (2.4.8) with respect to η and taking into account the second equality (2.3.5) we see that

$$\begin{aligned}\varphi_\eta(t,x;y,\eta) &= (\xi^*_\eta - \Phi_{xx}\cdot x^*_\eta)^T(x-x^*) \\ &\quad + O(|x-x^*|^2), \quad \forall (t;y,\eta) \in \mathfrak{D}_i,\end{aligned} \tag{2.4.9}$$

$$\Phi_{x\eta} = \xi^*_\eta - \Phi_{xx}\cdot x^*_\eta\,. \tag{2.4.10}$$

Note that we derived formulae (2.4.8)–(2.4.10) using only the assumptions (2.4.5), (2.4.4). This will allow us further on to use these formulae even in situations when we do not know *a priori* whether (2.4.3) holds, in particular in the proofs of Lemma 2.4.7 and Corollary 2.4.8.

In view of (2.4.10) the nondegeneracy condition can now be written in the form

$$\det(\xi^*_\eta - \Phi_{xx}\cdot x^*_\eta) \neq 0\,,$$

and this is a restriction on the matrix function Φ_{xx}.

LEMMA 2.4.4. *Let C_0 be a symmetric $n\times n$-matrix such that $\operatorname{Im} C_0$ is strictly positive. Then the matrix $C_1 = \xi^*_\eta - C_0\cdot x^*_\eta$ is nonsingular.*

PROOF. By (2.3.6) $\operatorname{Re} C_1^T\cdot x^*_\eta - (x^*_\eta)^T\cdot \operatorname{Re} C_1 = 0$. Therefore

$$\begin{aligned}(\operatorname{Re} C_1^T &- i\operatorname{Im} C_1^T)\cdot(\operatorname{Im} C_0)^{-1}\cdot(\operatorname{Re} C_1 + i\operatorname{Im} C_1) \\ &= \operatorname{Re} C_1^T\cdot(\operatorname{Im} C_0)^{-1}\cdot \operatorname{Re} C_1 + \operatorname{Im} C_1^T\cdot(\operatorname{Im} C_0)^{-1}\cdot \operatorname{Im} C_1\,.\end{aligned}$$

The real symmetric matrix on the right-hand side of this equality is nonnegative, and for any vector $\vec{c}$ from its kernel we have $x^*_\eta\vec{c} = 0$, $\xi^*_\eta\vec{c} = 0$. This implies $\vec{c} = 0$ because the transformation $(y,\eta)\to(x^*,\xi^*)$ is nondegenerate. Thus, the matrix C_1 is invertible. □

Lemma 2.4.4 and formula (2.4.10) immediately imply

COROLLARY 2.4.5. *Let φ be a phase function defined on a neighbourhood $\mathcal{O}$ of the set $\mathfrak{C}_i$, satisfying the conditions (2.4.3), (2.4.4) and such that the matrix function $\operatorname{Im}\Phi_{xx}$ is strictly positive. Then φ is nondegenerate (and, consequently, $\varphi\in\mathfrak{F}_i$).*

REMARK 2.4.6. In Lemma 2.4.4 and Corollary 2.4.5 the conditions $\operatorname{Im} C_0 > 0$ and $\operatorname{Im}\Phi_{xx} > 0$ can be weakened: it is sufficient for quadratic forms associated with these matrices to be strictly positive only on the subspace of vectors orthogonal to the covector ξ^*. Moreover, it is sufficient for them to be strictly positive only on the image of the linear mapping x^*_η.

Lemma 2.4.4 is the decisive step in the effective construction of nondegenerate phase functions associated with our canonical transformation (2.3.10), because the main problem with global constructions is the fulfilment of the condition $\det \varphi_{x\eta} \neq 0$ on $\mathfrak{C}_i$.

It only remains to explain how to satisfy the condition (2.4.3) (note that condition (2.4.3) is dependent on the choice of a particular neighbourhood $\mathcal{O}$). This problem is dealt with by the following

LEMMA 2.4.7. *Let φ be a phase function defined on a conic neighbourhood $\widetilde{\mathcal{O}} \subset (T_-, T_+) \times M_x \times O$ of the set $\mathfrak{C}_i$ and satisfying*

$$(2.4.11) \qquad \varphi = 0, \quad \varphi_x = \xi^*, \quad \det \varphi_{x\eta} \neq 0 \quad \text{on } \mathfrak{C}_i\,.$$

Then there exists a smaller neighbourhood $\mathcal{O}$ such that the restriction of φ to $\mathcal{O}$ is a phase function from the class $\mathfrak{F}_i$.

PROOF. In order to prove the lemma (i.e., to prove that φ satisfies the requirements of Definition 2.4.1) we only have to establish (2.4.3) on some neighbourhood O. Indeed, condition (2.4.4) is contained in (2.4.11) and nondegeneracy will follow from (2.4.3), (2.4.11).

Let us fix an arbitrary $(t; y, \eta) \in \mathfrak{D}_i$ and consider the x-solutions of the equation (more precisely, system of equations) $\varphi_\eta = 0$. By (2.4.9) $x = x^*$ is a solution, and, moreover, in view of (2.4.11) this solution is unique for x sufficiently close to x^*. We satisfy this uniqueness condition by reducing sufficiently the original neighbourhood $\widetilde{\mathcal{O}}$ in the x-variables.

Let us now prove that in a sufficiently small neighbourhood of $\mathfrak{C}_i$ the equation $\varphi_\eta = 0$ does not have solutions with $(t; y, \eta) \notin \mathfrak{D}_i$, i.e., with $t \notin \mathfrak{T}_i(y, \eta)$.

One way of doing this would be to consider the extended manifold $\widehat{M}$ (see 1.1.2), extended Hamiltonian trajectories $(\widehat{x}^*(t; y, y), \widehat{\xi}^*(t; y, y))$, and the extended phase function $\widehat{\varphi}(t, \widehat{x}; y, \eta)$. Then arguments from the first part of the proof show that $\widehat{x} = \widehat{x}^*$ is the unique solution of the equation $\widehat{\varphi}_\eta = 0$, and, because the trajectories are transversal to the boundary and t is close to $\mathfrak{T}_i(y, \eta)$, but not in $\mathfrak{T}_i(y, \eta)$, this would imply that $\widehat{x}^* \notin M$. However, for methodological reasons we do not want to resort to the notion of an extended manifold $\widehat{M}$ without utmost necessity (not least because this requires a detailed description of the domains of definition of our trajectories and phase functions), so we give below another argument which does not use any extensions.

Set

$$(2.4.12) \quad \mathfrak{C}_i' := \{ (t, x'; y, \eta) \,:\, t = t_i^*(y, \eta),\ x' = x^{*\prime}(t_i^*(y, \eta), y, \eta),\ (y, \eta) \in O \}$$

for $0 < i \leqslant \mathbf{r}$, and $\mathfrak{C}_0' := \mathfrak{C}_{\mathbf{r}+1}' := \varnothing$. According to this definition the set $\mathfrak{C}_i'$ is a subset of $(-T_-, T_+) \times \partial M_x \times O$. Further in this book we shall also view the set $\mathfrak{C}_i'$ as a subset of $(-T_-, T_+) \times M_x \times O$. Clearly, $\mathfrak{C}_i' \cup \mathfrak{C}_{i+1}'$ is the boundary of the set $\mathfrak{C}_i \subset (T_-, T_+) \times M_x \times O$.

For definiteness we shall consider the equation $\varphi_\eta = 0$ in the neighbourhood of $\mathfrak{C}_i'$, i.e., we shall consider the situation when t lies on the left-hand side of the set $\mathfrak{T}_i(y, \eta)$ and is close to $\mathfrak{T}_i(y, \eta)$. In other words, $i \neq 0$ and

$$(2.4.13) \qquad T_- < t < t_i^*(y, \eta), \qquad t_i^*(y, \eta) - t \ll 1\,.$$

In the remaining part of the proof of Lemma 2.4.6 we shall denote for brevity

$$\mathbf{x}^* = (\mathbf{x}^{*\prime}, 0) = x^*(t_i^*(y,\eta), y, \eta) \in \partial M_x, \quad \mathbf{t}^* = t_i^*(y,\eta), \quad \mathbf{h}_\xi = h_\xi(\mathbf{x}^*, \xi^*(\mathbf{t}^*; y, \eta)).$$

Note that

$$\mathbf{h}_{\xi_n} > 0. \tag{2.4.14}$$

Let us fix an arbitrary point $(y,\eta) \in O$ and search for (t,x)-solutions of the equation $\varphi_\eta = 0$ which are close to $(\mathbf{t}^*, \mathbf{x}^*)$. We shall denote these solutions by $(\mathbf{t}, \mathbf{x})$. The identity $\varphi_\eta(\mathbf{t}, \mathbf{x}; y, \eta) = 0$ can be rewritten as

$$\mathbf{F}_{\eta t}\,(\mathbf{t} - \mathbf{t}^*) \;+\; \mathbf{F}_{\eta x}\,(\mathbf{x} - \mathbf{x}^*) \;=\; O(\rho^2), \qquad \rho \to 0, \tag{2.4.15}$$

where

$$\mathbf{F}_{\eta t} = \varphi_{\eta t}(\mathbf{t}^*, \mathbf{x}^*; y, \eta), \quad \mathbf{F}_{\eta x} = \varphi_{\eta x}(\mathbf{t}^*, \mathbf{x}^*; y, \eta), \quad \rho = \sqrt{(\mathbf{t} - \mathbf{t}^*)^2 + |\mathbf{x} - \mathbf{x}^*|^2}.$$

Differentiating (2.4.9) with respect to t, setting $t = \mathbf{t}^*$, $x = \mathbf{x}^*$, and taking into account (1.3.3), we obtain $\mathbf{F}_{\eta t} = -\mathbf{F}_{\eta x}\,\mathbf{h}_\xi$. So, multiplying (2.4.15) by the matrix $(\mathbf{F}_{\eta x})^{-1}$ we get

$$(\mathbf{x} - \mathbf{x}^*) \;-\; \mathbf{h}_\xi\,(\mathbf{t} - \mathbf{t}^*) \;=\; O(\rho^2). \tag{2.4.16}$$

In particular, the nth component of (2.4.16) is

$$\mathbf{x}_n \;-\; \mathbf{h}_{\xi_n}\,(\mathbf{t} - \mathbf{t}^*) \;=\; O(\rho^2). \tag{2.4.17}$$

Recall that we are looking for $\mathbf{t} < \mathbf{t}^*$ (see (2.4.13)) which are close to $\mathbf{t}^*$. With account of (2.4.14) this means that

$$-\,\mathbf{h}_{\xi_n}\,(\mathbf{t} - \mathbf{t}^*) \;=\; |\mathbf{h}_{\xi_n}|\,|\mathbf{t} - \mathbf{t}^*|. \tag{2.4.18}$$

Combining (2.4.17), (2.4.18) and the inequality $\mathbf{x}_n \geqslant 0$, we obtain

$$\mathbf{t} - \mathbf{t}^* \;=\; O(\rho^2). \tag{2.4.19}$$

The substitution of (2.4.19) into (2.4.16) gives

$$\mathbf{x} - \mathbf{x}^* \;=\; O(|\mathbf{x} - \mathbf{x}^*|^2). \tag{2.4.20}$$

For sufficiently small ρ the only $\mathbf{x}$ satisfying (2.4.20) is $\mathbf{x} = \mathbf{x}^*$. But then $\rho = |\mathbf{t} - \mathbf{t}^*|$ and (2.4.19) turns into $\mathbf{t} - \mathbf{t}^* = O((\mathbf{t} - \mathbf{t}^*)^2)$. Clearly, for sufficiently small ρ the only $\mathbf{t}$ satisfying this identity is $\mathbf{t} = \mathbf{t}^*$, which contradicts our assumption $\mathbf{t} < \mathbf{t}^*$. So, reducing the original neighbourhood $\widetilde{\mathcal{O}}$ near $\mathfrak{C}_i'$ in the (t,x)-variables (and, similarly, near $\mathfrak{C}_{i+1}'$) we arrange that the equation $\varphi_\eta = 0$ has no solutions with $(t; y, \eta) \notin \mathfrak{D}_i$.

By switching from the original neighbourhood $\widetilde{\mathcal{O}}$ to some smaller neighbourhood $\mathcal{O}$ we have satisfied the condition (2.4.3). □

In the proof of Lemma 2.4.7 we have implicitly used the fact that the set $\mathfrak{C}_i$ is closed in $(T_-, T_+) \times M_x \times O$. This has allowed us to proceed from the construction of a neighbourhood of an arbitrary fixed point $(t, x; y, \eta) \in \mathfrak{C}_i$ to the neighbourhood of the whole set $\mathfrak{C}_i$.

Lemma 2.4.7 shows that when we construct effectively a phase function from the class $\mathfrak{F}_i$ we need not worry about the fulfilment of the condition (2.4.3): this condition can always be satisfied later by reducing (if necessary) the initial domain of definition of the phase function, as long as the conditions (2.4.11) are fulfilled on $\mathfrak{C}_i$.

An important consequence of Lemmas 2.4.4 and 2.4.7 is that the set $\mathfrak{F}_i$ is not empty. This fact is established by the following

COROLLARY 2.4.8. *For any canonical transformation of the form* (2.3.10) *there exists a nondegenerate phase function associated with this transformation.*

PROOF. Take an arbitrary symmetric $n \times n$-matrix function $C \in C^\infty(\mathfrak{D}_i)$ which is positively homogeneous in η of degree 1 and has strictly positive imaginary part. Then, by Lemma 2.4.4, we have

$$\det(\xi^*_\eta(t;y,\eta) - C(t;y,\eta)\cdot x^*_\eta(t;y,\eta)) \neq 0, \qquad \forall (t;y,\eta) \in \mathfrak{D}_i. \tag{2.4.21}$$

Let us take an arbitrary phase function with Taylor's expansion (2.4.8), $\Phi_{xx} \equiv C$. Formulae (2.4.8), (2.4.10), (2.4.21) imply (2.4.11). Reducing (if necessary) the initial domain of definition of our phase function, we obtain, according to Lemma 2.4.7, the required phase function $\varphi \in \mathfrak{F}_i$. □

REMARK 2.4.9. The matrix function $C(t;y,\eta)$ appearing in the proof of Corollary 2.4.8 is not a tensor. Under the transformation of coordinates $x \to \widetilde{x}$ it changes in a more complicated way:

$$\widetilde{C} = \sum_k \xi^*_k\,(\partial^2 x_k/\partial \widetilde{x}^2) + (\partial x/\partial \widetilde{x})^T\, C\,(\partial x/\partial \widetilde{x}), \tag{2.4.22}$$

where all the derivatives are evaluated at the point $\widetilde{x} = \widetilde{x}^*(t;y,\eta)$. So by C we really understand a set of local coordinate representations which match in the sense (2.4.22). The matrix C admits a geometric interpretation in terms of a (generally speaking, nonlinear) connection of the manifold M. For example, if we take a linear connection with Christoffel symbols Γ^k_{ij}, then the matrix with elements $\sum_k \Gamma^k_{ij}(x^*)\,\xi^*_k$ behaves in the same way as C. For us, however, the geometric interpretation of C is not important; the essential fact is that the condition (2.4.21) is invariant under changes of coordinates x, y.

The following lemma implies that the set $\mathfrak{F}_i$ is connected or, more exactly, that any two phase functions φ_0 and φ_1 from $\mathfrak{F}_i$ can be continuously transformed one into another in $\mathfrak{F}_i$ on a sufficiently small neighbourhood of $\mathfrak{C}_i$.

LEMMA 2.4.10. *Let* $\varphi^{(0)}, \varphi^{(1)} \in \mathfrak{F}_i$ *be phase functions with domains of definition* $\mathcal{O}^{(0)}$, $\mathcal{O}^{(1)}$, *respectively, and* $\mathcal{O} = \mathcal{O}^{(0)} \cap \mathcal{O}^{(1)}$ *be their common domain of definition. Then there exists a family of phase functions* $\psi^{(s)} \in \mathfrak{F}_i$ *smoothly depending on the parameter* $s \in [0,1]$ *and with domain of definition* $\mathcal{O}$ (*for each fixed* s), *such that* $\psi^{(0)} \equiv \varphi^{(0)}$ *and* $\psi^{(1)} \equiv \varphi^{(1)}$ *on* $\mathcal{O}$.

PROOF. Let us take some real-valued function $b \in C^\infty(\mathcal{O})$ homogeneous in η of degree 1, such that

$$b(t,x^*;y,\eta) = 0, \quad b_x(t,x^*;y,\eta) = 0, \quad b_{xx}(t,x^*;y,\eta) > 0, \qquad \forall (t;y,\eta) \in \mathfrak{D}_i,$$

and $b > 0$ outside the set $\mathfrak{C}_i$. Set

$$\psi^{(s)} = (1-s)\,\varphi^{(0)} + s\,\varphi^{(1)} + i\,s\,(1-s)\,b, \qquad 0 \leqslant s \leqslant 1.$$

Obviously, $\psi^{(0)} \equiv \varphi^{(0)}$ and $\psi^{(1)} \equiv \varphi^{(1)}$ on $\mathcal{O}$, and $\psi^{(s)}$ satisfies the conditions (2.4.3), (2.4.4) for all $s \in [0,1]$. Besides,

$$\partial_{xx}\psi^{(s)}\big|_{x=x^*} = (1-s)\,\partial_{xx}\varphi^{(0)}\big|_{x=x^*} + s\,\partial_{xx}\varphi^{(1)}\big|_{x=x^*} + i\,s\,(1-s)\,b_{xx}\big|_{x=x^*}.$$

Since $\partial_{xx}\varphi^{(0)}\big|_{x=x^*} \geqslant 0$ and $\partial_{xx}\varphi^{(1)}\big|_{x=x^*} \geqslant 0$, it follows from Corollary 2.4.4 that $\psi^{(s)}$ is nondegenerate for all $0 < s < 1$. Nondegeneracy of $\psi^{(s)}$ for $s = 0,1$ follows from the nondegeneracy of $\varphi^{(0)}$, $\varphi^{(1)}$. This completes the proof. □

REMARK 2.4.11. Examination of the proof of Lemma 2.4.10 shows that not only is $\mathfrak{F}_i$ connected, but it is *contractible* as a topological space. This means that if we fix an arbitrary phase function $\psi \in \mathfrak{F}_i$ (point in this topological space) then there exists a continuous mapping f from $\mathfrak{F}_i \times [0,1]$ to $\mathfrak{F}_i$ given explicitly by the formula

$$f(\psi,s) = (1-s)\,\psi + s\,\varphi + i\,s\,(1-s)\,b$$

(b is the same as in the proof of Lemma 2.4.10) such that $f(\psi,0) = \psi$ and $f(\psi,1) = \varphi$. Of course, some explanations are required here on how to view $\mathfrak{F}_i$ as a topological space. The most natural way of doing this is to identify all phase functions with the same Taylor expansions in powers of $x - x^*(t;y,\eta)$ (irrespective of their domains of definition) and use the topology generated by the standard $C^\infty(\mathfrak{D}_i)$-seminorms of these Taylor coefficients. A consequence of the contractibility of $\mathfrak{F}_i$ is the fact that $\mathfrak{F}_i$ is *simply connected*, i.e. any continuous curve in $\mathfrak{F}_i$ can be contracted to a point. All these nice properties — connectedness, simple connectedness, contractibility etc. — are really consequences of the simple fact that the set $\mathring{\mathfrak{F}}_i := \{\varphi \in \mathfrak{F}_i : \operatorname{Im}\Phi_{xx} > 0 \text{ on } \mathfrak{C}_i\}$ (the "interior" of $\mathfrak{F}_i$) is *convex* in the sense that for any two phase functions $\varphi^{(0)},\varphi^{(1)} \in \mathring{\mathfrak{F}}_i$ with domains of definition $\mathcal{O}^{(0)}$, $\mathcal{O}^{(1)}$, respectively, the phase function $\psi^{(s)} := (1-s)\,\varphi^{(0)} + s\,\varphi^{(1)}$ with domain of definition $\mathcal{O} = \mathcal{O}^{(0)} \cap \mathcal{O}^{(1)}$ belongs to $\mathring{\mathfrak{F}}_i$ for all $s \in [0,1]$.

Sometimes it is convenient to construct a phase function locally (i.e. not in a neighbourhood of the whole $\mathfrak{C}_i$, but in a neighbourhood of a smaller set $\widetilde{\mathfrak{C}}_i$), and then it is necessary to know whether this phase function can be extended up to a phase function of the class $\mathfrak{F}_i$ (defined on a neighbourhood of $\mathfrak{C}_i$). The following lemma gives a positive answer to this question.

LEMMA 2.4.12. *Let $\widetilde{\mathfrak{C}}_i$ be a closed (in $(T_-, T_+) \times M_x \times O$) conic subset of $\mathfrak{C}_i$, and $\widetilde{\mathcal{O}}$ be a conic neighbourhood of $\widetilde{\mathfrak{C}}_i$ in $(T_-, T_+) \times M_x \times O$. Let $\widetilde{\varphi}$ be a phase function defined on $\widetilde{\mathcal{O}}$ and satisfying the conditions*

$$\widetilde{\varphi} = 0, \quad \widetilde{\varphi}_x = \xi^* \quad \text{on} \quad \mathfrak{C}_i \cap \widetilde{\mathcal{O}}, \qquad \det\widetilde{\varphi}_{x\eta} \neq 0 \quad \text{on} \quad \widetilde{\mathfrak{C}}_i. \tag{2.4.23}$$

Then there exists a phase function $\varphi \in \mathfrak{F}_i$ with domain of definition $\mathcal{O}$, and a smaller conic neighbourhood $\widetilde{\mathcal{O}}_1 \subset \widetilde{\mathcal{O}}$ of $\widetilde{\mathfrak{C}}_i$, such that $\varphi = \widetilde{\varphi}$ on $\mathcal{O} \cap \widetilde{\mathcal{O}}_1$.

PROOF. Let $\psi \in \mathfrak{F}_i$ be some phase function with domain of definition $\mathcal{P}$, such that the matrix function $\operatorname{Im}\psi_{xx}$ is strictly positive on $\mathfrak{C}_i$ (such a phase function exists; see the proof of Lemma 2.4.8).

Let $\widetilde{\mathcal{O}}_2 \subset \widetilde{\mathcal{O}} \cap \mathcal{P}$ be some small conic neighbourhood of the set $\widetilde{\mathfrak{C}}_i$ such that

$$\det\widetilde{\varphi}_{x\eta}(t,x;y,\eta) \neq 0 \quad \text{on} \quad \widetilde{\mathcal{O}}_2. \tag{2.4.24}$$

Such a neighbourhood $\widetilde{\mathcal{O}}_2$ exists because $\det\widetilde{\varphi}_{x\eta}(t,x;y,\eta)\neq 0$ on $\widetilde{\mathfrak{C}}_i$ (see the third condition (2.4.23)) and $\widetilde{\mathfrak{C}}_i$ is closed in $(T_-\,,T_+)\times M_x\times O$.

Let us take some real-valued function $\chi\in C^\infty((T_-\,,T_+)\times M_x\times O)$ positively homogeneous in η of degree 0, such that $0\leqslant\chi\leqslant 1$ on $(T_-\,,T_+)\times M_x\times O$, $\chi=0$ on some small (smaller than $\widetilde{\mathcal{O}}_2$) conic neighbourhood $\widetilde{\mathcal{O}}_1$ of $\widetilde{\mathfrak{C}}_i$, and $\operatorname{supp}(1-\chi)\subset\widetilde{\mathcal{O}}_2$.

Set $\varphi:=\chi\psi+(1-\chi)\widetilde{\varphi}$ with initial domain of definition $\mathcal{P}$. Obviously, $\varphi=\widetilde{\varphi}$ on $\widetilde{\mathcal{O}}_1$. Direct substitution with account of (2.4.23) shows that the phase function φ satisfies the first two conditions (2.4.11). Let us show that it satisfies the third condition (2.4.11). Consider an arbitrary point $(t,x;y,\eta)\in\mathfrak{C}_i$. Then we have the following two possibilities:

$\chi(t,x;y,\eta)\neq 0$, in which case the inequality $\det\varphi_{x\eta}\neq 0$ follows from the formulae $\operatorname{Im}\varphi_{xx}\geqslant\chi\operatorname{Im}\psi_{xx}>0$, (2.4.10) and Lemma 2.4.4;

$\chi(t,x;y,\eta)=0$, in which case $(t,x;y,\eta)\in\widetilde{\mathcal{O}}_2$ and $\varphi_{x\eta}=\widetilde{\varphi}_{x\eta}$, so the inequality $\det\varphi_{x\eta}\neq 0$ follows from (2.4.24).

According to Lemma 2.4.7 the initial domain of definition $\mathcal{P}$ of the phase function φ can be reduced (if necessary) to some $\mathcal{O}$, such that the restriction of φ to $\mathcal{O}$ is a phase function from the class $\mathfrak{F}_i$. □

We already know that there are infinitely many nondegenerate phase functions associated with the transformation (2.3.10). The choice of a particular phase function from the class $\mathfrak{F}_i$ is a matter of convenience. It often makes sense to choose a phase function which is locally linear with respect to x. Lemma 2.4.13 and Corollary 2.4.14 (see below) show that such a choice is always possible.

LEMMA 2.4.13. *Let* $(t_0,x_0;y_0,\eta_0)\in\mathfrak{C}_i$, *and* x *be a coordinate system in a neighbourhood of the point* $x_0=x^*(t_0;y_0,\eta_0)$ *such that* $\det\xi^*_\eta(t_0;y_0,\eta_0)\neq 0$. *Then there exists a phase function* $\varphi\in\mathfrak{F}_i$ *such that in a small conic neighbourhood of the point* $(t_0,x_0;y_0,\eta_0)$

$$\varphi(t,x;y,\eta)=\langle x-\widehat{x}^*(t;y,\eta)\,,\,\widehat{\xi}^*(t;y,\eta)\,\rangle\,. \tag{2.4.25}$$

In (2.4.25) the "wide hat" is placed over x^*, ξ^* to show that it might be necessary to consider these Hamiltonian trajectories on the extended cotangent bundle $T'\widehat{M}$; see 1.1.2. In other words, it might be necessary to consider $(x^*(t;y,y)$, $\xi^*(t;y,y))$ for $t\notin\mathfrak{T}_i(y,\eta)$. Such a necessity arises when $x_0\in\partial M$. But of course the choice of a particular extension of the manifold M and of a particular extension of the differential operator A does not influence the result, in the sense that our oscillatory integrals will differ only by a C^∞-term depending on this choice.

Recall that according to Lemma 2.3.2 we can always choose local coordinates in which $\det\xi^*_\eta(t_0;y_0,\eta_0)\neq 0$. However, in such coordinates ∂M is not necessarily given by the equation $x_n=0$; see Remark 2.3.3.

PROOF OF LEMMA 2.4.13. Denote $\widetilde{\mathfrak{C}}_i:=\{(t_0,x_0;y_0,\lambda\eta_0)\,:\,\lambda>0\}$, and let $\Omega\subset\widehat{M}$ be the coordinate patch corresponding to our chosen local coordinate system x, with $x_0\in\Omega$. Let $\widetilde{\mathcal{O}}$ be a small conic neighbourhood of $\widetilde{\mathfrak{C}}_i$ in $(T_-\,,T_+)\times M_x\times O$, such that for any point $(t,x;y,\eta)\in\widetilde{\mathcal{O}}$ we have

$x \in \Omega$ and $\widehat{x}^*(t;y,\eta) \in \Omega$. Set $\widetilde{\varphi}(t,x;y,\eta) := \langle\, x - \widehat{x}^*(t;y,\eta)\,,\, \widehat{\xi}^*(t;y,\eta)\,\rangle$. Obviously, the phase function $\widetilde{\varphi}$ with domain of definition $\widetilde{\mathcal{O}}$ satisfies all the conditions of Lemma 2.4.12. Applying Lemma 2.4.12 we conclude the proof of Lemma 2.4.13. □

Lemmas 2.3.2, 2.3.3, Remark 2.3.4, and Lemma 2.4.13 immediately imply

COROLLARY 2.4.14. *For any fixed point from the set* $\mathfrak{C}_i$ *there exists a phase function* $\varphi \in \mathfrak{F}_i$ *with domain of definition* $\mathcal{O}$, *and a conic neighbourhood of this point* $\widetilde{\mathcal{O}} \subset \mathcal{O}$, *such that on* $\widetilde{\mathcal{O}}$ *this phase function can be written down in the form* (2.4.25) *in some local coordinates* $\widehat{x}$.

An important question is whether one can choose a real phase function. Globally the answer to this question is negative: the class $\mathfrak{F}_i$ may not contain any real phase functions (see [**LapSaVa**]). However, Corollary 2.4.14 shows that one can always choose a phase function from $\mathfrak{F}_i$ which is locally real, i.e., is real near a given point $(t_0, x_0; y_0, \eta_0) \in \mathfrak{C}_i$.

REMARK 2.4.15. Up till now we did not mention one simple property of phase functions $\varphi \in \mathfrak{F}_i$:

$$\varphi_y = -\eta \quad \text{on } \mathfrak{C}_i\,. \tag{2.4.26}$$

Formula (2.4.26) is obtained by differentiating (2.4.5) with respect to y, and using (2.4.4′) and the first equality (2.3.5). We will hardly ever need to use formula (2.4.26) because further on we will mostly consider our oscillatory integrals as distributions with respect to the variables t or x, but very rarely as distributions with respect to the variable y.

2. The eikonal equation. The Hamiltonian $h(x,\xi)$, which was defined initially for real $x \in M$ and real $\xi \in T'_x M$ and is an algebraic function in ξ, can be extended analytically to complex $\xi \neq 0$ with sufficiently small imaginary part. Let φ be a phase function from the class $\mathfrak{F}_i$. Consider the function

$$\mathfrak{e}(t,x;y,\eta) := \varphi_t(t,x;y,\eta) + h(x, \varphi_x(t,x;y,\eta))\,. \tag{2.4.27}$$

The expression (2.4.27) is well defined for $(t,x;y,\eta)$ lying in some conic neighbourhood of $\mathfrak{C}_i$. Without loss of generality we shall assume that this neighbourhood coincides with our usual neighbourhood $\mathcal{O}$ introduced in the previous subsection.

It follows from (2.4.6), (2.4.4) and (2.3.16) that

$$\mathfrak{e} = 0 \quad \text{on } \mathfrak{C}_i\,. \tag{2.4.28}$$

An interesting question is whether the stronger statement

$$\mathfrak{e} = 0 \quad \text{on } \mathcal{O} \tag{2.4.29}$$

is true.

The equation $\mathfrak{e} = 0$ is called the *eikonal equation*. In the traditional construction of oscillatory integrals associated with Hamiltonian flows the phase φ is chosen to satisfy (2.4.29), i.e., to satisfy the eikonal equation on $\mathcal{O}$. However, an examination of our definition of the class of phase functions $\mathfrak{F}_i$ shows that for our phase functions the eikonal equation does not necessarily hold on $\mathcal{O}$. And this fact has a profound meaning: by sacrificing the eikonal equation outside the set $\mathfrak{C}_i$ we have achieved the globality of our construction.

The point of this subsection is to show that though we cannot guarantee the fulfilment of the eikonal equation on $\mathcal{O}$, we can guarantee that the function $\mathfrak{e}$ has a second order zero on $\mathfrak{C}_i$.

THEOREM 2.4.16. *For any phase function* $\varphi \in \mathfrak{F}_i$

$$\mathfrak{e}_t = 0\,, \quad \mathfrak{e}_x = 0\,, \quad \mathfrak{e}_y = 0\,, \quad \mathfrak{e}_\eta = 0 \qquad \text{on } \mathfrak{C}_i\,. \tag{2.4.30}$$

PROOF OF THEOREM 2.4.16. Out of the four equalities (2.4.30) it is sufficient to prove only the second because the three others are an immediate consequence of the second and of (2.4.28). This fact is established by setting $x = x^*(t; y, \eta)$ and differentiating the identity (2.4.28) with respect to t, y and η.

Let us prove that $(\mathfrak{e}_x)|_{\mathfrak{C}_i} = 0$. Direct differentiation of (2.4.27) with account of (2.4.4′) and (2.4.7) gives

$$(\mathfrak{e}_x)|_{\mathfrak{C}_i} = \Phi_{xt} + h_x(x^*, \xi^*) + \Phi_{xx}\, h_\xi(x^*, \xi^*)\,, \tag{2.4.31}$$

where $\Phi_{xt}(t; y, \eta) := \varphi_{xt}(t, x^*; y, \eta)$. Formulae (2.4.8), (1.3.3) imply

$$\Phi_{xt} = -\,h_x(x^*, \xi^*) - \Phi_{xx}\, h_\xi(x^*, \xi^*)\,. \tag{2.4.32}$$

Substituting (2.4.32) into (2.4.31) we obtain $(\mathfrak{e}_x)|_{\mathfrak{C}_i} = 0$. □

It may seem (from the proof given above) that the origin of Theorem 2.4.16 is accidental, i.e., that the function $\mathfrak{e}$ has a second order zero on $\mathfrak{C}_i$ only due to some exotic relations between the derivatives. In order to expose better the origin of Theorem 2.4.16 we sketch below another, purely variational, proof of this theorem. This proof is remarkable in the sense that it does not require explicit calculations of the second derivatives.

VARIATIONAL PROOF OF THEOREM 2.4.16. Let $(x^*(t), \xi^*(t))$ be a billiard trajectory. We simplify notation by dropping references to the starting point (y, η), and will do the same with the phase function $\varphi(t, x) \in \mathfrak{F}_i$ and the function $\mathfrak{e}(t, x) := \varphi_t(t, x) + h(x, \varphi_x(t, x))$. Let (z, ζ) be an arbitrary point on this trajectory, i.e., $(z, \zeta) = (x^*(\tau), \xi^*(\tau))$ for some τ. We need to prove that

$$\mathfrak{e}(\tau, x) = O(|x - z|^2)\,, \qquad x \to z\,. \tag{2.4.33}$$

Without loss of generality we shall assume that $z \notin \partial M$; this is sufficient in view of the smoothness of all the functions involved. Let $\tau_1 < \tau < \tau_2$ be some moments of time which are sufficiently close to τ, so that our trajectory does not experience reflections on the time interval $[\tau_1, \tau_2]$. Set $z_1 := x^*(\tau_1)$, $z_2 := x^*(\tau_2)$.

It is well known (and can easily be checked directly) that the Hamiltonian trajectory $(x^*(t), \xi^*(t))$ is a stationary point of the functional

$$L := \int_{t=\tau_1}^{t=\tau_2} \frac{\langle \xi(t), \mathrm{d}x(t) \rangle}{h(x(t), \xi(t))} \tag{2.4.34}$$

under the conditions

$$x(\tau_1) = z_1\,, \qquad x(\tau_2) = z_2\,. \tag{2.4.35}$$

Here $x(t)$ is an arbitrary smooth curve on the manifold M, $\xi(t)$ is an arbitrary smooth covector function, $t \in [\tau_1, \tau_2]$ is an arbitrary variable used for parametrization, and d denotes the full differential (so, $\mathrm{d}x = \dot{x}(t)\, dt$).

Note that in the variational problem (2.4.34), (2.4.35) we can allow $\xi(t)$ to be complex-valued. This is possible because our Hamiltonian $h(x,\xi)$ is analytic for complex $\xi \neq 0$ with sufficiently small imaginary part and because the first variation of a functional is a linear form.

From now on let us consider trajectories of a special type: $x(t)$ is an arbitrary smooth curve on the manifold M satisfying (2.4.35) and $\xi(t) = \varphi_x(t, x(t))$. Obviously, the trajectory $(x^*(t), \xi^*(t))$ is a stationary point of the functional (2.4.34) on this special subclass of trajectories $(x(t), \xi(t))$.

Substituting $\xi(t) = \varphi_x(t, x(t))$ into (2.4.34), using the equality

$$\langle \varphi_x(t, x(t)), \mathrm{d}x(t)\rangle = \mathrm{d}\varphi(t, x(t)) - \varphi_t(t, x(t))\, dt\,,$$

and integrating by parts in $\mathrm{d}\varphi$ with account of the equalities (2.4.35) and $\varphi(\tau_1, z_1) = \varphi(\tau_2, z_2) = 0$, we obtain

$$\begin{aligned} L &= \int_{t=\tau_1}^{t=\tau_2} \frac{\langle \varphi_x(t, x(t)), \mathrm{d}x(t)\rangle}{h(x(t), \varphi_x(t, x(t)))} \\ &= \int_{t=\tau_1}^{t=\tau_2} \frac{\mathrm{d}\varphi(t, x(t))}{h(x(t), \varphi_x(t, x(t)))} - \int_{\tau_1}^{\tau_2} \frac{\varphi_t(t, x(t))\, dt}{h(x(t), \varphi_x(t, x(t)))} \\ &= \int_{\tau_1}^{\tau_2} \big(\varphi(t, x(t))\, p(t, x(t), \dot{x}(t)) + q(t, x(t))\big)\, dt\,, \end{aligned} \tag{2.4.36}$$

where

$$p(t, x(t), \dot{x}(t)) = -\frac{d}{dt}\frac{1}{h(x(t), \varphi_x(t, x(t)))}, \qquad q(t, x(t)) = -\frac{\varphi_t(t, x(t))}{h(x(t), \varphi_x(t, x(t)))}.$$

Let $\delta x(t) := x(t) - x^*(t)$ be the variation of the curve $x(t)$, and let δL be the corresponding first variation of the functional L. Let us compute δL using the last formula (2.4.36). The term $\varphi(t, x(t))\, p(t, x(t), \dot{x}(t))$ does not give a contribution to δL because $\varphi(t, x^*(t)) = p(t, x^*(t), \dot{x}^*(t)) = 0$ (the product of φ and p has a second order zero with respect to δx). So

$$\delta L = \int_{\tau_1}^{\tau_2} \langle q_x(t, x^*(t)), \delta x(t)\rangle\, dt\,. \tag{2.4.37}$$

According to the variational principle $\delta L = 0$. As the vector function $\delta x(t)$, $t \in [\tau_1, \tau_2]$, can be chosen arbitrarily (the only restriction being $\delta x(\tau_1) = \delta x(\tau_2) = 0$) the integral (2.4.37) is zero for all $\delta x(t)$ if and only if $q_x(t, x^*(t)) \equiv 0$ for $t \in [\tau_1, \tau_2]$. In particular, $q_x(\tau, z) = 0$. But $q(\tau, z) = -1$. So we have

$$q(\tau, x) = -1 + O(|x - z|^2)\,, \qquad x \to z\,. \tag{2.4.38}$$

It remains only to note that formula (2.4.38) is equivalent to (2.4.33). □

REMARK 2.4.17. In our variational arguments we managed to avoid dealing with reflections. However, this is not essential. It can be checked that the variational principle formulated in this subsection remains true for billiard trajectories, with the following natural modification: at the moment of reflection the curve $x^*(t)$ is allowed to have a vertex (it is continuous and all the one-sided derivatives exist) and the covector function $\xi^*(t)$ is allowed to have a discontinuity (it has one-sided limits and one-sided derivatives). The requirement of continuity of the first

$n-1$ components of $\xi^*(t)$ at the moment of reflection, which was axiomatically introduced in 1.3.3, is a simple consequence of the variational principle.

Theorem 2.4.16 plays a fundamental role in the effective construction of the wave group, see Chapter 3.

2.5. Restriction of phase functions to the boundary

In this section we consider the restrictions of phase functions of the class $\mathfrak{F}_i$ to the set $x \in \partial M$ and describe properties of these restrictions. We need this because eventually (in Section 3.4) we will have to construct linear combinations of oscillatory integrals which satisfy the boundary conditions (1.1.2).

The results of this section concern only the case when the billiard trajectories really experience reflections. Here and below we use notation introduced in 2.3.2 and in Section 2.4. Let us only remind that the type of billiard trajectories is assumed to be fixed, that $0 < t_1^*(y,\eta) < t_2^*(y,\eta) < \cdots < t_{\mathbf{r}}^*(y,\eta) < T_+$ are the moments of reflection, $\mathbf{r} > 0$, and that

$$\mathfrak{C}_i \cap \{\, (t,x;y,\eta) \;:\; x \in \partial M \,\} \;=\; \mathfrak{C}'_i \cup \mathfrak{C}'_{i+1}\,,$$

where the $\mathfrak{C}'_i$ are the sets defined by (2.4.12).

Recall that according to the reflection law the first $n-1$ components of ξ are preserved at the moment of reflection, whereas ξ_n experiences a jump. So

$$\xi^*(t_i^*(y,\eta)-0,y,\eta) \;=\; \left(\xi^{*\prime}(y,\eta),\xi_n^-(y,\eta)\right), \tag{2.5.1$^-$}$$

$$\xi^*(t_i^*(y,\eta)+0,y,\eta) \;=\; \left(\xi^{*\prime}(y,\eta),\xi_n^+(y,\eta)\right), \tag{2.5.1$^+$}$$

$i = 1,2,\dots,\mathbf{r}$, with some $\xi^{*\prime} \in C^\infty(O)$, $\xi_n^\mp \in C^\infty(O)$. Here we omit the dependence of the quantities $\xi^{*\prime}$, $\xi_n^\mp$ on the number of the reflection i in order to simplify notation. Note that the smooth dependence of the quantities $\xi^{*\prime}$, $\xi_n^\mp$ on (y,η) is a consequence of the facts that the type of trajectories is fixed and O is a T_+-admissible set; see 2.3.2. Let us also set for brevity $\xi^\mp := (\xi^{*\prime}(y,\eta),\xi_n^\mp(y,\eta))$.

In this subsection it will often be convenient for us to treat $\xi^{*\prime}$, $\xi_n^\mp$, $\xi^\mp$ as quantities defined on $\mathfrak{C}'_i$. In this case $\xi^{*\prime}$, $\xi_n^\mp$, $\xi^\mp$ at a point $(t,x';y,\eta) \in \mathfrak{C}'_i$ will be understood as $\xi^{*\prime}(y,\eta)$, $\xi_n^\mp(y,\eta)$, $\xi^\mp(y,\eta)$, respectively.

LEMMA 2.5.1. *Let φ be a phase function defined on some conic neighbourhood $\mathcal{O} \subset (T_-, T_+) \times M_x \times O$ of the set $\mathfrak{C}_i$, and satisfying the conditions* (2.4.5), (2.4.4) *(in particular, φ can be from the class $\mathfrak{F}_i$). Let $\varphi' = \varphi'(t,x';y,\eta) := \varphi(t,x',0;y,\eta)$ be the restriction of this phase function to the set $\{x \in \partial M\}$. Then*

$$\varphi' = 0\,, \quad {\varphi'}_{x'} = \xi^{*\prime}, \quad {\varphi'}_t = -h(y,\eta) \qquad \text{on } \mathfrak{C}'_i \cup \mathfrak{C}'_{i+1}\,, \tag{2.5.2}$$

and

$$\det \begin{pmatrix} {\varphi'}_{x'\eta} \\ {\varphi'}_{t\eta} \end{pmatrix} = -\,h_{\xi_n}(x',0,\xi^+)\,\det \varphi_{x\eta} \quad \text{on } \mathfrak{C}'_i\,, \tag{2.5.3$^+$}$$

$$\det \begin{pmatrix} {\varphi'}_{x'\eta} \\ {\varphi'}_{t\eta} \end{pmatrix} = -\,h_{\xi_n}(x',0,\xi^-)\,\det \varphi_{x\eta} \quad \text{on } \mathfrak{C}'_{i+1}\,, \tag{2.5.3$^-$}$$

where

$$\begin{pmatrix} {\varphi'}_{x'\eta} \\ {\varphi'}_{t\eta} \end{pmatrix} = \begin{pmatrix} {\varphi'}_{x_1\eta_1} & \cdots & {\varphi'}_{x_1\eta_n} \\ \vdots & \ddots & \vdots \\ {\varphi'}_{x_{n-1}\eta_1} & \cdots & {\varphi'}_{x_{n-1}\eta_n} \\ {\varphi'}_{t\eta_1} & \cdots & {\varphi'}_{t\eta_n} \end{pmatrix}.$$

PROOF. The equalities (2.5.2) immediately follow from (2.4.5), (2.4.4′) and (2.4.6), respectively. To prove ($2.5.3^{\pm}$) we differentiate (2.4.9) with respect to t and obtain

$$\varphi_{\eta t} = -(\Phi_{x\eta})^T h_\xi(x^*, \xi^*) + O(|x - x^*|),$$

where $(y, \eta) \to (x^*(t; y, \eta), \xi^*(t; y, \eta))$ is the shift along the ith leg of the Hamiltonian billiards. Hence it follows that for any point $(t, x; y, \eta) \in \mathfrak{C}_i$

(2.5.4) $$\begin{pmatrix} \varphi_{x'\eta} \\ \varphi_{t\eta} \end{pmatrix} = C \cdot \Phi_{x\eta}$$

with

(2.5.5) $$C = C(y, \eta) = \begin{pmatrix} 1 & 0 & \cdots & 0 & 0 \\ 0 & 1 & \cdots & 0 & 0 \\ \vdots & \vdots & \ddots & \vdots & \vdots \\ 0 & 0 & \cdots & 1 & 0 \\ -h_{\xi_1} & -h_{\xi_2} & \cdots & -h_{\xi_{n-1}} & -h_{\xi_n} \end{pmatrix},$$

$h_{\xi_k} = h_{\xi_k}(x^*(t; y, \eta), \xi^*(t; y, \eta))$. It remains to note that on $\mathfrak{C}'_i \cup \mathfrak{C}'_{i+1}$ we have $\varphi_{x'\eta} = {\varphi'}_{x'\eta}$, $\varphi_{t\eta} = {\varphi'}_{t\eta}$. So formulae (2.5.4), (2.5.5), ($2.5.1^+$) and ($2.5.1^-$) (with $i+1$ instead of i) imply ($2.5.3^{\pm}$). □

Now, let φ be a phase function from the class $\mathfrak{F}_i$ with domain of definition $\mathcal{O} \subset (T_-, T_+) \times M_x \times O$, and let φ' be its restriction to the set $\{x \in \partial M\}$ with domain of definition $\mathcal{O}' := \{\mathcal{O} \cap \{x \in \partial M\}\} \subset (T_-, T_+) \times \partial M_x \times O$. Let $\mathcal{O}$ be a sufficiently small neighbourhood of $\mathfrak{C}_i$, so that $\mathcal{O}' = \mathcal{O}'_i \cup \mathcal{O}'_{i+1}$ where $\mathcal{O}'_i \supset \mathfrak{C}'_i$ and $\mathcal{O}'_{i+1} \supset \mathfrak{C}'_{i+1}$ are disjoint open sets corresponding to t close to $t^*_i(y, \eta)$ and to $t^*_{i+1}(y, \eta)$, respectively. Let us consider the behaviour of the function φ' on the set $\mathcal{O}'_i$ ($i \geqslant 1$); the set $\mathcal{O}'_{i+1}$ can be dealt with analogously. Recall that we are dealing with T_+-admissible billiard trajectories, so

(2.5.6) $$h_{\xi_n}(x', 0, \xi^+) > 0 \quad \text{on } \mathfrak{C}'_i.$$

Formulae (2.5.2), ($2.5.3^+$), (2.5.6) yield

(2.5.7) $${\varphi'}_{x'} = \xi^{*\prime}, \quad {\varphi'}_t = -h(y, \eta), \quad \det \begin{pmatrix} {\varphi'}_{x'\eta} \\ {\varphi'}_{t\eta} \end{pmatrix} \neq 0 \qquad \text{on } \mathfrak{C}'_i.$$

From (2.4.3) and the definition of φ' as a restriction of the phase function $\varphi \in \mathfrak{F}_i$ to the set $\{x \in \partial M\}$ we immediately obtain

(2.5.8) $${\varphi'}_\eta(t, x'; y, \eta) = 0 \quad \text{if and only if} \quad (t, x'; y, \eta) \in \mathfrak{C}'_i.$$

Comparing formulae (2.5.8), (2.5.7) with (2.4.3), (2.4.4′) we see that φ' possesses properties similar to those of φ if we replace x_n by t. Thus, it is natural to introduce the following definition (cf. Definition 2.4.1).

DEFINITION 2.5.2. We say that $\varphi' = \varphi'(t, x'; y, \eta)$ is a nondegenerate *boundary* phase function associated with the ith reflection of our Hamiltonian billiards if φ' is defined on a conic neighbourhood $\mathcal{O}' \subset (T_-, T_+) \times \partial M_x \times O$ of the set $\mathfrak{C}'_i$ and satisfies the conditions (2.5.8), (2.5.7). By $\mathfrak{F}'_i$ we denote the set of all nondegenerate boundary phase functions associated with the ith reflection of our billiard trajectories, $1 \leqslant i \leqslant \mathbf{r}$.

Similarly to Remark 2.4.2 we have

REMARK 2.5.3. Condition (2.5.8) in Definition 2.5.2 can be replaced by the two weaker conditions:

$$\varphi' = 0 \quad \text{on } \mathfrak{C}'_i, \tag{2.5.9}$$

$$\varphi'_\eta \neq 0 \quad \text{on } \mathcal{O}' \setminus \mathfrak{C}'_i. \tag{2.5.10}$$

In this case we will automatically get $\varphi_\eta = 0$ on $\mathfrak{C}_i$: this fact is established by differentiating the identities $\varphi(t^*(y,\eta), x^*(t^*(y,\eta)); y, \eta) = 0$, $x^*_n(t^*(y,\eta)); y, \eta) = 0$ with respect to η_k, and taking into account the second equality (2.3.5).

Further on, to avoid confusion, we shall often call our original phase functions from the class $\mathfrak{F}_i$ nondegenerate *standard* phase functions associated with the ith leg of the canonical transformation (2.3.10). This will allow us to distinguish them from the nondegenerate boundary phase functions associated with the ith reflection (class $\mathfrak{F}'_i$), as well as from the nondegenerate boundary layer phase functions (class $\mathfrak{F}^{\mathrm{bl}}_i$) which will be introduced in 2.6.4.

Properties of boundary phase functions are similar to those of standard ones, so for $\varphi' \in \mathfrak{F}'_i$ we could give a list of results analogous to the results of Section 2.4. However, we shall refrain from doing this for the following reason: there is a simple relation between boundary phase functions and standard phase functions which makes a separate consideration of boundary phase functions unnecessary. Indeed, Lemma 2.5.1 and Definition 2.5.2 imply that the restriction of any phase function of the classes $\mathfrak{F}_{i-1}$ or $\mathfrak{F}_i$ to the set $\{x \in \partial M\}$ (with the additional condition that t is close to $t^*_i(y,\eta)$) is a phase function of the class $\mathfrak{F}'_i$. Conversely, we shall prove in the next section (Lemma 2.6.1) that any phase function of the class $\mathfrak{F}'_i$ can be extended up to phase functions of the classes $\mathfrak{F}_{i-1}$, $\mathfrak{F}_i$.

2.6. Extension of phase functions from the boundary

1. Standard extension.

LEMMA 2.6.1. *Let $\varphi' \in \mathfrak{F}'_i$ be a boundary phase function with domain of definition $\mathcal{O}' \subset (T_-, T_+) \times \partial M_x \times O$, $i \geqslant 1$. Then there exist standard phase functions $\varphi_{i-1} \in \mathfrak{F}_{i-1}$ and $\varphi_i \in \mathfrak{F}_i$ whose restrictions to the set $\{x \in \partial M\}$ coincide with φ' in some small conic neighbourhood of $\mathfrak{C}'_i$.*

PROOF. For the sake of definiteness we will find the phase function $\varphi_i \in \mathfrak{F}_i$; the phase function $\varphi_{i-1} \in \mathfrak{F}_{i-1}$ can be found in the same manner.

It is sufficient to define φ_i only for small x_n and for t close to $t^*_i(y,\eta)$ (moment of ith reflection); in view of Lemma 2.4.12, after that φ_i can be extended up to a phase function of the class $\mathfrak{F}_i$. So we must construct φ_i in such a way as to satisfy (2.4.23) with $\widetilde{\varphi} = \varphi_i$, $\widetilde{\mathfrak{C}}_i = \mathfrak{C}'_i$.

We shall construct φ_i in the form

$$\varphi_i(t,x;y,\eta) \;=\; \varphi'(t,x';y,\eta) \;+\; x_n\,\widetilde{\varphi}(t,x;y,\eta) \tag{2.6.1}$$

(thus, φ_i will automatically coincide with φ' for $x \in \partial M$). Then the first two conditions (2.4.23) for φ_i are equivalent to the equalities

$$\varphi_i(t,x^*;y,\eta) \;=\; \varphi'(t,x^{*\prime};y,\eta) \;+\; x_n^*\,\widetilde{\varphi}(t,x^*;y,\eta) \;=\; 0\,, \tag{2.6.2}$$

$$(\varphi_i)_{x'}(t,x^*;y,\eta) \;=\; \varphi'_{\,x'}(t,x^{*\prime};y,\eta) \;+\; x_n^*\,\widetilde{\varphi}_{x'}(t,x^*;y,\eta) \;=\; \xi^{*\prime}\,, \tag{2.6.3}$$

$$(\varphi_i)_{x_n}(t,x^*;y,\eta) \;=\; \widetilde{\varphi}(t,x^*;y,\eta) \;+\; x_n^*\,\widetilde{\varphi}_{x_n}(t,x^*;y,\eta) \;=\; \xi_n^* \tag{2.6.4}$$

with $(x^*,\xi^*) = \big(x^*(t;y,\eta),\xi^*(t;y,\eta)\big)$ (shift along the ith leg of the Hamiltonian billiards).

Let us note that x_n^* has a first order zero at the point $t=t_i^*$: as $t\to t_i^*+0$,

$$x_n^*(t;y,\eta) \;=\; (t-t_i^*)\,h_{\xi_n}(x^{*\prime},0,\xi^*)|_{t=t_i^*+0} \;+\; O((t-t_i^*)^2)\,. \tag{2.6.5}$$

Formulae (2.6.5), (2.5.6) imply that for any smooth function $f(t;y,\eta)$ which is equal to zero at $t=t_i^*$, the function $(x_n^*)^{-1}f$ is well defined and smooth. By (2.5.7),

$$\varphi'(t,x^{*\prime};y,\eta)|_{t=t_i^*} = 0\,, \quad \big(\varphi'_{\,x'}(t,x^{*\prime};y,\eta) \;-\; \xi^{*\prime}\big)\big|_{t=t_i^*} = 0\,. \tag{2.6.6}$$

Moreover, due to (2.5.7) and (2.3.17) we have

$$\begin{aligned}
\varphi'(t,x^{*\prime};y,\eta) \;&=\; (t-t_i^*)\big(\frac{d}{dt}\varphi'(t,x^{*\prime};y,\eta)\,\big)\big|_{t=t_i^*+0} \;+\; O((t-t_i^*)^2)\\
&=\; (t-t_i^*)\,\big(-h(y,\eta) \;+\; \langle\,\xi^{*\prime},h_{\xi'}(x^{*\prime},0,\xi^*)\,\rangle|_{t=t_i^*+0}\big) \;+\; O((t-t_i^*)^2)\\
&=\; -\,(t-t_i^*)\,\big(\xi_n^*\,h_{\xi_n}(x^{*\prime},0,\xi^*)\big)\big|_{t=t_i^*+0} \;+\; O((t-t_i^*)^2)\,,
\end{aligned}$$

and therefore, with account of (2.6.4),

$$\big((x_n^*)^{-1}\,\varphi'(t,x^{*\prime};y,\eta) \;+\; \xi_n^*\big)\big|_{t=t_i^*+0} = 0\,. \tag{2.6.7}$$

For $t>t_i^*$, let

$$\widetilde{\varphi}(t,x^*;y,\eta) \;=\; -\,(x_n^*)^{-1}\,\varphi'(t,x^{*\prime};y,\eta)\,, \tag{2.6.8}$$

$$\widetilde{\varphi}_{x'}(t,x^*;y,\eta) \;=\; (x_n^*)^{-1}\,\big(\xi^{*\prime} \;-\; \varphi'_{\,x'}(t,x^{*\prime};y,\eta)\big)\,, \tag{2.6.9}$$

$$\widetilde{\varphi}_{x_n}(t,x^*;y,\eta) \;=\; (x_n^*)^{-1}\,\big(\xi_n^* \;+\; (x_n^*)^{-1}\varphi'(t,x^{*\prime};y,\eta)\big)\,. \tag{2.6.10}$$

In view of (2.6.6), (2.6.7) the right-hand sides of (2.6.8)–(2.6.10) are well defined and smooth up to $t=t_i^*+0$.

Let us take some smooth function $\widetilde{\varphi}$ satisfying (2.6.8)–(2.6.10). Direct substitution shows that the equalities (2.6.8)–(2.6.10) imply (2.6.2)–(2.6.4), so the phase function (2.6.1) satisfies the first two conditions (2.4.23). The third condition (2.4.23) (nondegeneracy of φ_i) is an immediate consequence of formulae (2.5.3$^+$), (2.5.6) and of the third condition (2.5.7). □

2. Matching sets of phase functions.

DEFINITION 2.6.2. Consider a set of standard phase functions

$$\varphi_i \in \mathfrak{F}_i\,, \qquad i = 0, 1, 2, \ldots, \mathbf{r}, \tag{2.6.11}$$

with domains of definition $\mathcal{O}_i \subset (T_-\,, T_+) \times M_x \times O$, respectively. The set (2.6.11) is said to be *matching* if there exists a set of boundary phase functions $\varphi'_i \in \mathfrak{F}'_i$, $i = 1, 2, \ldots, \mathbf{r}$, with disjoint domains of definition $\mathcal{O}'_i \subset (T_-\,, T_+) \times \partial M_x \times O$, such that

$$\mathcal{O}_i \cap \{x \in \partial M\} = \mathcal{O}'_i \cup \mathcal{O}'_{i+1} \tag{2.6.12}$$

for $i = 0, 1, 2, \ldots, \mathbf{r}$ (here $\mathcal{O}'_0 = \mathcal{O}'_{\mathbf{r}+1} := \varnothing$), and

$$\varphi_{i-1}(t, x', 0; y, \eta) = \varphi_i(t, x', 0; y, \eta) = \varphi'_i(t, x'; y, \eta) \quad \text{on } \mathcal{O}'_i \tag{2.6.13}$$

for $i = 1, 2, \ldots, \mathbf{r}$.

In other words, formulae (2.6.12), (2.6.13) mean that that the phase functions from the set (2.6.11) are glued together at the boundary ∂M_x, in the sense that the restrictions to $\{x \in \partial M\}$ of consecutive phase functions from the set (2.6.11) coincide.

Note that equalities (2.5.3$^{\pm}$), (2.6.13) imply the following useful formula:

$$h_{\xi_n}(x', 0, \xi^{*\prime}, \xi_n^-) \det(\varphi_{i-1})_{x\eta} = h_{\xi_n}(x', 0, \xi^{*\prime}, \xi_n^+) \det(\varphi_i)_{x\eta} \quad \text{on} \quad \mathfrak{C}'_i, \tag{2.6.14}$$

$i = 1, 2, \ldots, \mathbf{r}$. Formula (2.6.14) holds for any matching set of phase functions.

The following lemma states that any standard phase function associated with a particular leg of our Hamiltonian billiards can be extended up to a matching set of phase functions.

LEMMA 2.6.3. *Given an integer* $0 \leqslant i_0 \leqslant \mathbf{r}$ *and a standard phase function* $\varphi_{i_0} \in \mathfrak{F}_{i_0}$ *with domain of definition* $\widetilde{\mathcal{O}}_{i_0}$, *there exists a matching set of phase functions* (2.6.11) *with domains of definition* $\mathcal{O}_i$ *such that* $\mathcal{O}_{i_0} \subset \widetilde{\mathcal{O}}_{i_0}$ *and the phase function* φ_{i_0} *from this set coincides with our original phase function* φ_{i_0} *on* $\mathcal{O}_{i_0}$.

PROOF. Applying to the phase function φ_{i_0} Lemma 2.5.1 we see that the restriction of φ_{i_0} to $\{x \in \partial M\}$ is a combination of two boundary phase functions $\varphi'_{i_0} \in \mathfrak{F}'_{i_0}$ and $\varphi'_{i_0+1} \in \mathfrak{F}'_{i_0+1}$. Applying Lemma 2.6.1 to φ'_{i_0} and φ'_{i_0+1} we extend these boundary phase functions up to matching phase functions $\varphi_{i_0-1} \in \mathfrak{F}_{i_0-1}$ and $\varphi_{i_0+1} \in \widetilde{\mathfrak{F}}_{i_0+1}$. This process of consecutive application of Lemmas 2.5.1 and 2.6.1 can be continued until we get full matching sets of $\mathbf{r}+1$ phase functions φ_i and $\mathbf{r}$ boundary phase functions φ'_i. It remains only to note that during the process it might be necessary to reduce repeatedly the initial domains of definition of our phase functions φ_i and φ'_i (because we want the domains of definition of boundary phase functions to be disjoint (see Definition 2.6.2), and because each application of Lemma 2.6.1 implies a possible reduction of the domains of definition). □

An obvious consequence of Corollary 2.4.8 and Lemma 2.6.3 is the fact that a set of matching phase functions (2.6.11) exists. Moreover, there are infinitely many such sets.

We conclude this subsection with a lemma stating that the set of all matching sets of phase functions is connected.

LEMMA 2.6.4. *Let* $\varphi_i^{(0)} \in \mathfrak{F}_i$, $i = 0, 1, 2, \ldots, \mathbf{r}$, *and* $\varphi_i^{(1)} \in \mathfrak{F}_i$, $i = 0, 1, 2, \ldots, \mathbf{r}$, *be two matching sets of phase functions with domains of definition* $\mathcal{O}_i^{(0)}$, $\mathcal{O}_i^{(1)}$, *respectively. Let* $\mathcal{O}_i = \mathcal{O}_i^{(0)} \cap \mathcal{O}_i^{(1)}$ *be the common domains of definition of* $\varphi_i^{(0)}$ *and* $\varphi_i^{(1)}$. *Then there exists a family of matching sets of phase functions* $\psi_i^{(s)} \in \mathfrak{F}_i$, $i = 0, 1, 2, \ldots, \mathbf{r}$, *smoothly depending on the parameter* $s \in [0, 1]$ *and with domains of definition* $\mathcal{O}_i$ *(for each fixed* s*), such that* $\psi_i^{(0)} \equiv \varphi_i^{(0)}$ *and* $\psi_i^{(1)} \equiv \varphi_i^{(1)}$ *on* $\mathcal{O}_i$.

THE PROOF of this lemma repeats that of Lemma 2.4.10. □

REMARK 2.6.5. Similarly to Remark 2.4.11, it is easy to show that the set of all matching sets of phase functions is contractible as a topological space.

3. Simplicity of phase functions. The following lemma states that all the phase functions from $\mathfrak{F}_i$ are simple in the sense of Definition 2.2.4. (As mentioned in Remark 2.2.5, this result is not valid for a general canonical transformation.)

LEMMA 2.6.6. *Any nondegenerate standard phase function associated with the canonical transformation* (2.3.10) *is simple.*

PROOF. Consider an arbitrary phase function $\varphi_{i_0} \in \mathfrak{F}_{i_0}$ associated with the i_0th leg of our Hamiltonian billiards, $0 \leqslant i_0 \leqslant \mathbf{r}$. Let us extend our given phase function φ_{i_0} up to a matching set of phase functions (2.6.11); this can always be done according to Lemma 2.6.3. Then the reasoning presented in 1.5.3 (and based on the formula (2.6.14)) shows that there exists a continuous branch of $\arg\big(\det^2(\varphi_{i_0})_{x\eta}\big)$ for $(t, x; y, \eta) \in \mathfrak{C}_{i_0}$. Obviously, this branch admits a continuous extension to a small neighbourhood $\mathcal{O}_{i_0} \subset (T_-, T_+) \times M_x \times O$ of $\mathfrak{C}_{i_0}$. □

It might seem unnatural that we have given the proof of Lemma 2.6.6 in 1.5.3, long before we obtained Lemma 2.6.3 and formula (2.6.14) which form the basis of this proof. We were forced to do this in order to give the second definition of the Maslov index (Definition 1.5.9) among the main results of the book.

It will be convenient for us to specify the choice of a particular continuous branch of $\arg\big(\det^2(\varphi_{i_0})_{x\eta}\big)$.

DEFINITION 2.6.7. Let $\varphi_{i_0} \in \mathfrak{F}_{i_0}$ be a standard phase function associated with the i_0th leg of our Hamiltonian billiards, $0 \leqslant i_0 \leqslant \mathbf{r}$. Then the choice of the continuous branch $\arg_0\big(\det^2(\varphi_{i_0})_{x\eta}\big)$ of the argument of $\det^2(\varphi_{i_0})_{x\eta}$ is uniquely specified by the following conditions: having taken a matching set of phase functions (2.6.11) containing the original phase function φ_{i_0} we require

$$\arg_0\big(\det^2(\varphi_0)_{x\eta}\big)\big|_{t=0,\, x=y} = 0\,,$$
$$\arg_0\big(\det^2(\varphi_i)_{x\eta}\big)\big|_{\mathfrak{C}'_i} = \arg_0\big(\det^2(\varphi_{i-1})_{x\eta}\big)\big|_{\mathfrak{C}'_i}\,, \qquad i = 1, 2, \ldots, \mathbf{r}\,.$$

The existence and uniqueness of the continuous branch of the argument specified by Definition 2.6.7 follow from the construction given in 1.5.3. Of course, $\arg_0\big(\det^2(\varphi_i)_{x\eta}\big)\big|_{\mathfrak{C}_i} = \arg_0 f(t; y, \eta)$, where $\arg_0 f(t; y, \eta)$ was defined in 1.5.3.

It is easy to check (using Lemma 2.6.4) that the choice of the continuous branch $\arg_0\big(\det^2(\varphi_{i_0})_{x\eta}\big)$ specified by Definition 2.6.7 does not depend on the particular matching set of phase functions.

Similarly to Definition 2.2.4 we shall call a nondegenerate boundary phase function $\varphi' \in C^\infty(\mathcal{O}_i')$ *simple* if there exists a continuous branch of the argument of $\det^2\begin{pmatrix}\varphi'_{x'\eta}\\ \varphi'_{t\eta}\end{pmatrix}$. Lemmas 2.6.1, 2.6.6 and formulae (2.5.3$^\pm$) imply that all the boundary phase functions $\varphi' \in \mathfrak{F}_i'$ are simple. It will be convenient for us to specify the choice of a particular continuous branch of $\arg\left(\det^2\begin{pmatrix}\varphi'_{x'\eta}\\ \varphi'_{t\eta}\end{pmatrix}\right)$.

DEFINITION 2.6.8. Let $\varphi' \in \mathfrak{F}_i'$ be a boundary phase function associated with the ith reflection of our Hamiltonian billiards, $1 \leqslant i \leqslant \mathbf{r}$. Then the choice of the continuous branch $\arg_0\left(\det^2\begin{pmatrix}\varphi'_{x'\eta}\\ \varphi'_{t\eta}\end{pmatrix}\right)$ of the argument of $\det^2\begin{pmatrix}\varphi'_{x'\eta}\\ \varphi'_{t\eta}\end{pmatrix}$ is uniquely specified by the following condition: having taken a standard phase function $\varphi_{i-1} \in \mathfrak{F}_{i-1}$ such that $\varphi_{i-1}|_{\partial M_x} = \varphi'$ we require

$$\arg_0\left(\det^2\begin{pmatrix}\varphi'_{x'\eta}\\ \varphi'_{t\eta}\end{pmatrix}\right)\Bigg|_{\mathfrak{C}_i'} = \arg_0\big(\det^2(\varphi_{i-1})_{x\eta}\big)\big|_{\mathfrak{C}_i'}\,.$$

4. Boundary layer phase functions. Let us now have a more careful look at the procedure of extension of a boundary phase function at the ith reflection which was described in subsection 1. The crucial step which determines the type of extension is the choice of a particular real ξ_n-root of the equation

$$h(x^{*\prime}(t_i^*(y,\eta);y,\eta),0,\xi^{*\prime}(t_i^*(y,\eta);y,\eta),\xi_n) = h(y,\eta)\,. \tag{2.6.15}$$

Having chosen this root $\xi_n = \xi_n(y,\eta)$ we then use

$$\big(x^{*\prime}(t_i^*(y,\eta);y,\eta),0,\ \xi^{*\prime}(t_i^*(y,\eta);y,\eta),\xi_n(y,\eta)\big)$$

as a starting point for the reflected Hamiltonian trajectory; after that we build upon the reflected trajectories the "reflected" phase functions $\varphi \in \mathfrak{F}_i$ in accordance with Definition 2.4.1 and the matching condition at the boundary. We did not previously deliberate on the choice of the ξ_n-root of the equation (2.6.15) because we originally considered only *real* ξ_n-roots, and the choice of these real ξ_n-roots was declared fixed throughout this chapter (see 2.3.2).

But (2.6.15) is equivalent to the algebraic equation

$$A_{2m}(x^{*\prime}(t_i^*(y,\eta);y,\eta),0,\xi^{*\prime}(t_i^*(y,\eta);y,\eta),\xi_n) = A_{2m}(y,\eta). \tag{2.6.16}$$

The equation (2.6.16) has exactly $2m$ ξ_n-roots some of which are real and some of which may be *complex*, and there is no mathematical reason to disregard the complex ones. The boundary layer phase functions which we are about to introduce are the phase functions associated with the complex ξ_n-roots of (2.6.16).

If all the ξ_n-roots of (2.6.16) are real, then there will be no boundary layer phase functions. So in this subsection we shall be concerned with the case when the equation (2.6.16) has complex ξ_n-roots. Let us choose one of the roots with positive imaginary part and denote it as $\xi_n = \xi_n^+(y,\eta)$, $\operatorname{Im}\xi_n^+ > 0$. We need to show that $\xi_n^+(y,\eta)$ is a well-defined smooth "function" on O. In view of the simplicity of all the ξ_n-roots of the equation (2.6.16) (see Definition 1.3.25) the "function" $\xi_n^+(y,\eta)$ is well defined and smooth locally, but in principle it might happen that the movement along a closed curve in O turns one complex ξ_n-root into another. However, such a situation cannot occur because we have imposed the condition of simple connectedness of O (see 2.3.2). Note that in the case $m \leqslant 2$ the latter condition is not needed because the equation (2.6.16) has at most one ξ_n-root in the upper complex half-plane.

In this subsection the choice of the sequential number of the reflection i, as well as the choice of the complex ξ_n-root of the equation (2.6.16) will be assumed to be fixed. We shall denote for brevity

$$\mathbf{x}^* = (\mathbf{x}^{*\prime}, 0) := (x^{*\prime}(t_i^*(y,\eta); y,\eta), 0) \in \partial M_x, \quad \xi^+ := (\xi^{*\prime}(t_i^*(y,\eta); y,\eta), \xi_n^+(y,\eta)).$$

Similarly to Section 2.5 we shall often consider ξ^+ as a quantity defined on $\mathfrak{C}_i'$. This is possible because there is a natural bijection between the sets O and $\mathfrak{C}_i'$.

We are now prepared to define the notion of a complex reflected trajectory.

For the sake of simplicity let us assume first that we are dealing with the analytic case. This means, in particular, that we allow only analytic changes of local coordinates, and our Hamiltonian is extended to an analytic function

$$h(x,\xi) := (A_{2m}(x,\xi))^{1/(2m)} \tag{2.6.17}$$

of complex variables (x,ξ), where the choice of branch is specified by the condition

$$h(x,\xi^+) > 0 \quad \text{on } \mathfrak{C}_i'. \tag{2.6.18}$$

Obviously this function h is analytic in (x,ξ) for complex x sufficiently close to $\mathbf{x}^*(y,\eta)$ and complex ξ sufficiently close to $\xi^+(y,\eta)$. Let us stress once again that now $\xi^+(y,\eta)$ is complex, and not real as when we dealt with standard extensions of phase functions. A complex reflected trajectory $(x^*(t;y,\eta), \xi^*(t;y,\eta))$ is defined as the solution of the system of equations

$$\dot{x}^* = h_\xi(x^*,\xi^*), \qquad \dot{\xi}^* = -h_x(x^*,\xi^*) \tag{2.6.19}$$

with initial condition $(x^*,\xi^*) = (\mathbf{x}^*,\xi^+)$ at $t = t_i^*$. Obviously this complex trajectory is well defined and analytic in t for t sufficiently close to t_i^*.

Let us now drop the condition of analyticity. The difficulty is that now we cannot define the complex reflected trajectories as solutions of (2.6.19) because the Hamiltonian $h(x,\xi)$ is not defined for complex x.

In principle this difficulty can be overcome by using the so-called *almost analytic* extension to complex x. This means that one introduces a smooth function $\widetilde{h}$ of $\operatorname{Re} x$, $\operatorname{Im} x$, ξ such that for any real $z \in M$ and any $l \in \mathbb{N}$

$$\widetilde{h}(x,\xi) = \sum_{|\alpha| \leqslant l} \frac{1}{\alpha!} \left(\partial_x^\alpha h\right)\big|_{x=z} (x-z)^\alpha + O\big(|x-z|^{l+1}\big) \quad \text{as } x \to z.$$

However, in such a construction it is hard to assign an invariant meaning to complex x, that is, to identify the complex quantity x with a point from some manifold. So we develop below a slightly different approach.

We shall view the Hamiltonian $h(x,\xi)$ as a formal Taylor expansion in powers of $x - \mathbf{x}^*$; each term of this expansion is, of course, analytic in x and in ξ. We define the complex reflected trajectory as a formal Taylor expansion in powers of $t - t_i^*$ which formally satisfies (2.6.19) and the necessary initial condition. In this case in place of (2.6.19) we shall write

$$\dot{x}^* \simeq h_\xi(x^*,\xi^*), \qquad \dot{\xi}^* \simeq -h_x(x^*,\xi^*), \tag{2.6.20}$$

where the sign $\simeq$ stands for the equality of formal Taylor expansions. Note that our complex Taylor expansions behave under changes of local coordinates x in the same way as real Taylor expansions, so one can assign to them an invariant meaning. Note also that from Euler's identity it follows that the vector $\dot{x}^*|_{t=t_i^*}$ is nonreal, which means that the rays x^* are indeed complex (they leave the real manifold M).

Having defined the notion of a complex reflected trajectory we can now define the notion of a boundary layer phase function. By analogy with Definition 2.4.1 and Remark 2.4.2 we introduce

DEFINITION 2.6.9. We say that the nondegenerate phase function φ is a *boundary layer* phase function associated with the ith reflection of our Hamiltonian billiards if φ is defined on a conic neighbourhood $\mathcal{O} \subset (T_-, T_+) \times M_x \times O$ of the set $\mathfrak{C}_i'$ and satisfies the following three conditions:

$$\varphi_\eta(t,x;y,\eta) = 0 \quad \text{if and only if} \quad (t,x;y,\eta) \in \mathfrak{C}_i', \tag{2.6.21}$$

$$\varphi(t,x^*;y,\eta) \simeq 0, \tag{2.6.22}$$

$$\varphi_x(t,x^*;y,\eta) \simeq \xi^*, \tag{2.6.23}$$

where $(x^*,\xi^*) = (x^*(t;y,\eta),\xi^*(t;y,\eta))$ is a complex reflected trajectory corresponding to the ith reflection, $\operatorname{Im}\xi_n^*|_{t=t_i^*} = \xi_n^+ > 0$. By $\mathfrak{F}_i^{\mathrm{bl}}$ we denote the set of all nondegenerate boundary layer phase functions associated with the ith reflection of our Hamiltonian billiards, $1 \leqslant i \leqslant \mathbf{r}$, and with the particular choice of the complex root ξ_n^+.

REMARK 2.6.10. The condition (2.6.21) is a consequence of conditions (2.6.22), (2.6.23) and $\det \varphi_{x\eta}|_{\mathfrak{C}_i'} \neq 0$ (nondegeneracy) if the neighbourhood $\mathcal{O}$ is chosen to be sufficiently small.

Of course, boundary layer phase functions defined above are a special case of phase functions introduced in Section 2.2.

Let us list the basic properties of boundary layer phase functions. We will do this briefly because the properties of boundary layer phase functions (as well as their proofs) mostly repeat those of standard phase functions.

Similarly to formula (2.4.28) and Theorem 2.4.16 we have

THEOREM 2.6.11. *For any phase function* $\varphi \in \mathfrak{F}_i^{\mathrm{bl}}$

$$\mathfrak{e}|_{x=x^*} \simeq 0, \quad \mathfrak{e}_t|_{x=x^*} \simeq 0, \quad \mathfrak{e}_x|_{x=x^*} \simeq 0, \quad \mathfrak{e}_y|_{x=x^*} \simeq 0, \quad \mathfrak{e}_\eta|_{x=x^*} \simeq 0,$$

where $\mathfrak{e}$ *is defined by* (2.4.27).

Similarly to Lemma 2.5.1 we have

LEMMA 2.6.12. *Let* φ *be a phase function defined on some conic neighbourhood* $\mathcal{O} \subset (T_-, T_+) \times M_x \times O$ *of the set* $\mathfrak{C}_i'$, $1 \leqslant i \leqslant \mathbf{r}$, *and satisfying the conditions* (2.6.22), (2.6.23) (*in particular,* φ *can be from the class* $\mathfrak{F}_i^{\mathrm{bl}}$). *Let* $\varphi' = \varphi'(t,x';y,\eta) := \varphi(t,x',0;y,\eta)$ *be the restriction of this phase function to the set* $\{x \in \partial M\}$. *Then*

$$\varphi' = 0, \quad \varphi'_{x'} = \xi^{*\prime}, \quad \varphi'_t = -h(y,\eta) \qquad \text{on } \mathfrak{C}_i', \tag{2.6.24}$$

and

$$\det \begin{pmatrix} {\varphi'}_{x'\eta} \\ {\varphi'}_{t\eta} \end{pmatrix} = -\, h_{\xi_n}(x',0,\xi^+) \det \varphi_{x\eta} \quad \text{on } \mathfrak{C}'_i \,. \tag{2.6.25}$$

An immediate consequence of Lemma 2.6.12 and Definition 2.5.2 is that for any phase function $\varphi \in \mathfrak{F}_i^{\mathrm{bl}}$ we have

$$\varphi|_{x\in\partial M} \in \mathfrak{F}'_i \,. \tag{2.6.26}$$

Recall (see Section 2.5) that by $\mathfrak{F}'_i$ we denote the class of nondegenerate *boundary* phase functions (not to be confused with *boundary layer* phase functions!).

Similarly to Lemma 2.6.1 we have

LEMMA 2.6.13. *Let $\varphi' \in \mathfrak{F}'_i$ be a boundary phase function with domain of definition $\mathcal{O}' \subset (T_-\,, T_+) \times \partial M_x \times O$, $1 \leqslant i \leqslant \mathbf{r}$. Then there exists a boundary layer phase function $\varphi \in \mathfrak{F}_i^{\mathrm{bl}}$ whose restriction to the set $\{x \in \partial M\}$ coincides with φ' in some small conic neighbourhood of $\mathfrak{C}'_i$.*

PROOF. The construction of the required phase function φ repeats that from the proof of Lemma 2.6.1. It remains only to show that (2.6.21) holds in a sufficiently small neighbourhood of $\mathfrak{C}'_i$. Formulae

$$\operatorname{Im} \varphi|_{x\in\partial M} = \operatorname{Im}\varphi' \geqslant 0\,, \qquad \operatorname{Im} \varphi_{x_n}|_{\mathfrak{C}'_i} = \operatorname{Im}\xi_n^+ > 0$$

imply that in a sufficiently small neighbourhood of $\mathfrak{C}'_i$ we have $\operatorname{Im}\varphi \geqslant \mathrm{const}\, x_n$, $\mathrm{const} > 0$. But $\{\varphi_\eta = 0\} \subset \{\varphi = 0\}$, so we should search for solutions of the equation $\varphi_\eta = 0$ only for $x \in \partial M$. In this case $\varphi = \varphi'$, so (2.6.21) follows from the condition (2.5.8) of Definition 2.5.2. □

Now we can state (similarly to Corollary 2.4.8) that the class $\mathfrak{F}_i^{\mathrm{bl}}$ is nonempty. Indeed, we already know that the class $\mathfrak{F}'_i$ is nonempty (see the last paragraph in Section 2.5), and in view of Lemma 2.6.13 any phase function from the class $\mathfrak{F}'_i$ can be extended up to a phase function from the class $\mathfrak{F}_i^{\mathrm{bl}}$.

Similarly to Lemma 2.4.10 and Remark 2.4.11 it can be shown that the class $\mathfrak{F}_i^{\mathrm{bl}}$ is connected and, moreover, contractible (this follows from the connectedness and contractibility of the class $\mathfrak{F}'_i$).

DEFINITION 2.6.14 (extension of Definition 2.6.2). Let $0 < \mathbf{i} \leqslant \mathbf{r}$. Consider a set of phase functions

$$\varphi_i \in \mathfrak{F}_i\,, \quad i = 0,1,2,\dots,\mathbf{i}-1\,, \qquad \varphi_{\mathbf{i}} \in \mathfrak{F}_{\mathbf{i}}^{\mathrm{bl}}\,, \tag{2.6.27}$$

with domains of definition $\mathcal{O}_i \subset (T_-\,, T_+) \times M_x \times O$, $i = 0,1,2,\dots,\mathbf{i}$, respectively. The set (2.6.27) is said to be *matching* if there exists a set of boundary phase functions $\varphi'_i \in \mathfrak{F}'_i$, $i = 1,2,\dots,\mathbf{i}$, with disjoint domains of definition $\mathcal{O}'_i \subset (T_-\,, T_+) \times \partial M_x \times O$, such that

$$\mathcal{O}_i \cap \{x \in \partial M\} = \mathcal{O}'_i \cup \mathcal{O}'_{i+1}$$

for $i = 0,1,2,\dots,\mathbf{i}$ (here $\mathcal{O}'_0 = \mathcal{O}'_{\mathbf{i}+1} := \varnothing$), and

$$\varphi_{i-1}(t,x',0;y,\eta) = \varphi_i(t,x',0;y,\eta) = \varphi'_i(t,x';y,\eta) \quad \text{on} \quad \mathcal{O}'_i$$

for $i = 1,2,\dots,\mathbf{i}$.

Similarly to Lemma 2.6.3 it can be shown that a matching set (2.6.27) can be constructed starting from any phase function $\varphi_{i_0} \in \mathfrak{F}_{i_0}$, $0 \leqslant i_0 \leqslant \mathbf{i}-1$, or $\varphi_{\mathbf{i}} \in \mathfrak{F}_{\mathbf{i}}^{\mathrm{bl}}$.

Similarly to Lemma 2.6.4 and Remark 2.6.5 it can be shown that the set of all matching sets of the type (2.6.27) is connected and, moreover, contractible as a topological space.

Let us show now that any nondegenerate boundary layer phase function $\varphi \in \mathfrak{F}_i^{\mathrm{bl}}$ associated with the canonical transformation (2.3.10) is simple. Unfortunately, we cannot do this by repeating directly the reasoning from the proof of Lemma 2.6.6 because it relied on the fact that $h_{\xi_n}(x^*|_{t=t_i^*}, \xi^*|_{t=t_i^*})$ is real (see formula (2.6.14)), which is not necessarily true when we are dealing with a complex reflected trajectory. We overcome this difficulty by choosing the set O to be simply connected; see 2.3.3.

Note that for $m \leqslant 2$ simple connectedness is not needed. Indeed, let φ' be the restriction of the boundary layer phase function φ to the set $\{x \in \partial M\}$. In view of formulae (2.6.25), (2.6.26) and the fact that boundary phase functions are simple (see the end of subsection 3), the simplicity of the boundary layer phase function φ is equivalent to the existence of a continuous branch of

$$\arg\left(\left(h_{\xi_n}(x^*|_{t=t_i^*}, \xi^*|_{t=t_i^*})\right)^2\right), \tag{2.6.28}$$

where (x^*, ξ^*) is the complex reflected trajectory. Note that our choice of the coordinate x_n is fixed (see 1.1.2), so the expression $h_{\xi_n}(x^*|_{t=t_i^*}, \xi^*|_{t=t_i^*})$ does not depend on the choice of coordinates x'. The existence of a continuous branch of (2.6.28) is equivalent to the existence of a continuous branch of

$$\arg\left(h_{\xi_n}(x^*|_{t=t_i^*}, \xi^*|_{t=t_i^*})\right). \tag{2.6.29}$$

As $A_{2m}(x,\xi) = (h(x,\xi))^{2m}$ and $h(x^*|_{t=t_i^*}, \xi^*|_{t=t_i^*}) > 0$, the existence of a continuous branch of (2.6.29) is equivalent to the existence of a continuous branch of

$$\arg\left(A'_{2m}(x^*|_{t=t_i^*}, \xi^*|_{t=t_i^*})\right) \tag{2.6.30}$$

(here the prime stands for differentiation with respect to ξ_n). When $m \leqslant 2$ the latter can be handled by an elementary factorization of the polynomial A'_{2m} in terms of the ξ_n-roots of the algebraic equation (2.6.16).

REMARK 2.6.15. In this subsection when dealing with the case $m \geqslant 3$ we have used the simple connectedness of the set $O \subset T'\overset{\circ}{M}$. We have used this condition on two separate occasions, and for two different reasons. The first time we needed it to guarantee that when we move along a closed curve in O, one complex ξ_n-root of (2.6.16) cannot turn *into* another. The second time we needed it to guarantee that when we move along a closed curve in O, pairs of different complex ξ_n-roots of (2.6.16) cannot make full rotations *around* one another.

It will be convenient for us to specify in an arbitrary fixed way the choice of a particular continuous branch of (2.6.29).

DEFINITION 2.6.16. By $\arg_0\Big(h_{\xi_n}\big(x^*|_{t=t_i^*}\,,\,\xi^*|_{t=t_i^*}\big)\Big)$ we shall denote a particular continuous branch of the argument (2.6.29).

Similarly to Definitions 2.6.7, 2.6.8 we introduce

DEFINITION 2.6.17. Let $\varphi \in \mathfrak{F}_i^{\mathrm{bl}}$ be a boundary layer phase function associated with the ith reflection of our Hamiltonian billiards, $1 \leqslant i \leqslant \mathbf{r}$. Then the choice of the continuous branch $\arg_0(\det^2\varphi_{x\eta})$ of the argument of $\det^2\varphi_{x\eta}$ is uniquely specified by the following condition: having taken a standard phase function $\varphi_{i-1} \in \mathfrak{F}_{i-1}$ such that $\varphi_{i-1}|_{\partial M_x} = \varphi|_{\partial M_x}$ we require

$$\arg_0(\det^2\varphi_{x\eta})\big|_{\mathfrak{C}_i'} = \arg_0(\det^2(\varphi_{i-1})_{x\eta})\big|_{\mathfrak{C}_i'} - 2\arg_0\Big(h_{\xi_n}\big(x^*|_{t=t_i^*}\,,\,\xi^*|_{t=t_i^*}\big)\Big)\,.$$

The existence of the continuous branch $\arg_0(\det^2\varphi_{x\eta})$ specified by Definition 2.6.17 follows from formulae (2.5.3$^-$) (with i replaced by $i-1$) and (2.6.25).

Finally, let us remark once again on the necessity of introducing boundary layer phase functions. Apart from abstract mathematical reasons, boundary layer phase functions are necessary for the effective construction of the Schwartz kernel of the wave group; see Chapter 3. For a second order operator this does not require the use of boundary layer phase functions (and, in fact, there are no boundary layer phase functions in this case), but in the general case boundary layer phase functions are necessary to satisfy the m boundary conditions (1.1.2). At each reflection we need a total of exactly m different phase functions (standard or boundary layer) to satisfy these boundary conditions.

2.7. Standard oscillatory integrals associated with homogeneous canonical transformations

Oscillatory integrals described in this section are a special (but most important for us) case of those introduced in Section 2.2. The only difference with Section 2.2 is that now we use a special type of phase functions introduced in Section 2.4, namely, nondegenerate standard phase functions associated with the ith leg of the canonical transformation (2.3.10) (class $\mathfrak{F}_i$). The use of special properties of phase functions from the class $\mathfrak{F}_i$ will enable us to study the corresponding oscillatory integrals in greater detail.

1. Definition of a standard oscillatory integral. As before, in this section we deal only with the transformation (2.3.10) though most of the results are valid for a general transformation (2.3.1) as well (see Remark 2.4.3). Throughout this section we will be using notation introduced in Sections 2.1–2.6.

DEFINITION 2.7.1. We say that a global oscillatory integral

$$\mathcal{I}_{\varphi,a}(t,x,y) = \int e^{i\varphi(t,x;y,\eta)}\, a(t,x;y,\eta)\, \varsigma(t,x;y,\eta)\, d_\varphi(t,x;y,\eta)\, đ\eta \tag{2.7.1}$$

with phase function φ and amplitude $a \in S^l$ is a *standard oscillatory integral* associated with the ith leg of the canonical transformation (2.3.10) if

1. φ and a are defined in a conic neighbourhood $\mathcal{O} \subset (T_-\,, T_+) \times M_x \times O$ of the set $\mathfrak{C}_i$.
2. φ is a nondegenerate standard phase function associated with the ith leg of this transformation, i.e., $\varphi \in \mathfrak{F}_i$.

3. $\det \varphi_{x\eta} \neq 0$ on $\mathcal{O}$.
4. $\operatorname{supp} a \subset (T_-, T_+) \times M_x \times \mathbf{O}$, where $\mathbf{O}$ is some conically compact conic subset of O.

In formula (2.7.1)

$$(2.7.2) \qquad d_\varphi(t,x;y,\eta) = (\det{}^2 \varphi_{x\eta})^{1/4} = |\det \varphi_{x\eta}|^{1/2} \, e^{i \arg_0(\det^2 \varphi_{x\eta})/4},$$

where the continuous branch $\arg_0(\det^2 \varphi_{x\eta})$ is chosen in accordance with Definition 2.6.7. The function ς in formula (2.7.1) is a cut-off around the set $\mathfrak{C}_i$ (see Section 2.2).

Let us make some comments concerning the above definition.

1. In view of Lemma 2.6.6 the phase function φ appearing in Definition 2.7.1 is simple.

2. Recall (see Section 2.2) that condition (3) of Definition 2.7.1 can always be fulfilled by choosing the neighbourhood $\mathcal{O}$ to be small enough.

3. In Definition 2.7.1 and further on, whenever we say that a conic set $\mathbf{O} \subset T'M_y$ is conically compact, we mean that the set $\mathbf{O} \cap \{h(y,\eta) = 1\}$ is compact. Obviously, condition (4) of Definition 2.7.1 is equivalent to the inclusion $\mathbf{O}_a \subset O$, where

$$\mathbf{O}_a = \overline{\operatorname{cone hull}\{(y,\eta) \in O : \ (t,x;y,\eta) \in \operatorname{supp} a \text{ for some } (t,x;y,\eta) \in \mathcal{O}\}}$$

is the closure in $T'M_y$ of the conic hull of the projection of $\operatorname{supp} a$ on O.

4. As $\mathfrak{C}_i$ is closed in $(T_-, T_+) \times M_x \times O$, condition (4) of Definition 2.7.1 guarantees that the set $\mathfrak{C}_i \cap \operatorname{supp} a$ is conically separated from the set $\partial\mathcal{O}$; here

$$\mathfrak{C}_i = \{(t,x;y,\eta) \in \mathcal{O} : \varphi_\eta = 0\} = C_\varphi,$$

and $\partial\mathcal{O}$ is the boundary of $\mathcal{O}$ in $(T_-, T_+) \times M_x \times T'M_y$. Therefore we can choose the cut-off ς as was explained in Section 2.2. Then the oscillatory integral (2.7.1) is well defined,

$$(2.7.3) \quad \mathcal{I}_{\varphi,a} \in C^\infty((T_-, T_+) \times M_y; \mathcal{D}'(\mathring{M}_x)) \cap C^\infty(M_x \times M_y; \mathcal{D}'(T_-, T_+)),$$

and $\mathcal{I}_{\varphi,a}$ is a half-density in x and in y. The projection of $\mathfrak{C}_i$ on $(T_-, T_+) \times M_x \times M_y$ is

$$\{(t,x;y) : (y,\eta) \in O, \ t \in \mathfrak{T}_i, \ x = x^*(t;y,\eta) \text{ for some } \eta \in T'_yM\},$$

so (see 2.1.1) $\mathcal{I}_{\varphi,a}$ is smooth outside the intersection of this set with

$$\{(t,x;y) : (t,x;y,\eta) \in \operatorname{cone supp} a \text{ for some } \eta \in T'_yM\}.$$

DEFINITION 2.7.2. A distribution $\mathcal{I}(t,x,y)$ from the class (2.7.3) which can be written modulo $C^\infty((T_-, T_+) \times M_x \times M_y)$ as a standard oscillatory integral (2.7.1) is called a *standard Lagrangian distribution* of order l associated with the ith leg of the canonical transformation (2.3.10).

Below we mostly omit the word "standard". This should not cause misunderstanding because this section is totally devoted to standard oscillatory integrals.

A natural question is whether for a given Lagrangian distribution $\mathcal{I}$ its amplitude and phase function are uniquely defined. Certainly, in this question, as everywhere in this book, we consider $\mathcal{I}$ to be given modulo C^∞ (i.e., up to the addition of a half-density from $C^\infty((T_-, T_+) \times M_x \times M_y)$) and a to be given modulo $S^{-\infty}$ (i.e., up to the addition of a function from $S^{-\infty}$). The answer to this question turns out to be negative. Moreover, a Lagrangian distribution $\mathcal{I}$ can be represented by an oscillatory integral with an arbitrary phase function φ associated with the canonical transformation (see subsection 4), and for each φ there exists an infinite number of amplitudes a such that $\mathcal{I} = \mathcal{I}_{\varphi,a}$ modulo C^∞.

Nevertheless, some uniqueness properties are attained if we restrict ourselves to the use of amplitudes independent of x. In subsection 2 for a given oscillatory integral $\mathcal{I}_{\varphi,a}$ with fixed phase function we reduce the amplitude a to a certain special amplitude independent of x. The restriction of this special amplitude to $\mathfrak{D}_i$ is determined uniquely, modulo $S^{-\infty}$, by the Lagrangian distribution $\mathcal{I}$ and by the phase function φ (subsection 3). Moreover, its leading term is independent of φ (subsection 4).

In subsection 5 we consider pseudodifferential operators as a special case of our oscillatory integrals, and in subsection 6 we describe the action of a pseudodifferential operator on an oscillatory integral.

2. Reduction of the amplitude. Further on in this section we denote by $\widehat{\mathfrak{D}}_i$ the projection of $\mathcal{O}$ on $(T_-, T_+) \times O$, that is

$$\widehat{\mathfrak{D}}_i = \{(t; y, \eta) \ : \ (t, x; y, \eta) \in \mathcal{O} \ \text{ for some } \ x \in M_x\} . \tag{2.7.4}$$

Obviously, this set $\widehat{\mathfrak{D}}_i \subset (T_-, T_+) \times O$ is a conic neighbourhood of the set $\mathfrak{D}_i$; see (2.4.1). The introduction of the set $\widehat{\mathfrak{D}}_i$ is convenient for the following reason: whenever we deal with a function of the variables $(t; y, \eta) \in \widehat{\mathfrak{D}}_i$ we can consider it as a function of the variables $(t, x; y, \eta) \in \mathcal{O}$, which possesses, of course, the special property of being independent of x.

LEMMA 2.7.3. *Let $a(t, x; y, \eta) \in C^\infty(\mathcal{O})$ be a function positively homogeneous in η of degree l, satisfying condition (4) of Definition 2.7.1, and let $\varphi \in \mathfrak{F}_i$. Then there exist a function $a^*(t; y, \eta) \in C^\infty(\widehat{\mathfrak{D}}_i)$ positively homogeneous in η of degree l and a covector field*

$$\vec{g}\,(t, x; y, \eta) = \{g_1(t, x; y, \eta), \dots, g_n(t, x; y, \eta)\} \in C^\infty(\mathcal{O})$$

positively homogeneous in η of degree l such that

$$\operatorname{supp} a^* \subset (T_-, T_+) \times \mathbf{O}, \qquad \operatorname{supp} \vec{g} \subset (T_-, T_+) \times M_x \times \mathbf{O}, \tag{2.7.5}$$

and

$$a(t, x; y, \eta) = a^*(t; y, \eta) + \langle \varphi_\eta(t, x; y, \eta), \vec{g}\,(t, x; y, \eta)\rangle . \tag{2.7.6}$$

Moreover, for a given function a^* (satisfying $a^*|_{\mathfrak{D}_i} = a|_{\mathfrak{C}_i}$ and the first condition (2.7.5)) the covector field $\vec{g}$ can be constructed effectively in the form (E.7), (E.14), (E.19), (E.21)–(E.24) *with*

$$z = x - x^*, \qquad f = \varphi_\eta, \qquad u = (t; y, \eta), \qquad a|_{z=0} = a^*, \tag{2.7.7}$$

an arbitrary positive Hermitian covariant tensor $B = B(y) \in C^\infty(M_y)$, and some cut-off functions $\chi_k = \chi_k(t, x; y, \eta) \in C^\infty(\mathcal{O})$ positively homogeneous in η of degree 0 which are identically equal to 1 in some neighbourhoods of the set $\mathfrak{C}_i$.

The crucial point of Lemma 2.7.3 is the fact that the function a^* is independent of x. Note also that in the formulation of this lemma the words "covector" and "covariant tensor" are used in relation to the coordinates y.

PROOF OF LEMMA 2.7.3. Consider the function

$$a^*(t;y,\eta) := a(t, x^*(t;y,\eta); y, \eta) \tag{2.7.8}$$

which is a smooth function of the variables $(t;y,\eta)$ with initial domain of definition $\mathfrak{D}_i$. Let us smoothly extend the function $a^*(t;y,\eta)$ to $\widehat{\mathfrak{D}}_i$ preserving homogeneity (see (2.7.4)). Let us choose the extension in such a way that we do not increase the projection of the support of the function on O. For the sake of simplicity we shall retain for the extended function the old notation $a^*(t;y,\eta)$. The function $a^*(t;y,\eta) \in C^\infty(\widehat{\mathfrak{D}}_i)$ constructed in this way satisfies the first condition (2.7.5). Of course in the case $\partial M = \varnothing$ the above statements concerning extensions are unnecessary.

With our chosen function a^* the equality (2.7.6) is automatically satisfied for $(t,x;y,\eta) \in \mathfrak{C}_i$. So it remains to choose $\vec{g}$ in such a way as to satisfy (2.7.6) for $(t,x;y,\eta) \in \mathcal{O} \setminus \mathfrak{C}_i$.

The fact that $\vec{g}$ constructed in accordance with (E.7), (E.14), (E.19), (E.21)–(E.24), (2.7.7) satisfies (2.7.6) on the whole set $\mathcal{O}$ follows immediately from the results of Appendix E; see E.3; namely, formula (2.7.6) is a special case of formula (E.26). Finally, the second condition (2.7.5) holds because in our explicit construction $\vec{g}$ is a combination of derivatives of a and a^*. □

The expression for $\vec{g}$ given by (E.7), (E.14), (E.19), (E.21)–(E.24), (2.7.7) possesses important invariance properties: it behaves as a function (i.e., does not change) under changes of local coordinates x, and as a covector under changes of local coordinates y. These invariance properties follow from the results of E.5. In Lemma 2.7.3 we placed an arrow over g to stress the fact that it is a covector field. Obviously, φ_η behaves as a function under changes of local coordinates x, and as a vector under changes of local coordinates y. Consequently the expression $\langle \varphi_\eta, \vec{g} \rangle$ appearing in (2.7.6) is independent of the choice of local coordinates x and y.

COROLLARY 2.7.4. *Suppose that the amplitude $a(t,x;y,\eta)$ in the oscillatory integral (2.7.1) is positively homogeneous in η of degree l. Then the oscillatory integral (2.7.1) coincides modulo C^∞ with an oscillatory integral with the same phase function and amplitude*

$$\mathbf{a} := a^* + i\, d_\varphi^{-1} \operatorname{div}_\eta(d_\varphi\, \vec{g}) = a^* + i\, d_\varphi^{-1} \sum_{j=1}^{n} \partial_{\eta_j}(d_\varphi\, g_j), \tag{2.7.9}$$

where a^* and $\vec{g}$ are the function and the covector field from Lemma 2.7.3.

PROOF OF COROLLARY 2.7.4. Let us substitute (2.7.6) into the oscillatory integral (2.7.1), replace $\varphi_\eta e^{i\varphi}$ by $-i\nabla_\eta(e^{i\varphi})$ and integrate by parts with respect to η. This transforms (2.7.1) into

$$\begin{aligned} \mathcal{I}_{\varphi,a}(t,x,y) = &\int e^{i\varphi(t,x;y,\eta)}\, \mathbf{a}(t,x;y,\eta)\, \varsigma(t,x;y,\eta)\, d_\varphi(t,x;y,\eta)\, đ\eta \\ &+ \int e^{i\varphi(t,x;y,\eta)}\, r(t,x;y,\eta)\, d_\varphi(t,x;y,\eta)\, đ\eta, \end{aligned} \tag{2.7.10}$$

where $\mathbf{a}$ is given by formula (2.7.9), and $r := i\langle \vec{g}, \nabla_\eta \varsigma\rangle$. Note that parasitic integrated terms related to $\partial\mathcal{O}$ do not appear on the right-hand side of (2.7.10) because $\operatorname{supp} \vec{g}$ is conically separated from the set $(T_-, T_+) \times M_x \times \partial\mathcal{O}$.

Since for sufficiently large $|\eta|$ we have $\varsigma = 1$ in a neighbourhood of the set $\mathfrak{C}_i$, the function r is identically zero in a neighbourhood of $\mathfrak{C}_i$. Consequently the second term on the right-hand side of (2.7.10) is a C^∞-half-density (see the statement following Definition 2.1.5). □

It is easy to see that the expression $i\,d_\varphi^{-1}\operatorname{div}_\eta(d_\varphi\,\vec{g})$ appearing in (2.7.9) is invariant under changes of local coordinates x and y. Note also that the two terms on the right-hand side of (2.7.9) have different degrees of homogeneity: a^* is positively homogeneous in η of degree l (as the original amplitude a), whereas $i\,d_\varphi^{-1}\operatorname{div}_\eta(d_\varphi\,\vec{g})$ is positively homogeneous in η of degree $l-1$.

Corollary 2.7.4 is one of the central points of this section. It allows us to eliminate the variable x from the leading homogeneous term of the amplitude because a^* depends only on $(t;y,\eta)$. This opens the way to the complete elimination of the variable x from the amplitude of an oscillatory integral.

THEOREM 2.7.5. *Any Lagrangian distribution* (2.7.1) *of order l can be written modulo C^∞ in the form*

$$(2.7.11)\qquad \mathcal{I}_{\varphi,\mathfrak{a}}(t,x,y) = \int e^{i\varphi(t,x;y,\eta)}\,\mathfrak{a}(t;y,\eta)\,\varsigma(t,x;y,\eta)\,d_\varphi(t,x;y,\eta)\,d\!\!{}^-\!\eta$$

with an amplitude $\mathfrak{a}\in S^l$ independent of x. This amplitude $\mathfrak{a}$ is defined on the set $\widehat{\mathfrak{D}}_i$ given by (2.7.4), *and* $\operatorname{supp}\mathfrak{a}\subset(T_-,T_+)\times\mathbf{O}$, *where $\mathbf{O}$ is the same as in Definition* 2.7.1.

The condition $\operatorname{supp}\mathfrak{a}\subset(T_-,T_+)\times\mathbf{O}$ guarantees that (2.7.11) is an oscillatory integral associated with the canonical transformation (2.3.10) in the sense of Definition 2.7.1.

PROOF OF THEOREM 2.7.5. Applying Corollary 2.7.4 to the leading (of degree l) homogeneous term of the amplitude a we represent (modulo C^∞) our original oscillatory integral as a sum of two: one with amplitude independent of x and positively homogeneous in η of degree l, and the other with amplitude dependent on x and of the class S^{l-1}. Treating the latter in a similar way and repeating this procedure infinitely many times we obtain a full expansion for the required amplitude $\mathfrak{a}(t;y,\eta)$ into positively homogeneous in η terms. Using an analogue of Lemma 2.1.2 we produce the amplitude $\mathfrak{a}(t;y,\eta)$ itself. □

Dealing with amplitudes independent of x is very convenient, so further on in this book in all our constructions we will always reduce amplitudes to such a form.

3. Symbol of a Lagrangian distribution. Theorem 2.7.5 enables us to introduce

DEFINITION 2.7.6. The restriction of the amplitude $\mathfrak{a}(t;y,\eta)$ (appearing in (2.7.11)) to the set $\mathfrak{D}_i$ is called the *symbol* of the Lagrangian distribution $\mathcal{I}(t,x,y)$.

Definition 2.7.6 does not look perfect and requires some comments. First, it is not *a priori* clear whether the symbol is uniquely defined. Uniqueness is established by the following argument and Proposition 2.7.7.

The symbol and the phase function determine the Lagrangian distribution uniquely modulo C^∞. That is, the choice of different amplitudes $\mathfrak{a}(t;y,\eta)\in C^\infty(\widehat{\mathfrak{D}}_i)$

which coincide on $\mathfrak{D}_i$ changes our oscillatory integral (2.7.11) only by a C^∞-half-density. Indeed, the difference between two such amplitudes is an amplitude a which has an infinite order zero on $\mathfrak{C}_i$. For the corresponding oscillatory integral we can carry out the iterative reduction procedure described above with $a^* \equiv 0$ and

$$\vec{g} = \frac{a}{\varphi_\eta^T B \overline{\varphi_\eta}} B \overline{\varphi_\eta},$$

which will reduce this oscillatory integral to an oscillatory integral with $S^{-\infty}$-amplitude. Here we use the fact that the function $\varphi_\eta^T B \overline{\varphi_\eta}$ has a strictly second order zero on the set $\mathfrak{C}_i$; that is, in a neighbourhood of any given point $(t_0, x_0; y_0, \eta_0) \in \mathfrak{C}_i$ we have $\varphi_\eta^T B \overline{\varphi_\eta} \geqslant C \operatorname{dist}^2\big((t,x;y,\eta);\mathfrak{C}_i\big)$ with some constant $C = C(t_0, x_0; y_0, \eta_0) > 0$.

Conversely, we have

PROPOSITION 2.7.7. *The symbol $\mathfrak{a}$ is determined by the Lagrangian distribution $\mathcal{I}$ and the phase function φ uniquely modulo $S^{-\infty}$.*

THE PROOF of this proposition is based on the stationary phase formula and is rather technical, so we have moved it to Appendix C.

Another hitch with Definition 2.7.6 is that the symbol of a Lagrangian distribution may depend not only on the distribution itself but also on the phase function φ. Examples show that the symbol does actually depend on φ. However, this dependence on φ occurs only in lower order terms. This fact will be established in the next subsection.

In order to proceed further we will need to derive a more explicit expression for the symbol $\mathfrak{a}(t;y,\eta)$. It will be convenient for us to introduce the linear operator $\mathfrak{S}$ mapping the original amplitude $a(t,x;y,\eta)$ into the corresponding symbol $\mathfrak{a}(t;y,\eta)$. The operator $\mathfrak{S}$ depends, of course, on the phase function φ.

DEFINITION 2.7.8. The linear operator L is said to be positively homogeneous in η of degree $p \in \mathbb{R}$ if for any $q \in \mathbb{R}$ and any function f positively homogeneous in η of degree q the function Lf is positively homogeneous in η of degree $p+q$.

According to the construction carried out in the previous subsection, the operator $\mathfrak{S}$ is not positively homogeneous in η, but it admits an asymptotic expansion into a series of positively homogeneous (in η) terms:

$$\mathfrak{S} \sim \sum_{r=0}^{\infty} \mathfrak{S}_{-r}, \tag{2.7.12}$$

where the operators $\mathfrak{S}_{-r}$ are positively homogeneous in η of degree $-r$. Lemma 2.7.3, Corollary 2.7.4, and the iterative procedure described in the proof of Theorem 2.7.5 allow us to give an explicit expression for the operators $\mathfrak{S}_{-r}$: combining formulae (E.7), (E.14), (E.16), (E.19), (E.21)–(E.23), (2.7.7), and (2.7.9) we get **[Va14]**

$$\mathfrak{S}_0\, a = a\big|_{x=x^*}, \tag{2.7.13}$$

and for $r \geqslant 1$

$$\mathfrak{S}_{-r}\, a = \mathfrak{S}_0\, (\mathfrak{P}_{-1,r})^r\, a \tag{2.7.14}$$

where

$$\mathfrak{P}_{-1,r}\, a \;=\; i\, d_\varphi^{-1}\, \mathrm{div}_\eta \left(d_\varphi \left(\sum_{k=1}^{2r} \sum_{|\alpha|=k-1} \frac{(-\varphi_\eta)^\alpha}{\alpha!\, k} \left((\varphi_{x\eta})^{-1}\partial_x\right)^\alpha \right) (\varphi_{x\eta})^{-1}\partial_x a \right).$$

Here $\partial_x = \mathrm{grad}_x$ (column), $(\varphi_{x\eta})^{-1}\partial_x = (\partial_{f_1}, \ldots, \partial_{f_n})$ is the column of first order linear differential operators ∂_{f_j} in x, and

$$\left((\varphi_{x\eta})^{-1}\partial_x\right)^\alpha \;=\; \partial_f^\alpha \;=\; \partial_{f_1}^{\alpha_1} \cdots \partial_{f_n}^{\alpha_n}\,.$$

It is shown in Appendix E that the operators ∂_{f_j} commute, so the order of terms in the last formula does not matter.

The operator $\mathfrak{P}_{-1,r}$ is positively homogeneous in η of degree -1, which is indicated by the first subscript. In our definition of $\mathfrak{P}_{-1,r}$ we have restricted summation to $k \leqslant 2r$ because terms with $k > 2r$ disappear when we set $x = x^*$. The cut-off factors $\chi_k(t,x;y,\eta)$ present in the original formula (E.23) have also disappeared because they are identically equal to 1 in some neighbourhoods of the set $\mathfrak{C}_i$. And g'' did not contribute to (2.7.14) because according to (E.16), (2.7.7) it has an infinite order zero on $\mathfrak{C}_i$.

The operators (2.7.13), (2.7.14) can be rewritten in the form

$$\mathfrak{S}_{-r} \;=\; \mathfrak{S}_0\, \mathfrak{L}_{-r}(t, \partial_x; y, \eta, \partial_\eta)\,, \qquad r = 0, 1, 2, \ldots, \tag{2.7.15}$$

where the $\mathfrak{L}_{-r}$ are differential operators positively homogeneous in η of degree $-r$ with coefficients independent of x. Note that $\mathfrak{L}_{-r} \neq (\mathfrak{P}_{-1,r})^r$ as the coefficients of $\mathfrak{P}_{-1,r}$ depend on x. Obviously,

$$\mathfrak{L}_0 = 1 \tag{2.7.16}$$

(identity operator), whereas for $r \geqslant 1$ the differential operators $\mathfrak{L}_{-r}$ are of the form

$$\mathfrak{L}_{-r} \;=\; \sum_{\substack{0 \leqslant |\beta| \leqslant r \\ 1 \leqslant |\alpha| \leqslant 2r - |\beta|}} C_{r,\alpha,\beta}\, \partial_x^\alpha\, \partial_\eta^\beta\,, \tag{2.7.17}$$

with coefficients $C_{r,\alpha,\beta} = C_{r,\alpha,\beta}(t; y, \eta) \in C^\infty(\mathfrak{D}_i)$ positively homogeneous in η of degree $|\beta| - r$. The coefficients $C_{r,\alpha,\beta}$ are determined by the phase function $\varphi \in \mathfrak{F}_i$ and can be explicitly derived from (2.7.14).

The following lemma gives a simple formula for the coefficients $C_{1,\alpha,\beta}$ with $|\alpha| = 2$. We will not need the other coefficients because further on (in the next subsection, as well as in Sections 3.3 and 3.4) we will have to apply the operator $\mathfrak{L}_{-1}$ only to functions which have a second order zero on $\mathfrak{C}_i$.

LEMMA 2.7.9. *The operator* $\mathfrak{L}_{-1}$ *has the form*

$$\mathfrak{L}_{-1} \;=\; -\frac{i}{2}\, \mathrm{tr}\left(x_\eta^* \cdot \Phi_{x\eta}^{-1} \cdot \partial_{xx}\right) \;+\; \sum_{\substack{0 \leqslant |\beta| \leqslant 1 \\ |\alpha| = 1}} C_{1,\alpha,\beta}\, \partial_x^\alpha\, \partial_\eta^\beta\,. \tag{2.7.18}$$

Note that when $\mathfrak{L}_{-1}$ acts on functions with a second order zero on $\mathfrak{C}_i$, the second term on the right-hand side of (2.7.18) gives a zero contribution. It is also easy to see that on functions with a second order zero on $\mathfrak{C}_i$, the first term on the right-hand side of (2.7.18) is invariant under changes of local coordinates x and

y (all the matrices involved behave as tensors under changes of coordinates). The latter is not surprising because the full operator (2.7.14) is invariant.

PROOF OF LEMMA 2.7.9. Formula (2.7.18) is an immediate consequence of formulae (2.7.14), (2.4.9), (2.4.10). □

Let us now make the following useful observation. Consider the oscillatory integral (2.7.1) with amplitude $a(t,x;y,\eta) \in S^l$ which has a first or a second order zero on the set $\mathfrak{C}_i$. Formula (2.7.18) shows that in both cases the order of the amplitude can be decreased, generally speaking, only by one: $\mathfrak{a}(t;y,\eta) \in S^{l-1}$. Moreover, if $a(t,x;y,\eta)$ has a zero of order $2N-1$ or $2N$ on the set $\mathfrak{C}_i$ $(N \in \mathbb{N})$, then formula (2.7.17) shows that in both cases the order of the amplitude can be decreased, generally speaking, only by N: $\mathfrak{a}(t;y,\eta) \in S^{l-N}$. This happens because in the process of integrating by parts we have to differentiate the remaining factors $(x-x^*)$ with respect to η.

There is, however, one special situation when the order of the amplitude can be decreased exactly by the order of the zero of $a(t,x;y,\eta)$. This is the case $x^* \equiv y$. Obviously, in this situation the derivatives of $(x-x^*)$ with respect to η are zero. This special case $x^* \equiv y$ will be considered in subsection 5, and it will be shown that oscillatory integrals of this type are in fact pseudodifferential operators.

4. Change of the phase function. Now we prove the following theorem which demonstrates that, in a sense, all phase functions associated with a canonical transformation are equivalent.

THEOREM 2.7.10. *Let $\mathcal{I}(t,x,y)$ be a Lagrangian distribution of order l associated with the ith leg of the canonical transformation* (2.3.10). *Then for any phase function $\varphi \in \mathfrak{F}_i$ there exists an amplitude $a \in S^l$ possessing the properties described in Definition* 2.7.1 *and such that $\mathcal{I}(t,x,y) = \mathcal{I}_{\varphi,a}(t,x,y)$ modulo a C^∞-half-density.*

PROOF. According to Definition 2.7.2 and Theorem 2.7.5 there exist a phase function $\varphi^{(0)} \in \mathfrak{F}_i$ and an amplitude $\mathfrak{a}^{(0)}(t;y,\eta) \in S^l$ such that $\mathcal{I}(t,x,y) = \mathcal{I}_{\varphi^{(0)},\mathfrak{a}^{(0)}}(t,x,y)$ modulo C^∞, i.e.,

$$\mathcal{I}_{\varphi^{(0)},\mathfrak{a}^{(0)}}(t,x,y) - \mathcal{I}(t,x,y) \in C^\infty((T_-,T_+)\times M_x \times M_y). \tag{2.7.19}$$

Let $\varphi \in \mathfrak{F}_i$ be the phase mentioned in the formulation of the theorem, and let $\mathfrak{a}(t;y,\eta) \in S^l$ be the corresponding amplitude which we want to construct. The amplitude $\mathfrak{a}$ must be constructed in such a way that the condition

$$\mathcal{I}_{\varphi,\mathfrak{a}}(t,x,y) - \mathcal{I}(t,x,y) \in C^\infty((T_-,T_+)\times M_x \times M_y) \tag{2.7.20}$$

is satisfied.

Let $\mathcal{O}^{(0)}$, $\mathcal{O}$ be the respective domains of definition of the phase functions $\varphi^{(0)}, \varphi$. Without loss of generality we shall assume that $\mathcal{O}^{(0)} = \mathcal{O}$. This can always be achieved by restriction to their common domain of definition $\mathcal{O}^{(0)} \cap \mathcal{O}$.

According to Lemma 2.4.10 there is a family of phase functions $\psi(s;t,x;y,\eta) \in \mathfrak{F}_i$ smoothly depending on the parameter $s \in [0,1]$ and with domain of definition $\mathcal{O}$ (for each fixed s) such that $\psi(0;t,x;y,\eta) = \varphi^{(0)}(t,x;y,\eta)$, $\psi(1;t,x;y,\eta) = \varphi(t,x;y,\eta)$. Without loss of generality we shall assume that $\det\psi_{x\eta} \neq 0$ on $[0,1]\times\mathcal{O}$ and that there exists a smooth branch of $\arg(\det^2\psi_{x\eta})$ on $[0,1]\times\mathcal{O}$.

This can always be achieved by reducing the original neighbourhood $\mathcal{O}$ of the set $\mathfrak{C}_i$.

The amplitude $\mathfrak{a}^{(0)}$ is defined on $\widehat{\mathfrak{D}}_i$ (projection of $\mathcal{O}$ on $(T_-, T_+)\times O$; see (2.7.4)) and satisfies the condition

$$\operatorname{supp}\mathfrak{a}^{(0)} \subset (T_-, T_+)\times \mathbf{O}, \tag{2.7.21}$$

where $\mathbf{O}$ is some conically compact conic subset of O. The amplitude $\mathfrak{a}$ which we are about to construct will also be defined on $\widehat{\mathfrak{D}}_i$ and will also satisfy the condition $\operatorname{supp}\mathfrak{a}\subset (T_-, T_+)\times \mathbf{O}$.

The idea of the proof of Theorem 2.7.10 is to construct a family of amplitudes $\mathfrak{b}(s;t;y,\eta)\in S^l$ smoothly depending on the parameter $s\in[0,1]$ and with domain of definition $[0,1]\times\widehat{\mathfrak{D}}_i$ in such a way that

$$\operatorname{supp}\mathfrak{b} \subset [0,1]\times(T_-, T_+)\times \mathbf{O}, \tag{2.7.22}$$

$$\mathfrak{b}(0;t;y,\eta) - \mathfrak{a}^{(0)}(t;y,\eta) \in S^{-\infty}, \tag{2.7.23}$$

$$\frac{\partial}{\partial s}\mathcal{I}_{\psi,\mathfrak{b}}(s;t,x,y) \in C^\infty([0,1]\times(T_-, T_+)\times M_x\times M_y), \tag{2.7.24}$$

and set $\mathfrak{a}(t;y,\eta):=\mathfrak{b}(1;t;y,\eta)$. Then (2.7.20) will follow from (2.7.19), (2.7.22)–(2.7.24).

The technical realization of this idea is described below.

We have

$$\mathcal{I}_{\psi,\mathfrak{b}}(s;t,x,y) = \int e^{i\psi(s;t,x;y,\eta)}\,\mathfrak{b}(s;t;y,\eta)\,\varsigma(t,x;y,\eta)\,d_\psi(s;t,x;y,\eta)\,đ\eta. \tag{2.7.25}$$

Differentiating (2.7.25) with respect to s, we obtain

$$\frac{\partial}{\partial s}\mathcal{I}_{\psi,\mathfrak{b}}(s;t,x,y) = \mathcal{I}_{\psi,b}(s;t,x,y), \tag{2.7.26}$$

where

$$b = \frac{\partial}{\partial s}\mathfrak{b} + f_1\,\mathfrak{b} + f_2\,\mathfrak{b}, \tag{2.7.27}$$

$$f_1\equiv f_1(t,x;y,\eta) = i\frac{\partial}{\partial s}\psi, \qquad f_2\equiv f_2(t,x;y,\eta) = d_\psi^{-1}\frac{\partial}{\partial s}d_\psi. \tag{2.7.28}$$

Let us examine closer the functions (2.7.28). By Liouville's formula for any nonsingular smooth matrix function $C(s)$ we have

$$\left(\det C(s)\right)^{-1}\frac{\partial}{\partial s}\det C(s) = \operatorname{tr}\left(C^{-1}(s)\,\frac{\partial}{\partial s}C(s)\right);$$

therefore, with account of (2.2.4),

$$f_2 = \frac{1}{2}\operatorname{tr}\left(\psi_{x\eta}^{-1}\cdot\frac{\partial}{\partial s}\psi_{x\eta}\right).$$

Since the canonical transformation itself does not depend on s, it follows from the latter formula and (2.4.10) that

$$f_2 = -\frac{1}{2}\operatorname{tr}\left(\Psi_{x\eta}^{-1}\cdot\frac{\partial}{\partial s}\Psi_{xx}\cdot x^*_\eta\right) + O(|x-x^*|),\quad \forall(t;y,\eta)\in\mathfrak{D}_i \tag{2.7.29}$$

(we are using notation introduced in Section 2.4 with ψ and Ψ instead of φ and Φ). Besides, the equality (2.4.8) implies

(2.7.30)
$$f_1 = \frac{i}{2}\left\langle\left(\frac{\partial}{\partial s}\Psi_{xx}\right)(x-x^*), x-x^*\right\rangle + O(|x-x^*|^3), \qquad \forall (t;y,\eta)\in\mathfrak{D}_i .$$

Condition (2.7.23) means that

$$\mathfrak{b}_{l-k}(0;t;y,\eta) = \mathfrak{a}^{(0)}_{l-k}(t;y,\eta) \quad \text{on } \widehat{\mathfrak{D}}_i , \qquad k=0,1,2,\ldots, \tag{2.7.31}$$

where $\mathfrak{a}^{(0)}_l$, $\mathfrak{a}^{(0)}_{l-1}$, ... and $\mathfrak{b}_l$, $\mathfrak{b}_{l-1}$, ... are the terms of the amplitudes $\mathfrak{a}^{(0)}$ and $\mathfrak{b}$, respectively, that are positively homogeneous in η.

Condition (2.7.24) holds if the corresponding symbol is identically zero in all its homogeneous terms. In view of (2.7.26), (2.7.27), (2.7.12), and (2.7.15) this means

$$\frac{\partial}{\partial s}\mathfrak{b}_{l-k} = \sum_{r=0}^{k} L_{-r}\,\mathfrak{b}_{l+r-k} \quad \text{on } \mathfrak{D}_i , \qquad k=0,1,2,\ldots, \tag{2.7.32}$$

where

$$L_{-r} \equiv L_{-r}(s;t;y,\eta,\partial_\eta) := -(\mathfrak{S}_{-r-1}f_1 + \mathfrak{S}_{-r}f_2)(\,\cdot\,) \tag{2.7.33}$$

(here the operator $\mathfrak{S}$ is associated with the phase ψ). In writing (2.7.32), (2.7.33) we used the fact that the functions f_1 and f_2 are positively homogeneous in η of degrees 1 and 0, respectively, and that $f_1|_{x=x^*} = 0$ (see (2.7.30)). Of course, the differential operators L_{-r} are positively homogeneous in η of degree $-r$.

Direct substitution of formulae (2.7.15), (2.7.16), (2.7.18), (2.7.29), (2.7.30) into (2.7.33) shows that L_0 is an operator of multiplication by a function, and, moreover, this function is

(2.7.34)
$$\begin{aligned} L_0 &= \left(\frac{i}{2}\operatorname{tr}\left(x^*_\eta\cdot\Psi^{-1}_{x\eta}\cdot\partial_{xx}f_1\right) - f_2\right)\Bigg|_{x=x^*} \\ &= -\frac{1}{2}\operatorname{tr}\left(x^*_\eta\cdot\Psi^{-1}_{x\eta}\cdot\frac{\partial}{\partial s}\Psi_{xx}\right) + \frac{1}{2}\operatorname{tr}\left(\Psi^{-1}_{x\eta}\cdot\frac{\partial}{\partial s}\Psi_{xx}\cdot x^*_\eta\right) = 0. \end{aligned}$$

Formula (2.7.34) is the central point of our proof.

Further on it will be convenient for us to extend the coefficients of the differential operators L_{-r} from $\mathfrak{D}_i$ to $\widehat{\mathfrak{D}}_i$. We extend L_0 as identically zero (in order to preserve (2.7.34)), and we extend the L_{-r}, $r\geqslant 1$, in an arbitrary smooth way.

With account of (2.7.34) we can now rewrite (2.7.32), (2.7.31) as

$$\begin{aligned} \frac{\partial}{\partial s}\mathfrak{b}_l &= 0, & \mathfrak{b}_l|_{s=0} &= \mathfrak{a}^{(0)}_l, \\ \frac{\partial}{\partial s}\mathfrak{b}_{l-1} &= L_{-1}\mathfrak{b}_l, & \mathfrak{b}_{l-1}|_{s=0} &= \mathfrak{a}^{(0)}_{l-1}, \\ \frac{\partial}{\partial s}\mathfrak{b}_{l-2} &= L_{-1}\mathfrak{b}_{l-1} + L_{-2}\mathfrak{b}_l, & \mathfrak{b}_{l-2}|_{s=0} &= \mathfrak{a}^{(0)}_{l-2}, \\ &\ldots & &\ldots \end{aligned}$$

The unique solution of this infinite system is

$$\mathfrak{b}_l = \mathfrak{a}_l^{(0)}, \tag{2.7.35}$$

$$\mathfrak{b}_{l-k} = \mathfrak{a}_{l-k}^{(0)} + \int_0^s \left(\sum_{r=1}^{k} L_{-r} \mathfrak{b}_{l+r-k} \right) ds, \qquad k = 1, 2, \dots \tag{2.7.36}$$

(the latter is a recursive formula). Formulae (2.7.35), (2.7.36), (2.7.21) imply

$$\operatorname{supp} \mathfrak{b}_{l-k} \subset (T_-, T_+) \times \mathbf{O}, \qquad k = 0, 1, 2, \dots .$$

Using a version of Lemma 2.1.2 we can produce a function $\mathfrak{b}(s; t; y, \eta) \in S^l$ such that $\mathfrak{b} \sim \sum_{k=0}^{\infty} \mathfrak{b}_{l-k}$ and $\operatorname{supp} \mathfrak{b} \subset (T_-, T_+) \times \mathbf{O}$. Then $\mathfrak{a}(t; y, \eta) = \mathfrak{b}(1; t; y, \eta)$ is the required amplitude. □

THEOREM 2.7.11. *The leading homogeneous term of the symbol of a Lagrangian distribution is independent of the choice of the phase function and is uniquely determined by the Lagrangian distribution itself.*

PROOF. The equality (2.7.35) means precisely that the leading term does not change when one changes the phase function, and by Proposition 2.7.7 this term is uniquely determined for a fixed phase function. □

Theorem 2.7.11 enables us to define the following more invariant object.

DEFINITION 2.7.12. The leading homogeneous term $\mathfrak{a}_l$ (of degree l) of the symbol $\mathfrak{a}$ is called the *principal symbol* of the Lagrangian distribution $\mathcal{I}$ (of order l).

5. Pseudodifferential operators. We can now see that the situation with Lagrangian distributions is similar to that in the theory of pseudodifferential operators. One can define the full symbol of a Lagrangian distribution, but it depends on the choice of the phase function like the full symbol of a pseudodifferential operator depends on the choice of coordinates. However, its leading part — the principal symbol — is correctly defined and determined only by the distribution itself. This analogy is not accidental: the Schwartz kernel of a pseudodifferential operator from Ψ_0^l can be regarded as a Lagrangian distribution associated with the identity transformation.

Indeed, we can always assume the local phase functions $\langle x - y, \eta \rangle$ introduced in 2.1.3 to be defined not on subsets of $M_x \times M_y \times \mathbb{R}^n$ but on some subsets of $M_x \times T'M_y$ (we write η instead of θ to match notation introduced in different sections). Then these local phase functions satisfy the conditions of Definition 2.4.1 with fixed $t = 0$ and $x^* \equiv y$, $\xi^* \equiv \eta$, i.e., they are associated with the identity transformation. Moreover, these phase functions are nondegenerate and simple in the sense of Definitions 2.2.3, 2.2.4.

Let $\mathfrak{F}^0$ be the class of global phase functions $\varphi(x; y, \eta) \in C^\infty(M_x \times T'M_y)$ such that

$$\varphi_\eta(x; y, \eta) \neq 0 \quad \text{for} \quad x \neq y,$$

and in a neighbourhood of the diagonal $\{x = y\}$

$$\varphi(x; y, \eta) = \langle x - y, \eta \rangle + O(|x - y|^2), \tag{2.7.37}$$

assuming that the local x- and y-coordinates are the same. Obviously, the phase functions $\varphi \in \mathfrak{F}^0$ are nondegenerate and simple, and are associated with the identity transformation. By Theorem 2.7.11 (with fixed $t = 0$) we can transform local oscillatory integrals with phase functions $\langle x - y, \eta \rangle$ into oscillatory integrals with a global phase function $\varphi \in \mathfrak{F}^0$. Summing up the latter, we see that the Schwartz kernel of an arbitrary pseudodifferential operator from the class Ψ_0^l can be represented by one oscillatory integral with this global phase function $\varphi \in \mathfrak{F}^0$. In the same manner a global oscillatory integral with phase function $\varphi \in \mathfrak{F}^0$ can be transformed into a sum of oscillatory integrals with phase functions of the form $\langle x - y, \eta \rangle$. Thus, we have proved

PROPOSITION 2.7.13. *Let* $\varphi \in \mathfrak{F}^0$. *Then for any pseudodifferential operator from* Ψ_0^l *there exists an amplitude* $a(x; y, \eta) \in S^l$ *with* supp a *lying in some conically compact conic subset of* $\overset{\circ}{M}_x \times T'\overset{\circ}{M}_y$ *such that the Schwartz kernel of this pseudodifferential operator coincides, modulo a* C_0^∞*-half-density, with the oscillatory integral* (2.7.1), $t = 0$. *Conversely, any such oscillatory integral defines the Schwartz kernel of a pseudodifferential operator from* Ψ_0^l.

Recall (see 2.1.3) that in considering phase functions of the form $\varphi_0 = \langle x - y, \eta \rangle$ we always assume that the coordinate systems in x and in y are the same. Certainly, we have $d_{\varphi_0} \equiv 1$ in these coordinates. This implies that the dual symbol of a pseudodifferential operator coincides with the symbol of the corresponding Lagrangian distribution. We had to write here "dual" because in 2.1.3 we defined (in accordance with traditional notation) the symbol independent of x as the dual one.

REMARK 2.7.14. If we assume the function $\varphi_0 = \langle x - y, \eta \rangle$ to be defined on a subset of $M_x \times T'M_y$ and change coordinates x or y, then the equality $d_{\varphi_0} \equiv 1$ may fail. Moreover, if we take different coordinate systems in x and in y, then even the restriction $d_{\varphi_0}|_{x=y}$ may differ from 1. This follows from the fact that d_{φ_0} behaves as a (1/2)-density in x and as a $(-1/2)$-density in y.

6. Action of a pseudodifferential operator on an oscillatory integral.

LEMMA 2.7.15. *Let* $\mathcal{I}(t, x, y)$ *be a Lagrangian distribution of order* l *associated with the canonical transformation* (2.3.10), *and let* $A(x, D_x)$ *be a differential operator of order* p *acting on* M. *Let* $\mathfrak{a}_l(t; y, \eta)$ *and* $A_p(x, \xi)$ *be the principal symbols of* $\mathcal{I}$ *and* A, *respectively. Then* $A\mathcal{I}$ *is a Lagrangian distribution of order* $l+p$ *associated with the canonical transformation* (2.3.10), *and its principal symbol is* $A_p(x^*(t; y, \eta), \xi^*(t; y, \eta))\, \mathfrak{a}_l(t; y, \eta)$.

PROOF. Lemma 2.7.15 is proved by direct substitution: if $\mathcal{I} = \mathcal{I}_{\varphi, a}$ (modulo C^∞), then $A\mathcal{I}$ is (modulo C^∞) an oscillatory integral with the same phase function φ and amplitude

$$d_\varphi^{-1}\, e^{-i\varphi} A\,(e^{i\varphi}\, a\, d_\varphi)\,. \tag{2.7.38}$$

Taking the leading term in (2.7.38) (positively homogeneous in η of degree $l + p$) and setting $x = x^*$ we obtain the required principal symbol. □

LEMMA 2.7.16. *Lemma* 2.7.15 *remains true if* A *is a pseudodifferential operator from* Ψ_0^p.

PROOF. Let $(t_0, x_0; y_0, \eta_0)$ be an arbitrary point from $\mathfrak{C}_i$ (of course, $x_0 = x^*(t_0; y_0, \eta_0)$), and z_0 an arbitrary point from $\overset{\circ}{M}$. It is sufficient to prove Lemma 2.7.16 in the case when the support of the amplitude of the oscillatory integral representing (modulo C^∞) the Lagrangian distribution $\mathcal{I}$ lies in a small conic neighbourhood of the point $(t_0, x_0; y_0, \eta_0)$, and the support of the Schwartz kernel of the pseudodifferential operator A lies in a small neighbourhood of the point (z_0, z_0); the general case is handled by appropriate partitions of unity.

The case $x_0 \neq z_0$ is trivial ($A\mathcal{I} \equiv 0$), so further on we assume $x_0 = z_0$.

In view of Theorem 2.7.10, Corollary 2.4.14, and Theorem 2.7.5 there exist local coordinates x in a neighbourhood of the point x_0 in which the Lagrangian distribution $\mathcal{I}$ can be written modulo C^∞ in the form

$$\int e^{i\langle x - x^*(t;y,\eta), \xi^*(t;y,\eta)\rangle}\, \mathfrak{a}(t; y, \eta)\, \varsigma(t, x; y, \eta)\, d_\varphi(t; y, \eta)\, đ\eta\,. \tag{2.7.39}$$

The pseudodifferential operator A can be written in the form

$$(Av)(z) = \int e^{i\langle z - x, \xi\rangle}\, \sigma_A(z, \xi)\, v(x)\, dx\, đ\xi \tag{2.7.40}$$

(see 2.1.3 and 2.7.5). In (2.7.40) we use the same local coordinates for x and z as used for x in (2.7.39). Acting by (2.7.40) on (2.7.39) we obtain (modulo C_0^∞) an oscillatory integral with phase function $\langle z - x^*(t; y, \eta), \xi^*(t; y, \eta)\rangle$ and amplitude

$$\sigma_A(z, \xi^*(t; y, \eta))\, \mathfrak{a}(t; y, \eta)\,. \tag{2.7.41}$$

Taking the leading term in (2.7.41) (positively homogeneous in η of degree $l + p$) and setting $z = x^*$ we obtain the required principal symbol. □

2.8. Boundary layer oscillatory integrals associated with homogeneous canonical transformations

Boundary layer oscillatory integrals introduced in this section are a special case of global oscillatory integrals from Section 2.2. Properties of boundary layer oscillatory integrals are similar to those of standard oscillatory integrals described in Section 2.7, the difference being that the singularities of boundary layer oscillatory integrals lie on the boundary of the manifold.

Below we continue using notation introduced in Sections 2.4–2.7.

Similarly to Definitions 2.7.1, 2.7.2 we introduce the following two definitions.

DEFINITION 2.8.1. We say that a global oscillatory integral (2.7.1) with phase function φ and amplitude $a \in S^l$ is a *boundary layer oscillatory integral* associated with the ith reflection of our Hamiltonian billiards if

1. φ and a are defined in a conic neighbourhood $\mathcal{O} \subset (T_-, T_+) \times M_x \times O$ of the set $\mathfrak{C}'_i$.
2. φ is a nondegenerate boundary layer phase function associated with the ith reflection of our Hamiltonian billiards, i.e., $\varphi \in \mathfrak{F}_i^{\mathrm{bl}}$.
3. $\det \varphi_{x\eta} \neq 0$ on $\mathcal{O}$.
4. $\operatorname{supp} a \subset (T_-, T_+) \times M_x \times \mathbf{O}$, where $\mathbf{O}$ is some conically compact conic subset of O.

In (2.7.1) d_φ is defined by (2.7.2), where the continuous branch $\arg_0(\det^2\varphi_{x\eta})$ is chosen in accordance with Definition 2.6.17. The function ς in formula (2.7.1) is a cut-off around the set $\mathfrak{C}_i'$ (see Section 2.2).

DEFINITION 2.8.2. A distribution $\mathcal{I}(t,x,y)$ from the class (2.7.3) which can be written modulo $C^\infty((T_-, T_+) \times M_x \times M_y)$ as a boundary layer oscillatory integral (2.7.1) is called a *boundary layer Lagrangian distribution* of order l associated with the ith reflection of our Hamiltonian billiards.

Below, as in the previous section, we denote by $\widehat{\mathfrak{D}}_i$ the projection of $\mathcal{O}$ on $(T_-, T_+) \times O$; see (2.7.4). Obviously, this set $\widehat{\mathfrak{D}}_i \subset (T_-, T_+) \times O$ is a conic neighbourhood of the set $\{(t;y,\eta) : (y,\eta) \in O,\ t = t_i^*(y,\eta)\}$.

Similarly to Lemma 2.7.3 we have

LEMMA 2.8.3. *Let $a(t,x;y,\eta) \in C^\infty(\mathcal{O})$ be a function positively homogeneous in η of degree l satisfying condition (4) of Definition 2.8.1, and let $\varphi \in \mathfrak{F}_i^{\mathrm{bl}}$. Then there exist a function $a^*(t;y,\eta) \in C^\infty(\widehat{\mathfrak{D}}_i)$ positively homogeneous in η of degree l and a covector field*

$$\vec{g}\,(t,x;y,\eta) = \{g_1(t,x;y,\eta), \dots, g_n(t,x;y,\eta)\} \in C^\infty(\mathcal{O})$$

positively homogeneous in η of degree l such that (2.7.5), (2.7.6) hold. Moreover, the function a^* and the covector field $\vec{g}$ can be constructed effectively in the form (E.7), (E.14), (E.19), (E.21)–(E.23), (E.31), (E.33) with

$$z = x - \mathbf{x}^*, \quad f = \varphi_\eta, \quad u = (y,\eta), \quad v = t - t_i^*(y,\eta), \quad a_0 = a^*, \tag{2.8.1}$$

an arbitrary positive Hermitian covariant tensor $B = B(y) \in C^\infty(M_y)$, and some cut-off functions $\chi_k = \chi_k(t,x;y,\eta) \in C^\infty(\mathcal{O})$ positively homogeneous in η of degree 0 which are identically equal to 1 in some neighbourhoods of the set $\mathfrak{C}_i'$.

Recall that by $t_i^*(y,\eta)$ we denote the moment of the ith reflection, and by $\mathbf{x}^* = (\mathbf{x}^{*\prime}, 0) := x^*(t_i^*(y,\eta); y,\eta) \in \partial M_x$ the point of the ith reflection.

THE PROOF of Lemma 2.8.3 follows immediately from the results of Appendix E (see E.4); namely, formula (2.7.6) in this case is a special case of formula (E.30). We were able to apply the results of E.4 because for a boundary layer phase function, condition (E.29) is fulfilled: it follows from the fact that, for the corresponding complex reflected trajectory , $\operatorname{Im} \dot{x}^*|_{t=t_i^*} \neq 0$; see the discussion following formula (2.6.20).

COROLLARY 2.8.4. *Corollary 2.7.4 remains true for boundary layer oscillatory integrals, with a^* and $\vec{g}$ from Lemma 2.8.3.*

THE PROOF of Corollary 2.8.3 repeats that of Corollary 2.7.4.

THEOREM 2.8.5. *Theorem 2.7.5 remains true for boundary layer Lagrangian distributions.*

THE PROOF of Theorem 2.8.5 repeats that of Theorem 2.7.5.

Similarly to Definition 2.7.6 we introduce

DEFINITION 2.8.6. The jet of the amplitude $\mathfrak{a}(t;y,\eta)$ (appearing in (2.7.11)) at $t=t_i^*(y,\eta)$ is called the *symbol* of the boundary layer Lagrangian distribution $\mathcal{I}(t,x,y)$.

Here by a *jet* we understand the equivalence class of all amplitudes $\mathfrak{a}(t;y,\eta)\in S^l$ defined in some conic neighbourhoods of the set $\{t=t_i^*(t;y,\eta)\}\subset(T_-,T_+)\times O$ which differ by amplitudes with an infinite order zero at $t=t_i^*(t;y,\eta)$. In other words, the jet of the function $\mathfrak{a}$ at $t=t_i^*$ is uniquely determined by the set of Taylor coefficients $\frac{1}{k!}\,\partial_t^k\mathfrak{a}\big|_{t=t_i^*}$, $k=0,1,2,\dots$.

In accordance with traditional terminology we should have called our jet an ∞-jet, but we omit the "infinity" for brevity.

It may seem that our two definitions of the symbol, namely Definition 2.7.6 for standard oscillatory integrals and Definition 2.8.6 for boundary layer ones, are completely different. However, in reality Definition 2.7.6 also implicitly incorporates the notion of a jet: when we defined the symbol of a standard oscillatory integral we automatically defined all its partial derivatives with respect to t, y, η. In Definition 2.7.6 we did not have to mention the notion of a jet explicitly because in the case of a standard oscillatory integral the set $\{\varphi_\eta=0\}$ is more massive than in the case of a boundary layer oscillatory integral. More precisely, the dimension of the manifold $\{\varphi_\eta=0\}$ in the standard and boundary layer cases is $2n+1$ and $2n$, respectively. As a result, in the standard case the values of the amplitude $\mathfrak{a}(t;y,\eta)$ on $\{\varphi_\eta=0\}$ uniquely determine all its partial derivatives.

The symbol and the phase function determine the boundary layer Lagrangian distribution uniquely modulo C^∞. That is, the choice of different representatives $\mathfrak{a}(t;y,\eta)$ of the corresponding equivalence class changes our boundary layer oscillatory integral (2.7.11) only by a C^∞-half-density. This fact is established by the same argument as in 2.7.3 with $\mathfrak{C}_i$ replaced by $\mathfrak{C}_i'$.

Similarly to Proposition 2.7.7 we have

PROPOSITION 2.8.7. *The symbol* $\mathfrak{a}$ *is uniquely determined modulo* $S^{-\infty}$ *by the boundary layer Lagrangian distribution* $\mathcal{I}$ *and the boundary layer phase function* φ.

THE PROOF of this proposition is given in Appendix C.

As in 2.7.3, it will be convenient for us to introduce the linear operator $\mathfrak{S}$ mapping the original amplitude $a(t,x;y,\eta)$ of a boundary layer oscillatory integral into the corresponding symbol $\mathfrak{a}(t;y,\eta)$. It can be checked that all the formulae (2.7.12)–(2.7.18) remain true for boundary layer oscillatory integrals, with the only difference that in this case $x^*=x^*(t;y,\eta)$ is the complex reflected ray corresponding to our boundary layer phase function φ, $\mathfrak{S}_0=(\,\cdot\,)|_{x=x^*}$ is the operator of restriction to this complex ray, and $\Phi_{x\eta}=\mathfrak{S}_0\varphi_{x\eta}$. According to the convention introduced in 2.6.4 we view our complex reflected ray x^* as a formal Taylor expansion in powers of $t-t^*$, consequently it is more appropriate to write formulae (2.7.13)–(2.7.18) with the sign $\simeq$ instead of the sign $=$; recall that in our notation the sign $\simeq$ stands for the equality of formal Taylor expansions. The fact that we deal with formal (that is, not necessarily convergent) infinite Taylor expansions does not create any difficulties because the symbol of a boundary layer Lagrangian distribution is defined modulo $O(|t-t_i^*|^\infty)$.

The procedure of the reduction of the amplitude described in this section allows us to give an alternative representation for the operator of restriction to the complex reflected ray:

$$\mathfrak{S}_0\, a \simeq \left. \left(I - \varphi_\eta^T \left(\sum_{k=1}^{\infty} \sum_{|\alpha|=k-1} \frac{(-\varphi_\eta)^\alpha}{\alpha!\, k} \left((\varphi_{x\eta})^{-1} \partial_x \right)^\alpha \right) (\varphi_{x\eta})^{-1} \partial_x a \right) \right|_{x=\mathbf{x}^*}$$

where I is the identity operator. This representation has the advantage that it does not require the explicit solution of the complex Hamiltonian system of equations (2.6.20); x^* enters the above formula implicitly as the complex x-root of the equation $\varphi_\eta = 0$, and ξ^* as the value of φ_x on the complex set $\{\varphi_\eta = 0\}$.

THEOREM 2.8.8. *Let $\mathcal{I}(t,x,y)$ be a boundary layer Lagrangian distribution of order l associated with the* ith *reflection of our Hamiltonian billiards. Then for any phase function $\varphi \in \mathfrak{F}_i^{\mathrm{bl}}$ there exists an amplitude $a \in S^l$ possessing properties described in Definition* 2.8.1 *and such that $\mathcal{I}(t,x,y) = \mathcal{I}_{\varphi,a}(t,x,y)$ modulo a C^∞-half-density.*

THE PROOF of Theorem 2.8.8 repeats that of Theorem 2.7.10.

Similarly to Theorem 2.7.11 we have

THEOREM 2.8.9. *The leading homogeneous term of the symbol of a boundary layer Lagrangian distribution is independent of the choice of the phase function and is uniquely determined by the Lagrangian distribution itself.*

This enables us to introduce, similarly to Definition 2.7.12, the following

DEFINITION 2.8.10. The leading homogeneous term $\mathfrak{a}_l$ (of degree l) of the symbol $\mathfrak{a}$ is called the *principal symbol* of the boundary layer Lagrangian distribution $\mathcal{I}$ (of order l).

Of course, in Theorem 2.8.9 and Definition 2.8.10 the leading homogeneous term of the symbol is understood as a jet, that is, a positively homogeneous in η amplitude $\mathfrak{a}_l(t;y,\eta)$ defined modulo $O(|t-t_i^*|^\infty)$.

It remains to note the obvious fact that Lemma 2.7.15 (action of a differential operator) remains true for boundary layer Lagrangian distributions. Lemma 2.7.16 also has an analogue for boundary layer Lagrangian distributions, but a trivial one: acting on a boundary layer Lagrangian distribution with a pseudodifferential operator from Ψ_0^p we obtain a C^∞-half-density.

2.9. Boundary oscillatory integrals associated with homogeneous canonical transformations

The boundary oscillatory integrals described in this section appear naturally when we take boundary traces of standard or boundary layer oscillatory integrals. The exposition in this section is brief because it goes along the same lines as in Sections 2.7, 2.8.

Below we continue using notation introduced in Sections 2.4–2.8.

DEFINITION 2.9.1. We say that an oscillatory integral

$$\mathcal{I}_{\varphi',a}(t,x',y) = \int e^{i\varphi'(t,x';y,\eta)}\, a(t,x';y,\eta)\, \varsigma(t,x';y,\eta)\, d_{\varphi'}(t,x';y,\eta)\, \text{đ}\eta \tag{2.9.1}$$

with phase function φ' and amplitude $a \in S^l$ is a *boundary oscillatory integral* associated with the ith reflection of our Hamiltonian billiards if

1. φ' and a are defined in a conic neighbourhood $\mathcal{O}' \subset (T_-\,,\,T_+)\times\partial M_x\times O$ of the set $\mathfrak{C}_i'$.

2. φ' is a nondegenerate boundary phase function associated with the ith reflection of our Hamiltonian billiards, i.e., $\varphi'\in\mathfrak{F}_i'$.

3. $\det\begin{pmatrix}\varphi'_{x'\eta}\\ \varphi'_{t\eta}\end{pmatrix}\neq 0$ on $\mathcal{O}'$.

4. $\operatorname{supp} a\subset (T_-\,,\,T_+)\times\partial M_x\times\mathbf{O}$, where $\mathbf{O}$ is some conically compact conic subset of O.

In formula (2.9.1)

$$
\begin{aligned}
d_{\varphi'}(t,x';y,\eta) &= \left(\det{}^2\begin{pmatrix}\varphi'_{x'\eta}\\ \varphi'_{t\eta}\end{pmatrix}\right)^{1/4}\\
&= \left|\det\begin{pmatrix}\varphi'_{x'\eta}\\ \varphi'_{t\eta}\end{pmatrix}\right|^{1/2}\exp\left(\frac{i}{4}\arg_0\left(\det{}^2\begin{pmatrix}\varphi'_{x'\eta}\\ \varphi'_{t\eta}\end{pmatrix}\right)\right),
\end{aligned}
\tag{2.9.2}
$$

where the continuous branch $\arg_0\left(\det^2\begin{pmatrix}\varphi'_{x'\eta}\\ \varphi'_{t\eta}\end{pmatrix}\right)$ is chosen in accordance with Definition 2.6.8. The function ς in formula (2.9.1) is a cut-off around the set $\mathfrak{C}_i'$ (see Section 2.2).

It is easy to see that the oscillatory integral (2.9.1) is well defined,

$$
\mathcal{I}_{\varphi',a}\in C^\infty(\partial M_x\times M_y;\,\mathcal{D}'(T_-\,,\,T_+))\,, \tag{2.9.3}
$$

and $\mathcal{I}_{\varphi',a}$ is a half-density in x' and in y (recall that our coordinate x_n is fixed).

DEFINITION 2.9.2. A distribution $\mathcal{I}(t,x',y)$ from the class (2.9.3) which can be written modulo $C^\infty((T_-\,,\,T_+)\times\partial M_x\times M_y)$ as a boundary oscillatory integral (2.9.1) is called a *boundary Lagrangian distribution* of order l associated with the ith reflection of our Hamiltonian billiards.

Set $a^*(y,\eta):=a(t_i^*(y,\eta),\mathbf{x}^{*\prime}(y,\eta);y,\eta)\in C^\infty(O)$. Obviously, $\operatorname{supp} a^*\subset\mathbf{O}$.

LEMMA 2.9.3. *Let* $a(t,x';y,\eta)\in C^\infty(\mathcal{O}')$ *be a function positively homogeneous in* η *of degree* l *satisfying condition* (4) *of Definition* 2.9.1, *and let* $\varphi'\in\mathfrak{F}_i'$. *Then there exists a covector field*

$$
\vec{g}\,(t,x';y,\eta)=\{g_1(t,x';y,\eta),\dots,g_n(t,x';y,\eta)\}\in C^\infty(\mathcal{O}')
$$

positively homogeneous in η *of degree* l *such that*

$$
\operatorname{supp}\vec{g}\subset(T_-\,,\,T_+)\times\partial M_x\times\mathbf{O}\,, \tag{2.9.4}
$$

and

$$
a(t,x';y,\eta)\;=\;a^*(y,\eta)\;+\;\langle\varphi'_\eta(t,x';y,\eta),\vec{g}\,(t,x';y,\eta)\,\rangle\,. \tag{2.9.5}
$$

Moreover, the covector field $\vec{g}$ *can be constructed effectively in the form* (E.7), (E.14), (E.19), (E.21)–(E.24) *with*

$$
z=(x'-\mathbf{x}^{*\prime},\,t-t_i^*)\,,\qquad f=\varphi'_\eta\,,\qquad u=(y,\eta)\,, \tag{2.9.6}
$$

an arbitrary positive Hermitian covariant tensor $B=B(y)\in C^\infty(M_y)$, *and some cut-off functions* $\chi_k=\chi_k(t,x';y,\eta)\in C^\infty(\mathcal{O}')$ *positively homogeneous in* η *of degree* 0, *which are identically equal to* 1 *in some neighbourhoods of the set* $\mathfrak{C}_i'$.

THE PROOF of Lemma 2.9.3 follows immediately from the results of Appendix E (see E.3); namely, formula (2.9.5) is a special case of formula (E.26).

Lemma 2.9.3 allows us to eliminate the variables x' and t from the leading homogeneous term of the amplitude of the oscillatory integral (2.9.1) (cf. Corollary 2.7.4), which eventually leads to the following analogue of Theorem 2.7.5.

THEOREM 2.9.4. *Any boundary Lagrangian distribution* (2.9.1) *of order l can be written modulo C^∞ in the form*

$$\mathcal{I}_{\varphi',\mathfrak{a}}(t,x',y) = \int e^{i\varphi'(t,x';y,\eta)}\,\mathfrak{a}(y,\eta)\,\varsigma(t,x';y,\eta)\,d_{\varphi'}(t,x';y,\eta)\,đ\eta \tag{2.9.7}$$

with an amplitude $\mathfrak{a}\in S^l$ independent of x' and t. This amplitude $\mathfrak{a}$ is defined on O and $\operatorname{supp}\mathfrak{a}\subset\mathbf{O}$, *where $\mathbf{O}$ is the same as in Definition* 2.9.1.

DEFINITION 2.9.5. The amplitude $\mathfrak{a}(y,\eta)$ (appearing in (2.9.7)) is called the *symbol* of the boundary Lagrangian distribution $\mathcal{I}(t,x',y)$.

Obviously, the symbol and the phase function determine the boundary Lagrangian distribution uniquely modulo C^∞. Conversely, we have

PROPOSITION 2.9.6. *The symbol $\mathfrak{a}$ is uniquely determined modulo $S^{-\infty}$ by the boundary Lagrangian distribution $\mathcal{I}$ and the boundary phase function φ'.*

THE PROOF of this proposition is given in Appendix C.

As in 2.7.3, we can introduce the linear operator $\mathfrak{S}'$ mapping the original amplitude $a(t,x';y,\eta)$ of the boundary oscillatory integral into the corresponding symbol $\mathfrak{a}(y,\eta)$. Similarly to (2.7.12) this operator admits an asymptotic expansion into a series of positively homogeneous (in η) terms:

$$\mathfrak{S}' \sim \sum_{r=0}^{\infty}\mathfrak{S}'_{-r}, \tag{2.9.8}$$

where

$$\mathfrak{S}'_0\,a = a|_{x=\mathbf{x}^{*\prime},\,t=t_i^*} \tag{2.9.9}$$

(cf. (2.7.13)). We will not need explicit formulae for the operators $\mathfrak{S}'_{-r}$, $r\geqslant 1$.

Similarly to Theorems 2.7.10 and 2.7.11 it is easy to prove the following two theorems.

THEOREM 2.9.7. *Let $\mathcal{I}(t,x',y)$ be a boundary Lagrangian distribution of order l associated with the* ith *reflection of our Hamiltonian billiards. Then for any phase function $\varphi'\in\mathfrak{F}'_i$ there exists an amplitude $a\in S^l$ possessing properties described in Definition* 2.9.1 *and such that $\mathcal{I}(t,x',y) = \mathcal{I}_{\varphi',a}(t,x',y)$ modulo a C^∞-half-density.*

THEOREM 2.9.8. *The leading homogeneous term of the symbol of a boundary Lagrangian distribution is independent of the choice of the phase function and is uniquely determined by the Lagrangian distribution itself.*

This enables us to introduce, similarly to Definition 2.7.12, the following

DEFINITION 2.9.9. The leading homogeneous term $\mathfrak{a}_l$ (of degree l) of the symbol $\mathfrak{a}$ is called the *principal symbol* of the boundary Lagrangian distribution $\mathcal{I}$ (of order l).

Consider a differential operator $B(x', D_x)$ of order p, and suppose that the restriction $(B\cdot)|_{\partial M_x}$ acts from the space of half-densities on M_x into the space of half-densities on ∂M_x. In particular, B can be the identity operator. The latter acts into the space of half-densities on ∂M_x because in 1.1.2 we agreed to specify once and for all the choice of the "normal" coordinate. Denote by $B_p(x', \xi)$ the principal symbol of $(B\cdot)|_{\partial M_x}$.

LEMMA 2.9.10. *Let $\mathcal{I}(t, x, y)$ be a standard Lagrangian distribution of order l associated with the ith leg of the canonical transformation* (2.3.10), $i = 1, \dots, \mathbf{r}-1$, *and let $\mathfrak{a}_l(t; y, \eta)$ be its principal symbol. Then $(B\mathcal{I})|_{\partial M_x}$ is a sum of two boundary Lagrangian distributions of order $l + p$: a distribution associated with the ith reflection with principal symbol*

$$\frac{B_p\left(x^{*\prime}|_{t=t_i^*}, \xi^*|_{t=t_i^*}\right) \mathfrak{a}_l|_{t=t_i^*}}{\sqrt{h_{\xi_n}\left(x^*|_{t=t_i^*}, \xi^*|_{t=t_i^*}\right)}}, \tag{2.9.10}$$

and a distribution associated with the $(i+1)$st *reflection with principal symbol*

$$\frac{B_p\left(x^{*\prime}|_{t=t_{i+1}^*}, \xi^*|_{t=t_{i+1}^*}\right) \mathfrak{a}_l|_{t=t_{i+1}^*}}{\sqrt{-h_{\xi_n}\left(x^*|_{t=t_{i+1}^*}, \xi^*|_{t=t_{i+1}^*}\right)}}. \tag{2.9.11}$$

Here

$$(x^*(t; y, \eta), \xi^*(t; y, \eta)), \qquad t_i^*(y, \eta) \leqslant t \leqslant t_{i+1}^*(y, \eta), \qquad (y, \eta) \in O,$$

are the corresponding Hamiltonian trajectories, and $t_i^*(y, \eta)$ and $t_{i+1}^*(y, \eta)$ are the moments of the ith and $(i+1)$th reflections, respectively. The square roots in (2.9.10), (2.9.11) *are chosen to be positive.*

The same result holds for $i = 0$ and $i = \mathbf{r}$, only in these cases $(B\mathcal{I})|_{\partial M_x}$ is one boundary Lagrangian distribution of order $l+p$ corresponding to the first and last reflection, respectively (not a sum of two boundary Lagrangian distributions). The principal symbol of this boundary Lagrangian distribution is given by formula (2.9.11) *in the case $i = 0$, and by* (2.9.10) *in the case $i = \mathbf{r}$.*

LEMMA 2.9.11. *Let $\mathcal{I}(t, x, y)$ be a boundary layer Lagrangian distribution of order l associated with the ith reflection, $i = 1, 2, \dots, \mathbf{r}$, and let $\mathfrak{a}_l(t; y, \eta)$ be its principal symbol. Then $(B\mathcal{I})|_{\partial M_x}$ is a boundary Lagrangian distribution of order $l + p$ associated with the ith reflection. The principal symbol of this boundary Lagrangian distribution is given by formula* (2.9.10). *Here*

$$(x^*(t; y, \eta), \xi^*(t; y, \eta)), \qquad (y, \eta) \in O,$$

are the corresponding complex Hamiltonian trajectories, and

$$\sqrt{h_{\xi_n}\left(x^*|_{t=t_i^*}, \xi^*|_{t=t_i^*}\right)} = \left|h_{\xi_n}\left(x^*|_{t=t_i^*}, \xi^*|_{t=t_i^*}\right)\right|^{1/2} \exp\left(\frac{i}{2}\arg_0\left(h_{\xi_n}\left(x^*|_{t=t_i^*}, \xi^*|_{t=t_i^*}\right)\right)\right),$$

where the continuous branch $\arg_0\left(h_{\xi_n}\left(x^*|_{t=t_i^*}, \xi^*|_{t=t_i^*}\right)\right)$ *is chosen in accordance with Definition* 2.6.16.

Theorems 2.9.10 and 2.9.11 are proved by direct substitution (cf. Theorem 2.7.15). The square roots in (2.9.10), (2.9.11) appear as a consequence of Lemmas 2.5.1, 2.6.12 and Definitions 2.7.1, 2.8.1, 2.9.1, 2.6.7, 2.6.8, 2.6.17 which establish a relation between d_φ and $d_{\varphi'}$.

2.10. Parameter-dependent oscillatory integrals

When we introduced the notion of a standard oscillatory integral associated with our canonical transformation (Definition 2.7.1), we required the support of the amplitude to lie in a tube separated from ∂M_y. Further on in this book we will need to consider the case when the distance from this tube to ∂M_y (which will be characterized by the parameter d) tends to zero. For this reason we introduce in this section the notion of parameter-dependent oscillatory integrals and give their basic properties.

We will not have to develop the theory of parameter-dependent oscillatory integrals to the same extent as for oscillatory integrals without a parameter because we will not consider reflections from the boundary in the parameter-dependent case. The singularities of our parameter-dependent oscillatory integrals will stay at a distance $\gtrsim d$ from ∂M_x. This simplifies matters substantially.

In this section $d \in (0, d_0]$ is a parameter and $d_0 < 1$ is a small fixed number.

Subsections 1–5 deal with parameter-dependent time-dependent oscillatory integrals (and for the sake of brevity we shall omit the word "time-dependent" in these subsections), whereas subsection 6 deals with parameter-dependent time-independent oscillatory integrals.

1. Phase functions. In this subsection we define the phase functions which we use in our parameter-dependent oscillatory integrals. These phase functions themselves do not depend on the parameter d.

Let $\widehat{M}$ be some (fixed) smooth extension of the manifold M; see 1.1.2. By $\widehat{h} \in C^\infty(T'\widehat{M})$ we shall denote some (fixed) extension of our original Hamiltonian $h \in C^\infty(T'M)$, and by $(\widehat{x}^*(t;\widehat{y},\widehat{\eta}), \widehat{\xi}^*(t;\widehat{y},\widehat{\eta}))$ the corresponding Hamiltonian trajectories. We shall denote

$$\widehat{\mathfrak{D}} := \mathbb{R} \times T'\widehat{M}_{\widehat{y}},$$
$$\widehat{\mathfrak{C}} := \{ (t, \widehat{x}; \widehat{y}, \widehat{\eta}) \ : \ (t; \widehat{y}, \widehat{\eta}) \in \widehat{\mathfrak{D}} \,,\ \widehat{x} = \widehat{x}^*(t;\widehat{y},\widehat{\eta}) \}\,.$$

We shall say that a phase function $\widehat{\varphi} \in C^\infty(\mathbb{R} \times \widehat{M}_{\widehat{x}} \times T'\widehat{M}_{\widehat{y}})$ belongs to the class $\widehat{\mathfrak{F}}$ if the following conditions are satisfied:

$$\widehat{\varphi}_{\widehat{\eta}}(t,\widehat{x};\widehat{y},\widehat{\eta}) = 0 \quad \text{if and only if} \quad (t,\widehat{x};\widehat{y},\widehat{\eta}) \in \widehat{\mathfrak{C}},$$
$$\widehat{\varphi}_{\widehat{x}}(t,\widehat{x}^*;\widehat{y},\widehat{\eta}) = \widehat{\xi}^*, \qquad \forall\,(t;\widehat{y},\widehat{\eta}) \in \widehat{\mathfrak{D}},$$
$$\det \widehat{\varphi}_{\widehat{x}\widehat{\eta}}(t,\widehat{x}^*;\widehat{y},\widehat{\eta}) \neq 0, \qquad \forall\,(t;\widehat{y},\widehat{\eta}) \in \widehat{\mathfrak{D}}.$$

The properties of phase functions $\widehat{\varphi} \in \widehat{\mathfrak{F}}$ are listed in Section 2.4 (note that in this case everything is facilitated by the fact that the manifold $\widehat{M}$ has no boundary). In particular, any phase function $\widehat{\varphi} \in \widehat{\mathfrak{F}}$ is simple in some conic neighbourhood

$\widehat{\mathcal{O}} \subset \mathbb{R} \times \widehat{M_{\widehat{x}}} \times T'\widehat{M_{\widehat{y}}}$ of the set $\widehat{\mathfrak{C}}$. In accordance with Definition 2.6.7 we shall specify the choice of the continuous branch $\arg_0(\det^2 \widehat{\varphi}_{\widehat{x}\widehat{\eta}})$ of the argument $\arg(\det^2 \widehat{\varphi}_{\widehat{x}\widehat{\eta}})$ by the condition

$$\arg_0\left(\det^2 \widehat{\varphi}_{\widehat{x}\widehat{\eta}}\right)\big|_{t=0,\, \widehat{x}=\widehat{y}} = 0\,.$$

Finally, by $\check{\mathfrak{F}}$ we shall denote the class of phase functions

$$\varphi \in C^\infty(\mathbb{R} \times M_x \times T'M_y)$$

which are restrictions to the set $\{\widehat{x} \in M,\ \widehat{y} \in M\}$ of phase functions

$$\widehat{\varphi} \in C^\infty(\mathbb{R} \times \widehat{M_{\widehat{x}}} \times T'\widehat{M_{\widehat{y}}})$$

from the class $\widehat{\mathfrak{F}}$. Naturally, by $\arg_0(\det^2 \varphi_{x\eta})$ we shall denote the restriction of $\arg_0(\det^2 \widehat{\varphi}_{\widehat{x}\widehat{\eta}})$ to the set $\{\widehat{x} \in M,\ \widehat{y} \in M\}$. Similarly, $\mathfrak{C} := \widehat{\mathfrak{C}} \cap \{\widehat{x} \in M,\ \widehat{y} \in M\}$.

Phase functions from $\check{\mathfrak{F}}$ are the ones we are going to use in our parameter-dependent oscillatory integrals. Clearly, a restriction of a function $\varphi \in \check{\mathfrak{F}}$ to an appropriate set $\mathcal{O} \subset \mathbb{R} \times M_x \times T'\overset{\circ}{M}_y$ gives a phase function from $\mathfrak{F}_0$.

2. Amplitudes. Further on ε_1 and ε_2 are arbitrary fixed positive constants satisfying the conditions

$$\varepsilon_1 \leqslant 1/2\,, \tag{2.10.1}$$

$$\varepsilon_1 < \varepsilon_2\,. \tag{2.10.2}$$

Throughout this section we shall denote

$$M_y(d) = \{d \leqslant y_n \leqslant 4d\} \subset M_y\,,$$

$$\mathbf{O}^+(d) = T'M_y(d) \cap \{h_{\xi_n}(y,\eta) \geqslant -d^{\varepsilon_1}\}\,, \tag{2.10.3$^+$}$$

$$\mathbf{O}^-(d) = T'M_y(d) \cap \{h_{\xi_n}(y,\eta) \leqslant d^{\varepsilon_1}\}\,. \tag{2.10.3$^-$}$$

The set $M_y(d)$ is a narrow strip stretching "parallel" to the boundary of the manifold M_y, and the sets $\mathbf{O}^\pm(d)$ are conic subsets of $T'M_y(d)$. Obviously, $\mathbf{O}^+(d) \cup \mathbf{O}^-(d) = T'M_y(d)$.

Let us choose a constant $T_{\varepsilon_1} > 0$ such that $x_n^*(t;y,\eta) \geqslant d/2$ for all

$$(t;y,\eta) \in (\mathbf{O}^+(d) \times [-T_{\varepsilon_1} d, T_{\varepsilon_1} d^{1-\varepsilon_1}]) \cup (\mathbf{O}^-(d) \times [-T_{\varepsilon_1} d^{1-\varepsilon_1}, T_{\varepsilon_1} d])\,.$$

Set

$$T_-^+(d) = -T_{\varepsilon_1} d\,, \qquad T_+^+(d) = T_{\varepsilon_1} d^{1-\varepsilon_1}\,, \tag{2.10.4$^+$}$$

$$T_-^-(d) = -T_{\varepsilon_1} d^{1-\varepsilon_1}\,, \qquad T_+^-(d) = T_{\varepsilon_1} d\,. \tag{2.10.4$^-$}$$

In this subsection we shall introduce two types of amplitudes:

1. Amplitudes of type $^+$, which will be defined for $t \in (T_-^+(d), T_+^+(d))$.
2. Amplitudes of type $^-$, which will be defined for $t \in (T_-^-(d), T_+^-(d))$.

In all subsequent formulae the type $^+$ or $^-$ will always be indicated by the respective superscript.

Denote

$$\mathbf{D}^\pm := \{(d,t) :\ d \in (0, d_0],\ t \in (T_-^\pm(d), T_+^\pm(d))\}\,.$$

We shall consider amplitudes $a^\pm(d,t,x;y,\eta)$ defined on $\mathbf{D}^\pm \times M_x \times T'M_y$ which for any fixed $d \in (0,d_0]$ are infinitely smooth as functions of $(t,x;y,\eta)$ and satisfy the following conditions.

1. For each $d \in (0,d_0]$ we have

$$\begin{aligned}&\operatorname{supp} a^\pm(d,\cdot\,,\,\cdot\,;\,\cdot\,,\,\cdot\,)\\ &\qquad\subset\,(T^\pm_-(d),T^\pm_+(d))\times M_x\times(\mathbf{O}^\pm(d)\cap\{h(y,\eta)\geqslant d^{-1-\varepsilon_2}\}).\end{aligned}\tag{2.10.5}$$

2. In any local coordinates x, y for any nonnegative integer p and any multiindices a, β, γ we have

$$\left|\partial_t^p\partial_x^\alpha\partial_y^\beta\partial_\eta^\gamma\,a^\pm\right|\,\leqslant\,c^\pm_{p\alpha\beta\gamma}\,d^{-|\beta|-|\gamma|\varepsilon_1}\,|\eta|^{l-|\gamma|}\tag{2.10.6}$$

uniformly over $\mathbf{D}^\pm \times K_x \times T'K_y$. Here and throughout this section K_x, K_y are arbitrary compact sets in the coordinate patches $\Omega_x \subset M_x$, $\Omega_y \subset M_y$, respectively. The constants $c^\pm_{p\alpha\beta\gamma}$ depend, of course, on the choice of local coordinates and compact sets.

Amplitudes satisfying the two conditions stated above will be called *parameter-dependent amplitudes of the type* $(0,0;1,\varepsilon_1)^\pm$. The class of such amplitudes will be denoted by $S^l(0,0;1,\varepsilon_1)^\pm$. In this notation the four numbers 0, 0, 1, and ε_1 indicate the extra negative powers of the small parameter d which we get upon differentiating with respect to t, x, y, and η.

Note that the amplitudes from the class $S^l(0,0;1,\varepsilon_1)^\pm$ behave differently in the variables x and y: condition (2.10.6) allows these amplitudes to be highly irregular in y, that is, each differentiation in y gives an extra factor of d^{-1}. The irregular behaviour with respect to y is unavoidable because formulae (2.10.3$^\pm$), (2.10.5) imply $\operatorname{supp} a^\pm(d,\cdot\,,\,\cdot\,;\,\cdot\,,\,\cdot\,)\subset\{d\leqslant y_n\leqslant 4d\}$.

Our definition of the class of amplitudes $S^l(0,0;1,\varepsilon_1)^\pm$ is invariant under changes of local coordinates x and y. For x this fact is obvious, whereas for y it requires some explanation. Let $y=y(\widetilde{y})$ be a change of local coordinates. Then the dual coordinates change as

$$\eta\,=\,\eta(\widetilde{y},\widetilde{\eta})=\sum_{k=1}^n\widetilde{\eta}_k\left.\frac{\partial\widetilde{y}_k}{\partial y}\right|_{y=y(\widetilde{y})}.$$

Consequently

$$\begin{aligned}\frac{\partial\widetilde{a}^\pm}{\partial\widetilde{y}_p}\;&=\;\sum_{j=1}^n\left.\frac{\partial a^\pm}{\partial y_j}\right|_{y=y(\widetilde{y}),\,\eta=\eta(\widetilde{y},\widetilde{\eta})}\times\frac{\partial y_j}{\partial\widetilde{y}_p}\\ &\quad+\sum_{j,k,i=1}^n\left.\frac{\partial a^\pm}{\partial\eta_j}\right|_{y=y(\widetilde{y}),\,\eta=\eta(\widetilde{y},\widetilde{\eta})}\times\widetilde{\eta}_k\times\left.\frac{\partial^2\widetilde{y}_k}{\partial y_j\partial y_i}\right|_{y=y(\widetilde{y}),\,\eta=\eta(\widetilde{y},\widetilde{\eta})}\times\frac{\partial y_i}{\partial\widetilde{y}_p}\,.\end{aligned}$$

The right-hand side of this formula can be estimated through (2.10.6), and we see that it is bounded by $\text{const}\,d^{-1}\,|\widetilde{\eta}|^l$ as long as $\varepsilon_1\leqslant 1$, which is true by (2.10.1). Similarly one can consider higher order and mixed derivatives.

REMARK 2.10.1. It is useful to note that any parameter-dependent amplitude from our class $S^l(0,0;1,\varepsilon_1)^\pm$ is uniformly contained in Hörmander's class $S^l_{\rho,\delta}$ (see, for example, [**Sh**]) with

$$\rho\,=\,1\,-\,\frac{\varepsilon_1}{1+\varepsilon_2}\,,\qquad\delta\,=\,\frac{1}{1+\varepsilon_2}\,;\tag{2.10.7}$$

here uniformity means that the constants appearing in the estimates of derivatives in the standard definition do not depend on d. Note that formulae (2.10.1), (2.10.2), (2.10.7) imply

$$1-\rho \leqslant \delta < \rho . \tag{2.10.8}$$

The left inequality (2.10.8) guarantees that $S^{\,l}_{\rho,\delta}$ is invariantly defined as a class of functions on $\mathbb{R}\times M_x\times T'M_y$. The right inequality is a usual condition in microlocal analysis which ensures that the amplitude possesses some basic good properties (see [**Sa9**] where the classes $S^{\,l}_{\rho,\delta}$ are discussed in greater detail). In view of the first observation, our previous statement about the invariance of $S^{\,l}(0,0;1,\varepsilon_1)^{\pm}$ becomes quite natural (though it does not directly follow from (2.10.8)).

We denote $S^{\,-\infty,\pm}=\bigcap_{l\in\mathbb{R}}S^{\,l}(0,0;1,\varepsilon_1)^{\pm}$. Here we have dropped the indices 0, 0, 1, ε_1 in $S^{\,-\infty,\pm}$ because this class does not actually depend on them.

We shall use the standard notation

$$b(d,t,x;y,\eta)\sim\sum_{k=0}^{\infty}b_k(d,t,x;y,\eta)$$

for describing asymptotic convergence with respect to two independent asymptotic parameters $d\to+0$ and $|\eta|\to+\infty$. This means that for sufficiently large q the remainder

$$b(d,t,x;y,\eta)\ -\ \sum_{k=0}^{q}b_k(d,t,x;y,\eta)$$

(and its every derivative with respect to $(t,x;y,\eta)$) is uniformly bounded by a constant times an arbitrary positive power of d times an arbitrary negative power of $|\eta|$.

Similarly to Lemma 2.1.2 we have

LEMMA 2.10.2. *For any sequence of functions* $a^{\pm}_{l-k}(d,t,x;y,\eta)$, $k=0,1,\dots,$ *satisfying the condition*

$$d^{k(1+\varepsilon_1)}a^{\pm}_{l-k}(d,t,x;y,\eta)\in S^{\,l-k}(0,0;1,\varepsilon_1)^{\pm} \tag{2.10.9}$$

there exists a function $a^{\pm}(d,t,x;y,\eta)\in S^{\,l}(0,0;1,\varepsilon_1)^{\pm}$ *such that*

$$a^{\pm}(d,t,x;y,\eta)\sim\sum_{k=0}^{\infty}a^{\pm}_{l-k}(d,t,x;y,\eta). \tag{2.10.10}$$

This function is determined uniquely modulo $S^{\,-\infty,\pm}$.

3. Cut-offs. Let $\widehat{\varsigma}(t,\widehat{x};\widehat{y},\widehat{\eta})\in C^\infty(\mathbb{R}\times\widehat{M_{\widehat{x}}}\times T'\widehat{M_{\widehat{y}}})$ be a function positively homogeneous in $\widehat{\eta}$ of degree 0 and such that

1. $\operatorname{supp}\widehat{\varsigma}$ is a conically compact subset of $\widehat{\mathcal{O}}$.
2. $\widehat{\varsigma}=1$ on some small conic neighbourhood of $\widehat{\mathfrak{C}}$.

(See subsection 1 for definitions of $\widehat{\mathcal{O}}$ and $\widehat{\mathfrak{C}}$.) By $\varsigma(t,x;y,\eta)$ we shall denote the restriction of the function $\widehat{\varsigma}(t,\widehat{x};\widehat{y},\widehat{\eta})$ to the set $\{\widehat{x}\in M,\ \widehat{y}\in M\}$.

By $\varsigma_+\in C^\infty(\mathbb{R}_+)$ we shall denote a function such that $\varsigma_+(z)=0$ for $z\leqslant 1/8$ and $\varsigma_+(z)=1$ for $z\geqslant 1/4$.

In our oscillatory integrals the full cut-off will be $\varsigma_+(x_n/d)\,\varsigma(t,x;y,\eta)$. The term $\varsigma_+(x_n/d)\in C_0^\infty(\overset{\circ}{M}_x)$ is introduced in order to detach the support of the oscillatory integral from ∂M_x. The condition $\varsigma_+(z) = 1$ for $z \geqslant 1/4$ is necessary to ensure that $\varsigma_+(x_n/d) \equiv 1$ in a conic neighbourhood of $\mathfrak{C} \cap \operatorname{supp} a^\pm$.

4. Oscillatory integrals. For $t \in (T_-^\pm(d), T_+^\pm(d))$ we shall consider parameter-dependent oscillatory integrals of the form

$$\begin{aligned} &\mathcal{I}_{\varphi,a^\pm}(d,t,x,y) \\ &\quad = \varsigma_+(x_n/d) \int e^{i\varphi(t,x;y,\eta)}\, a^\pm(d,t,x;y,\eta)\, \varsigma(t,x;y,\eta)\, d_\varphi(t,x;y,\eta)\, {đ}\eta \end{aligned} \tag{2.10.11}$$

with phase function $\varphi \in \check{\mathfrak{F}}$ and amplitude $a^\pm \in S^l(0,0;1,\varepsilon_1)^\pm$. In the above formula

$$d_\varphi(t,x;y,\eta) \;=\; (\det{}^2\varphi_{x\eta})^{1/4} \;=\; |\det \varphi_{x\eta}|^{1/2}\, e^{i\arg_0(\det^2\varphi_{x\eta})/4},$$

where the continuous branch $\arg_0(\det^2\varphi_{x\eta})$ is chosen as described in subsection 1 and the cut-offs ς and ς_+ are as in subsection 3.

We shall call oscillatory integrals (2.10.11) *parameter-dependent oscillatory integrals of the type* $(0,0;1,\varepsilon_1)^\pm$.

5. Properties of parameter-dependent oscillatory integrals.

DEFINITION 2.10.3. Let $f^\pm(d,t,x,y)$ be a half-density in (x,y) defined for $d \in (0,d_0]$, $t \in (T_-^\pm(d), T_+^\pm(d))$, $x \in M_x$, $y \in M_y$. We shall say that $f^\pm$ is of the class $d^\infty((T_-^\pm(d), T_+^\pm(d)) \times M_x \times M_y)$ if

1. For each fixed $d \in (0,d_0]$, it is infinitely smooth.
2. $\operatorname{supp} f^\pm(d,\cdot,\cdot,\cdot) \subset (T_-^\pm(d), T_+^\pm(d)) \times \overset{\circ}{M}_x \times \overset{\circ}{M}_y$, $\forall d \in (0,d_0]$.
3. For any $r,s \in \mathbb{N}$ and any differential operators $F_x(x,D_x)$, $F_y(y,D_y)$ the half-density $d^{-r} D_t^s F_x F_y f^\pm$ is uniformly bounded over $d \in (0,d_0]$, $t \in (T_-^\pm(d), T_+^\pm(d))$, $x \in M_x$, $y \in M_y$.

Roughly speaking, $f^\pm \in d^\infty((T_-^\pm(d), T_+^\pm(d)) \times M_x \times M_y)$ means that any given partial derivative of $f^\pm$ with respect to (t,x,y) (including $f^\pm$ itself) vanishes faster than any given power of d as $d \to 0$.

Repeating the arguments from 2.1.1 with account of (2.10.5), (2.10.6), it is easy to see that a change of the cut-offs ς, ς_+ changes the oscillatory integral (2.10.11) by a $d^\infty((T_-^\pm(d), T_+^\pm(d)) \times M_x \times M_y)$-term.

Similarly to Definition 2.7.2 we introduce the following

DEFINITION 2.10.4. A distribution $\mathcal{I}^\pm(d,t,x,y)$ which can be written modulo $d^\infty((T_-^\pm(d), T_+^\pm(d)) \times M_x \times M_y)$ as a parameter-dependent oscillatory integral (2.10.11) is called a *parameter-dependent Lagrangian distribution* of order l of the type $(0,0;1,\varepsilon_1)^\pm$.

Going through the arguments presented in the previous sections one can check that the theory developed for standard oscillatory integrals without a parameter remains mostly true for parameter-dependent oscillatory integrals of the type $(0,0;1,\varepsilon_1)^\pm$. The difference is that all the operations with parameter-dependent oscillatory integrals are carried out modulo $d^\infty((T_-^\pm(d), T_+^\pm(d)) \times M_x \times M_y)$. For the sake of brevity we shall not repeat all these arguments. We state below the basic properties of parameter-dependent oscillatory integrals which we will use later in Section 3.5.

Note that the derivation of properties listed below does not require operations in the variable y which plays the role of a parameter. This is important because, as we noted above in subsection 2, our amplitudes are highly irregular in y. As to x, our amplitudes behave nicely in this variable in the sense that we do not get negative powers of the small parameter d when we differentiate with respect to x. We still have, of course, a slight irregularity in the variable η: each differentiation with respect to η gives a factor $d^{-\varepsilon_1}$, but this is more than compensated by the additional factor $|\eta|^{-1} \lesssim d^{1+\varepsilon_2}$; see (2.10.6), (2.10.5), (2.10.2).

Similarly to Theorem 2.7.5 we have

THEOREM 2.10.5. *Any parameter-dependent Lagrangian distribution of order* l *of the type* $(0,0;1,\varepsilon_1)^\pm$ *can be written modulo* $d^\infty((T_-^\pm(d), T_+^\pm(d)) \times M_x \times M_y)$ *in the form*

$$\begin{aligned} &\mathcal{I}_{\varphi,\mathfrak{a}^\pm}(d,t,x,y) \\ &\quad = \varsigma_+(x_n/d) \int e^{i\varphi(t,x;y,\eta)}\, \mathfrak{a}^\pm(d,t;y,\eta)\, \varsigma(t,x;y,\eta)\, d_\varphi(t,x;y,\eta)\, đ\eta \end{aligned} \tag{2.10.12}$$

with an amplitude $\mathfrak{a}^\pm \in S^l(0,0;1,\varepsilon_1)^\pm$ *independent of* x.

Similarly to Definition 2.7.6 we introduce

DEFINITION 2.10.6. The amplitude $\mathfrak{a}^\pm$ (appearing in (2.10.12) and independent of x) is called the *symbol* of the parameter-dependent Lagrangian distribution $\mathcal{I}^\pm(d,t,x,y)$.

Similarly to Proposition 2.7.7 we have

PROPOSITION 2.10.7. *The symbol* $\mathfrak{a}^\pm$ *is uniquely determined modulo* $S^{-\infty,\pm}$ *by the parameter-dependent Lagrangian distribution* $\mathcal{I}^\pm$ *and the phase function* φ.

LEMMA 2.10.8. *The symbol of a parameter-dependent Lagrangian distribution can be computed in accordance with our standard formulae* (2.7.12)–(2.7.18), *with asymptotic convergence being understood as described above in subsection* 2.

Let us illustrate Lemma 2.10.8 by the following argument. Let $a^\pm(d,t,x;y,\eta) \in S^l(0,0;1,\varepsilon_1)^\pm$. Then by (2.7.14), (2.10.6) we have

$$d^{k\varepsilon_1}\, \mathfrak{S}_{-k}\, a^\pm(d,t,x;y,\eta) \in S^{l-k}(0,0;1,\varepsilon_1)^\pm. \tag{2.10.13}$$

Set $a_{l-k}^\pm := \mathfrak{S}_{-k}\, a^\pm$. Then formula (2.10.13) implies (2.10.9), and consequently we have asymptotic convergence (2.10.10).

Similarly to Theorem 2.7.10 we have

THEOREM 2.10.9. *Let* $\mathcal{I}^\pm(d,t,x,y)$ *be a parameter-dependent Lagrangian distribution of order* l *of the type* $(0,0;1,\varepsilon_1)^\pm$. *Then for any phase function* $\varphi \in \check{\mathfrak{F}}$ *there exists an amplitude* $a^\pm \in S^l(0,0;1,\varepsilon_1)^\pm$ *such that*

$$\mathcal{I}^\pm(d,t,x,y) = \mathcal{I}_{\varphi,a^\pm}(d,t,x,y)$$

modulo a d^∞*-half-density.*

Similarly to Lemma 2.7.15 we have

LEMMA 2.10.10. *Let $\mathcal{I}^{\pm}(d,t,x,y)$ be a parameter-dependent Lagrangian distribution of order l of the type $(0,0;1,\varepsilon_1)^{\pm}$, and let $A(x,D_x)$ be a differential operator of order p acting on M. Then $A\mathcal{I}^{\pm}$ is a parameter-dependent Lagrangian distribution of order $l+p$ of the type $(0,0;1,\varepsilon_1)^{\pm}$, and the amplitude of the corresponding oscillatory integral can be evaluated in accordance with* (2.7.38).

REMARK 2.10.11. Continuing our analogy with $S^l_{\rho,\delta}$ (see Remark 2.10.1), let us note that in the case of oscillatory integrals with amplitudes from $S^l_{\rho,\delta}$ and more or less general phase functions the condition (2.10.8) is insufficient for developing a proper theory, and should be replaced by the stronger condition

$$1-\rho \leqslant \delta < 1/2. \tag{2.10.14}$$

In particular, (2.10.14) ensures that the symbol of a Lagrangian distribution associated with a Hamiltonian flow is well defined within the class $S^l_{\rho,\delta}$; this fact is established by a simple analysis of formulae (2.7.16), (2.7.17). But in our case (2.10.14) does not necessarily hold. We managed to overcome this difficulty only because we produced results which did not require operations in the irregular spatial variable y.

6. Parameter-dependent pseudodifferential operators. Let us consider time-independent amplitudes $a(d,x;y,\eta)$ defined on $(0,d_0]\times M_x\times T'M_y$ which are infinitely smooth in $(x;y,\eta)$ and satisfy the following three conditions:

$$\operatorname{supp} a(d,\,\cdot\,;\,\cdot\,,\,\cdot\,) \subset \{h(y,\eta)\geqslant d^{-1-\varepsilon_2}\}, \qquad \forall d\in(0,d_0], \tag{2.10.15}$$

$$\begin{aligned}&\operatorname{supp} a(d,\,\cdot\,;\,\cdot\,,\,\cdot\,)\cap\{x=y\}\\ &\qquad\subset M_x(d)\times T'M_y(d), \qquad \forall d\in(0,d_0],\end{aligned} \tag{2.10.16}$$

and in any local coordinates x, y for any nonnegative multiindices a, β, γ

$$\left|\partial_x^\alpha\partial_y^\beta\partial_\eta^\gamma\, a\right| \leqslant c_{\alpha\beta\gamma}\, d^{-|\alpha|-|\beta|-|\gamma|\varepsilon_1}\,|\eta|^{l-|\gamma|} \tag{2.10.17}$$

uniformly over $(0,d_0]\times K_x\times T'K_y$.

We shall call such amplitudes *parameter-dependent time-independent amplitudes of the type* $(1;1,\varepsilon_1)$. The class of these amplitudes will be denoted by $S^l(1;1,\varepsilon_1)$. In this notation the three numbers 1, 1, and ε_1 indicate the extra negative powers of the small parameter d which we get differentiating with respect to x, y, and η.

Note that according to (2.10.17) amplitudes from the class $S^l(1;1,\varepsilon_1)$ may behave highly irregularly both in x and in y. This is the basic difference with the class of time-dependent amplitudes $S^l(0,0;1,\varepsilon_1)^{\pm}$ introduced in subsection 2.

Our definition of the class of time-independent amplitudes $S^l(1;1,\varepsilon_1)$ is, of course, invariant under changes of local coordinates x and y.

Naturally, we denote $S^{-\infty}=\bigcap_{l\in\mathbb{R}} S^l(1;1,\varepsilon_1)$. A time-independent version of Lemma 2.10.2 holds for our amplitudes.

Let $\varsigma(x;y,\eta)\in C^\infty(M_x\times T'M_y)$ be a function positively homogeneous in η of degree 0 and such that

1. $\operatorname{supp}\varsigma$ lies in a small conic neighbourhood of the set $\{x=y\}$.
2. $\varsigma=1$ on some small conic neighbourhood of the set $\{x=y\}$.

Let $\varsigma_+\in C^\infty(\mathbb{R})$ be the same as in subsection 3.

We shall consider time-independent oscillatory integrals of the form

(2.10.18)
$$\mathcal{I}_{\varphi,a}(d,x,y) = \varsigma_+(x_n/d)\left(\int e^{i\varphi(x;y,\eta)}\,a(d,x;y,\eta)\,\varsigma(x;y,\eta)\,d_\varphi(x;y,\eta)\,đ\eta\right)\varsigma_+(y_n/d)$$

with phase function $\varphi \in \mathfrak{F}^0$ (see 2.7.5) and amplitude $a \in S^l(1;1,\varepsilon_1)$. We shall call oscillatory integrals (2.10.18) *parameter-dependent time-independent oscillatory integrals of the type* $(1;1,\varepsilon_1)$.

Similarly to Definition 2.10.3 we define the class $d^\infty(M_x \times M_y)$. By analogy with Definition 2.10.4 we introduce

DEFINITION 2.10.12. A distribution $\mathcal{I}(d,x,y)$ which can be written modulo $d^\infty(M_x \times M_y)$ as a parameter-dependent time-independent oscillatory integral (2.10.18) is called a *parameter-dependent time-independent Lagrangian distribution* of order l of the type $(1;1,\varepsilon_1)$.

DEFINITION 2.10.13. An operator $P : C^\infty(\overset{\circ}{M}_y) \to C_0^\infty(\overset{\circ}{M}_x)$ the Schwartz kernel of which is a parameter-dependent time-independent Lagrangian distribution of order $\mathbf{l}$ of the type $(1;1,\varepsilon_1)$ is called a *parameter-dependent pseudodifferential operator* of order $\mathbf{l}$ of the type $(1;1,\varepsilon_1)$.

Note that for each fixed $d \in (0,d_0]$ a parameter-dependent pseudodifferential operator of order $\mathbf{l}$ belongs to the class $\Psi_0^{\mathbf{l}}$ introduced in 2.1.3; see also 2.7.5.

One can check that all the theory developed for pseudodifferential operators without a parameter remains true for parameter-dependent oscillatory integrals of the type $(1;1,\varepsilon_1)$, with all the operations being carried out modulo $d^\infty(M_x \times M_y)$. This statement requires some explanation because previously (subsection 5) we dealt only with properties the derivation of which did not require operations with respect to "highly irregular" spatial variables. In the case of a parameter-dependent pseudodifferential operator of the type $(1;1,\varepsilon_1)$ both x and y are highly irregular in the sense that each differentiation with respect to x or y gives a factor of d^{-1}. This irregularity does not, however, cause problems in the case of a parameter-dependent pseudodifferential operator because the phase function in this case is more special. The difference can be illustrated by considering the procedure of evaluation of the symbol of a Lagrangian distribution (i.e., the exclusion of the variable x from the amplitude). In the time-dependent case this procedure requires us to perform up to $2r$ differentiations with respect to x whenever we want to gain a factor of $|\eta|^{-r}$; see formulae (2.7.15), (2.7.17). In the case of the Schwartz kernel of a pseudodifferential operator we have to perform only up to r differentiations with respect to x whenever we want to gain a factor of $|\eta|^{-r}$; this fact can be established by analyzing formula (2.7.14) with account of the special structure of φ ($x^* \equiv y$ does not depend on η).

We state below the basic properties of parameter-dependent pseudodifferential operators which we will later use (mostly implicitly) in Sections 3.5 and 5.3.

For parameter-dependent time-independent oscillatory integrals of the type $(1;1,\varepsilon_1)$ we have natural analogues of Theorems 2.10.5, 2.10.9, Proposition 2.10.7, Lemmas 2.10.8, 2.10.10, and Definition 2.10.6. We do not state them explicitly for the sake of brevity.

If P is a parameter-dependent pseudodifferential operator of order $\mathbf{l}$ of the type $(1;1,\varepsilon_1)$, then P^* is a parameter-dependent pseudodifferential operator of order $\mathbf{l}$ of the type $(1;1,\varepsilon_1)$. If P, Q are parameter-dependent pseudodifferential operators of orders $\mathbf{l}$, $\mathbf{m}$ of the type $(1;1,\varepsilon_1)$, then their composition PQ is a parameter-dependent pseudodifferential operator of order $\mathbf{l}+\mathbf{m}$ of the type $(1;1,\varepsilon_1)$. Usual rules for the calculation of symbols (see 2.1.3) apply in these cases.

Let us now consider amplitudes $a^{\pm}(d,x;y,\eta) \in S^l(1;1,\varepsilon_1)$ satisfying the additional condition

$$\operatorname{supp} a(d,\,\cdot\,;\,\cdot\,,\,\cdot\,) \cap \{x=y\} \subset M_x(d) \times \mathbf{O}^{\pm}(d) \tag{2.10.19}$$

(cf. (2.10.16)), where the sets $\mathbf{O}^{\pm}(d)$ are defined by formulae ($2.10.3^{\pm}$). Such amplitudes will be called *parameter-dependent time-independent amplitudes of the type* $(1;1,\varepsilon_1)^{\pm}$, and the class of these amplitudes will be denoted by $S^l(1;1,\varepsilon_1)^{\pm}$. The respective oscillatory integrals, Lagrangian distributions and pseudodifferential operators will be said to be of the type $(1;1,\varepsilon_1)^{\pm}$. It is easy to check that all the operations described in this subsection retain the type $(1;1,\varepsilon_1)^{\pm}$.

Note that a distribution $\mathcal{I}^{\pm}(d,x,y)$ is a parameter-dependent time-independent Lagrangian distribution of the type $(1;1,\varepsilon_1)^{\pm}$ if and only if it is the restriction to $t=0$ of a parameter-dependent time-dependent Lagrangian distribution of the type $(0,0;1,\varepsilon_1)^{\pm}$. This fact is not completely obvious when one compares the oscillatory integrals (2.10.11) and (2.10.18) because conditions (2.10.6) and (2.10.17) prescribe a different type of behaviour of the amplitudes with respect to the variable x; the difficulty can be overcome by excluding the variable x, which is possible according to Theorem 2.10.5 and its time-independent analogue.

REMARK 2.10.14. Let us return for the last time to our analogy with $S^l_{\rho,\delta}$ (see Remarks 2.10.1 and 2.10.11). It is known **[Sh]** that in the case of pseudodifferential operators with amplitudes from $S^l_{\rho,\delta}$ the condition (2.10.8) is sufficient for developing a proper theory, and there is no necessity for replacing them by the stronger condition (2.10.14). In view of this observation it is not surprising that in this subsection we were able to handle the situation when amplitudes are highly irregular both in x and in y.

7. The price to pay. The remarkable thing about the theory of parameter-dependent Lagrangian distributions developed in subsections 1–6 is that all operations are carried out modulo $d^\infty(M_x \times M_y)$, that is, modulo smooth terms rapidly decreasing as $d \to 0$. Unfortunately, we have to pay a certain price for such nice properties. This price is the exclusion of η with $|\eta| \lesssim d^{-1-\varepsilon_2}$ (see (2.10.5) and (2.10.15)). In this subsection we produce estimates concerning the approximation of a given distribution containing all η by a parameter-dependent distribution containing only η with $|\eta| \gtrsim d^{-1-\varepsilon_2}$.

DEFINITION 2.10.15. Let $f(d,x,y)$ be a half-density defined for $d \in (0,d_0]$, $x \in M_x$, $y \in M_y$. We shall say that f is of the class $d^l(M_x \times M_y)$ if

1. For each $d \in (0,d_0]$ it is infinitely smooth with respect to $(x,y) \in M_x \times M_y$.
2. For each $d \in (0,d_0]$ we have $\operatorname{supp} f(d,\,\cdot\,,\,\cdot\,) \subset \overset{\circ}{M}_x \times \overset{\circ}{M}_y$.
3. In any local coordinates x, y for any multiindices α, β we have

$$\left|\partial_x^\alpha \partial_y^\beta f\right| \leqslant c_{\alpha\beta}\, d^{(l-|\alpha|-|\beta|)(1+\varepsilon_2)} \tag{2.10.20}$$

uniformly over $(0,d_0] \times K_x \times K_y$.

We denote $d^{\infty}(M_x \times M_y) = \bigcap_{l\in\mathbb{R}} d^l(M_x \times M_y)$, $d^{-\infty}(M_x \times M_y) = \bigcup_{l\in\mathbb{R}} d^l(M_x \times M_y)$. Obviously, the class $d^{\infty}(M_x \times M_y)$ defined in this manner coincides with the one introduced in subsection 6 (time-independent analogue of Definition 2.10.3).

Let us consider the Dirac function $\delta(x-y)$; more precisely, by $\delta(x-y)$ we understand the Schwartz kernel of the identity operator acting on $C_0^\infty(\mathring{M})$-half-densities. Let $\chi_0 \in C_0^\infty(\mathbb{R}_+)$ be a function such that $\operatorname{supp}\chi_0 \subset [1,4]$. In the remainder of this section we discuss the question how well the distribution $\chi_0(x_n/d)\,\delta(x-y)$ can be approximated by our parameter-dependent oscillatory integrals. This question is motivated by the construction which we eventually will have to perform in Section 5.3 (see Lemma 5.3.2).

The distribution $\delta(x-y)$ can be viewed as the restriction to $\mathring{M}_x \times \mathring{M}_y$ of the Schwartz kernel of a pseudodifferential operator of the class Ψ^0 acting on the extended manifold $\widehat{M}$. So according to the results of Section 2.7 (and, in particular, 2.7.5) for any phase function $\varphi \in \mathfrak{F}^0$ (see the definition at the beginning of subsection 6) there exists a symbol $\mathfrak{c}(y,\eta) \in C^\infty(T'M_y)$ of the class S^0 such that

$$\delta(x-y) - \int e^{i\varphi(x;y,\eta)}\,\mathfrak{c}(y,\eta)\,\varsigma(x;y,\eta)\,\chi_+(h(y,\eta))\,d_\varphi(x;y,\eta)\,đ\eta \in C^\infty(M_x \times M_y)\,.$$

Here ς is as defined in subsection 6, and $\chi_+ \in C^\infty(\mathbb{R}_+)$ is a function such that $\chi_+(z)=0$ for $z \leqslant 1$ and $\chi_+(z)=1$ for $z \geqslant 2$.

Multiplying the latter formula by $\chi_0(x_n/d)\,\varsigma_+(y_n/d)$ we get

$$\chi_0(x_n/d)\,\delta(x-y) - \widetilde{\mathcal{I}}(d,x,y) \in d^0(M_x \times M_y)\,, \tag{2.10.21}$$

where

$$\widetilde{\mathcal{I}}(d,x,y) = \chi_0(x_n/d)\left(\int e^{i\varphi(x;y,\eta)}\,\mathfrak{c}(y,\eta)\,\varsigma(x;y,\eta)\,\chi_+(h(y,\eta))\,d_\varphi(x;y,\eta)\,đ\eta\right)\varsigma_+(y_n/d)\,.$$

The distribution $\widetilde{\mathcal{I}}(d,x,y)$ is not of the type described in Definition 2.10.12 because the support of the amplitude is too close to $\eta=0$. In order to rectify this, set

$$a(d,x;y,\eta) := \chi_0(x_n/d)\,\mathfrak{c}(y,\eta)\,\chi_+(d^{1+\varepsilon_2}h(y,\eta)) \in S^0(1;1,\varepsilon_1)$$

and consider the oscillatory integral $\mathcal{I}_{\varphi,a}(d,x,y)$ defined by formula (2.10.18) with our particular choice of amplitude. The distribution $\widetilde{\mathcal{I}}(d,x,y) - \mathcal{I}_{\varphi,a}(d,x,y)$ is represented by an integral in which integration is carried out over the finite ball $h(y,\eta) \leqslant 2d^{-1-\varepsilon_2}$, so elementary estimates give

$$\widetilde{\mathcal{I}}(d,x,y) - \mathcal{I}_{\varphi,a}(d,x,y) \in d^{-n}(M_x \times M_y)\,. \tag{2.10.22}$$

Combining (2.10.21) and (2.10.22) we get

$$\chi_0(x_n/d)\,\delta(x-y) - \mathcal{I}_{\varphi,a}(d,x,y) \in d^{-n}(M_x \times M_y) \subset d^{-\infty}(M_x \times M_y)\,.$$

Thus, we have approximated the given distribution $\chi_0(x_n/d)\,\delta(x-y)$ by a parameter-dependent Lagrangian distribution of the type described in subsection 6 with a $d^{-\infty}(M_x \times M_y)$-error. The appearance of this $d^{-\infty}(M_x \times M_y)$-error is inevitable, and is caused by the inherent limitations of our approach.

CHAPTER 3

Construction of the Wave Group

Consider the time-dependent unitary operator $\mathbf{U}(t) : L_2(M) \to L_2(M)$ defined as

$$\mathbf{U}(t) := \exp(-it\mathcal{A}^{1/(2m)}) = \sum_{k=1}^{\infty} \exp(-it\lambda_k)\, v_k(x) \int_{M_y} (\,\cdot\,)\, \overline{v_k(y)}\, dy\,. \tag{3.0.1}$$

Here λ_k are the eigenvalues and v_k are the orthonormalized eigenfunctions of the problem (1.1.1), (1.1.2), and $t \in \mathbb{R}$ is a parameter. The operator (3.0.1) is called the *wave group*. The wave group plays a central role in our book: information on the singularities of the Schwartz kernel of the operator $\mathbf{U}(t)$ will allow us to derive with high accuracy (by use of Fourier Tauberian theorems) asymptotics of the counting function $N(\lambda)$ and of the spectral function $e(\lambda, y, y)$.

The aim of this chapter is to construct effectively modulo C^∞ the Schwartz kernel of the wave group.

On a manifold with boundary it is difficult to achieve this aim fully. So we will be forced to restrict ourselves slightly by performing some microlocalization in the variables (x, ξ), (y, η), and some localization in the variable t. Microlocalization in (x, ξ), (y, η) will be introduced by studying the operator

$$\mathbf{U}_P(t) := \mathbf{U}(t)\, P \tag{3.0.2}$$

instead of the original operator $\mathbf{U}(t)$. Here $P \in \Psi_0^1$ is a pseudodifferential operator satisfying certain admissibility conditions. Localization in t will be introduced by doing our constructions on a time interval $(T_-\,, T_+)$ which may be finite.

In Section 3.1 we describe some general properties of distributions characterizing Schwartz kernels of operators of the type $\mathbf{U}_P(t)$. We prove the abstract Theorem 3.1.1 which plays a basic role in this chapter. This theorem will eventually allow us to justify our constructions, i.e., to prove that the oscillatory integrals being constructed really represent modulo C^∞ the Schwartz kernel of the operator $\mathbf{U}_P(t)$.

In Section 3.2 we demonstrate how to apply Theorem 3.1.1 to oscillatory integrals.

In Sections 3.3 and 3.4 we give an effective (in terms of oscillatory integrals) construction of the Schwartz kernel of the operator $\mathbf{U}_P(t)$ for manifolds without and with boundary, respectively.

Finally, in Section 3.5 we repeat our analysis for the parameter-dependent case, when the support of the Schwartz kernel of P is close to the boundary.

3.1. Characteristic properties of distributions associated with the wave group

1. Function spaces. In this subsection we define precisely the spaces of our distributions. Some of the notation is traditional and was used earlier (Chapter 2) without any comment, but we feel it necessary at this stage to give exact definitions in order to prepare the ground for the formulation and subsequent proof of Theorem 3.1.1.

We denote by $C^\infty(M)$ the vector space of infinitely differentiable (up to the boundary!) complex-valued half-densities $v(x)$ on M equipped with the usual C^∞-topology defined by the semi-norms

$$v \to \sum_{p=1}^{q} \sum_{|\alpha|\leqslant k} \max_{x\in M^{(p)}} |\partial_x^\alpha(\chi_p v)|,$$

where k runs through all nonnegative integers, and $1 = \sum_{p=1}^q \chi_p(x)$, $\chi_p \in C^\infty(M)$, $\operatorname{supp}\chi_p \subset M^{(p)}$, is some partition of unity on M with local coordinates x in coordinate patches $M^{(p)}$. We denote by $C_{\bar{B}}^\infty(M)$ the subspace of $C^\infty(M)$ consisting of all the half-densities which satisfy the boundary conditions

$$\left.(B^{(j)} A^r \overline{v})\right|_{\partial M} = 0, \qquad j = 1, 2, \ldots, m, \qquad r = 0, 1, 2, \ldots;$$

the topology on $C_{\bar{B}}^\infty(M)$ is taken to be the same as on $C^\infty(M)$. Following Hörmander [**Hö3,** Vol. 3, App. B.2] we denote by $\dot{\mathcal{D}}'(M)$, $\dot{\mathcal{D}}'_{\bar{B}}(M)$ the spaces dual to $C^\infty(M)$, $C_{\bar{B}}^\infty(M)$, respectively, i.e., spaces of linear continuous functionals on $C^\infty(M)$, $C_{\bar{B}}^\infty(M)$. The value of the functional (distribution) u on the test half-density v will be denoted by $\langle u, v\rangle_x$ with the subscript x emphasizing the variable in which the distribution is acting. We shall say that a distribution $u \in \dot{\mathcal{D}}'_{\bar{B}}(M)$ has order $s \in \mathbb{R}$ if the sequence $\lambda_k^{-s}\langle u, \overline{v_k}\rangle_x$, $k = 1, 2, \ldots$, is bounded. Since we *a priori* know the classical rough asymptotic formula $\lambda_k \sim ck^{1/n}$, $c > 0$ (see e.g. [**RoShSo**]), and rough estimates for $\partial_x^\alpha v_k(x)$ in terms of powers of k (these follow from standard embedding theorems), it is easy to show that distributions from the class $\dot{\mathcal{D}}'_{\bar{B}}(M)$ have finite orders.

Let $T_- < T_+$ be finite or $\mp\infty$. In accordance with traditional notation we denote by $\mathcal{D}'(T_-, T_+)$ the vector space of linear continuous functionals (distributions) on $C_0^\infty(T_-, T_+)$; see [**Hö3,** Vol. 1] for details. The value of the distribution $f \in \mathcal{D}'(T_-, T_+)$ on the test function $g \in C_0^\infty(T_-, T_+)$ will be denoted by $\langle f, g\rangle_t$.

Throughout this section we shall always assume that our distributions behave as half-densities in the variables x, y.

We will be considering distributions of the class $\dot{\mathcal{D}}'_{\bar{B}}(M_x)$ depending on the additional parameter $y \in M_y$ (the subscripts x and y are used to distinguish between the two copies of the manifold M). We will write $u_0(x,y) \in C^\infty(M_y; \dot{\mathcal{D}}'_{\bar{B}}(M_x))$ if

1. For any $v \in C_{\bar{B}}^\infty(M_x)$ we have $\langle u_0, v\rangle_x \in C^\infty(M_y)$.
2. If $w_k \in C_{\bar{B}}^\infty(M_x)$ and $w_k \to 0$ in $C_{\bar{B}}^\infty(M_x)$ as $k \to \infty$, then $\langle u_0, w_k\rangle_x \to 0$ in $C^\infty(M_y)$.

In other words, $u_0(x,y) \in C^\infty(M_y; \dot{\mathcal{D}}'_{\bar{B}}(M_x))$ means that u_0 defines a continuous linear map from $C_{\bar{B}}^\infty(M_x)$ to $C^\infty(M_y)$.

We will also be considering distributions of the class $\dot{\mathcal{D}}'_{\bar{B}}(M_x)$ depending on the parameters $t \in (T_-\,, T_+)$ and $y \in M_y$. We will write

$$u(t,x,y) \in C^\infty((T_-\,, T_+) \times M_y; \dot{\mathcal{D}}'_{\bar{B}}(M_x))$$

if

1. For any $v \in C^\infty_{\bar{B}}(M_x)$ we have $\langle u, v\rangle_x \in C^\infty((T_-\,, T_+) \times M_y)$.
2. If $w_k \in C^\infty_{\bar{B}}(M_x)$ and $w_k \to 0$ in $C^\infty_{\bar{B}}(M_x)$ as $k \to \infty$, then $\langle u, w_k\rangle_x \to 0$ in $C^\infty((T_-\,, T_+) \times M_y)$.

This means that u defines a continuous linear map from $C^\infty_{\bar{B}}(M_x)$ to $C^\infty((T_-\,, T_+) \times M_y)$. Note that convergence in $C^\infty((T_-\,, T_+) \times M_y)$ is understood as convergence of any given partial derivative uniformly over $[t_-\,, t_+] \times M_y$, where $[t_-\,, t_+]$ is an arbitrary closed bounded interval in $(T_-\,, T_+)$.

Similarly, we will be considering distributions of the class $\mathcal{D}'(T_-\,, T_+)$ depending on the parameters $(x, y) \in M_x \times M_y$. We will write

$$u(t,x,y) \in C^\infty(M_x \times M_y; \mathcal{D}'(T_-\,, T_+))$$

if

1. For any $g \in C^\infty_0(T_-\,, T_+)$ we have $\langle u, g\rangle_t \in C^\infty(M_x \times M_y)$.
2. If $g_k \in C^\infty_0(T_-\,, T_+)$ and $g_k \to 0$ in $C^\infty_0(T_-\,, T_+)$ as $k \to \infty$, then $\langle u, g_k\rangle_t \to 0$ in $C^\infty(M_x \times M_y)$.

This means that u defines a continuous linear map from $C^\infty_0(T_-\,, T_+)$ to $C^\infty(M_x \times M_y)$.

By $C^\infty_B(M)$ we shall denote the subspace of $C^\infty(M)$ consisting of all the u which satisfy the boundary conditions

$$(B^{(j)} A^r u)\Big|_{\partial M} = 0\,, \qquad j = 1, 2, \ldots, m, \qquad r = 0, 1, 2, \ldots\,.$$

By $C^\infty_{B_x}(M_x \times M_y)$ and $C^\infty_{B_x}((T_-\,, T_+) \times M_x \times M_y)$ we shall denote the subspaces of $C^\infty(M_x \times M_y)$ and $C^\infty((T_-\,, T_+) \times M_x \times M_y)$, respectively, consisting of all the u which satisfy the boundary conditions

$$(B^{(j)}_x A^r_x u)\Big|_{\partial M_x} = 0\,, \qquad j = 1, 2, \ldots, m, \qquad r = 0, 1, 2, \ldots\,.$$

The subscript in $B^{(j)}_x$ and A_x indicates that the operators $B^{(j)}$ and A act in the variable x.

We shall use the notation "$f = O(|\lambda|^{-\infty})$ as $\lambda \to -\infty$" to describe the fact that the function $f(\lambda) \in C^\infty(\mathbb{R}^1)$ vanishes faster than any given negative power of $|\lambda|$ as $\lambda \to -\infty$. More generally, we shall use this notation for "functions" $f(\lambda, x, y)$ depending on additional parameters $x \in M_x$, $y \in M_y$ to describe the fact that f as well as any given derivative of f with respect to x, y vanishes faster than any given negative power of $|\lambda|$ as $\lambda \to -\infty$ uniformly over $M_x \times M_y$.

2. Main result. The main result of this section is the following abstract theorem which plays a fundamental role in our book, allowing us to avoid the consideration of an ill-posed Cauchy problem for the equation $D_t^{2m} u = A_x u$. A version of this theorem first appeared in [**Va3**]; see also [**Va4**], [**Va7**], [**Va11**]–[**Va13**], [**SaVa2**].

THEOREM 3.1.1. *Let* $T_- < 0 < T_+$ *be finite or* $\mp\infty$, *and let* $u(t,x,y) \in C^\infty(M_x \times M_y; \mathcal{D}'(T_-, T_+))$. *Then*

$$u(t,x,y) - \exp(-it\mathcal{A}_x^{1/(2m)})u_0(x,y) \in C^\infty((T_-, T_+) \times M_x \times M_y) \tag{3.1.1}$$

for some $u_0(x,y) \in C^\infty(M_y; \dot{\mathcal{D}}'_{\overline{B}}(M_x))$ *if and only if the following three conditions are fulfilled:*

$$D_t^{2m}u - A_x u \in C^\infty((T_-, T_+) \times M_x \times M_y), \tag{3.1.2}$$

$$(B_x^{(j)}u)\big|_{\partial M_x} \in C^\infty((T_-, T_+) \times \partial M_x \times M_y), \qquad j = 1, 2, \ldots, m, \tag{3.1.3}$$

$$\mathcal{F}^{-1}_{t\to\lambda}[gu] = O(|\lambda|^{-\infty}) \quad as \quad \lambda \to -\infty, \qquad \forall g \in C_0^\infty(T_-, T_+). \tag{3.1.4}$$

Before proceeding to the proof of Theorem 3.1.1 let us make several remarks and give some explanations.

Formulae (3.1.1)–(3.1.4) are understood in the sense of distributions in t. In particular, the expression $\mathcal{F}^{-1}_{t\to\lambda}[gu]$ is the $C^\infty(\mathbb{R} \times M_x \times M_y)$-"function" defined as $\mathcal{F}^{-1}_{t\to\lambda}[gu] = (2\pi)^{-1}\langle\, u(t,x,y), \exp(it\lambda)g(t)\,\rangle_t$.

The subscript in $\mathcal{A}_x$ indicates that the operator $\mathcal{A}$ acts in the variable x. The operator $\exp(-it\mathcal{A}_x^{1/(2m)})$ originally defined on $C_{\overline{B}}^\infty(M_x)$ is extended to $\dot{\mathcal{D}}'_{\overline{B}}(M_x)$ in the standard way (by duality). That is, the value of the distribution $\exp(-it\mathcal{A}_x^{1/(2m)})u_0(x,y)$ on the test half-density $v(x) \in C_{\overline{B}}^\infty(M_x)$ is defined as

$$\begin{aligned} &\langle\, \exp(-it\mathcal{A}_x^{1/(2m)})u_0(x,y), v(x)\,\rangle_x = \langle\, u_0(x,y), \exp(-it\,\overline{\mathcal{A}}_x^{1/(2m)})\, v(x)\,\rangle_x \\ &\quad = \sum_{k=1}^{\infty} \exp(-it\lambda_k)\, \langle\, u_0(x,y), \overline{v_k(x)}\,\rangle_x \int_{M_x} v_k(x)\, v(x)\, dx\,. \end{aligned} \tag{3.1.5}$$

The expression $\exp(-it\mathcal{A}_x^{1/(2m)})u_0(x,y)$ can then be considered as a distribution in t. Indeed, the value of the distribution $\exp(-it\mathcal{A}_x^{1/(2m)})u_0(x,y)$ on the test function $g \in C_0^\infty(T_-, T_+)$ is defined as

$$\begin{aligned} &\langle\, \exp(-it\mathcal{A}_x^{1/(2m)})u_0(x,y), g(t)\,\rangle_t \\ &\quad = \sum_{k=1}^{\infty} v_k(x)\, \langle\, u_0(x,y), \overline{v_k(x)}\,\rangle_x \int_{T_-}^{T_+} \exp(-it\lambda_k)\, g(t)\, dt\,. \end{aligned} \tag{3.1.6}$$

Elementary integration by parts

$$\int_{M_x} v_k(x)\, v(x)\, dx = \lambda_k^{-s} \int_{M_x} v_k(x)\, (\overline{A}_x^{s} v(x))\, dx\,,$$

$$\int_{T_-}^{T_+} \exp(-it\lambda_k)\, g(t)\, dt = (i\lambda_k)^{-s} \int_{T_-}^{T_+} \exp(-it\lambda_k)\, (\partial_t^s g(t))\, dt\,,$$

$\forall s \in \mathbb{N}$, proves that the quantities $\int_{M_x} v_k(x)\, v(x)\, dx$, $\int_{T_-}^{T_+} \exp(-it\lambda_k)\, g(t)\, dt$ vanish faster than any given negative power of λ_k as $k \to \infty$, and, consequently, faster than any given negative power of k. On the other hand, the quantities $v_k(x)$, $\langle\, u_0(x,y), \overline{v_k(x)}\,\rangle_x$ grow not faster than some positive powers of k (see the previous subsection). This argument shows that for fixed $(t,y) \in (T_-, T_+) \times M_y$

and $(x,y) \in M_x \times M_y$ the series (3.1.5) and (3.1.6) are absolutely convergent. Moreover, it is easy to check that

$$\begin{aligned}(3.1.7)\qquad &\exp(-it\mathcal{A}_x^{1/(2m)})u_0(x,y)\\ &\in C^\infty((T_-, T_+)\times M_y; \dot{\mathcal{D}}'_{\widetilde{B}}(M_x)) \cap C^\infty(M_x \times M_y; \mathcal{D}'(T_-, T_+)).\end{aligned}$$

Note that if

$$u_0(x,y) = \delta(x-y) = \sum_{k=1}^{\infty} v_k(x)\,\overline{v_k(y)},$$

then $\exp(-it\mathcal{A}_x^{1/(2m)})u_0(x,y)$ is the Schwartz kernel of the wave group.

3. Proof of Theorem 3.1.1. If (3.1.1) holds, then (3.1.2)–(3.1.4) are obviously fulfilled. So we have to prove only that formulae (3.1.2)–(3.1.4) imply (3.1.1) with some u_0. It is convenient to split this proof into several parts which we shall consider as separate lemmas.

LEMMA 3.1.2. *Let $b_{jr}(t,x') \in C^\infty((T_-, T_+)\times \partial M)$, $j = 1,2,\ldots,m$, $r = 0,1,2,\ldots$, be a set of "functions" (functions in t and half-densities in x'). Then there exists a $w(t,x) \in C^\infty((T_-, T_+)\times M)$ (function in t and half-density in x) such that*

$$(3.1.8)\qquad (B^{(j)}A^r w)\Big|_{\partial M} = b_{jr}, \qquad j = 1,2,\ldots,m, \qquad r = 0,1,2,\ldots .$$

PROOF OF LEMMA 3.1.2. In order to simplify notation let us reenumerate our boundary "functions", boundary operators and their orders with one index $k = j + mr$:

$$b_{j+mr} := b_{jr}, \qquad B^{(j+mr)} := B^{(j)}A^r, \qquad m_{j+mr} := m_j + 2mr,$$

$j = 1,2,\ldots,m$, $r = 0,1,2,\ldots$. Recall that $0 \leqslant m_1 < m_2 < \cdots < m_m \leqslant 2m-1$ are the orders of $B^{(1)}, B^{(2)}, \ldots, B^{(m)}$.

Each operator $B^{(k)}$, $k = 1,2,\ldots$, can be represented in the form $B^{(k)} = c_k(x')\partial_{x_n}^{m_k} - \widetilde{B}^{(k)}$, $c_k(x') \neq 0$, where $\widetilde{B}^{(k)}$ is a "boundary" differential operator of order m_k without the leading conormal derivative (see Conditions 1.1.3 and 1.1.1 in subsection 1.1.4):

$$\widetilde{B}^{(k)} = \sum_{p=0}^{m_k-1} \widetilde{B}^{(kp)}\,\partial_{x_n}^p .$$

Here the $B^{(kp)}$ are differential operators in x' of order $\leqslant m_k - p$. Without loss of generality we shall assume that $c_k(x') = 1$; the general case is reduced to this one by an obvious renormalization of the operators $B^{(k)}$ and of the "functions" $b_k(t,x')$. Recall that near ∂M we are using special local coordinates $x = (x', x_n)$ with specified coordinate x_n; see 1.1.2.

Let us construct w as a formal Taylor expansion in x_n,

$$(3.1.9)\qquad w(t,x) \sim \sum_{k=1}^{\infty} w_k(t,x')\,\frac{x_n^{m_k}}{m_k!} .$$

Substituting (3.1.9) into (3.1.8) we obtain an infinite system of differential equations:

$$w_1 = b_1 \,, \tag{3.1.10}$$

$$w_k = \sum_{l=1}^{k-1} \left(\widetilde{B}^{(km_l)} \, w_l \right) + b_k \,, \qquad k = 2, 3, \dots \,. \tag{3.1.11}$$

Due to its triangular structure this system can be solved explicitly: (3.1.10) gives w_1 and (3.1.11) gives a recurrent procedure for the determination of w_k, $k = 2, 3, \dots$.

It remains to note that given an arbitrary formal Taylor expansion (3.1.9) one can construct an infinitely smooth "function" w with such Taylor coefficients. (This simple statement is proved analogously to [**Sh,** Proposition 3.5]; basically we are using here Lemma 2.1.2 of our book, but with the asymptotic parameter $x_n \to 0$ instead of $|\theta| \to \infty$.)

Lemma 3.1.2 has been proved. □

LEMMA 3.1.3. *Let $t_- < 0 < t_+$ be finite real numbers, and let $a(t,x) \in C^\infty([t_-\,, t_+] \times M)$ be a "function" (function in t and half-density in x) which satisfies*

$$\left. (B^{(j)} A^r a) \right|_{\partial M} = 0 \,, \qquad j = 1, 2, \dots, m, \qquad r = 0, 1, 2, \dots \,. \tag{3.1.12}$$

Then there exists a $w(t,x) \in C^\infty([t_-\,, t_+] \times M)$ (function in t and half-density in x) such that

$$D_t^{2m} w = A w + a \,, \tag{3.1.13}$$

$$\left. (B^{(j)} A^r w) \right|_{\partial M} = 0 \,, \qquad j = 1, 2, \dots, m, \qquad r = 0, 1, 2, \dots \tag{3.1.14}$$

PROOF OF LEMMA 3.1.3. Set

$$w(t,x) = \sum_{k=1}^{\infty} v_k(x) \, w_k(t) \,, \tag{3.1.15}$$

$$w_k(t) = \sum_{l=1}^{m} w_{kl}^+(t) + \sum_{l=m+1}^{2m} w_{kl}^-(t) \,, \tag{3.1.16}$$

$$w_{kl}^\pm(t) = -\frac{i \exp(-it\lambda_{kl})}{2m\lambda_{kl}^{2m-1}} \int_{t_\pm}^{t} \exp(i\tau\lambda_{kl}) \, a_k(\tau) \, d\tau \,, \tag{3.1.17}$$

$$a_k(t) = \int_M a(t,x) \, \overline{v_k(x)} \, dx \,, \tag{3.1.18}$$

$$\lambda_{kl} = \lambda_k \exp(i\pi(l-1)m^{-1}) \,, \qquad l = 1, 2, \dots, 2m \,. \tag{3.1.19}$$

Recall that by λ_k and v_k, $k = 1, 2, \dots$, we denote the eigenvalues and the orthonormalized eigenfunctions of the problem (1.1.1), (1.1.2). Due to the boundary conditions (3.1.12) the quantity $a_k(t)$ defined by (3.1.18) vanishes faster than any given negative power of k as $k \to \infty$ uniformly over $[t_-\,, t_+]$; the same is true for any given derivative of $a_k(t)$ with respect to t. An elementary analysis of formulae (3.1.16), (3.1.17), (3.1.19) shows that this rapid decay property is inherited by the terms of the series (3.1.15): this series converges absolutely, uniformly over $[t_-, t_+] \times M$, as well as the series of any given derivatives with respect to t, x. Thus, (3.1.15) defines an infinitely smooth "function" which can be differentiated under the $\sum_{k=1}^{\infty}$ sign.

Straightforward substitution shows that the constructed w satisfies (3.1.13) and (3.1.14). □

REMARK 3.1.4. Lemmas 3.1.2, 3.1.3 and their proofs remain true if the "functions" b_{jr} and a depend smoothly on the additional parameter $y \in M_y$. In this case the resulting w will also smoothly depend on y.

LEMMA 3.1.5. *Let* $u(t,x,y) \in C^\infty(M_x \times M_y; \mathcal{D}'(T_-, T_+))$ *and* $u_0(x,y) \in C^\infty(M_y; \dot{\mathcal{D}}'_{\bar{B}}(M_x))$ *be such that* (3.1.1) *holds. Then* $u(t,x,y)$ *can be viewed as a distribution of the class* $C^\infty((T_-, T_+) \times M_y; \dot{\mathcal{D}}'_{\bar{B}}(M_x))$.

PROOF OF LEMMA 3.1.5. The required result follows immediately from formula (3.1.7). □

The next two simple results follow from the definition of the spaces C_{B_x}; see subsection 1.

LEMMA 3.1.6. *Under the conditions of Lemma* 3.1.5 *assume, in addition, that*

$$(B_x^{(j)} A_x^r u)\Big|_{\partial M_x} = 0 \quad \text{at} \quad t = 0, \qquad j = 1, 2, \dots, m, \qquad r = 0, 1, 2, \dots . \tag{3.1.20}$$

Then

$$u(0,x,y) - u_0(x,y) \in C^\infty_{B_x}(M_x \times M_y). \tag{3.1.21}$$

LEMMA 3.1.7. *If* $u_0(x,y) \in C^\infty_{B_x}(M_x \times M_y)$, *then*

$$\exp(-it\mathcal{A}_x^{1/(2m)}) u_0(x,y) \in C^\infty_{B_x}((T_-, T_+) \times M_x \times M_y).$$

Suppose now that we have proved the following

LEMMA 3.1.8. *Let* $T_- < 0 < T_+$ *be finite or* $\mp\infty$ *and let*

$$u(t,x,y) \in C^\infty(M_x \times M_y; \mathcal{D}'(T_-, T_+)). \tag{3.1.22}$$

Let the condition (3.1.4) *be fulfilled as well as*

$$D_t^{2m} u - A_x u = 0, \tag{3.1.23}$$

$$(B_x^{(j)} u)\Big|_{\partial M_x} = 0, \qquad j = 1, 2, \dots, m. \tag{3.1.24}$$

Then

$$u(t,x,y) - \exp(-it\mathcal{A}_x^{1/(2m)}) u_0(x,y) \in C^\infty_{B_x}((T_-, T_+) \times M_x \times M_y) \tag{3.1.25}$$

for some $u_0(x,y) \in C^\infty(M_y; \dot{\mathcal{D}}'_{\bar{B}}(M_x))$.

Let us show that Lemma 3.1.8 implies Theorem 3.1.1.

Suppose that the distribution (3.1.22) satisfies the conditions (3.1.2)–(3.1.4). It follows from Lemma 3.1.2 and Remark 3.1.4 that we can turn (3.1.3) into (3.1.24) by adding to u a $C^\infty((T_-, T_+) \times M_x \times M_y)$-term. Such an operation does not spoil (3.1.2) and (3.1.4), and it has no influence on the formula (3.1.1) which we are proving.

Take now arbitrary finite real numbers t_-, t_+ such that

$$T_- < t_- < 0 < t_+ < T_+ . \tag{3.1.26}$$

Formulae (3.1.2), (3.1.24) imply that the "function" $a(t,x,y) := D_t^{2m}u - A_x u$ satisfies the conditions of Lemma 3.1.3. It follows from Lemma 3.1.3 and Remark 3.1.4 that we can turn (3.1.2) into (3.1.23) by adding to u a $C^\infty([t_-,t_+]\times M_x \times M_y)$-term which satisfies (3.1.14) in the variable x. Such an operation does not spoil (3.1.24) and (3.1.4), and it does not influence the formula (3.1.1) which we are proving, only the time interval becomes smaller ((t_-, t_+) instead of (T_-, T_+)).

Applying Lemma 3.1.8 we conclude that

$$u(t,x,y) - \exp(-it\mathcal{A}_x^{1/(2m)})u_0(x,y) \in C^\infty_{B_x}((t_-, t_+) \times M_x \times M_y) \tag{3.1.27}$$

for some $u_0(x,y) \in C^\infty(M_y; \dot{\mathcal{D}}'_{\overline{B}}(M_x))$, dependent, generally speaking, on our choice of the numbers t_-, t_+. But Lemmas 3.1.6, 3.1.7 imply that $u_0(x,y)$ can be chosen to be independent of t_-, t_+: say, we can take $u_0(x,y) = u(0,x,y)$. As t_-, t_+ are arbitrary numbers satisfying (3.1.26), formula (3.1.27) implies (3.1.1).

Thus we have reduced the proof of Theorem 3.1.1 to the proof of Lemma 3.1.8. So further on we assume the conditions of Lemma 3.1.8 to be fulfilled and prove (3.1.25).

Consider the distribution

$$u_k(t,y) = \int_{M_x} u(t,x,y)\,\overline{v_k(x)}\,dx \in C^\infty(M_y; \mathcal{D}'(T_-, T_+)) . \tag{3.1.28}$$

Formulae (3.1.23), (3.1.24) imply

$$D_t^{2m} u_k = \lambda_k^{2m} u_k , \tag{3.1.29}$$

and consequently for each fixed point y and fixed local coordinates the distribution (3.1.28) is, in fact, a function

$$u_k(t,y) = \sum_{l=1}^{2m} u_{kl}(t,y) , \tag{3.1.30}$$

$$u_{kl}(t,y) = c_{kl}(y)\exp(-it\lambda_{kl}) , \tag{3.1.31}$$

where the λ_{kl} are given by formula (3.1.19).

Let $\|a_{lp}\|$ be the symmetric $2m$ by $2m$ matrix with elements

$$a_{lp} = \exp(i\pi(l-1)(p-1)m^{-1}) , \qquad l,p = 1,2,\ldots,2m,$$

and $\|b_{lp}\| = \|a_{lp}\|^{-1}$. We have the useful formula

$$u_{kl}(t,y) = \sum_{p=1}^{2m} b_{lp}\,(-\lambda_k)^{1-p}\,D_t^{p-1}u_k(t,y) . \tag{3.1.32}$$

Formula (3.1.32) allows us to express the u_{kl} in terms of derivatives of u_k. In particular, formulae (3.1.28), (3.1.32) imply $u_{kl} \in C^\infty(M_y; \mathcal{D}'(T_-, T_+))$, which in turn implies $c_{kl} \in C^\infty(M_y)$.

We shall need the following elementary

LEMMA 3.1.9. *Let $\theta \in C_0^\infty(\mathbb{R})$ be a function such that $\theta \geqslant 0$, $\theta \not\equiv 0$, and let $r, \mathbf{r} \geqslant 0$ be integers. Then there exists a positive constant $\mathbf{c}_{r\mathbf{r}}$ such that for any $t_0 \in \mathbb{R}$, any $\lambda > 0$, and any solution $\mathbf{u} \not\equiv 0$ of the ordinary differential equation $D_t^{2m}\mathbf{u} = \lambda^{2m}\mathbf{u}$ the function g defined by*

$$g(t) = \operatorname{const} \theta(\lambda(t - t_0))\, \overline{\mathbf{u}(t)}\,, \qquad \int \mathbf{u}(t)\, g(t)\, dt = 1\,,$$

satisfies

$$|\mathbf{u}^{(r)}(t_0)\, g^{(\mathbf{r})}(t)| \leqslant \mathbf{c}_{r\mathbf{r}}\, \lambda^{1+r+\mathbf{r}}\,, \qquad \forall t \in \mathbb{R}\,,$$

where the superscript indicates the order of the derivative.

PROOF OF LEMMA 3.1.9. It is sufficient to prove the lemma for $t_0 = 0$ (because the differential equation has constant coefficients) and $\lambda = 1$ (by homogeneity). After this observation the proof becomes trivial: one just has to write $\mathbf{u}(t)$ as a linear combination of $2m$ linearly independent solutions. □

DEFINITION 3.1.10. Consider a sequence of "functions" (more precisely, functions in the variable t and half-densities in the variable y)

$$w_k(t,y) \in C^\infty((T_-\,, T_+) \times M_y)\,, \qquad k = 1, 2, \ldots,$$

and let $1 = \sum_{p=1}^q \chi_p(y)$, $\chi_p \in C^\infty(M_y)$, $\operatorname{supp}\chi_p \subset M_y^{(p)}$, be some partition of unity on M_y with local coordinates y in coordinate patches $M_y^{(p)}$. We will say that this sequence *increases slowly* if for any real numbers t_-, t_+ satisfying (3.1.26), any multiindex $\alpha \geqslant 0$ and any integer $r \geqslant 0$ there exists a natural number s such that $\lambda_k^{-s}\partial_t^r\partial_y^\alpha(\chi_p(y)\, w_k(t,y)) \to 0$ as $k \to \infty$ uniformly over $t \in [t_-\,, t_+]$, $y \in M_y^{(p)}$, $p = 1, 2, \ldots, q$. We will say that this sequence *decreases rapidly* if for any real numbers t_-, t_+ satisfying (3.1.26), any multiindex $\alpha \geqslant 0$, any integer $r \geqslant 0$ and any natural number s we have $\lambda_k^s\partial_t^r\partial_y^\alpha(\chi_p(y)\, w_k(t,y)) \to 0$ as $k \to \infty$ uniformly over $t \in [t_-\,, t_+]$, $y \in M_y^{(p)}$, $p = 1, 2, \ldots, q$.

LEMMA 3.1.11. *The sequence $u_k(t,y)$, $k = 1, 2, \ldots,$ defined by (3.1.28) increases slowly.*

PROOF OF LEMMA 3.1.11. Suppose that the statement of Lemma 3.1.11 is false. Then there exist real numbers t_-, t_+ satisfying (3.1.26), a multiindex $\alpha \geqslant 0$, an integer $r \geqslant 0$, a coordinate patch $M_y^{(p)}$ with local coordinates y, a cut-off function $\chi_p(y) \in C^\infty(M_y)$ with $\operatorname{supp}\chi_p \subset M_y^{(p)}$, and sequences $k_s \in \mathbb{N}$, $t_s \in [t_-\,, t_+]$, $y_s \in \operatorname{supp}\chi_p$, $s = 1, 2, \ldots,$ such that $k_s \to +\infty$ as $s \to +\infty$ and

$$\left|\lambda_{k_s}^{-s}\left(\partial_t^r\partial_y^\alpha(\chi_p(y)\, u_{k_s}(t,y))\right)\big|_{t=t_s,\, y=y_s}\right| \geqslant 1\,. \tag{3.1.33}$$

Denote by $\mathbf{u}(t,x,y)$ the distribution defined by the formula

$$\langle \mathbf{u}(t,x,y), g(t)\rangle_t = \partial_y^\alpha\left(\chi_p(y)\, \langle u(t,x,y), (-\partial_t)^r g(t)\rangle_t\right).$$

Clearly, (3.1.22) implies

$$\mathbf{u}(t,x,y) \in C^\infty(M_x \times M_y; \mathcal{D}'(T_-\,, T_+))\,. \tag{3.1.34}$$

Denote also

$$\mathbf{u}_k(t,y) = \int_{M_x} \mathbf{u}(t,x,y)\,\overline{v_k(x)}\,dx \equiv \partial_t^r \partial_y^\alpha (\chi_p(y)\,u_k(t,y)) .$$

Set

$$g_s(t) = \text{const}_s\,\theta(\lambda_{k_s}(t-t_s))\,\overline{\mathbf{u}_{k_s}(t,y_s)}\,, \qquad \int \mathbf{u}_{k_s}(t,y_s)\,g_s(t)\,dt = 1\,,$$

where θ is the same as in Lemma 3.1.9. Then

(3.1.35)
$$\int_{M_x} \langle \mathbf{u}(t,x,y_s), g_s(t)\rangle_t\,\overline{v_{k_s}(x)}\,dx = 1\,.$$

Formula (3.1.33) and Lemma 3.1.9 imply that for any integer $\mathbf{r} \geqslant 0$ we have

$$|g_s^{(\mathbf{r})}(t)| \leqslant \mathbf{c}_{r\mathbf{r}}\,\lambda_{k_s}^{1+r+\mathbf{r}-s}\,, \qquad \forall t \in \mathbb{R}\,,$$

which means that $g_s \to 0$ in $C_0^\infty(T_-\,,T_+)$ as $s \to \infty$. Therefore by (3.1.34) $\langle \mathbf{u}(t,x,y), g_s(t)\rangle_t \to 0$ in $C^\infty(M_x \times M_y)$ as $s \to \infty$. But the latter contradicts (3.1.35).

Lemma 3.1.11 has been proved. □

LEMMA 3.1.12. *For any* $l = 1,2,\dots,2m$ *the sequence* $u_{kl}(t,y)$, $k = 1,2,\dots$, *defined by* (3.1.28), (3.1.30), *and* (3.1.31) *increases slowly.*

PROOF OF LEMMA 3.1.12. The required result follows immediately from formula (3.1.32) and Lemma 3.1.11. □

Lemma 3.1.12 allows us to represent our distribution $u(t,x,y)$ in the form of a series of smooth "functions"

(3.1.36)
$$u(t,x,y) = \sum_{k=1}^{\infty} \sum_{l=1}^{2m} u_{kl}(t,y)\,v_k(x)\,.$$

The above series correctly defines $u(t,x,y)$ as a distribution of the class $C^\infty(M_x \times M_y; \mathcal{D}'(T_-\,,T_+))$, as well as $C^\infty((T_-\,,T_+) \times M_y;\, \dot{\mathcal{D}}'_{\bar{B}}(M_x))$.

LEMMA 3.1.13. *For any* $l \neq 1,\, m+1$ *the sequence* $u_{kl}(t,y)$, $k = 1,2,\dots$, *decreases rapidly.*

PROOF OF LEMMA 3.1.13. Assume that $m \geqslant 2$ (otherwise no number $l = 1,2,\dots,2m$ satisfies the condition $l \neq 1,\, m+1$). Let us consider first the case $2 \leqslant l \leqslant m$. Having fixed l and arbitrary s, r, α, χ_p (in the notation of Definition 3.1.10) set

$$\mu_k(t_-\,,t_+) := \max_{t\in[t_-,t_+],\ y\in M_y^{(p)}} \lambda_k^s\,|\partial_t^r \partial_y^\alpha(\chi_p(y)\,u_{kl}(t,y))|$$

where t_-, t_+ are arbitrary real numbers satisfying (3.1.26). Choose a positive $\varepsilon < T_+ - t_+$. Due to the exponential behaviour of the "function" $u_{kl}(t,y)$ in the variable t (see (3.1.31)) we have

(3.1.37)
$$\mu_k(t_-\,,t_+) \leqslant \exp(-\varepsilon\,\lambda_k \sin(\pi/m))\,\mu_k(t_-\,,t_+ + \varepsilon)\,.$$

According to Lemma 3.1.12 and Definition 3.1.10 there exists a natural number $\widetilde{s}$ such that

$$\lambda_k^{-\widetilde{s}} \mu_k(t_-\,, t_+ + \varepsilon) \to 0 \quad \text{as} \quad k \to \infty\,. \tag{3.1.38}$$

Combining (3.1.37) and (3.1.38) we get

$$\mu_k(t_-\,, t_+) \leqslant c\,\lambda_k^{\widetilde{s}} \exp(-\varepsilon\,\lambda_k \sin(\pi/m)) \tag{3.1.39}$$

where $c = \max_{k\in\mathbb{N}}(\lambda_k^{-\widetilde{s}}\mu_k(t_-\,, t_+{+}\varepsilon))$. As the right-hand side of (3.1.39) contains an exponential term which is obviously stronger than the term $\lambda_k^{s+\widetilde{s}}$, it follows from (3.1.39) that $\mu_k(t_-\,, t_+)$ vanishes as $k \to \infty$. This means rapid decrease in the sense of Definition 3.1.10.

The case $m+2 \leqslant l \leqslant 2m$ is handled similarly by estimating $\mu_k(t_-\,, t_+)$ through $\mu_k(t_- - \varepsilon\,, t_+)$, $0 < \varepsilon < t_- - T_-$.

Lemma 3.1.13 has been proved. □

LEMMA 3.1.14. *The sequence $u_{k\,m+1}(t,y)$, $k = 1, 2, \ldots$, decreases rapidly.*

PROOF OF LEMMA 3.1.14. As in the proof of the previous lemma let us fix arbitrary s, α, χ_p, and let us also fix an arbitrary $g \in C_0^\infty(T_-\,, T_+)$, $g \not\equiv 0$. It follows from (3.1.4) that

$$\lambda_k^s \partial_y^\alpha \left(\chi_p(y) \int_{M_x} \mathcal{F}^{-1}_{t\to\lambda}[g(t)\,u(t,x,y)]\,\overline{v_k(x)}\,dx\right) \to 0 \text{ as } \lambda \to -\infty \tag{3.1.40}$$

uniformly over $y \in M_y^{(p)}$, $k \in \mathbb{N}$; here uniformity over k is established by integration by parts in the variable x. Formulae (3.1.36), (3.1.31) and Lemmas 3.1.12, 3.1.13 allow us to rewrite (3.1.40) in the form

$$\lambda_k^s \partial_y^\alpha \left(\chi_p(y) \sum_{l=1,m+1} \mathcal{F}^{-1}_{t\to(\lambda-\lambda_{kl})}[g(t)]\,c_{kl}(y))\right) \to 0 \text{ as } \lambda \to -\infty\,. \tag{3.1.41}$$

Let us fix some $\mu \in \mathbb{R}$ such that $\mathcal{F}^{-1}_{t\to\mu}[g(t)] \neq 0$, and let us relate λ and k by the condition $\lambda = \mu - \lambda_k$. Then (3.1.41) implies

$$g_{k1}(y) + g_{k\,m+1}(y) \to 0 \quad \text{as} \quad k \to \infty\,, \tag{3.1.42}$$

where

$$g_{kl}(y) = \lambda_k^s\,\mathcal{F}^{-1}_{t\to(\mu-\lambda_k-\lambda_{kl})}[g(t)]\,\left(\partial_y^\alpha(\chi_p(y)\,c_{kl}(y))\right), \qquad l = 1, m+1\,. \tag{3.1.43}$$

Let us now examine formula (3.1.43) in the case $l = 1$. As $\lambda_{k1} = \lambda_k$ the sequence $\mathcal{F}^{-1}_{t\to(\mu-\lambda_k-\lambda_{k1})}[g(t)]$ decreases rapidly (the usual property of the Fourier transform of a C_0^∞-function), and the presence of the slowly increasing factor $\lambda_k^s\partial_y^\alpha(\chi_p(y)\,c_{kl}(y))$ (see Lemma 3.1.12) on the right-hand side of (3.1.43) cannot spoil this rapid decrease. Thus,

$$g_{k1}(y) \to 0 \quad \text{as} \quad k \to \infty\,. \tag{3.1.44}$$

Formulae (3.1.42)–(3.1.44) and $\lambda_{k\,m+1} = -\lambda_k$ imply that

$$g_{k\,m+1}(y) = \lambda_k^s\,\mathcal{F}^{-1}_{t\to\mu}[g(t)]\,\left(\partial_y^\alpha(\chi_p(y)\,c_{k\,m+1}(y))\right) \to 0 \quad \text{as} \quad k \to \infty\,.$$

Dividing the latter formula by the constant $\mathcal{F}^{-1}_{t\to\mu}[g(t)]$ we obtain the required rapid decrease property.

Lemma 3.1.14 has been proved. □

Now it only remains to rearrange (3.1.36) as

$$u(t,x,y) = \exp(-it\mathcal{A}_x^{1/(2m)})u_0(x,y) + \sum_{k=1}^{\infty}\sum_{l=2}^{2m} u_{kl}(t,y)\,v_k(x)\,, \tag{3.1.45}$$

$$u_0(x,y) = \sum_{k=1}^{\infty} c_{k1}(y)\,v_k(x)\,. \tag{3.1.46}$$

By Lemmas 3.1.13, 3.1.14 the infinite sum on the right-hand side of (3.1.45) defines a $C^\infty_{B_x}((T_-\,,\,T_+)\times M_x\times M_y)$-"function", and by Lemma 3.1.12 formula (3.1.46) defines a distribution of the class $C^\infty(M_y;\dot{\mathcal{D}}'_{\bar B}(M_x))$.

Lemma 3.1.8, and with it our Theorem 3.1.1, have been proved.

4. Choice of the distribution $u_0(x,y)$. In this subsection we assume that formulae (3.1.2)–(3.1.4) hold.

Theorem 3.1.1 states that (3.1.1) holds for some $u_0(x,y)\in C^\infty(M_y;\dot{\mathcal{D}}'_{\bar B}(M_x))$, but it does not give an explicit formula for $u_0(x,y)$. The natural desire is to take $u(0,x,y)$ (which is well defined by Lemma 3.1.5) for $u_0(x,y)$. Unfortunately, this may not be justified due to the following reason. Denote $w(x,y)=u(0,x,y)-u_0(x,y)$, where $u_0(x,y)$ is the proper initial distribution. Then (3.1.1) implies that w is of the class $C^\infty(M_x\times M_y)$, but not necessarily of the class $C^\infty_{B_x}(M_x\times M_y)$. Consequently, we cannot apply Lemma 3.1.7 and are unable to guarantee the smoothness of $\exp(-it\mathcal{A}_x^{1/(2m)})w(x,y)$.

This difficulty with the choice of $u_0(x,y)$ disappears if we consider distributions $u(t,x,y)$ satisfying the additional condition (3.1.20). Then, by Lemmas 3.1.6, 3.1.7, we can take $u(0,x,y)$ for $u_0(x,y)$, and, moreover, all possible $u_0(x,y)$ are described by (3.1.21). In all our applications condition (3.1.20) will be automatically satisfied.

3.2. Representation of the wave group by means of oscillatory integrals: sufficient conditions

In the remaining part of this chapter (Sections 3.2–3.5) we will be considering distributions

$$\mathbf{u}_P(t,x,y)\in C^\infty((T_-\,,\,T_+)\times M_y;\dot{\mathcal{D}}'_{\bar B}(M_x))\cap C^\infty(M_x\times M_y;\mathcal{D}'(\mathbb{R})) \tag{3.2.1}$$

of a special type. Namely, let P be a pseudodifferential operator of class Ψ^1_0, and let $\mathbf{U}_P(t)$ be the time-dependent operator defined by (3.0.2), (3.0.1). The distribution to be studied is the Schwartz kernel $\mathbf{u}_P$ of $\mathbf{U}_P(t)$:

$$\mathbf{U}_P(t) = \int_M \mathbf{u}_P(t,x,y)\,(\,\cdot\,)\,dx\,. \tag{3.2.2}$$

Of course the distribution $\mathbf{u}_P$ can be alternatively defined as

$$\mathbf{u}_P(t,x,y) = \exp(-it\mathcal{A}_x^{1/(2m)})\,p(x,y)\,, \tag{3.2.3}$$

where p is the Schwartz kernel of P; the right-hand side of (3.2.3) is understood in the sense of (3.1.5) or (3.1.6). Note that according to notation introduced in 2.1.3 $P \in \Psi_0^1$ implies that $\operatorname{supp} p$ is compact in $\overset{\circ}{M}_x \times \overset{\circ}{M}_y$, i.e., $\operatorname{supp} p$ is separated from the boundary. This allows us to integrate by parts in x, guaranteeing the inclusion (3.2.1).

We will attempt to approximate $\mathbf{u}_P$ by the distribution

$$u_P(t,x,y) = \sum_\alpha \mathcal{I}_{\varphi^{(\alpha)},a^{(\alpha)}}(t,x,y), \tag{3.2.4}$$

which is a finite sum of standard and boundary layer oscillatory integrals associated with our branching Hamiltonian billiards. These oscillatory integrals were described in Sections 2.7 and 2.8, respectively; they have the form

$$\begin{aligned}&\mathcal{I}_{\varphi^{(\alpha)},a^{(\alpha)}}(t,x,y)\\ &\quad= \int e^{i\varphi^{(\alpha)}(t,x;y,\eta)} a^{(\alpha)}(t,x;y,\eta)\,\varsigma^{(\alpha)}(t,x;y,\eta)\,d_{\varphi^{(\alpha)}}(t,x;y,\eta)\,đ\eta,\end{aligned} \tag{3.2.5}$$

and are assumed to be defined on some time interval (T_-, T_+). By α we denote a summation index which takes a finite number of values.

THEOREM 3.2.1. *Let $T_- < 0 < T_+$ be finite or $\mp\infty$, and let the oscillatory integrals (3.2.5) be such that*

$$D_t^{2m}\mathcal{I}_{\varphi^{(\alpha)},a^{(\alpha)}} - A(x,D_x)\mathcal{I}_{\varphi^{(\alpha)},a^{(\alpha)}} \in C^\infty((T_-, T_+)\times M_x \times M_y), \quad \forall\alpha, \tag{3.2.6}$$

$$(B^{(j)}(x',D_x)u_P)\big|_{\partial M_x} \in C^\infty((T_-,T_+)\times \partial M_x \times M_y), \quad j=1,2,\dots,m, \tag{3.2.7}$$

$$u_P(0,x,y) - p(x,y) \in C_0^\infty(\overset{\circ}{M}_x \times \overset{\circ}{M}_y). \tag{3.2.8}$$

Then

$$u_P(t,x,y) - \mathbf{u}_P(t,x,y) \in C^\infty((T_-, T_+)\times M_x \times M_y). \tag{3.2.9}$$

PROOF. Theorem 3.2.1 is an immediate consequence of Theorem 3.1.1 and Remark 3.1.4. We only have to check the fulfilment of (3.1.4).

Let us show that each oscillatory integral in the sum (3.2.4) satisfies the condition (3.1.4), i.e., that

$$\int e^{i(\varphi^{(\alpha)}(t,x;y,\eta)+t\lambda)} g(t)\,a^{(\alpha)}(t,x;y,\eta)\,\varsigma^{(\alpha)}(t,x;y,\eta)\,d_{\varphi^{(\alpha)}}(t,x;y,\eta)\,dt\,đ\eta \tag{3.2.10}$$

is $O(|\lambda|^{-\infty})$ as $\lambda \to -\infty$. Here $g \in C_0^\infty(T_-, T_+)$ and $a^{(\alpha)} \in S^l$.

According to the definitions of standard and boundary layer oscillatory integrals (see Chapter 2) we have $\varphi_t^{(\alpha)} = -h(y,\eta)$ on the set $C_{\varphi^{(\alpha)}} = \{\varphi_\eta^{(\alpha)} = 0\}$. Moreover, without loss of generality we can assume that $\operatorname{supp}\varsigma^{(\alpha)}$ is small enough, so that

$$\operatorname{Re}\varphi_t^{(\alpha)} \leqslant -h(y,\eta)/2 \quad \text{on } \operatorname{supp}\varsigma^{(\alpha)}. \tag{3.2.11}$$

Set $L = (\varphi_t^{(\alpha)} + \lambda)^{-1}D_t$. Then $L\,e^{i(\varphi^{(\alpha)}+t\lambda)} = e^{i(\varphi^{(\alpha)}+t\lambda)}$, and integrating (3.2.10) by parts in t we get

$$\int e^{i(\varphi^{(\alpha)}+t\lambda)} \left(L^T\right)^k \left(g\,a^{(\alpha)}\,\varsigma^{(\alpha)}\,d_{\varphi^{(\alpha)}}\right) dt\,đ\eta, \tag{3.2.12}$$

where $L^T = D_t(\varphi_t^{(\alpha)} + \lambda)^{-1}$ and $k = 1, 2, \dots$. Elementary analysis of the integral (3.2.12) with account of (3.2.11) shows that for $k > n + l$ this integral converges absolutely and admits the estimate $O(|\lambda|^{n+l-k})$ as $\lambda \to -\infty$. Since k can be taken arbitrarily large, the required result follows. □

3.3. Representation of the wave group by means of oscillatory integrals: effective construction for manifolds without boundary

In this section we consider the case $\partial M = \varnothing$ and construct effectively (in terms of oscillatory integrals) the distribution $\mathbf{u}_P(t, x, y)$ with $P \in \Psi^{\mathbf{1}}$. Clearly, for manifolds without boundary $\Psi^{\mathbf{1}} = \Psi_0^{\mathbf{1}}$.

As in 1.3.1 we denote by Φ^t the Hamiltonian flow generated by the Hamiltonian (1.1.14). All the constructions in this section are carried out without any restrictions on t, i.e., for $t \in \mathbb{R}$.

The main result of this section is the following

THEOREM 3.3.1. *For any standard phase function $\varphi(t, x; y, \eta)$ associated with the canonical transformation Φ^t there exists a symbol $\mathfrak{a}(t; y, \eta) \in S^{\mathbf{1}}$ such that the oscillatory integral $\mathcal{I}_{\varphi,\mathfrak{a}}(t, x, y)$ coincides with the Schwartz kernel $\mathbf{u}_P(t, x, y)$ of the operator $\exp(-itA^{1/(2m)})P$ modulo a smooth half-density.*

PROOF. In view of Theorem 2.7.10 it is sufficient to prove Theorem 3.3.1 only for one standard phase function φ associated with Φ^t. So further on φ is fixed.

We will prove Theorem 3.3.1 by applying Theorem 3.2.1 with $T_\mp = \mp\infty$. This means we must satisfy the equation

$$D_t^{2m}\mathcal{I}_{\varphi,\mathfrak{a}} - A(x, D_x)\mathcal{I}_{\varphi,\mathfrak{a}} \in C^\infty(\mathbb{R} \times M_x \times M_y) \tag{3.3.1}$$

and the initial condition

$$\mathcal{I}_{\varphi,\mathfrak{a}}(0, x, y) - p(x, y) \in C^\infty(M_x \times M_y), \tag{3.3.2}$$

where p is the Schwartz kernel of the pseudodifferential operator P.

By Proposition 2.7.13 and Theorem 2.7.5 there exists a symbol $\mathfrak{p}(y, \eta) \in S^{\mathbf{1}}$ such that the Schwartz kernel of the pseudodifferential operator P coincides with the oscillatory integral $\mathcal{I}_{\varphi_0,\mathfrak{p}}$ where $\varphi_0 := \varphi|_{t=0}$. Therefore (3.3.2) is equivalent to

$$\mathfrak{a}_{\mathbf{l}-s}|_{t=0} = \mathfrak{p}_{\mathbf{l}-s}, \qquad s = 0, 1, \dots, \tag{3.3.3}$$

where $\mathfrak{a}_{\mathbf{l}-s}$ and $\mathfrak{p}_{\mathbf{l}-s}$ are the homogeneous terms of the symbols $\mathfrak{a}$ and $\mathfrak{p}$, respectively, with the subscript $\mathbf{l} - s$ indicating the degree of homogeneity in η.

Now we have to extend $\mathfrak{a}$ to $t \neq 0$ in such a way that (3.3.1) would be fulfilled. Substituting (2.7.11) into (3.3.1) and differentiating under the integral sign we see that modulo C^∞ the left-hand side of (3.3.1) is an oscillatory integral of the form (2.7.1) with the same phase function φ and amplitude

$$a = e^{-i\varphi}\, d_\varphi^{-1}\, D_t^{2m}(e^{i\varphi}\,\mathfrak{a}\, d_\varphi) - \mathfrak{a}\, e^{-i\varphi}\, d_\varphi^{-1}\, A(x, D_x)(e^{i\varphi}\, d_\varphi)\ ;$$

note that we did not include the derivatives of the cut-off ς into the above formula because the corresponding terms give a C^∞-contribution; see the remark following

Definition 2.1.5. By Leibniz's formula we have

$$D_t^{2m}(e^{i\varphi}\,\mathfrak{a}\,d_\varphi) = \sum_{j=0}^{2m}\sum_{k=0}^{j} \frac{2m!}{k!\,(j-k)!\,(2m-j)!}\,(D_t^{2m-j}e^{i\varphi})\,(D_t^{j-k}\mathfrak{a})\,D_t^k d_\varphi\,,$$

and in an arbitrary local coordinate system

$$A(x,D_x)(e^{i\varphi}\,d_\varphi) = \sum_\alpha \frac{1}{\alpha!}\,(A^{(\alpha)}(x,D_x)e^{i\varphi})\,D_x^\alpha d_\varphi\,,$$

where $A^{(\alpha)}(x,\xi) = \partial_\xi^\alpha A(x,\xi)$ and

$$A(x,\xi) = A_{2m}(x,\xi) + A_{2m-1}(x,\xi) + \cdots + A_1(x,\xi) + A_0(x)$$

is the full symbol of the differential operator A in these coordinates. Obviously,

$$e^{-i\varphi}\,D_t^{2m-j}e^{i\varphi} \in S^{2m-j}\,, \qquad e^{-i\varphi}\,A^{(\alpha)}(x,D_x)e^{i\varphi} \in S^{2m-|\alpha|}\,.$$

Moreover, the "functions" $d_\varphi^{-1}D_t^k d_\varphi$, $d_\varphi^{-1}D_x^\alpha d_\varphi$ and the operators D_t^{j-k} are positively homogeneous in η of degree 0 (see Definition 2.7.8). Therefore

$$a = \sum_{s=0}^{2m} L_{2m-s}\,\mathfrak{a}\,, \tag{3.3.4}$$

where $L_{2m-s} = L_{2m-s}(t,x,D_t;y,\eta)$ are linear differential operators of order s positively homogeneous in η of degree $2m-s$. In particular, L_{2m} is an operator of multiplication by a function

$$L_{2m} = L_{2m}(t,x;y,\eta) = \varphi_t^{2m} - A_{2m}(x,\varphi_x)\,, \tag{3.3.5}$$

and L_{2m-1} is a first order differential operator

$$L_{2m-1} = L_{2m-1}^{(0)}(t,x;y,\eta) + L_{2m-1}^{(1)}(t,x;y,\eta)\,D_t\,, \tag{3.3.6}$$

where

$$\begin{aligned} L_{2m-1}^{(0)} = &-A_{2m-1}(x,\varphi_x) \\ &- i\Bigg(2m\,\varphi_t^{2m-1}\,(d_\varphi)^{-1}\,(d_\varphi)_t + m\,(2m-1)\,\varphi_t^{2m-2}\,\varphi_{tt} \\ &\qquad - (d_\varphi)^{-1}\sum_{j=1}^n (A_{2m})_{\xi_j}(x,\varphi_x)\,(d_\varphi)_{x_j} \\ &\qquad - \frac{1}{2}\sum_{j,k=1}^n \varphi_{x_jx_k}\,(A_{2m})_{\xi_j\xi_k}(x,\varphi_x)\Bigg)\,, \end{aligned} \tag{3.3.7}$$

$$L_{2m-1}^{(1)} = 2m\,\varphi_t^{2m-1} \tag{3.3.8}$$

(the subscripts $_t$, $_x$, $_\xi$ denote derivatives with respect to t, x, ξ). The operators L_{2m-s} are, of course, invariant under changes of local coordinates x and y; these operators map functions of the variables $(t;y,\eta)$ into functions of the variables $(t,x;y,\eta)$.

The equation (3.3.1) is equivalent to

$$\mathfrak{S}\, a \in S^{-\infty}, \tag{3.3.9}$$

where $\mathfrak{S}$ is the linear operator mapping the amplitude of a standard oscillatory integral into the corresponding symbol; see 2.7.3. Substituting (2.7.12) and (3.3.4) into (3.3.9), we rewrite (3.3.9) in equivalent form

$$\sum_{k=0}^{\infty} \mathcal{L}_{2m-k}\, \mathfrak{a} \sim 0, \tag{3.3.10}$$

where

$$\mathcal{L}_{2m-k} = \sum_{r+s=k} \mathfrak{S}_{-r}\, L_{2m-s} \tag{3.3.11}$$

(in this sum r may take all nonnegative integer values, and s may take the values $0, 1, 2, \ldots, 2m$). The operators $\mathcal{L}_{2m-k}$ are positively homogeneous in η of degree $2m-k$. In view of (2.7.15) the operators $\mathcal{L}_{2m-k}$ are differential operators in t and η, independent of x, that is, $\mathcal{L}_{2m-k} = \mathcal{L}_{2m-k}(t, D_t; y, \eta, D_\eta)$. Exclusion of the variable x was the main aim of the transformations carried out above.

Let us now examine closer the operators $\mathcal{L}_{2m}$ and $\mathcal{L}_{2m-1}$.

According to (3.3.11), (2.7.13), and (3.3.5) we have

$$\mathcal{L}_{2m} = L_{2m}(t, x^*; y, \eta) = h^{2m}(y, \eta) - A_{2m}(x^*, \xi^*) = 0. \tag{3.3.12}$$

According to to (3.3.11), (2.7.13), (2.7.15) we have

$$\mathcal{L}_{2m-1}\, \mathfrak{a} = L_{2m-1}(t, x^*, D_t; y, \eta)\, \mathfrak{a} + \mathfrak{S}_0\, \mathfrak{L}_{-1}(t, \partial_x; y, \eta, \partial_\eta)\, (L_{2m}(t, x; y, \eta)\, \mathfrak{a}),$$

where $\mathfrak{L}_{-1}$ is the differential operator from (2.7.15). By Theorem 2.4.16 we know that the function $L_{2m}(t, x; y, \eta)$ has a second order zero on the set $\{x = x^*\}$; consequently we can use Lemma 2.7.9 for the evaluation of the operator $\mathfrak{S}_0\, \mathfrak{L}_{-1}\, L_{2m}$ which turns out to be an operator of multiplication by a function. With account of (3.3.6), (3.3.8) this yields

$$\mathcal{L}_{2m-1} = \mathcal{L}_{2m-1}(t, D_t; y, \eta) = \mathcal{L}^{(0)}_{2m-1}(t; y, \eta) - 2m\, h^{2m-1}(y, \eta)\, D_t, \tag{3.3.13}$$

where

$$\mathcal{L}^{(0)}_{2m-1}(t; y, \eta) = L^{(0)}_{2m-1}(t, x^*; y, \eta) - \frac{i}{2} \operatorname{tr}\left(x^*_\eta \cdot \Phi^{-1}_{x\eta} \cdot (L_{2m})_{xx}\big|_{x=x^*}\right). \tag{3.3.14}$$

Note that the operator $\mathcal{L}_{2m-1}$ does not contain differentiations in η, so it is an ordinary differential operator; this crucial fact is a consequence of Theorem 2.4.16.

Let us expand our unknown amplitude $\mathfrak{a}$ into a sum of terms positively homogeneous in η, substitute this expansion in (3.3.9) and collect together terms with the same degree of homogeneity. Then, with account of (3.3.12), (3.3.13), we see that (3.3.9) turns into the following recurrent system of ordinary differential equations in t with respect to the unknown functions $\mathfrak{a}_{\mathfrak{l}-s}(t;y,\eta)$:

$$
\begin{aligned}
\mathcal{L}_{2m-1}\,\mathfrak{a}_{\mathfrak{l}} &= 0\,,\\
\mathcal{L}_{2m-1}\,\mathfrak{a}_{\mathfrak{l}-1} &= -\mathcal{L}_{2m-2}\,\mathfrak{a}_{\mathfrak{l}}\,,\\
\mathcal{L}_{2m-1}\,\mathfrak{a}_{\mathfrak{l}-2} &= -\mathcal{L}_{2m-2}\,\mathfrak{a}_{\mathfrak{l}-1} - \mathcal{L}_{2m-3}\,\mathfrak{a}_{\mathfrak{l}}\,,\\
\dots &= \dots\\
\mathcal{L}_{2m-1}\,\mathfrak{a}_{\mathfrak{l}-s} &= -\sum_{k=1}^{s}\mathcal{L}_{2m-k-1}\,\mathfrak{a}_{\mathfrak{l}-s+k}\,,\\
\dots &= \dots
\end{aligned}
\tag{3.3.15}
$$

Integrating the equations (3.3.15) successively with initial conditions (3.3.3) we obtain the (unique) homogeneous functions $\mathfrak{a}_{\mathfrak{l}-s}$, $s = 0, 1, \dots$. By Lemma 2.1.2 there exists an amplitude $\mathfrak{a} \in S^1$ with given asymptotic expansion $\sum_{s=0}^{\infty}\mathfrak{a}_{\mathfrak{l}-s}$. $\square$

According to Theorem 3.3.1 $\mathbf{u}_P(t,x,y)$ is a standard Lagrangian distribution associated with the Hamiltonian flow Φ^t. The following theorem gives an explicit formula for its principal symbol.

THEOREM 3.3.2. *The principal symbol of the Lagrangian distribution* $\mathbf{u}_P$ *is*

$$
\mathfrak{a}_{\mathfrak{l}}(t;y,\eta) = P_{\mathfrak{l}}(y,\eta)\exp\Big(-\frac{i}{2m}\,h^{1-2m}(y,\eta)\int_0^t A_{\mathrm{sub}}\left(x^*(\tau;y,\eta),\xi^*(\tau;y,\eta)\right)d\tau\Big),
\tag{3.3.16}
$$

where $P_{\mathfrak{l}}$ *is the principal symbol of the operator* P *and* A_{sub} *is the subprincipal symbol of the operator* A.

PROOF. The ordinary differential operator $\mathcal{L}_{2m-1}$ appearing on the left-hand sides of (3.3.15) is given by formula (3.3.13). Integration of the first equation (3.3.15) with the first initial condition (3.3.3) gives

$$
\mathfrak{a}_{\mathfrak{l}}(t;y,\eta) = \exp\Big(\frac{i}{2m}\,h^{1-2m}(y,\eta)\int_0^t \mathcal{L}^{(0)}_{2m-1}(\tau;y,\eta)\,d\tau\Big)\,\mathfrak{p}_{\mathfrak{l}}(y,\eta)\,.
$$

By the results of 2.7.5, 2.1.3 and Theorem 2.7.11 $\mathfrak{p}_{\mathfrak{l}}$ is the principal symbol of the pseudodifferential operator P. So in order to prove (3.3.16) we must show that

$$
\mathcal{L}^{(0)}_{2m-1}(t;y,\eta) = -A_{\mathrm{sub}}\left(x^*(t;y,\eta),\xi^*(t;y,\eta)\right).
\tag{3.3.17}
$$

The explicit formula for $\mathcal{L}^{(0)}_{2m-1}$ given by (3.3.14), (3.3.7), (3.3.5) is rather complicated, and, moreover, seemingly depends on the choice of the phase function φ. However, we can significantly simplify the computation of the function $\mathcal{L}^{(0)}_{2m-1}$ by making the following observation. If we set $\mathfrak{a}(t;y,\eta) \equiv 1$, then $\mathcal{L}^{(0)}_{2m-1}(t;y,\eta)$ is the principal symbol of the Lagrangian distribution $D_t^{2m}\mathcal{I}_{\varphi,\mathfrak{a}} - A(x,D_x)\mathcal{I}_{\varphi,\mathfrak{a}}$; consequently (see Theorem 2.7.11) the function $\mathcal{L}^{(0)}_{2m-1}$ does not depend on the choice of the phase function. Thus, it is sufficient to prove (3.3.17) in the case when locally the phase function φ has the form $\langle x - x^*, \xi^*\rangle$ for some coordinates x such that $\det \xi^*_\eta \neq 0$ (see Lemma 2.4.13). So further on we carry out our computations locally and with this special (linear in x) phase function.

We have $\varphi_x = \xi^*$, and, in view of (2.3.17), $\varphi_t = -h(y,\eta) - \langle x - x^*, h_x(x^*,\xi^*)\rangle$. Since

$$\begin{aligned}
\varphi_t^{2m} &= \Big(-h(y,\eta) - \langle x - x^*, h_x(x^*,\xi^*)\rangle\Big)^{2m} \\
&= h^{2m}(y,\eta) + 2m\, h^{2m-1}(y,\eta)\,\langle x - x^*, h_x(x^*,\xi^*)\rangle \\
&\quad + m\,(2m-1)\,h^{2m-2}(y,\eta)\,\langle x - x^*, h_x(x^*,\xi^*)\rangle^2 + O(|x - x^*|)^3, \\
A_{2m}(x,\varphi_x) &= h^{2m}(x,\xi^*) = h^{2m}(y,\eta) + \langle x - x^*, (h^{2m})_x(x^*,\xi^*)\rangle \\
&\quad + \frac{1}{2}\,\langle (h^{2m})_{xx}(x^*,\xi^*)(x - x^*), x - x^*\rangle + O(|x - x^*|)^3,
\end{aligned}$$

substituting these expressions in (3.3.5) we obtain

$$L_{2m} = -m\, h^{2m-1}(y,\eta)\,\langle h_{xx}(x^*,\xi^*)(x - x^*), x - x^*\rangle + O(|x - x^*|^3).$$

Consequently

$$(L_{2m})_{xx}\big|_{x=x^*} = -2\,m\,h^{2m-1}(y,\eta)\,h_{xx}(x^*,\xi^*). \tag{3.3.18}$$

In the chosen coordinates and for the chosen phase function we have

$$\varphi_{xx} \equiv 0, \qquad (d_\varphi)_x \equiv 0; \tag{3.3.19}$$

the latter is true because $\varphi_{x\eta} \equiv \xi^*_\eta$ and $d_\varphi \equiv c\,|\det \xi^*_\eta|^{1/2}$, where c is some constant factor determined by our choice of the branch of $\arg d_\varphi$; see Definition 2.7.1. We also have

$$\begin{aligned}
\varphi_{tt}\big|_{x=x^*} &= \frac{d}{dt}\Big(-h(y,\eta) - \langle x - x^*, h_x(x^*,\xi^*)\rangle\Big)\Big|_{x=x^*} \\
&= \langle h_\xi(x^*,\xi^*), h_x(x^*,\xi^*)\rangle,
\end{aligned} \tag{3.3.20}$$

and by Liouville's formula
(3.3.21)

$$\begin{aligned}
(d_\varphi)^{-1}\,(d_\varphi)_t &= \frac{1}{2}\,(\det \xi^*_\eta)^{-1}\,\frac{d}{dt}(\det \xi^*_\eta) = \frac{1}{2}\,\mathrm{tr}\Big((\xi^*_\eta)^{-1}\cdot \frac{d}{dt}\xi^*_\eta\Big) \\
&= -\frac{1}{2}\,\mathrm{tr}\Big((\xi^*_\eta)^{-1}\cdot \frac{d}{d\eta}h_x(x^*,\xi^*)\Big) \\
&= -\frac{1}{2}\,\mathrm{tr}\Big((\xi^*_\eta)^{-1}\cdot h_{xx}(x^*,\xi^*)\cdot x^*_\eta + (\xi^*_\eta)^{-1}\cdot h_{x\xi}(x^*,\xi^*)\cdot \xi^*_\eta\Big) \\
&= -\frac{1}{2}\,\mathrm{tr}\Big((\xi^*_\eta)^{-1}\cdot h_{xx}(x^*,\xi^*)\cdot x^*_\eta + h_{x\xi}(x^*,\xi^*)\Big).
\end{aligned}$$

Substituting (3.3.19)–(3.3.21) in (3.3.7) and setting $x = x^*$ we obtain

$$\begin{aligned}
&L^{(0)}_{2m-1}(t,x^*;y,\eta) \\
&\quad = -A_{2m-1}(x^*,\xi^*) \\
&\qquad - imh^{2m-1}(y,\eta)\,\mathrm{tr}\Big((\xi^*_\eta)^{-1}\cdot h_{xx}(x^*,\xi^*)\cdot x^*_\eta + h_{x\xi}(x^*,\xi^*)\Big) \\
&\qquad - im(2m-1)h^{2m-2}(y,\eta)\langle h_\xi(x^*,\xi^*), h_x(x^*,\xi^*)\rangle.
\end{aligned} \tag{3.3.22}$$

Finally, substituting (3.3.22) and (3.3.18) in (3.3.14) we get

$$\begin{aligned}\mathcal{L}_{2m-1}(t;y,\eta) &= -A_{2m-1}(x^*,\xi^*)\\ &\quad - im\big(h^{2m-1}(y,\eta)\operatorname{tr} h_{x\xi}(x^*,\xi^*)\\ &\qquad + (2m-1)\,h^{2m-2}(y,\eta)\,\langle h_\xi(x^*,\xi^*), h_x(x^*,\xi^*)\rangle\big)\\ &= -A_{2m-1}(x^*,\xi^*) + (2i)^{-1}\operatorname{tr}(A_{2m})_{x\xi}(x^*,\xi^*).\end{aligned}$$

But according to (2.1.9) the latter expression is the subprincipal symbol of the operator $-A$. This proves (3.3.17). □

It is remarkable that the complicated expression (3.3.14), (3.3.7), (3.3.5) turns out to be the subprincipal symbol. In order to highlight the underlying reasons for this miraculous simplification we give an alternative argument below.

OUTLINE OF ALTERNATIVE PROOF OF (3.3.17). For any $t \in \mathbb{R}$ we have

$$U_P^*(t)\,U_P(t) \;=\; P^*\,P\,. \tag{3.3.23}$$

The operator on the right-hand side of (3.3.23) is a pseudodifferential operator with principal symbol $|P_1(y,\eta)|^2$. The operator on the left-hand side of (3.3.23) can be written down as a composition of two oscillatory integrals, and its Schwartz kernel is given (modulo C^∞) by the formula

$$\begin{aligned}\int e^{i\varphi(t,x;y,\eta)-i\overline{\varphi(t,x;z,\zeta)}}\,\mathfrak{a}(t;y,\eta)\,\overline{\mathfrak{a}(t;z,\zeta)}\\ \times\,\varsigma(t,x;y,\eta)\,\overline{\varsigma(t,x;z,\zeta)}\,d_\varphi(t,x;y,\eta)\,\overline{d_\varphi(t,x;z,\zeta)}\,dx\,đ\zeta\,đ\eta\,.\end{aligned} \tag{3.3.24}$$

If we fix local coordinates and make a change of variables $\alpha = (\eta+\zeta)/2$, $\beta = \zeta-\eta$, then the stationary phase method can be applied with respect to x and β, which reduces (3.3.24) to an integral in α. This integral has a phase function of the form $\langle z-y,\alpha\rangle + O(|z-y|^2)$; consequently it is the Schwartz kernel of a pseudodifferential operator. Explicit calculations show that the principal symbol of this operator is $|\mathfrak{a}_1(t;y,\eta)|^2$. Thus,

$$|\mathfrak{a}_1(t;y,\eta)|^2 \;=\; |P_1(y,\eta)|^2\,. \tag{3.3.25}$$

Let us take $P_1(y,\eta) \equiv 1$. Then (3.3.25) implies $\mathfrak{a}_1(t;y,\eta) = \exp(i\theta(t;y,\eta))$ where θ is a real-valued function. Substituting this expression in the first equation (3.3.15) we get $\mathcal{L}^{(0)}_{2m-1}(t;y,\eta) = 2m\,h^{2m-1}(y,\eta)\,\theta_t(t;y,\eta)$, which means that the function $\mathcal{L}^{(0)}_{2m-1}$ is real-valued. But the subprincipal symbol of a self-adjoint operator is also real-valued; consequently in order to prove (3.3.17) it is sufficient to prove

$$\operatorname{Re}\,\mathcal{L}^{(0)}_{2m-1} \;=\; -\,\operatorname{Re}\,A_{\mathrm{sub}}(x^*,\xi^*)\,. \tag{3.3.26}$$

If we now take a phase function φ which is locally real-valued, then an elementary examination of formulae (3.3.14), (3.3.7), and (3.3.5) leads to the conclusion that

$$\operatorname{Re}\,\mathcal{L}^{(0)}_{2m-1} \;=\; -\,\operatorname{Re}\,A_{2m-1}(x^*,\xi^*)\,.$$

But according to (2.1.9) the latter expression is the real part of the subprincipal symbol of the operator $-A$. This proves (3.3.26). □

The argument given above is based on the self-adjointness of the operator A, which means that the physical system described by the operator A preserves energy. This led us to the conclusion that the "generalized wave equation" $D_t^{2m}u = A_x u$ also preserves energy (formula (3.3.23)), which in turn implies that the modulus of the principal symbol of its solution does not depend on time (formula (3.3.25)). The latter formula is the crucial point of the above argument: once (3.3.25) is established, the proof of (3.3.17) becomes elementary. The only disadvantage is that one has to use the stationary phase method in order to substantiate (3.3.25).

REMARK 3.3.3. All the results of this section remain true when A is a pseudodifferential operator. The proofs of these results also remain the same, with the only difference that the expansion of the symbol of the operator A does not necessarily terminate at the zero degree term. The fact that A may be pseudodifferential does not cause any difficulties because throughout this section we assume $\partial M = \varnothing$.

Theorem 3.3.2 leads us to the following

COROLLARY 3.3.4. *Let A be a differential operator with real coefficients. Then $\mathfrak{a}_1$ is independent of t and coincides with P_1.*

PROOF. By Remark 2.1.16 in this case $A_{\text{sub}} = 0$. This equality and (3.3.16) immediately imply the required result. □

3.4. Representation of the wave group by means of oscillatory integrals: effective construction for manifolds with boundary

In this section, as in the previous one, we construct effectively (modulo C^∞) the Schwartz kernel $\mathbf{u}_P(t,x,y)$ of the operator $\exp(-it\mathcal{A}^{1/(2m)})P$, the difference being that now $\partial M \neq \varnothing$. Of course, the pseudodifferential operator P is assumed to be from the class Ψ_0^1.

1. Construction for small $|t|$ (without reflection). The support of the Schwartz kernel $p(x,y)$ of the pseudodifferential operator P is separated from the boundary, i.e., there exists a small positive δ such that

$$\operatorname{supp} p \subset \{(x,y) \in M_x \times M_y \,:\, x_n \geqslant \delta \text{ and } y_n \geqslant \delta\}. \tag{3.4.1}$$

Let us choose a small positive T_δ such that Hamiltonian trajectories originating from points (y,η) with $y_n \geqslant \delta$ do not reach the boundary on the time interval $[-T_\delta\,, T_\delta]$. We will show in this subsection that for $t \in (-T_\delta\,, T_\delta)$ the construction of the oscillatory integral representing the distribution $\mathbf{u}_P(t,x,y)$ modulo C^∞ is exactly the same as in the case of a manifold without boundary.

Let $O \subset T'\mathring{M}_y$ be a small conic neighbourhood of $\operatorname{cone\,supp} P$,

$$\operatorname{cone\,supp} P \subset O, \tag{3.4.2}$$

such that Hamiltonian trajectories originating from points $(y,\eta) \in O$ do not reach the boundary on the time interval $(-T_\delta\,, T_\delta)$. A neighbourhood O possessing these properties exists because $\operatorname{cone\,supp} P \subset \{(y,\eta) \in T'M_y \,:\, y_n \geqslant \delta\}$.

In accordance with our usual notation set $-T_\delta = T_-$, $T_\delta = T_+$, and

$$\mathfrak{O} := \{(t;y,\eta) \,:\, (y,\eta) \in O\,, t \in (-T_\delta\,, T_\delta)\},$$
$$\mathfrak{C} := \{(t,x;y,\eta) \,:\, (t;y,\eta) \in \mathfrak{O}\,, x = x^*(t;y,\eta)\}.$$

Clearly,

$$\mathfrak{C} \cap \{x \in \partial M_x\} = \varnothing . \tag{3.4.3}$$

Let

$$\mathcal{O} \subset (-T_\delta\,,\, T_\delta) \times M_x \times O \tag{3.4.4}$$

be a small conic neighbourhood of the set $\mathfrak{C}$ such that

$$\mathcal{O} \cap \{x \in \partial M_x\} = \varnothing . \tag{3.4.5}$$

A neighbourhood $\mathcal{O}$ possessing these properties exists due to (3.4.3). In view of (3.4.4) and $O \cap \{y \in \partial M_y\} = \varnothing$ we also have

$$\mathcal{O} \cap \{y \in \partial M_y\} = \varnothing . \tag{3.4.6}$$

Let us consider the oscillatory integral

$$u_P(t,x,y) = \mathcal{I}_{\varphi,\mathfrak{a}} = \int e^{i\varphi(t,x;y,\eta)}\,\mathfrak{a}(t;y,\eta)\,\varsigma(t,x;y,\eta)\,d_\varphi(t,x;y,\eta)\,đ\eta \tag{3.4.7}$$

which was constructed in Section 3.3. Recall that this oscillatory integral represents the distribution $\mathbf{u}_P(t,x,y)$ modulo C^∞ for a manifold without boundary. Though now $\partial M \neq \varnothing$ and it is not clear *a priori* whether (3.4.7) really represents the Schwartz kernel of the operator $\exp(-it\mathcal{A}^{1/(2m)})P$, the formal constructions of the phase function φ and of the amplitude $\mathfrak{a}$ described in Section 3.3 can be carried out for $t \in (-T_\delta\,,\, T_\delta)$ irrespective of the presence of the boundary. In particular, φ and $\mathfrak{a}$ are smooth functions defined on $\mathcal{O}$ and $\widehat{\mathfrak{O}}$, respectively; here

$$\widehat{\mathfrak{O}} = \{(t;y,\eta)\ :\ (t,x;y,\eta) \in \mathcal{O} \ \text{ for some } \ x \in M_x\} . \tag{3.4.8}$$

Moreover, in accordance with the construction of the amplitude $\mathfrak{a}$ (see (3.3.15), (3.3.3)) and Lemma 2.1.2 we can assume, without loss of generality, that

$$\operatorname{supp}\mathfrak{a} \subset (-T_\delta\,,\, T_\delta) \times \operatorname{cone\,supp} P . \tag{3.4.9}$$

Formulae (3.4.9), (3.4.2) guarantee the fulfilment of condition 4 from Definition 2.7.1. Thus, (3.4.7) is a standard oscillatory integral associated with the 0th leg of our canonical transformation in the sense of Definition 2.7.1.

Formulae (3.4.5), (3.4.6) and condition 4 from the definition of the cut-off ς (see Section 2.2) imply that

$$\operatorname{supp} u_P \cap \{x \in \partial M_x\} = \varnothing , \qquad \operatorname{supp} u_P \cap \{y \in \partial M_y\} = \varnothing . \tag{3.4.10}$$

In view of (3.4.10) the boundary conditions (3.2.7) are fulfilled *exactly*, i.e.,

$$(B_x^{(j)}(x',D_x)u_P)\big|_{\partial M_x} = 0 , \qquad j = 1,2,\ldots,m.$$

Moreover, due to (3.4.10), $\operatorname{supp}(u_P|_{t=0})$ is separated from the boundary of the manifold $M_x \times M_y$; consequently we have (3.2.8). So we can apply Theorem 3.2.1 the same way as we did in the previous section, which gives us (3.2.9).

2. The case of a large time interval. Further on in this section T_-, T_+ and O are as in 2.3.2 (in particular, O is assumed to be sufficiently small). The pseudodifferential operator P is assumed to satisfy (3.4.2). The latter is a restriction on P. Note, however, that if $\operatorname{cone\,supp} P$ is T_+-admissible, then we can partition P into a finite sum of pseudodifferential operators with small conic supports which satisfy (3.4.2). So all the main results remain valid if $\operatorname{cone\,supp} P$ is T_+-admissible and not necessarily small. In the future we shall often implicitly use this simple observation.

We shall construct the distribution $\mathbf{u}_P(t,x,y)$ modulo C^∞ for $t \in (T_-, T_+)$, with T_+ not necessarily small. This will force us to consider reflections from the boundary.

The conditions on O imposed in 2.3.2 and Lemma 1.3.29 imply that the number of types of billiard trajectories originating from points $(y,\eta) \in O$ is finite and does not depend on (y,η). Let us denote the set of these types by $\mathfrak{N}(O)$.

Suppose first that the set $\mathfrak{N}(O)$ is empty, which means that trajectories originating from points $(y,\eta) \in O$ do not reach the boundary on the time interval $(0, T_+)$. By the argument from subsection 1 we already know that the whole construction from Section 3.3 remains valid in this case. Note, that in principle the set $\mathfrak{N}(O)$ may be empty not only because the chosen time interval is small, but also due to some peculiarities of the set O and the Hamiltonian flow Φ^t.

In the remaining part of this section we assume $\mathfrak{N}(O) \neq \varnothing$.

3. Construction with one reflection. In this subsection we shall consider the case when there is only one reflection on the time interval $(0, T_+)$, i.e., when the set $\mathfrak{N}(O)$ consists only of multiindices of length 1.

Let us denote by $0 < t_1^*(y,\eta) < T_+$ the moment of reflection, and by $2q$ the number of real ξ_n-roots of the equation (1.3.30) with $\tau = t_1^*(y,\eta)$. Obviously, the set $\mathfrak{N}(O)$ contains in our case exactly q elements and these elements are the numbers $1, 2, \dots, q$.

Let us denote by $\zeta^\mp(y,\eta)$ the real ζ-roots of the algebraic equation (1.3.30), where the superscript $\mp$ indicates the sign of the derivative

$$\left(\frac{\partial}{\partial\zeta} A_{2m}(x^{*\prime}(t_1^*(y,\eta);y,\eta),0,\xi^{*\prime}(t_1^*(y,\eta);y,\eta),\zeta)\right)\Bigg|_{\zeta=\zeta^\mp(y,\eta)}$$

(note that the above derivative is nonzero according to Definition 1.3.25 of an admissible trajectory). By virtue of (1.1.14) the real roots $\zeta^-(y,\eta)$ correspond to incident trajectories, whereas the real roots $\zeta^+(y,\eta)$ correspond to reflected trajectories; this explains the convenience of separating our real roots into two groups. Let us enumerate the real roots $\zeta^\mp(y,\eta)$ in order of their growth:

$$\zeta_1^-(y,\eta) < \zeta_1^+(y,\eta) < \zeta_2^-(y,\eta) < \zeta_2^+(y,\eta) < \cdots < \zeta_q^-(y,\eta) < \zeta_q^+(y,\eta).$$

We have $\zeta_l^\mp \in C^\infty(O)$, $l = 1, 2, \dots, q$, because these roots are simple.

Let us denote by $\zeta_l^\mp(y,\eta)$, $l = q+1, q+2, \dots, m$, the complex roots of the algebraic equation (1.3.30) with negative and positive imaginary parts, respectively. According to Definition 1.3.25 these complex roots are simple. This fact and the simple connectedness of the set O imply that we can choose the numeration of our complex roots in such a way that $\zeta_l^\mp \in C^\infty(O)$, $l = q+1, q+2, \dots, m$.

By

$$(x^*(t;y,\eta),\xi^*(t;y,\eta))\,,\qquad t\in(T_-\,,\,t_1^*(y,\eta)]\,,\quad (y,\eta)\in O\,, \tag{3.4.11}$$

we shall denote the trajectory before reflection. This trajectory corresponds to a real ξ_n-root of the algebraic equation (1.3.30):

$$\xi_n^*(t_1^*(y,\eta)-0;y,\eta) \;=\; \zeta_{\mathbf{k}}^-(y,\eta) \tag{3.4.12}$$

for some $1\leqslant \mathbf{k}\leqslant q$. By

$$(x^{(l)*}(t;y,\eta),\xi^{(l)*}(t;y,\eta)),\qquad t\in[t_1^*(y,\eta)\,,\,T_+)\,,\quad (y,\eta)\in O\,,\quad l=1,2,\dots,q\,, \tag{3.4.13}$$

we shall denote the real reflected trajectories. By

$$(x^{(l)*}(t;y,\eta),\xi^{(l)*}(t;y,\eta)),\quad (y,\eta)\in O,\quad l=\,q+1,\,q+2,\dots,\,m, \tag{3.4.14}$$

we shall denote the complex reflected trajectories (see 2.6.4). Recall that we have agreed to view these complex trajectories as formal Taylor expansions in powers of $t-t_1^*$.

We shall enumerate the real and complex reflected trajectories in accordance with

$$\xi_n^{(l)*}(t_1^*(y,\eta)+0;y,\eta) \;=\; \zeta_l^+(y,\eta)\,,\qquad l=1,2,\dots,m\,. \tag{3.4.15}$$

Set

$$\begin{aligned}\mathfrak{O} &:= \{\,(t;y,\eta)\,:\,(y,\eta)\in O\,,\,t\in(T_-\,,\,t_1^*(y,\eta)]\,\}\,,\\ \mathfrak{C} &:= \{\,(t,x;y,\eta)\,:\,(t;y,\eta)\in\mathfrak{O}\,,\,x=x^*(t;y,\eta)\,\}\,,\end{aligned}$$

and

$$\begin{aligned}\mathfrak{O}^{(l)} &:= \{\,(t;y,\eta)\,:\,(y,\eta)\in O\,,\,t\in[t_1^*(y,\eta)\,,T_+)\,\}\,,\\ \mathfrak{C}^{(l)} &:= \{\,(t,x;y,\eta)\,:\,(t;y,\eta)\in\mathfrak{O}^{(l)}\,,\,x=x^{(l)*}(t;y,\eta)\,\}\,,\end{aligned}$$

$l=1,2,\dots,q$. Under this notation the sets $\mathfrak{O}^{(l)}$ are identical for all l; however, we will stick to this notation because subsequently (next subsection), when we consider the case of an arbitrary number of reflections, these sets may become different.

Set also

$$\mathfrak{C}' \;:=\; \{\,(t,x';y,\eta)\;:\;t=t_1^*(y,\eta)\,,\;x'=x^{*\prime}(t_1^*(y,\eta),y,\eta)\,,\;(y,\eta)\in O\,\}\,.$$

According to this definition the set $\mathfrak{C}'$ is a subset of $(T_-\,,\,T_+)\times\partial M_x\times O$. However we shall often view $\mathfrak{C}'$ as a subset of $(T_-\,,\,T_+)\times M_x\times O$. This will be necessary because further on in this subsection the set $\mathfrak{C}'$ will play two roles: the set where the derivative $\frac{\partial}{\partial\eta}$ of the boundary phase function $\varphi'(t,x';y,\eta)$ vanishes, as well as the set where the derivative $\frac{\partial}{\partial\eta}$ of the boundary layer phase functions $\varphi^{(l)}(t,x;y,\eta)$, $l>q$, vanishes.

Note that

$$\mathfrak{C}\cap\{t=0,\;x\in\partial M_x\} \;=\; \varnothing\,, \tag{3.4.16}$$

$$\mathfrak{C}^{(l)}\cap\{t=0\} \;=\; \varnothing\,,\qquad l=1,2,\dots,q\,, \tag{3.4.17}$$

$$\mathfrak{C}'\cap\{t=0\} \;=\; \varnothing\,. \tag{3.4.18}$$

Let us choose some standard phase functions $\varphi(t,x;y,\eta)$ and $\varphi^{(l)}(t,x;y,\eta)$, $l=1,2,\dots,q$, associated (in the sense of Definition 2.4.1) with the canonical transformations (3.4.11) and (3.4.13), respectively. These phase functions are defined on some conic neighbourhoods $\mathcal{O},\mathcal{O}^{(l)}\subset(T_-\,,\,T_+)\times M_x\times O$ of the sets $\mathfrak{C}\,,\ \mathfrak{C}^{(l)}$. The existence of the phase functions φ, $\varphi^{(1)}$, $\varphi^{(2)}$, ... , $\varphi^{(q)}$ is established by Corollary 2.4.8.

REMARK 3.4.1. In Chapter 2 we denoted by φ_1, φ_2, ... , φ_i, ... , $\varphi_{\mathbf{r}}$ the phase functions associated with *different* legs of the *same* billiard trajectory, and the subscript i was used to indicate that i reflections have already been experienced. However in the current subsection we denote by $\varphi^{(1)}$, $\varphi^{(2)}$, ... , $\varphi^{(l)}$, ... , $\varphi^{(q)}$ the phase functions associated with the *same* (after one reflection) leg of *different* billiard trajectories, and here the subscript l indicates the choice of the billiard trajectory. The way to bring these two types of notation into agreement is to view the superscripts of φ appearing in the current section as *multiindices*. In particular, the superscripts of φ appearing in this subsection should be viewed as multiindices of length 1. Under this convention the length of the multiindex will correspond to the subscript i of φ (the number of experienced reflections). Thus, the phase functions $\varphi^{(1)}$, $\varphi^{(2)}$, ... , $\varphi^{(l)}$, ... , $\varphi^{(q)}$ in the current subsection are q different variants of the phase function φ_1 from Section 2.4. Similarly, the phase function φ from this section corresponds to the phase function φ_0 from Section 2.4 because the empty multiindex has length zero. Our convention on the use of multiindices will become clearer in the next subsection when we introduce the concept of a *tree*. In fact, our seeming disaccord in notation happened only because we are trying to avoid as long as possible the use of notions from graph theory, in order to make our presentation simpler.

REMARK 3.4.2. For boundary phase functions we have a similar correspondence between the notation of this section and that of Chapter 2, but with a shift by 1: the length of the multiindex plus 1 equals the sequential number of the reflection.

Let us choose some boundary layer phase functions $\varphi^{(l)}(t,x;y,\eta)\,,\ l=q+1,$ $q+2,\dots,m$, associated (in the sense of Definition 2.6.9) with the transformations (3.4.14). These phase functions are defined on some conic neighbourhoods $\mathcal{O}^{(l)}\subset$ $(T_-\,,\,T_+)\times M_x\times O$ of the set $\mathfrak{C}'$. The existence of phase functions $\varphi^{(q+1)}$, $\varphi^{(q+2)}$, ... , $\varphi^{(m)}$ was established in 2.6.4.

We shall assume that the sets $\mathcal{O}$, $\mathcal{O}^{(l)}$, $l=1,2,\dots,m$, are so small that

$$\mathcal{O}\cap\{t=0,\ x\in\partial M_x\}=\varnothing\,, \tag{3.4.19}$$

$$\mathcal{O}^{(l)}\cap\{t=0\}=\varnothing\,,\qquad l=1,2,\dots,m\,. \tag{3.4.20}$$

Neighbourhoods possessing these properties exist due to (3.4.16)–(3.4.18). As in subsection 1 we note also that the set $\mathcal{O}$ automatically satisfies (3.4.6), and similarly

$$\mathcal{O}^{(l)}\cap\{y\in\partial M_y\}=\varnothing\,,\qquad l=1,2,\dots,m\,. \tag{3.4.21}$$

Let us denote by $\widehat{\mathfrak{D}}$, $\widehat{\mathfrak{D}}^{(l)}$, $l=1,2,\dots,m$, the projections of the sets $\mathcal{O}$, $\mathcal{O}^{(l)}$, $l=1,2,\dots,m$, on $(T_-\,,T_+)\times O$; the set $\widehat{\mathfrak{D}}$ is defined by formula (3.4.8), and analogous formulae define $\widehat{\mathfrak{D}}^{(l)}$.

We shall assume that the phase functions φ, $\varphi^{(1)}$, $\varphi^{(2)}$, ... , $\varphi^{(m)}$ are chosen in such a way that for each $l = 1, 2, \dots, m$ the pair φ and $\varphi^{(l)}$ *match* in the sense of Definitions 2.6.2, 2.6.14. This means that there exists a boundary phase function $\varphi'(t, x'; y, \eta)$ (see Definition 2.5.2) defined on some conic neighbourhood $\mathcal{O}' \subset (T_-, T_+) \times \partial M_x \times O$ of the set $\mathfrak{C}'$ such that

$$\begin{aligned}\mathcal{O} \cap \{x \in \partial M\} &= \mathcal{O}^{(1)} \cap \{x \in \partial M\} = \mathcal{O}^{(2)} \cap \{x \in \partial M\} = \cdots \\ &= \mathcal{O}^{(m)} \cap \{x \in \partial M\} = \mathcal{O}'\end{aligned} \tag{3.4.22}$$

and

$$\begin{aligned}\varphi(t, x', 0; y, \eta) &= \varphi^{(1)}(t, x', 0; y, \eta) = \varphi^{(2)}(t, x', 0; y, \eta) = \cdots \\ &= \varphi^{(m)}(t, x', 0; y, \eta) = \varphi'(t, x'; y, \eta) \quad \text{on } \mathcal{O}'.\end{aligned} \tag{3.4.23}$$

The existence of a set of phase functions φ, $\varphi^{(1)}$, $\varphi^{(2)}$, ... , $\varphi^{(m)}$ and φ' possessing properties (3.4.22), (3.4.23) is established by Lemma 2.6.3 and its analogue for boundary layer phase functions (see 2.6.4).

Let $\mathfrak{a} \in S^1(\mathfrak{D})$ be the amplitude constructed in accordance with (3.3.15), (3.3.3) and satisfying

$$\operatorname{supp} \mathfrak{a} \subset (T_-, T_+) \times \operatorname{cone supp} P \tag{3.4.24}$$

(cf. (3.4.9)). Let us extend this amplitude to $\widehat{\mathfrak{D}} \setminus \mathfrak{D}$ (i.e., to $t > t_1^*(y, \eta)$) in an arbitrary smooth manner preserving (3.4.24), and retain for the extended amplitude (of the class $S^1(\widehat{\mathfrak{D}})$) the notation $\mathfrak{a}$.

Let us consider the oscillatory integral $\mathcal{I}_{\varphi,\mathfrak{a}}$ (see (3.4.7)) with phase function φ and amplitude $\mathfrak{a}$ defined above. As in subsection 1 we note that formulae (3.4.24), (3.4.2) ensure that $\mathcal{I}_{\varphi,\mathfrak{a}}$ is a standard oscillatory integral associated with the canonical transformation (3.4.11) in the sense of Definition 2.7.1. Let us also consider the oscillatory integrals

$$\mathcal{I}_{\varphi^{(l)},\mathfrak{a}^{(l)}} := \int e^{i\varphi^{(l)}(t,x;y,\eta)} \mathfrak{a}^{(l)}(t; y, \eta)\, \varsigma^{(l)}(t, x; y, \eta)\, d_{\varphi^{(l)}}(t, x; y, \eta)\, đ\eta, \tag{3.4.25}$$

$l = 1, 2, \dots, m$, where the phase functions $\varphi^{(l)}$ are as defined above and the $\mathfrak{a}^{(l)} \in S^1(\widehat{\mathfrak{D}}^{(l)})$ are some amplitudes satisfying

$$\operatorname{supp} \mathfrak{a}^{(l)} \subset (T_-, T_+) \times \operatorname{cone supp} P. \tag{3.4.26}$$

Formulae (3.4.26), (3.4.2) ensure that the oscillatory integrals (3.4.25) are standard ($l \leqslant q$) or boundary layer ($l > q$) oscillatory integrals associated with the canonical transformations (3.4.13), (3.4.14) in the sense of Definitions 2.7.1, 2.8.1, respectively.

We shall construct the distribution approximating modulo C^∞ the Schwartz kernel $\mathbf{u}_P(t, x, y)$ of the operator $\exp(-it\mathcal{A}^{1/(2m)})P$ in the form

$$u_P(t, x, y) = \mathcal{I}_{\varphi,\mathfrak{a}} + \sum_{l=1}^{m} \mathcal{I}_{\varphi^{(l)},\mathfrak{a}^{(l)}}, \tag{3.4.27}$$

where the amplitudes $\mathfrak{a}^{(l)}$ are to be determined is such a way that $u_P(t, x, y)$ satisfies the conditions of Theorem 3.2.1.

Formulae (3.4.19), (3.4.6), (3.4.20) imply that $u_P(0,x,y) \equiv \mathcal{I}_{\varphi,\mathfrak{a}}(0,x,y)$ and that $\operatorname{supp}(u_P|_{t=0})$ is separated from the boundary of the manifold $M_x \times M_y$; consequently we have (3.2.8).

Thus, in order to apply Theorem 3.2.1 we have to fulfill only the conditions (3.2.6) and (3.2.7), i.e.,

$$D_t^{2m}\mathcal{I}_{\varphi^{(l)},\mathfrak{a}^{(l)}} - A(x,D_x)\mathcal{I}_{\varphi^{(l)},\mathfrak{a}^{(l)}} \in C^\infty((T_-,T_+) \times M_x \times M_y), \quad l = 1,2,\ldots,m, \tag{3.4.28}$$

and

$$\left.(B^{(j)}(x',D_x)u_P)\right|_{\partial M_x} \in C^\infty((T_-,T_+) \times \partial M_x \times M_y), \quad j = 1,2,\ldots,m. \tag{3.4.29}$$

Note that we did not write the equation

$$D_t^{2m}\mathcal{I}_{\varphi,\mathfrak{a}} - A(x,D_x)\mathcal{I}_{\varphi,\mathfrak{a}} \in C^\infty((T_-,T_+) \times M_x \times M_y)$$

among the equations to be satisfied because it is already satisfied by our choice of the amplitude $\mathfrak{a}$.

Consider first the equations (3.4.28) with $l \leqslant q$. The procedure described in the proof of Theorem 3.3.1 reduces each of these equations to a recurrent system of ordinary differential equations in t with respect to the unknown functions $\mathfrak{a}^{(l)}_{\mathfrak{l}-s}(t;y,\eta)$:

$$\begin{aligned}
\mathcal{L}^{(l)}_{2m-1}\,\mathfrak{a}^{(l)}_{\mathfrak{l}} &= 0, \\
\mathcal{L}^{(l)}_{2m-1}\,\mathfrak{a}^{(l)}_{\mathfrak{l}-1} &= -\mathcal{L}^{(l)}_{2m-2}\,\mathfrak{a}^{(l)}_{\mathfrak{l}}, \\
\mathcal{L}^{(l)}_{2m-1}\,\mathfrak{a}^{(l)}_{\mathfrak{l}-2} &= -\mathcal{L}^{(l)}_{2m-2}\,\mathfrak{a}^{(l)}_{\mathfrak{l}-1} - \mathcal{L}^{(l)}_{2m-3}\,\mathfrak{a}^{(l)}_{\mathfrak{l}}, \\
\cdots &= \cdots \\
\mathcal{L}^{(l)}_{2m-1}\,\mathfrak{a}^{(l)}_{\mathfrak{l}-s} &= -\sum_{k=1}^{s}\mathcal{L}^{(l)}_{2m-k-1}\,\mathfrak{a}^{(l)}_{\mathfrak{l}-s+k}, \\
\cdots &= \cdots
\end{aligned} \tag{3.4.30}$$

where $(t;y,\eta) \in \mathfrak{D}^{(l)}$. The operators $\mathcal{L}^{(l)}_{2m-k}$ appearing in (3.4.30) are defined by the same formulae as the operators $\mathcal{L}_{2m-k}$ appearing in (3.3.15), the only difference being that $\varphi^{(l)}$, $x^{(l)*}$, $\xi^{(l)*}$ are placed instead of φ, x^*, ξ^*; see proof of Theorem 3.3.1 for details.

Consider now the equations (3.4.28) with $l > q$. The procedure described in the proof of Theorem 3.3.1 works in this case as well, and reduces each of these equations to a recurrent system similar to (3.4.30):

$$
\begin{aligned}
\mathcal{L}^{(l)}_{2m-1}\,\mathfrak{a}^{(l)}_{\mathfrak{l}} &\simeq 0\,,\\
\mathcal{L}^{(l)}_{2m-1}\,\mathfrak{a}^{(l)}_{\mathfrak{l}-1} &\simeq -\,\mathcal{L}^{(l)}_{2m-2}\,\mathfrak{a}^{(l)}_{\mathfrak{l}}\,,\\
\mathcal{L}^{(l)}_{2m-1}\,\mathfrak{a}^{(l)}_{\mathfrak{l}-2} &\simeq -\,\mathcal{L}^{(l)}_{2m-2}\,\mathfrak{a}^{(l)}_{\mathfrak{l}-1} - \mathcal{L}^{(l)}_{2m-3}\,\mathfrak{a}^{(l)}_{\mathfrak{l}}\,,\\
\cdots &\simeq \cdots\\
\mathcal{L}^{(l)}_{2m-1}\,\mathfrak{a}^{(l)}_{\mathfrak{l}-s} &\simeq -\sum_{k=1}^{s}\mathcal{L}^{(l)}_{2m-k-1}\,\mathfrak{a}^{(l)}_{\mathfrak{l}-s+k}\,,\\
\cdots &\simeq \cdots
\end{aligned}
\tag{3.4.31}
$$

where the sign $\simeq$ stands for the equality of formal Taylor expansions in powers of $t-t^*_{\mathfrak{l}}$. Naturally, the operators $\mathcal{L}^{(l)}_{2m-k}$ in (3.4.31) are defined similarly to those in (3.4.30), with account of the fact that now the corresponding Hamiltonian trajectories are complex. The latter implies that the coefficients of the operators $\mathcal{L}^{(l)}_{2m-k}$ appearing in (3.4.31) are defined modulo $O(|t-t^*_{\mathfrak{l}}|^\infty)$.

Let us now consider the boundary conditions (3.4.29). Substituting (3.4.27) on the left-hand side of (3.4.29) we obtain a boundary oscillatory integral with phase function φ' (see (3.4.23)) and amplitude

$$
b^{(j)}(t,x';y,\eta) + \sum_{l=1}^{m} b^{(j,l)}(t,x';y,\eta)\,, \tag{3.4.32}
$$

where

$$
b^{(j)} = \mathfrak{a}\,e^{-i\varphi'}\,(d_{\varphi'})^{-1}\,\left(B^{(j)}(x',D_x)\,e^{i\varphi}\,d_\varphi\right)\Big|_{\partial M_x} = \sum_{s=0}^{m_j} M^{(j)}_{m_j-s}\,\mathfrak{a}\,,
$$

$$
b^{(j,l)} = \mathfrak{a}^{(l)}\,e^{-i\varphi'}\,(d_{\varphi'})^{-1}\,\left(B^{(j)}(x',D_x)\,e^{i\varphi^{(l)}}\,d_{\varphi^{(l)}}\right)\Big|_{\partial M_x} = \sum_{s=0}^{m_j} M^{(j,l)}_{m_j-s}\,\mathfrak{a}^{(l)}\,.
$$

Here $M^{(j)}_{m_j-s} = M^{(j)}_{m_j-s}(t,x';y,\eta)$, $M^{(j,l)}_{m_j-s} = M^{(j,l)}_{m_j-s}(t,x';y,\eta)$ are functions positively homogeneous in η of degree m_j-s, and $m_j = \operatorname{ord} B^{(j)}$. Condition (3.4.29) holds if and only if the symbol corresponding to the amplitude (3.4.32) is of class $S^{-\infty}$. Consequently, acting on (3.4.32) with the operator $\mathfrak{S}'$ (see (2.9.8)) and collecting terms with the same degree of homogeneity we reduce (3.4.29) to the following recurrent system:

$$
\begin{aligned}
\sum_{l=1}^{m}\mathcal{M}^{(j,l)}_{m_j}\mathfrak{a}^{(l)\prime}_{\mathfrak{l}} &= -\mathcal{M}^{(j)}_{m_j}\mathfrak{a}'_{\mathfrak{l}}\,,\\
\sum_{l=1}^{m}\mathcal{M}^{(j,l)}_{m_j}\mathfrak{a}^{(l)\prime}_{\mathfrak{l}-1} &= -\mathcal{M}^{(j)}_{m_j}\mathfrak{a}'_{\mathfrak{l}-1} - \mathcal{M}^{(j)}_{m_j-1}\mathfrak{a}'_{\mathfrak{l}} - \sum_{l=1}^{m}\mathcal{M}^{(j,l)}_{m_j-1}\mathfrak{a}^{(l)\prime}_{\mathfrak{l}}\,,\\
\cdots &= \cdots\\
\sum_{l=1}^{m}\mathcal{M}^{(j,l)}_{m_j}\mathfrak{a}^{(l)\prime}_{\mathfrak{l}-s} &= -\sum_{k=0}^{s}\mathcal{M}^{(j)}_{m_j-k}\mathfrak{a}'_{\mathfrak{l}-s+k} - \sum_{k=1}^{s}\sum_{l=1}^{m}\mathcal{M}^{(j,l)}_{m_j-k}\mathfrak{a}^{(l)\prime}_{\mathfrak{l}-s+k}\,,\\
\cdots &= \cdots
\end{aligned}
\tag{3.4.33}
$$

$j = 1, 2, \ldots, m$. In (3.4.33) $\mathfrak{a}'_{1-s} := \mathfrak{a}_{1-s}|_{t=t_1^*}$ are given, $\mathfrak{a}^{(l)\prime}_{1-s} := \mathfrak{a}^{(l)}_{1-s}|_{t=t_1^*}$ are unknown, and

$$\mathcal{M}^{(j)}_{m_j-k} = \mathcal{M}^{(j)}_{m_j-k}(y,\eta,D_\eta) = \sum_{r+s=k} \mathfrak{S}'_{-r}\, M^{(j)}_{m_j-s}\,,$$
$$\mathcal{M}^{(j,l)}_{m_j-k} = \mathcal{M}^{(j,l)}_{m_j-k}(y,\eta,D_\eta) = \sum_{r+s=k} \mathfrak{S}'_{-r}\, M^{(j,l)}_{m_j-s}$$

are differential operators positively homogeneous in η of degree $m_j - k$. According to Lemmas 2.9.10, 2.9.11 the $\mathcal{M}^{(j)}_{m_j}$, $\mathcal{M}^{(j,l)}_{m_j}$ are operators of multiplication by functions

$$\mathcal{M}^{(j)}_{m_j} = \mathcal{M}^{(j)}_{m_j}(y,\eta) = \frac{B^{(j)}_{m_j}\left(x^{*\prime}|_{t=t_1^*}\,,\,\xi^*|_{t=t_1^*}\right)}{\sqrt{-h_{\xi_n}\left(x^*|_{t=t_1^*}\,,\,\xi^*|_{t=t_1^*}\right)}}\,, \tag{3.4.34}$$

$$\mathcal{M}^{(j,l)}_{m_j} = \mathcal{M}^{(j,l)}_{m_j}(y,\eta) = \frac{B^{(j)}_{m_j}\left(x^{(l)*\prime}\big|_{t=t_1^*}\,,\,\xi^{(l)*}\big|_{t=t_1^*}\right)}{\sqrt{h_{\xi_n}\left(x^{(l)*}\big|_{t=t_1^*}\,,\,\xi^{(l)*}\big|_{t=t_1^*}\right)}}\,. \tag{3.4.35}$$

The square root in (3.4.34) and the square roots in (3.4.35), $l \leqslant q$, are chosen to be positive, whereas the (complex) square roots in (3.4.35), $l > q$, are chosen in accordance with Lemma 2.9.11 and Definition 2.6.16. In view of (3.4.34), (3.4.35) the system (3.4.33) is a recurrent system of linear *algebraic* equations with respect to the unknowns $\mathfrak{a}^{(l)\prime}_{1-s}$. Moreover, as the number $\nu = A_{2m}(y,\eta)$ is not an eigenvalue of the auxiliary one-dimensional spectral problem (1.1.7), (1.1.8) with

$$(x',\xi') = (x^{*\prime}(t_1^*(y,\eta);y,\eta), \xi^{*\prime}(t_1^*(y,\eta);y,\eta)),$$

the matrix

$$\mathcal{M}^+ = \mathcal{M}^+(y,\eta) := \left(\mathcal{M}^{(j,l)}_{m_j}\right)^m_{j,l=1} \tag{3.4.36}$$

is invertible; here j is the number of the row and l is the number of the column. Solving the linear algebraic system (3.4.33) iteratively for $s = 0, 1, 2, \ldots$, we obtain the (unique) homogeneous functions $\mathfrak{a}^{(l)\prime}_{1-s} \equiv \mathfrak{a}^{(l)}_{1-s}|_{t=t_1^*}$, $l = 1, 2, \ldots, m$, $s = 0, 1, 2, \ldots$. Note that (3.4.24) implies

$$\operatorname{supp}\left(\mathfrak{a}^{(l)}_{1-s}|_{t=t_1^*}\right) \subset \operatorname{cone\,supp} P\,, \qquad l = 1, 2, \ldots, m, \quad s = 0, 1, 2, \ldots\,. \tag{3.4.37}$$

Integrating the ordinary differential equations (3.4.30) ($l \leqslant q$) with initial conditions at $t = t_1^*$ derived from (3.4.33) we obtain the (unique) homogeneous functions $\mathfrak{a}^{(l)}_{1-s} \in C^\infty(\mathfrak{O}^{(l)})$, $l = 1, 2, \ldots, q$, $s = 0, 1, 2, \ldots$. In view of (3.4.37)

$$\operatorname{supp} \mathfrak{a}^{(l)}_{1-s} \subset (T_-\,, T_+) \times \operatorname{cone\,supp} P\,. \tag{3.4.38}$$

By Lemma 2.1.2 there exists an amplitude $\mathfrak{a}^{(l)} \in S^1(\mathfrak{O}^{(l)})$ with given asymptotic expansion $\sum_{j=0}^\infty \mathfrak{a}^{(l)}_{1-j}$ and satisfying (3.4.26). Let us extend this amplitude to $\widehat{\mathfrak{O}}^{(l)} \setminus \mathfrak{O}^{(l)}$ (i.e., to $t < t_1^*(y,\eta)$) in an arbitrary smooth manner preserving (3.4.26), and retain for the extended amplitude (of the class $S^1(\widehat{\mathfrak{O}}^{(l)})$) the notation $\mathfrak{a}^{(l)}$.

Integrating the ordinary differential equations (3.4.31) ($l > q$) with initial conditions at $t = t_1^*$ derived from (3.4.33) we obtain the homogeneous functions

$\mathfrak{a}^{(l)}_{1-s} \in C^\infty(\widehat{\mathfrak{D}}^{(l)})$, $l = q+1, q+2, \ldots, m$, $s = 0, 1, 2, \ldots$. These functions are defined modulo $O(|t-t_1^*|^\infty)$. In view of (3.4.37) we can choose the $\mathfrak{a}^{(l)}_{1-s}$ in such a way that they satisfy (3.4.38). By Lemma 2.1.2 there exists an amplitude $\mathfrak{a}^{(l)} \in S^1(\mathfrak{D}^{(l)})$ with given asymptotic expansion $\sum_{j=0}^\infty \mathfrak{a}^{(l)}_{1-j}$ and satisfying (3.4.26).

With the amplitudes $\mathfrak{a}$, $\mathfrak{a}^{(1)}$, $\mathfrak{a}^{(2)}$, ..., $\mathfrak{a}^{(m)}$ constructed above we have satisfied all the conditions of Theorem 3.2.1.

It remains to note that the matching conditions (3.4.22), (3.4.23) limiting our choice of phase functions can eventually be dropped because by virtue of Theorems 2.7.10 and 2.8.8 each of the oscillatory integrals appearing in (3.4.27) can be written with an arbitrary phase function of the same class.

Thus, we have proved

THEOREM 3.4.3. *In the case of one reflection for any set of phase functions* φ, $\varphi^{(1)}$, $\varphi^{(2)}$, ..., $\varphi^{(m)}$ *associated with our Hamiltonian billiards there exist amplitudes* $\mathfrak{a}$, $\mathfrak{a}^{(1)}$, $\mathfrak{a}^{(2)}$, ..., $\mathfrak{a}^{(m)}$ *of the class* S^1 *such that the distribution* (3.4.27) *coincides with the Schwartz kernel* $\mathbf{u}_P(t,x,y)$ *of the operator* $\exp(-it\mathcal{A}^{1/(2m)})P$ *modulo* $C^\infty((T_-, T_+) \times M_x \times M_y)$.

In Section 1.4 we introduced the reflection matrix $R(\nu; x', \xi')$, $\nu \in \mathbb{R}$, $(x', \xi') \in T^*\partial M$. Set for brevity

$$R(y,\eta) := R\big(A_{2m}(y,\eta); x^{*\prime}(t_1^*(y,\eta); y,\eta), \xi^{*\prime}(t_1^*(y,\eta); y,\eta)\big). \tag{3.4.39}$$

For $(y,\eta) \in O$ the matrix $R(y,\eta)$ is of size $q \times q$. We shall denote the elements of the matrix $R(y,\eta)$ by $R_{lk}(y,\eta)$; here l is the number of the row and k is the number of the column.

Replacing in our construction the roots ζ_l^+ by ζ_l^- we can define, similarly to the matrix function $\mathcal{M}^+$ (see (3.4.36)), the matrix function $\mathcal{M}^-$. It is easy to see that the reflection matrix $R(y,\eta)$ is in fact the upper left $q \times q$-block of the matrix $-(\mathcal{M}^+)^{-1} \cdot \mathcal{M}^-$.

THEOREM 3.4.4. *In the case of one reflection the principal symbols of the standard Lagrangian distributions* $\mathcal{I}_{\varphi,\mathfrak{a}}$ *and* $\mathcal{I}_{\varphi^{(l)},\mathfrak{a}^{(l)}}$, $l \leqslant q$, *appearing in* (3.4.27) *are given by formulae* (3.3.16), $(t;y,\eta) \in \mathfrak{D}$, *and*

$$\begin{aligned} \mathfrak{a}^{(l)}_1(t;y,\eta) &= R_{l\mathbf{k}}(y,\eta)\, P_1(y,\eta) \\ &\times \exp\Big(-\frac{i}{2m} h^{1-2m}(y,\eta) \int_0^t A_{\mathrm{sub}}\left(x^{(l)*}(\tau;y,\eta), \xi^{(l)*}(\tau;y,\eta)\right) d\tau\Big), \end{aligned}$$

$(t;y,\eta) \in \mathfrak{D}^{(l)}$, *respectively. Here the number* $\mathbf{k}$ *is from* (3.4.12).

Recall that by P_1 we denote the principal symbol of the operator P, and by A_{sub} the subprincipal symbol of the operator A. By $(x^{(l)*}(\tau;y,\eta), \xi^{(l)*}(\tau;y,\eta))$, $0 \leqslant \tau \leqslant t$, in Theorem 3.4.4 we understand the billiard trajectory of type l originating from the point (y,η), that is, the concatenation of (3.4.11) and (3.4.13).

The proof of Theorem 3.4.4 is the same as that of Theorem 3.3.2. The extra factor $R_{l\mathbf{k}}(y,\eta)$ appears because $\mathfrak{a}^{(l)}_1|_{t=t_1^*} = R_{l\mathbf{k}}\,(\mathfrak{a}_1|_{t=t_1^*})$.

As the reflection matrix $R(y,\eta)$ is unitary (see Section 1.4 and Proposition A.2.1) we have

$$\sum_{l=1}^q |\mathfrak{a}^{(l)}_1(t;y,\eta)|^2 = |P_1(y,\eta)|^2.$$

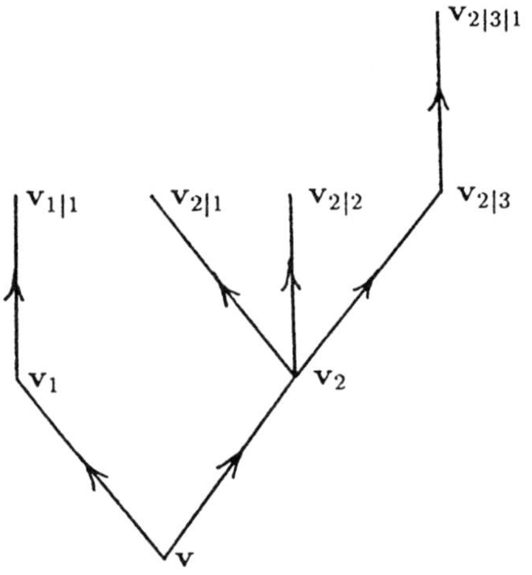

FIGURE 12. Example of a rooted tree.

The above formula and (3.3.25) illustrate the physical fact that the "generalized wave equation" $D_t^{2m}u = A_x u$ preserves energy.

4. Construction with an arbitrary number of reflections. In this subsection we shall finally consider the case when the number of reflections on the time interval $(0\,, T_+)$ is arbitrary, i.e., when the set $\mathfrak{N}(O)$ (see subsection 2) is nonempty and possibly contains multiindices of length greater than one.

This case is handled by the successive application of the procedure described in subsection 3. We formulate below the end results. It is convenient to do this using graph theory notation [**Big**].

In what follows we denote multiindices as $\mathfrak{n} = \mathfrak{n}_1|\mathfrak{n}_2|\dots|\mathfrak{n}_i$, $\mathfrak{m} = \mathfrak{m}_1|\mathfrak{m}_2|\dots|\mathfrak{m}_j$, and their concatenation as $\mathfrak{n}|\mathfrak{m} := \mathfrak{n}_1|\mathfrak{n}_2|\dots|\mathfrak{n}_i|\mathfrak{m}_1|\mathfrak{m}_2|\dots|\mathfrak{m}_j$. We allow the use of the empty multiindex, and view it as a multiindex of length zero. If $\mathfrak{n} = \varnothing$, then $\mathfrak{n}|\mathfrak{m} := \mathfrak{m}$; if $\mathfrak{m} = \varnothing$, then $\mathfrak{n}|\mathfrak{m} := \mathfrak{n}$.

Let us associate our branching Hamiltonian billiards with a rooted tree $\mathbf{T} = (\mathbf{V}, \mathbf{E})$. Here $\mathbf{V}$ is the vertex-set and $\mathbf{E}$ is the edge-set which are defined in the following way.

Individual vertices are denoted by $\mathbf{v}_\mathfrak{n}$, where $\mathfrak{n}$ is the multiindex enumerating the particular vertex. This multiindex takes values from the set of multiindices $\mathfrak{M}$. Thus, $\mathbf{V} = \{\mathbf{v}_\mathfrak{n} :\ \mathfrak{n} \in \mathfrak{M}\}$.

By $\mathring{\mathfrak{M}} \subset \mathfrak{M}$ we denote the set of all multiindices enumerating the *internal* vertices of our tree; this set includes the empty multiindex. The set $\mathfrak{M}\backslash\mathring{\mathfrak{M}}$ enumerates the leaves of our tree.

The enumeration of vertices is chosen in such a way that

1. If the vertex $\mathbf{v}_\mathfrak{m}$ is the son of the vertex $\mathbf{v}_\mathfrak{n}$, then $\mathfrak{m} = \mathfrak{n}|l$ for some natural number l.
2. $\mathfrak{M} \setminus \mathring{\mathfrak{M}} = \mathfrak{N}(O)$, where $\mathfrak{N}(O)$ is the set defined in subsection 2.

The above conditions uniquely define our rooted tree. In particular, the root of the tree is $\mathbf{v}$; this vertex is enumerated by an empty multiindex. The height of the tree is the maximal number of reflections which trajectories originating from points $(y, \eta) \in O$ can experience on the time interval $(0\,, T_+)$.

By $q_{\mathfrak{n}}$ we denote the number of sons of the (internal) vertex $\mathbf{v}_{\mathfrak{n}}$. The edge connecting the (internal) vertex $\mathbf{v}_{\mathfrak{n}}$ with its son $\mathbf{v}_{\mathfrak{n}|l}$ is denoted by $\mathbf{e}_{\mathfrak{n}|l}$. Thus, $\mathbf{E} = \{\mathbf{e}_{\mathfrak{n}|l} : \mathfrak{n} \in \mathring{\mathfrak{M}},\ l = 1, 2, \dots, q_{\mathfrak{n}}\}$.

We shall associate the internal vertices with reflections, the leaves with the moment $t = T_+$, and the edges with (real) legs of our billiards. In particular, the root $\mathbf{v}$ corresponds to the first reflection. The billiard trajectory of type $\mathfrak{n}$ corresponds to the path from the root $\mathbf{v}$ to the leaf $\mathbf{v}_{\mathfrak{n}}$; this path represents schematically the trajectory on the time interval between the moment of first reflection and the moment $t = T_+$. Under the chosen interpretation $q_{\mathfrak{n}}$ is the number of reflected trajectories corresponding to the given incident trajectory at the reflection $\mathbf{v}_{\mathfrak{n}}$.

By $\mathbf{t}^*_{\mathfrak{n}}(y, \eta)$ we denote the moment of the reflection $\mathbf{v}_{\mathfrak{n}}$. Under this notation $\mathbf{t}^*(y, \eta) \equiv t^*_1(y, \eta)$ is the moment of the first reflection.

Similarly to (3.4.12) we denote by $\mathbf{k}_{\mathfrak{n}}$ the sequential number of the incident trajectory at the reflection $\mathbf{v}_{\mathfrak{n}}$.

Denote

$$\mathfrak{T}_{\mathfrak{n}|l}(y, \eta) := \begin{cases} [\mathbf{t}^*_{\mathfrak{n}}(y, \eta)\,,\ \mathbf{t}^*_{\mathfrak{n}|l}(y, \eta)]\,, & \text{if} \quad \mathfrak{n}|l \notin \mathfrak{N}(O)\,, \\ [\mathbf{t}^*_{\mathfrak{n}}(y, \eta)\,,\ T_+)\,, & \text{if} \quad \mathfrak{n}|l \in \mathfrak{N}(O) \end{cases}$$

(cf. (2.3.13)). By

$$(3.4.40) \quad (x^{(\mathfrak{n}|l)*}(t; y, \eta), \xi^{(\mathfrak{n}|l)*}(t; y, \eta)), \qquad t \in \mathfrak{T}_{\mathfrak{n}|l}(y, \eta)\,, \quad (y, \eta) \in O\,, \quad l = 1, 2, \dots, q_{\mathfrak{n}}\,,$$

we denote the leg of the billiards corresponding to the edge $\mathbf{e}_{\mathfrak{n}|l}$ (cf. (3.4.13)). By

$$(3.4.41) \quad (x^{(\mathfrak{n}|l)*}(t; y, \eta), \xi^{(\mathfrak{n}|l)*}(t; y, \eta)), \quad (y, \eta) \in O, \quad l = q_{\mathfrak{n}} + 1,\ q_{\mathfrak{n}} + 2, \dots, m,$$

we denote the complex reflected trajectories corresponding to the reflection $\mathbf{v}_{\mathfrak{n}}$ (cf. (3.4.14), (3.4.15)). By $\varphi^{(\mathfrak{n}|l)}(t, x; y, \eta)$ we denote standard ($l \leqslant q_{\mathfrak{n}}$) and boundary layer ($l > q_{\mathfrak{n}}$) phase functions associated with the canonical transformations (3.4.40), (3.4.41).

We construct the distribution approximating modulo C^∞ the Schwartz kernel $\mathbf{u}_P(t, x, y)$ of the operator $\exp(-it\mathcal{A}^{1/(2m)})P$ in the form

$$(3.4.42) \qquad u_P(t, x, y) \;=\; \mathcal{I}_{\varphi, \mathfrak{a}} \;+\; \sum_{\mathfrak{n} \in \mathring{\mathfrak{M}}} \sum_{l=1}^{m} \mathcal{I}_{\varphi^{(\mathfrak{n}|l)}, \mathfrak{a}^{(\mathfrak{n}|l)}}\,,$$

cf. (3.4.27). Here the oscillatory integrals with $l \leqslant q_{\mathfrak{n}}$ are standard oscillatory integrals, whereas those with $l > q_{\mathfrak{n}}$ are boundary layer ones. The double sum in (3.4.42) may have more terms than the number of edges in our tree because in the general situation it contains boundary layer oscillatory integrals.

THEOREM 3.4.5 (extension of Theorem 3.4.3). *For any set of phase functions* φ, $\varphi^{(\mathfrak{n}|l)}$, $\mathfrak{n} \in \mathring{\mathfrak{M}}$, $l = 1, 2, \dots, m$, *associated with our Hamiltonian billiards there exist amplitudes* $\mathfrak{a}$, $\mathfrak{a}^{(\mathfrak{n}|l)}$, $\mathfrak{n} \in \mathring{\mathfrak{M}}$, $l = 1, 2, \dots, m$, *of the class* S^1 *such that the distribution* (3.4.42) *coincides with the Schwartz kernel* $\mathbf{u}_P(t, x, y)$ *of the operator* $\exp(-it\mathcal{A}^{1/(2m)})P$ *modulo* $C^\infty((T_-\,, T_+) \times M_x \times M_y)$.

Set $\mathfrak{O}^{(\mathfrak{n}|l)} := \{\,(t; y, \eta) : (y, \eta) \in O\,, t \in \mathfrak{T}_{\mathfrak{n}|l}(y, \eta)\,\}$.

Similarly to (3.4.39) we introduce the reflection matrix $R^{\mathfrak{n}}(y,\eta)$ (here $\mathfrak{n}$ is a superscript, not a power) corresponding to the reflection $\mathbf{v}_{\mathfrak{n}}$. We set $\mathfrak{R} = 1$ and inductively define $\mathfrak{R}_{\mathfrak{n}|l} = R^{\mathfrak{n}}_{l\mathbf{k}_{\mathfrak{n}}}\mathfrak{R}_{\mathfrak{n}}$, $\mathfrak{n} \in \overset{\circ}{\mathfrak{M}}$, $l = 1, 2, \ldots, q_{\mathfrak{n}}$. Thus, we have a scalar function $\mathfrak{R}_{\mathfrak{n}|l}(y,\eta)$ corresponding to each edge $\mathbf{e}_{\mathfrak{n}|l}$.

THEOREM 3.4.6 (extension of Theorem 3.4.4). *The principal symbols of the standard Lagrangian distributions* $\mathcal{I}_{\varphi,\mathfrak{a}}$ *and* $\mathcal{I}_{\varphi^{(\mathfrak{n}|l)},\mathfrak{a}^{(\mathfrak{n}|l)}}$, $l \leqslant q_{\mathfrak{n}}$, *appearing in* (3.4.42) *are given by formulae* (3.3.16), $(t; y, \eta) \in \mathfrak{O}$, *and*

$$\begin{aligned}\mathfrak{a}_1^{(\mathfrak{n}|l)}(t;y,\eta) &= \mathfrak{R}_{\mathfrak{n}|l}(y,\eta)\,P_1(y,\eta)\\ &\times \exp\Big(-\frac{i}{2m}h^{1-2m}(y,\eta)\int_0^t A_{\mathrm{sub}}\left(x^{(\mathfrak{n}|l)*}(\tau;y,\eta),\xi^{(\mathfrak{n}|l)*}(\tau;y,\eta)\right)d\tau\Big),\end{aligned}$$

$(t; y, \eta) \in \mathfrak{O}^{(\mathfrak{n}|l)}$, *respectively.*

5. The case of simple reflection. If the simple reflection condition (see Definition 1.3.28) is satisfied, then the general construction outlined in the previous subsection is significantly simplified. Namely, in this case $q_{\mathfrak{n}} = 1$, the set $\mathfrak{N}(O)$ contains only one multiindex

$$\mathfrak{N}(O) = \{\underbrace{1|\ldots|1}_{\mathbf{r}\ \text{times}}\}$$

($\mathbf{r}$ is the number of reflections), and the set $\overset{\circ}{\mathfrak{M}}$ contains $\mathbf{r}$ multiindices

$$\overset{\circ}{\mathfrak{M}} = \{\mathfrak{n} : \mathfrak{n} = \underbrace{1|\ldots|1}_{j-1\ \text{times}},\ j = 1,\ldots,\mathbf{r}\}$$

(recall that $\overset{\circ}{\mathfrak{M}}$ includes the empty multiindex). In fact our rooted tree in this case is a set of $\mathbf{r}$ consecutively connected edges.

In order to simplify our notation let us denote

$$\underbrace{1|\ldots|1}_{j-1\ \text{times}}|\,l := j;l, \qquad j = 1,\ldots,\mathbf{r}, \quad l = 1,\ldots,m.$$

Formula (3.4.42) can be rewritten in this notation as

$$u_P(t,x,y) = \mathcal{I}_{\varphi,\mathfrak{a}} + \sum_{j=1}^{\mathbf{r}}\sum_{l=1}^{m}\mathcal{I}_{\varphi^{(j;l)},\mathfrak{a}^{(j;l)}}. \tag{3.4.43}$$

Here the oscillatory integrals with $l = 1$ are standard oscillatory integrals, whereas those with $l > 1$ are boundary layer ones.

As usual, let us denote the moments of reflection by $t_j^*(y,\eta)$, $j = 1,\ldots,\mathbf{r}$.

The following is a special case of Theorem 3.4.6.

COROLLARY 3.4.7. *In the case of simple reflection the principal symbols of the standard Lagrangian distributions* $\mathcal{I}_{\varphi,\mathfrak{a}}$ *and* $\mathcal{I}_{\varphi^{(j;1)},\mathfrak{a}^{(j;1)}}$, *appearing in* (3.4.43) *are given by one formula*

$$e^{i(\mathfrak{f}_r(t;y,\eta)+\mathfrak{f}_s(t;y,\eta))}\,P_1(y,\eta) \tag{3.4.44}$$

for all $(t;y,\eta)\in(T_-\,,\,T_+)\times O$ *such that* $t\ne t_j^*(y,\eta)\,,\ j=1,\dots,\mathbf{r}\,.$ *Here* $\mathfrak{f}_r$ *and* $\mathfrak{f}_s$ *are the functions from Definition* 1.7.2.

In the above corollary the phrase "given by one formula" means that

1. If $t\in(T_-\,,\,t_1^*(y,\eta))$ then (3.4.44) is the principal symbol of $\mathcal{I}_{\varphi,\mathfrak{a}}$.
2. If $t\in(t_j^*(y,\eta)\,,\,t_{j+1}^*(y,\eta))\,,\ j=1,\dots,\mathbf{r}-1\,,$ then (3.4.44) is the principal symbol of $\mathcal{I}_{\varphi^{(j;1)},\mathfrak{a}^{(j;1)}}$.
3. If $t\in(t_{\mathbf{r}}^*(y,\eta)\,,\,T_+)$ then (3.4.44) is the principal symbol of $\mathcal{I}_{\varphi^{(\mathbf{r};1)},\mathfrak{a}^{(\mathbf{r};1)}}$.

The advantage of formula (3.4.44) is its clear meaning: recall that $\mathfrak{f}_r$ and $\mathfrak{f}_s$ are the phase shifts generated by the reflections and by the subprincipal symbol, respectively.

6. Action of a pseudodifferential operator on $\mathbf{u}_P$. Let $Q\in\Psi_0^{\mathbf{m}}$. Then Theorems 3.4.5, 2.7.16 imply that modulo C^∞

$$Q(x,D_x)\,\mathbf{u}_P(t,x,y)\;=\;\mathcal{I}_{\varphi,\mathfrak{b}}\;+\;\sum_{\mathfrak{n}\in\mathring{\mathfrak{M}}}\sum_{l=1}^{q_{\mathfrak{n}}}\mathcal{I}_{\varphi^{(\mathfrak{n}|l)},\mathfrak{b}^{(\mathfrak{n}|l)}} \tag{3.4.45}$$

(cf. (3.4.42)), with principal symbols

$$\begin{aligned}\mathfrak{b}_{1+\mathbf{m}}(t;y,\eta)\;&=\;Q_{\mathbf{m}}\big(x^*(t;y,\eta),\xi^*(t;y,\eta)\big)\,P_1(y,\eta)\\&\quad\times\,\exp\Big(-\frac{i}{2m}h^{1-2m}(y,\eta)\int_0^t A_{\mathrm{sub}}\left(x^*(\tau;y,\eta),\xi^*(\tau;y,\eta)\right)d\tau\Big)\,,\end{aligned}$$

$$\begin{aligned}\mathfrak{b}_{1+\mathbf{m}}^{(\mathfrak{n}|l)}(t;y,\eta)\;&=\;Q_{\mathbf{m}}\big(x^{(\mathfrak{n}|l)*}(t;y,\eta),\xi^{(\mathfrak{n}|l)*}(t;y,\eta)\big)\,\mathfrak{R}_{\mathfrak{n}|l}(y,\eta)\,P_1(y,\eta)\\&\quad\times\,\exp\Big(-\frac{i}{2m}h^{1-2m}(y,\eta)\int_0^t A_{\mathrm{sub}}\left(x^{(\mathfrak{n}|l)*}(\tau;y,\eta),\xi^{(\mathfrak{n}|l)*}(\tau;y,\eta)\right)d\tau\Big)\,.\end{aligned}$$

Note that (3.4.45) does not contain any boundary layer oscillatory integrals; they disappear because their singularities lie on the boundary. Of course, in the case of simple reflection the above formulae are simplified:

$$Q(x,D_x)\,\mathbf{u}_P(t,x,y)\;=\;\mathcal{I}_{\varphi,\mathfrak{b}}\;+\;\sum_{j=1}^{\mathbf{r}}\mathcal{I}_{\varphi^{(j;1)},\mathfrak{b}^{(j;1)}}$$

(cf. (3.4.43)), with principal symbols

$$\begin{aligned}\mathfrak{b}_{1+\mathbf{m}}(t;y,\eta)\;&=\;Q_{\mathbf{m}}\big(x^*(t;y,\eta),\xi^*(t;y,\eta)\big)e^{i\mathfrak{f}_s(t;y,\eta)}P_1(y,\eta)\,,\\ \mathfrak{b}_{1+\mathbf{m}}^{(j;1)}(t;y,\eta)\;&=\;Q_{\mathbf{m}}\big(x^*(t;y,\eta),\xi^*(t;y,\eta)\big)e^{i(\mathfrak{f}_r(t;y,\eta)+\mathfrak{f}_s(t;y,\eta))}P_1(y,\eta)\,,\end{aligned}$$

where $\big(x^*(t;y,\eta),\xi^*(t;y,\eta)\big)\,,\ t\in(T_-,T_+)\,,$ is the full billiard trajectory.

3.5. Construction of the wave group when the source is close to the boundary

In 3.4.1 we constructed the oscillatory integral approximating the Schwartz kernel $\mathbf{u}_P(t,x,y)$ of the operator $\exp(-it\mathcal{A}^{1/(2m)})P$ under the assumption that the trajectories originating from $\operatorname{cone\,supp}P$ do not reach the boundary. In this section we shall deal with the case when $\operatorname{cone\,supp}P$ and, consequently, the trajectories originating from $\operatorname{cone\,supp}P$ are close to the boundary. These trajectories may approach the boundary at awkward angles which makes the construction of compensating oscillatory integrals (as in 3.4.3, 3.4.4) difficult. So in this section we shall carry out the construction without reflections on the largest possible time interval.

The oscillatory integrals which we are about to construct will be parameter dependent of the type described in Section 2.10, with the distance from the tube of rays $x^*(t;y,\eta)$, $(y,\eta)\in\operatorname{cone\,supp}P$, to ∂M_x being characterized by the parameter $d\in(0,d_0]$, $d_0<1$. The length of the time interval will be of the order of $d^{1-\varepsilon_1}$ with $\varepsilon_1>0$, which is large compared to the distance from the boundary. Note that it is absolutely essential for our purposes to carry out the construction on a time interval of length $\gg d$ in order to achieve the necessary accuracy when we eventually (Section 5.3) apply a Fourier Tauberian theorem.

1. Parameter-dependent version of Theorem 3.1.1. Below we use the notation from Section 2.10, in particular, the d^∞-notation introduced in Definition 2.10.3 (see also 2.10.6). The $T^\pm_\pm(d)$ are defined in accordance with (2.10.4$^\pm$). By λ_k and v_k, $k=1,2,\dots$, we denote, as usual, the eigenvalues and the orthonormalized eigenfunctions of the problem (1.1.1), (1.1.2).

THEOREM 3.5.1. *Let $u^\pm(d,t,x,y)$ be a parameter-dependent distribution such that*

1. *For each $d\in(0,d_0]$ this distribution is of the class*

$$C^\infty((T^\pm_-(d)\,,\,T^\pm_+(d))\times M_y;\,\mathcal{E}'(\mathring{M}_x))\cap C^\infty(M_x\times M_y;\,\mathcal{D}'(T^\pm_-(d)\,,\,T^\pm_+(d)))\,.$$

2. *$u^\pm$ behaves as a function in the variables d, t and as a half-density in the variables x, y.*
3. *For each $d\in(0,d_0]$ we have $\operatorname{supp}u^\pm(d,\cdot,\cdot,\cdot,)\cap\{x\in\partial M_x\}=\varnothing$.*
4. *For any $s\in\mathbb{N}$ and any differential operator $F_y\equiv F_y(y,D_y)$ there exist $r,q\in\mathbb{N}$ such that the "function" $d^r\mathcal{A}_x^{-q}D_t^s\,F_y\,u^\pm$ is uniformly bounded over $d\in(0,d_0]$, $t\in(T^\pm_-(d),T^\pm_+(d))$, $x\in M_x$, $y\in M_y$.*
5. *There exists a constant $\varepsilon_3>0$ such that for any $r,s\in\mathbb{N}$ and any differential operator $F_y\equiv F_y(y,D_y)$ the "function" $d^{-r}D_t^s\,F_y\,\langle u^\pm\,,\,\overline{v_k}\rangle$ is uniformly bounded over $d\in(0,d_0]$, $t\in(T^\pm_-(d),T^\pm_+(d))$, $y\in M_y$, and k such that $\lambda_k\leqslant d^{-1-\varepsilon_3}$.*
6. *$D_t^{2m}u^\pm-A(x,D_x)u^\pm\in d^\infty((T^\pm_-(d),T^\pm_+(d))\times M_x\times M_y)$.*
7. *For any $g(t)\in C_0^\infty(-T_{\varepsilon_1}\,,\,T_{\varepsilon_1})$, $g_d(t):=g(t/d)$, any $r,s\in\mathbb{N}$, and any differential operator $F_y\equiv F_y(y,D_y)$ the "function" $d^{-r}\lambda^s\,F_y\,\mathcal{F}^{-1}_{t\to\lambda}[g_du^\pm]$ is uniformly bounded over $d\in(0,d_0]$, $\lambda\leqslant0$, $x\in M_x$, $y\in M_y$.*

Then

$$u^\pm(d,t,x,y)-\exp(-it\mathcal{A}_x^{1/(2m)})u^\pm(d,0,x,y)\in d^\infty((T^\pm_-(d)/2\,,\,T^\pm_+(d)/2)\times M_x\times M_y).$$

PROOF. The proof of Theorem 3.5.1 repeats that of Theorem 3.1.1 with additional control over the parameter d. We do not repeat these arguments for the sake of brevity. Note, however, that in the case of Theorem 3.5.1 the arguments involved in the proof are slightly simpler because the distribution in question is identically zero near ∂M_x; consequently the parts of the original proof of Theorem 3.1.1 dealing with the boundary conditions become unnecessary. □

2. Parameter-dependent version of Theorem 3.2.1. Further on in this section

$$P^{\pm} = \int_{M_y} p^{\pm}(d,x,y)\,(\,\cdot\,)\,dy$$

is a given parameter-dependent pseudodifferential operator of order 1 of the type $(1;1,\varepsilon_1)^{\pm}$ (see Definition 2.10.13 and the text following formula (2.10.19)), and

$$\mathbf{u}_{P^{\pm}}(d,t,x,y) = \exp(-it\mathcal{A}_x^{1/(2m)})\,p^{\pm}(d,x,y)\,.$$

THEOREM 3.5.2. *Let $\mathcal{I}_{\varphi,a^{\pm}}(d,t,x,y)$ be a parameter-dependent oscillatory integral of the type $(0,0;1,\varepsilon_1)^{\pm}$ (see (2.10.11)) such that*

$$D_t^{2m}\mathcal{I}_{\varphi,a^{\pm}} - A(x,D_x)\mathcal{I}_{\varphi,a^{\pm}} \in d^{\infty}((T_-^{\pm}(d)\,,\,T_+^{\pm}(d))\times M_x\times M_y)\,,$$

$$\mathcal{I}_{\varphi,a^{\pm}}(d,0,x,y) - p^{\pm}(d,x,y) \in d^{\infty}(M_x\times M_y)\,.$$

Then

$$\mathcal{I}_{\varphi,a^{\pm}}(d,t,x,y) - \mathbf{u}_{P^{\pm}}(d,t,x,y) \in d^{\infty}((T_-^{\pm}(d)/2\,,\,T_+^{\pm}(d)/2)\times M_x\times M_y)\,.$$

PROOF. Theorem 3.5.2 is proved by checking that the oscillatory integral in question satisfies all the conditions of Theorem 3.5.1. The arguments involved basically repeat those from the proof of Theorem 3.2.1 (with additional control over the parameter d). The only new element in the proof is concerned with establishing that condition (5) of Theorem 3.5.1 is satisfied. So we present below the arguments to this effect. We will show that condition (5) of Theorem 3.5.1 is satisfied for any ε_3 such that

$$0<\varepsilon_3<\varepsilon_2\,, \tag{3.5.1}$$

where ε_2 is the positive constant introduced at the beginning of 2.10.2.

Let l be the order of the parameter-dependent Lagrangian distribution $\mathcal{I}_{\varphi,a^{\pm}}$ (see Section 2.10 for details), and let q be the order of the differential operator F_y. It is easy to see that $D_t^s\,F_y\,\mathcal{I}_{\varphi,a^{\pm}}$ is a parameter-dependent Lagrangian distribution of order $l+s+q$ of the type $(0,0;1,\varepsilon_1)^{\pm}$. Consequently it is sufficient to establish the fulfilment of condition (5) of Theorem 3.5.1 in the case $s=1$, $F_y=I$.

We have (modulo d^{∞})

$$\mathcal{I}_{\varphi,a^{\pm}} = A(x,D_x)\,\mathcal{I}_{\varphi,a_{1,0}^{\pm}} + \mathcal{I}_{\varphi,a_{0,1}}\,,$$

where

$$a_{1,0}^{\pm} = \frac{a^{\pm}}{A_{2m}(x,\varphi_x)} \in S^{\,l-2m}(0,0;1,\varepsilon_1)^{\pm}\,,$$

$$a_{0,1}^{\pm} = a^{\pm} - e^{-i\varphi}\,(d_{\varphi})^{-1}\,A(x,D_x)\,(e^{i\varphi}\,a_{1,0}^{\pm}\,d_{\varphi}) \in S^{\,l-1}(0,0;1,\varepsilon_1)^{\pm}\,.$$

Iterating this procedure p times we get (modulo d^{∞})

$$\mathcal{I}_{\varphi,a^{\pm}} = \sum_{j=0}^{p} A^j(x,D_x)\,\mathcal{I}_{\varphi,a_{j,p-j}^{\pm}}\,, \tag{3.5.2}$$

with some amplitudes

$$a_{j,p-j}^{\pm} \in S^{\,l-2mj+j-p}(0,0;1,\varepsilon_1)^{\pm}\,. \tag{3.5.3}$$

Substituting (3.5.2) in $d^{-r}\langle\mathcal{I}_{\varphi,a^\pm,b}\,,\,\overline{v_k}\rangle$ and integrating by parts we obtain (modulo d^∞)

$$d^{-r}\langle\mathcal{I}_{\varphi,a^\pm}\,,\,\overline{v_k}\rangle = d^{-r}\sum_{j=0}^{p}\lambda_k^{2mj}\langle\mathcal{I}_{\varphi,a^\pm_{j,p-j}}\,,\,\overline{v_k}\rangle\,. \tag{3.5.4}$$

Now we make the following observation. Suppose we have a parameter-dependent Lagrangian distribution $\mathcal{I}$ of order $\ell<-n$. Then the corresponding oscillatory integral converges absolutely and we have the uniform pointwise estimate $|\mathcal{I}|\leqslant \text{const}\, d^{-(1+\varepsilon_2)(\ell+n)}$; here and further on in this subsection by const we denote various positive constants independent of d, t, y. This pointwise estimate implies an estimate for the $L_2(M_x)$-norm: $\|\mathcal{I}\|\leqslant \text{const}\, d^{-(1+\varepsilon_2)(\ell+n)}$. As $\|v_k\|=1$, by use of the Cauchy-Schwarz inequality we obtain

$$|\langle\mathcal{I}\,,\,\overline{v_k}\rangle|\leqslant \text{const}\, d^{-(1+\varepsilon_2)(\ell+n)} \tag{3.5.5}$$

with const independent of $k\in\mathbb{N}$.

Let us now return to (3.5.4), (3.5.3) and suppose that $p>l+n$. Set $\ell:=l-2mj+j-p$. We have $\ell\leqslant l-p<-n$, so we can apply (3.5.5) obtaining

$$|\langle\mathcal{I}_{\varphi,a^\pm_{j,p-j}}\,,\,\overline{v_k}\rangle|\leqslant \text{const}\, d^{(1+\varepsilon_2)(p+2mj-j-l-n)}\,. \tag{3.5.6}$$

Recall (see statement of condition (5) of Theorem 3.5.1) that

$$\lambda_k\leqslant d^{-1-\varepsilon_3}\,. \tag{3.5.7}$$

Substituting (3.5.7), (3.5.6) into (3.5.4) we get

$$\begin{aligned}&|d^{-r}\langle\mathcal{I}_{\varphi,a^\pm}\,,\,\overline{v_k}\rangle|\\ &\quad\leqslant \text{const}\; d^{-r-(1+\varepsilon_2)(l+n)}\sum_{j=0}^{p} d^{p(1+\varepsilon_2)+j(2m(\varepsilon_2-\varepsilon_3)-1-\varepsilon_2)}\,.\end{aligned} \tag{3.5.8}$$

But

$$\min_{0\leqslant j\leqslant p}\big(p(1+\varepsilon_2)+j(2m(\varepsilon_2-\varepsilon_3)-1-\varepsilon_2)\big) = p\varkappa\,,$$

where $\varkappa=\min\{1+\varepsilon_2, 2m(\varepsilon_2-\varepsilon_3)\}$. Clearly, by (3.5.1) $\varkappa>0$. So (3.5.8) implies

$$|d^{-r}\langle\mathcal{I}_{\varphi,a^\pm}\,,\,\overline{v_k}\rangle|\leqslant \text{const}\; d^{p\varkappa-r-(1+\varepsilon_2)(l+n)}\,.$$

For sufficiently large p the exponent on the right-hand side of this inequality becomes nonnegative, which proves the uniform boundedness of $d^{-r}\langle\mathcal{I}_{\varphi,a^\pm}\,,\,\overline{v_k}\rangle$. □

3. Effective construction of the wave group. Recall that according to the definitions and results of 2.10.6 the Schwartz kernel of our pseudodifferential operator $P^\pm$ can be written modulo d^∞ as a parameter-dependent time-independent oscillatory integral

(3.5.9)
$$\begin{aligned}p^\pm(d,x,y) &= \mathcal{I}_{\varphi_0,\mathfrak{p}^\pm}(d,x,y)\\ &= \varsigma_+(x_n/d)\left(\int e^{i\varphi_0(x;y,\eta)}\,\mathfrak{p}^\pm(d;y,\eta)\,\varsigma(x;y,\eta)\,d_{\varphi_0}(x;y,\eta)\,\text{đ}\eta\right)\varsigma_+(y_n/d)\,.\end{aligned}$$

Here $\varphi_0(x;y,\eta)\in\mathfrak{F}^0$ is a time-independent phase function (see 2.7.5) and $\mathfrak{p}^\pm(d;y,\eta)\in S^1(1;1,\varepsilon_1)^\pm$ is the symbol of the Lagrangian distribution $p^\pm(d,x,y)$.

THEOREM 3.5.3. *For any phase function* $\varphi(t,x;y,\eta) \in \check{\mathfrak{F}}$ *there exists a symbol* $\mathfrak{a}^\pm(d,t;y,\eta) \in S^1(0,0;1,\varepsilon_1)^\pm$ *such that the parameter-dependent oscillatory integral*

$$
\begin{aligned}
&\mathcal{I}_{\varphi,\mathfrak{a}^\pm}(d,t,x,y) \\
&\quad = \varsigma_+(x_n/d) \int e^{i\varphi(t,x;y,\eta)}\, \mathfrak{a}^\pm(d,t;y,\eta)\, \varsigma(t,x;y,\eta)\, d_\varphi(t,x;y,\eta)\, đ\eta
\end{aligned} \tag{3.5.10}
$$

coincides with the Schwartz kernel $\mathbf{u}_{P^\pm}(d,t,x,y)$ *of the operator* $\exp(-it\mathcal{A}^{1/(2m)})P^\pm$ *modulo* $d^\infty((T_-^\pm(d)/2\,,\,T_+^\pm(d)/2) \times M_x \times M_y)$.

The proof of Theorem 3.5.3 repeats that of Theorem 3.3.1. The only differences are that we apply Theorem 3.5.2 instead of Theorem 3.2.1, and we expand our amplitude $\mathfrak{a}^\pm$ into a series of nonhomogeneous terms (see (2.10.9), (2.10.10), (2.10.13)) instead of homogeneous ones.

CHAPTER 4

Singularities of the Wave Group

Let $\hat\rho$ be a function from $C_0^\infty(\mathbb{R})$, and $P \in \Psi_0^{\mathbf{l}}$, $Q \in \Psi_0^{\mathbf{m}}$ classical pseudodifferential operators with principal symbols $P_{\mathbf{l}}$, $Q_{\mathbf{m}}$ and subprincipal symbols P_{sub}, Q_{sub}, respectively. Let $\mathbf{u}_{P,Q}$ be the Schwartz kernel of the operator

$$\mathbf{U}_{P,Q}(t) \;:=\; Q^* \exp(-it\mathcal{A}^{(1/2m)})\, P \;=\; \int_{M_y} \mathbf{u}_{P,Q}(t,x,y)\,(\,\cdot\,)\,dy\,. \tag{4.0.1}$$

The objective of this chapter is to derive asymptotic formulae for expressions of the type

$$\mathcal{F}^{-1}_{t\to\lambda}\Big[\hat\rho(t)\,\mathbf{u}_{P,Q}(t,x,y)\Big] \tag{4.0.2}$$

and

$$\mathcal{F}^{-1}_{t\to\lambda}\Big[\hat\rho(t)\;\mathrm{Tr}\,\mathbf{U}_{P,Q}(t)\Big] \;=\; \int_M \mathcal{F}^{-1}_{t\to\lambda}\left[\hat\rho(t)\,\mathbf{u}_{P,Q}(t,y,y)\right]dy \tag{4.0.3}$$

as $\lambda\to+\infty$. As in Chapter 3, we set $\mathbf{U}_P = \mathbf{U}_{P,I}$ and $\mathbf{u}_P = \mathbf{u}_{P,I}$

In Section 4.1 we obtain general asymptotic formulae for (4.0.2) and (4.0.3) in the case when $\mathbf{u}_{P,Q}(t,x,y)$ is an arbitrary Lagrangian distribution associated with our canonical transformation (2.3.10), and not necessarily the Schwartz kernel of the operator (4.0.1).

In the remainder of the chapter (Sections 4.2.–4.5) $\mathbf{u}_{P,Q}(t,x,y)$ is the Schwartz kernel of the operator (4.0.1). Section 4.2 deals with the case when the support of $\hat\rho$ lies in a sufficiently small neighbourhood of the point $t=0$, whereas Section 4.3 deals with the case when the support of $\hat\rho$ is separated from the point $t=0$ and is not necessarily small. In Section 4.3 the pseudodifferential operator P is required to satisfy a certain admissibility condition, so Section 4.4 is devoted to the relaxation of this condition. Finally, Section 4.5 is concerned with the parameter-dependent situation, that is, situation when $\operatorname{cone\,supp} P$ is close to the boundary of $T'M$.

In Sections 4.2–4.4 δ is a small positive number such that

$$\operatorname{cone\,supp} P \,\subset\, \{x_n \geqslant \delta\}\,, \tag{4.0.4}$$

and $T_\delta>0$ is as defined in 3.4.1. Without loss of generality we assume that T_δ is so small that for trajectories originating from points (y,η) with $y_n\geqslant\delta$ there are no loops of length $\leqslant T_\delta$ (i.e. $x^*(t;y,\eta)\neq y$ for $t\in(0,T_\delta]$).

Throughout Sections 4.1–4.3, $T_-=-T_\delta$, T_+ and O are as in 2.3.2; in particular, O is sufficiently small and T_+-admissible.

4.1. Singularities of Lagrangian distributions

In this section we calculate the singularities with respect to t of the oscillatory integral

$$\mathcal{I}_{\varphi,\mathfrak{a}}(t,x,y) = \int e^{i\varphi(t,x;y,\eta)}\, \mathfrak{a}(t;y,\eta)\, \varsigma(t,x;y,\eta)\, d_\varphi(t,x;y,\eta)\, đ\eta\,,$$

defined in Section 2.7, and of its trace $\int_M \mathcal{I}_{\varphi,\mathfrak{a}}(t,y,y)\,dy$. Here $\varphi \in \mathfrak{F}_i$ is a phase function associated with the ith leg of our Hamiltonian billiards and $\mathfrak{a} \sim \sum_{k=0}^{\infty} \mathfrak{a}_{l-k}$, where the $\mathfrak{a}_{l-k}$ are positively homogeneous functions of degrees $l-k$. Further on we always assume that $\operatorname{cone\,supp}\varsigma$ is sufficiently small (such that $\varphi_t \neq 0$, $\varphi_x \neq 0$ and $\varphi_y \neq 0$ on $\operatorname{cone\,supp}\varsigma$) and that $x^*(t;y,\eta) \notin \partial M$, $\forall(t;y,\eta) \in \operatorname{cone\,supp}\mathfrak{a}$.

More exactly, we are going to calculate asymptotics of the inverse Fourier transforms $\mathcal{F}^{-1}_{t\to\lambda}$ of our oscillatory integrals as $\lambda \to \infty$. Obviously, this is equivalent to describing their singularities in t. Recall that by (2.7.3) all these integrals are distributions in t depending on the parameters (x,y). This fact underlies all further constructions.

For any $\hat\rho \in C_0^\infty(T_-,T_+)$ the function

$$\mathcal{F}^{-1}_{t\to\lambda}\Big[\hat\rho(t)\,\mathcal{I}_{\varphi,\alpha}(t,x,y)\Big] = (2\pi)^{-1}\int e^{i\lambda t}\,\hat\rho(t)\,\mathcal{I}_{\varphi,\alpha}(t,x,y)\,dt$$

is rapidly decreasing as $\lambda \to -\infty$ (see proof of Theorem 3.2.1). Therefore we have to consider only $\lambda \to +\infty$.

1. The main singularity. First we prove the elementary

PROPOSITION 4.1.1. *Let Ω be a compact subset of M such that $x^*(t;y,\eta) \notin \Omega$ for all $(t;y,\eta) \in \operatorname{cone\,supp}(\hat\rho\,\mathfrak{a})$. Then*

$$\mathcal{F}^{-1}_{t\to\lambda}\Big[\hat\rho(t)\,\mathcal{I}_{\varphi,\alpha}(t,x,y)\Big] = O(\lambda^{-\infty}), \qquad \lambda \to +\infty, \tag{4.1.1}$$

uniformly with respect to $x \in \Omega$ and $y \in M$.

PROOF. Under the conditions of Proposition 4.1.1,

$$\mathcal{I}_{\varphi,a} \in C^\infty((T_-,T_+)\times\widetilde\Omega\times M)$$

for some neighbourhood $\widetilde\Omega$ of Ω (see 2.7.1). Since $\hat\rho \in C_0^\infty(T_-,T_+)$, this immediately implies (4.1.1). □

THEOREM 4.1.2. *Let $\operatorname{cone\,supp}(\hat\rho\,\mathfrak{a})$ lie in a sufficiently small conic neighbourhood of the point $(t_0;y_0,\eta_0)$, let the phase function φ be real in a neighbourhood of $\operatorname{cone\,supp}(\hat\rho\,\mathfrak{a}\,\varsigma)$, and $x_0 = x^*(t_0;y_0,\eta_0)$. Then for x close to x_0 the equation $\varphi(t,x;y,\eta) = 0$ has the only t-solution $t_\star(x;y,\eta)$ close to t_0, and*

(4.1.2)

$$(2\pi)^{-1}\int e^{i\lambda t}\hat\rho(t)\mathcal{I}_{\varphi,\mathfrak{a}}(t,x,y)\,dt$$

$$\sim \lambda^{n+l-1}\int_{S_y^*M} e^{i\lambda t_\star}\, r_\star \sum_{j,k=0}^{\infty} \lambda^{-k-j}\, \mathcal{L}_{j,k}\,(\hat\rho\,\mathfrak{a}_{l-k})\big|_{t=t_\star}\, đ\tilde\eta\,, \qquad \lambda \to +\infty\,,$$

uniformly with respect to (x,y) *lying in a neighbourhood of* (x_0,y_0). *Here* $đ\widetilde{\eta} = (2\pi)^{-n} d\widetilde{\eta}$ *is as in* 1.1.10, $r_\star = r_\star(x;y,\eta) = -\left(\varphi_t(t_\star,x;y,\eta)\right)^{-1}$, *and* $\mathcal{L}_{j,k} = \mathcal{L}_{j,k}(x;y,\widetilde{\eta},D_t)$ *are differential operators defined as follows. Let* $r\in\mathbb{R}_+$ *be an independent variable,*

$$\mathcal{L}(x;y,\widetilde{\eta};D_t,D_r) := -2\,r_\star\,D_t\,D_r - r_\star^3\,\varphi_{tt}(t_\star,x;y,\widetilde{\eta})\,D_r^2\,, \tag{4.1.3}$$

and

$$\begin{aligned}\psi(t,x;y,\widetilde{\eta};r) := r\,\varphi(t,x;y,\widetilde{\eta}) &+ (t-t_\star)\\ &+ r_\star^{-1}(r-r_\star)(t-t_\star) - r_\star(t-t_\star)^2\,\varphi_{tt}(t_\star,x;y,\widetilde{\eta})/2\,.\end{aligned} \tag{4.1.4}$$

Then for $w\in C^\infty((T_-,T_+)\times S^*_yM)$

$$\begin{aligned}(\mathcal{L}_{j,k}w)(t;y,\widetilde{\eta}) := i^{-j} &\sum_{\substack{\mu,\nu\,:\,\nu-\mu=j,\\ 3\mu\leqslant 2\nu}} (\mu!\,\nu!)^{-1}\,2^{-\nu}\\ &\times\left(\mathcal{L}(x;y,\widetilde{\eta};D_t,D_r)\right)^\nu\left(r^{n+l-1-k}\,\psi^\mu(t,x;y,\widetilde{\eta};r)\right.\\ &\left.\left.\times\,\varsigma(t,x;y,\widetilde{\eta})\,d_\varphi(t,x;y,\widetilde{\eta})\,w(t;y,\widetilde{\eta})\right)\right|_{r=r_\star}.\end{aligned} \tag{4.1.5}$$

In particular,

$$\begin{aligned}\mathcal{L}_{0,0}\,(\hat\rho\,\mathfrak{a}_l) &= r_\star^{n+l-1}\,\varsigma\,d_\varphi\,\hat\rho\,\mathfrak{a}_l,\\ \mathcal{L}_{0,1}\,(\hat\rho\,\mathfrak{a}_{l-1}) &= r_\star^{n+l-2}\,\varsigma\,d_\varphi\,\hat\rho\,\mathfrak{a}_{l-1},\end{aligned} \tag{4.1.6}$$

$$\begin{aligned}\mathcal{L}_{1,0}\,(\hat\rho\,\mathfrak{a}_l) &= (n+l-1)\,r_\star^{n+l-1}D_t(\varsigma\,d_\varphi\,\hat\rho\,\mathfrak{a}_l)\\ &\quad+(2i)^{-1}(n+l)\,(n+l-1)\,r_\star^{n+l}\varphi_{tt}\,\varsigma\,d_\varphi\,\hat\rho\,\mathfrak{a}_l\,.\end{aligned} \tag{4.1.7}$$

PROOF. In the integral

$$(2\pi)^{-1}\int e^{i\varphi(t,x;y,\eta)+i\lambda t}\hat\rho(t)\,\mathfrak{a}(t;y,\eta)\,\varsigma(t,x;y,\eta)\,d_\varphi(t,x;y,\eta)\,đ\eta\,dt$$

we change the variables $\eta = \lambda r\widetilde{\eta}$, where $r\in\mathbb{R}_+$ and $\widetilde{\eta}$ are coordinates on the "sphere" S^*_yM. Then this integral takes the form

$$\begin{aligned}(2\pi)^{-1}\,\lambda^n\int &e^{i\lambda(r\varphi(t,x;y,\widetilde{\eta})+t)}\hat\rho(t)\mathfrak{a}(t;y,\lambda r\widetilde{\eta})\\ &\times\varsigma(t,x;y,\lambda r\widetilde{\eta})\,d_\varphi(t,x;y,\widetilde{\eta})\,r^{n-1}\,dr\,dt\,đ\widetilde{\eta}\,.\end{aligned} \tag{4.1.8}$$

Now we are going to apply the stationary phase formula (C.1) with respect to the variables t and r; this is justified by Corollary C.4 and Remark C.5.

The stationary points of the phase function $r\varphi+t$ are determined by the equations

$$\varphi(t,x;y,\widetilde{\eta}) = 0\,,\qquad r\,\varphi_t(t,x;y,\widetilde{\eta}) = -1\,.$$

The first equation has the only solution $t=t_\star$, and then the second one gives $r=-\varphi_t^{-1}$. Thus, there is a unique stationary point

$$t = t_\star(x;y,\widetilde{\eta})\,,\qquad r = r_\star(x;y,\widetilde{\eta}) = -\left(\varphi_t\left(t_\star(x;y,\widetilde{\eta}),x;y,\widetilde{\eta}\right)\right)^{-1}.$$

Let us denote $z=(t,r)$, $z_0=(t_\star,r_\star)$. Then, in the notation of Theorem C.1, we have

$$\begin{aligned}
f_{zz}(z_0) &= \begin{pmatrix} r_\star\,\varphi_{tt}(t_\star,x;y,\widetilde{\eta}) & -r_\star^{-1} \\ -r_\star^{-1} & 0 \end{pmatrix}, \\
\operatorname{sgn} f_{zz}(z_0) &= 0\,, \qquad |\det(\lambda f_{zz}(z_0)/2\pi)|^{-1/2} \;=\; 2\pi\,\lambda^{-1}\,r_\star\,, \\
\big(f_{zz}(z_0)\big)^{-1} &= \begin{pmatrix} 0 & -r_\star \\ -r_\star & -\,r_\star^3\,\varphi_{tt}(t_\star,x;y,\widetilde{\eta}) \end{pmatrix}, \\
\langle\,\big(f_{zz}(z_0)\big)^{-1} D_z\,,\,D_z\,\rangle &= \mathcal{L}(x;y,\widetilde{\eta};D_t,D_r)\,, \qquad f_0(z) \;=\; \psi(t,x;y,\widetilde{\eta},r)\,,
\end{aligned}$$

where $\mathcal{L}$ and ψ are defined by (4.1.3) and (4.1.4), respectively. The above equalities and (C.1) imply that the integral (4.1.8) coincides with

$$\begin{aligned}
&\lambda^{n-1}\int e^{i\lambda t_\star}\,r_\star\sum_{j<p}\lambda^{-j} \\
&\qquad\times\,L_j\Big(\hat\rho(t)\,\mathfrak{a}(t;y,\lambda r\widetilde{\eta})\,\varsigma(t,x;y,\lambda r\widetilde{\eta})\,d_\varphi(t,x;y,\widetilde{\eta})\,r^{n-1}\Big)\Big|_{\substack{t=t_\star\\ r=r_\star}}\,đ\widetilde{\eta}
\end{aligned}\tag{4.1.9}$$

modulo a term of order $O(\lambda^{l-p})$. Here $L_j=L_j(x;y,\widetilde{\eta};D_t,D_r)$ are the differential operators defined by (C.3). Since

$$\begin{aligned}
&L_j\,\Big(\hat\rho(t)\,\mathfrak{a}_{l-k}(t;y,\lambda r\widetilde{\eta})\,\varsigma(t,x;y,\widetilde{\eta})\,d_\varphi(t,x;y,\widetilde{\eta})\,r^{n-1}\Big)\Big|_{r=r_\star} \\
&\quad=\;\lambda^{l-k}\,L_j\Big(r^{n+l-1-k}\,\hat\rho(t)\,\mathfrak{a}_{l-k}(t;y,\widetilde{\eta})\,\varsigma(t,x;y,\widetilde{\eta})\,d_\varphi(t,x;y,\widetilde{\eta})\Big)\Big|_{r=r_\star} \\
&\quad=\;\lambda^{l-k}\,\mathcal{L}_{j,k}(\hat\rho\,\mathfrak{a}_{l-k})(t;y,\widetilde{\eta})\,,
\end{aligned}$$

replacing in (4.1.9) the amplitude $\mathfrak{a}$ by the sum of its homogeneous terms $\mathfrak{a}_{l-k}$ we obtain (4.1.2).

Formulae (4.1.6), (4.1.7) are obtained from (4.1.5) by straightforward calculations. Note that for the operator $\mathcal{L}_{1,0}$ one has to consider only the terms corresponding to $\nu=1$, $\mu=0$ and $\nu=2$, $\mu=1$; the term corresponding to $\nu=3$, $\mu=2$ disappears because the function ψ is linear in r and has a third order zero at $(t_\star,r_\star)$. □

REMARK 4.1.3. Since $h(y,\lambda r_\star\widetilde{\eta})\geqslant 1$ for sufficiently large λ, we can omit ς in (4.1.9) and, consequently, in (4.1.5)–(4.1.7) (see Remark C.7).

LEMMA 4.1.4. *Let $\vec{\mathfrak{b}}(y,\eta)=(\mathfrak{b}_1(y,\eta),\dots,\mathfrak{b}_n(y,\eta))$ be a smooth covector field positively homogeneous in η of degree $\varkappa$. Then*

$$(n+\varkappa-1)\int_{S_y^*M}\langle h_\eta,\vec{\mathfrak{b}}\rangle\,đ\widetilde{\eta}\;=\;\int_{S_y^*M}\operatorname{div}_\eta\vec{\mathfrak{b}}\,đ\widetilde{\eta}\,.$$

PROOF. Assume first that $\varkappa>1-n$. The vector h_η is normal to the $(n-1)$-dimensional surface $S_y^*M\subset T_y'M$, and $(2\pi)^n\,|h_\eta|\,đ\widetilde{\eta}$ is the Euclidean measure on S_y^*M. Applying the Gauss-Ostrogradskii formula and taking into account the fact that $\operatorname{div}_\eta\vec{\mathfrak{b}}$ is homogeneous of degree $\varkappa-1$, we obtain

$$(n+\varkappa-1)\int_{S_y^*M}\langle h_\eta,\vec{\mathfrak{b}}\rangle\,đ\widetilde{\eta}\;=(n+\varkappa-1)\int_{h(y,\eta)\leqslant 1}\operatorname{div}_\eta\vec{\mathfrak{b}}\,đ\eta\;=\int_{S_y^*M}\operatorname{div}_\eta\vec{\mathfrak{b}}\,đ\widetilde{\eta}\,.$$

If $\varkappa < 1-n$, then the lemma is proved by the same procedure, but with integration over the set $\{\eta : h(y,\eta) \geqslant 1\}$ instead of $\{\eta : h(y,\eta) \leqslant 1\}$. Finally, if $\varkappa = 1-n$, then we apply the lemma to the covector field $h^\varepsilon \vec{\mathfrak{b}}$ (homogeneous of degree $\varkappa+\varepsilon > 1-n$), and obtain the required formula by taking the limit as $\varepsilon \to 0$. $\square$

COROLLARY 4.1.5. *Let the conditions of Theorem* 4.1.2 *be fulfilled with* $t_0 = 0$, *and let* $\varphi = \langle x - x^*, \xi^* \rangle$ *in a neighbourhood of* $\operatorname{cone supp}(\varsigma\,\hat\rho\,\mathfrak{a})$. *Then*

(4.1.10)

$$\mathcal{F}^{-1}_{t\to\lambda}\Big[\hat\rho(t)\,\mathcal{I}_{\varphi,a}(t,y,y)\Big] = \lambda^{n+l-1}\,\hat\rho(0)\int_{S^*_yM} \mathfrak{a}_l\big|_{t=0}\,d\!\!{}^-\eta$$
$$+\ (n+l-1)\,\lambda^{n+l-2}\int_{S^*_yM}\Big(D_t(\hat\rho\,\mathfrak{a}_l)|_{t=0} - (2i)^{-1}\hat\rho(0)\{h,\,\mathfrak{a}_l|_{t=0}\}\Big)\,d\!\!{}^-\eta$$
$$+\ \lambda^{n+l-2}\,\hat\rho(0)\int_{S^*_yM}\Big(\mathfrak{a}_{l-1}|_{t=0} + (2i)^{-1}\operatorname{tr}\partial_{y\eta}\mathfrak{a}_l|_{t=0}\Big)\,d\!\!{}^-\eta\ +\ O(\lambda^{n+l-3})$$

as $\lambda \to +\infty$.

REMARK 4.1.6. The distribution $\mathcal{I}_{\varphi,a}(0,x,y)$ is the Schwartz kernel of a pseudodifferential operator. The integrand $\mathfrak{a}_{l-1}|_{t=0} + (2i)^{-1}\operatorname{tr}\partial_{y\eta}\mathfrak{a}_l|_{t=0}$ on the right-hand side of (4.1.10) coincides with the subprincipal symbol of this pseudodifferential operator written down via its dual symbol. This fact does not depend on the choice of the phase function. However, if we had taken a general phase function instead of $\langle x-x^*,\xi^*\rangle$, the explicit formula for the corresponding term would have become much more complicated.

PROOF OF COROLLARY 4.1.5. For small t the equation $\varphi(t,y;y,\eta)=0$ has the only t-solution $t_\star \equiv 0$. Since $x^*(0;y,\eta)\equiv y$, from (2.4.5), (2.4.6), (2.4.10) it follows that

$$\varphi(t_\star,y;y,\eta)\equiv 0\,,\quad r_\star^{-1} = -\varphi_t(t_\star,y;y,\eta)\equiv h(y,\eta)\,,\quad d_\varphi(t_\star,y;y,\eta)=1,$$

respectively (note that all these equalities hold for a general phase function φ as well). For our special phase function φ, in view of (3.3.20) and (3.3.21), we also have

$$\varphi_{tt}(0,y;y,\eta) = \langle h_y(y,\eta), h_\eta(y,\eta)\rangle\,,\qquad D_t\,d_\varphi(0,y;y,\eta) = -(2i)^{-1}\operatorname{tr} h_{y\eta}(y,\eta)\,.$$

Thus, at the point $t = t_\star = 0$, we have, for $(y,\eta)\in S^*_yM$,

$$\mathcal{L}_{0,0}(\hat\rho\,\mathfrak{a}_l) = \hat\rho\,\mathfrak{a}_l\,,\qquad \mathcal{L}_{0,1}(\hat\rho\,\mathfrak{a}_{l-1}) = \hat\rho\,\mathfrak{a}_{l-1}\,,$$
$$\mathcal{L}_{1,0}(\hat\rho\,\mathfrak{a}_l) = (n+l-1)\,D_t(\hat\rho\,\mathfrak{a}_l) - (2i)^{-1}(n+l-1)\,(\operatorname{tr}h_{y\eta})\,\hat\rho\,\mathfrak{a}_l$$
$$+\ (2i)^{-1}(n+l)\,(n+l-1)\,\langle h_y(y,\eta), h_\eta(y,\eta)\rangle\,\hat\rho\,\mathfrak{a}_l$$

(here we have omitted ς). Therefore, applying (4.1.2) we obtain

(4.1.11)

$$\begin{aligned}\mathcal{F}^{-1}_{t\to\lambda}\Big[\hat\rho(t)\,\mathcal{I}_{\varphi,a}(t,y,y)\Big] &= \lambda^{n+l-1}\,\hat\rho(0)\int_{S^*_yM}\mathfrak{a}_l|_{t=0}\,d\!\!\!{}^-\!\eta \;+\;(n+l-1)\,\lambda^{n+l-2}\int_{S^*_yM}D_t(\hat\rho\,\mathfrak{a}_l)|_{t=0}\,d\!\!\!{}^-\!\eta\\ &+\;(2i)^{-1}(n+l-1)\,\lambda^{n+l-2}\,\hat\rho(0)\int_{S^*_yM}\Big((n+l)\,\langle h_y,h_\eta\rangle\;-\;\operatorname{tr}h_{y\eta}\Big)\,\mathfrak{a}_l|_{t=0}\,d\!\!\!{}^-\!\eta\\ &+\;\lambda^{n+l-2}\,\hat\rho(0)\int_{S^*_yM}\mathfrak{a}_{l-1}|_{t=0}\,d\!\!\!{}^-\!\eta\;+\;O(\lambda^{n+l-3})\,,\qquad\lambda\to+\infty\,.\end{aligned}$$

By Lemma 4.1.4 we have

$$(n+l)\int_{S^*_yM}\langle h_y,h_\eta\rangle\,\mathfrak{a}_l\,d\!\!\!{}^-\!\eta\;=\;\int_{S^*_yM}\Big((\operatorname{tr}h_{y\eta})\,\mathfrak{a}_l\;+\;\langle h_y,\partial_\eta\mathfrak{a}_l\rangle\Big)\,d\!\!\!{}^-\!\eta\,,\tag{4.1.12}$$

and

$$(n+l-1)\int_{S^*_yM}\langle h_\eta,\partial_y\mathfrak{a}_l\rangle\,d\!\!\!{}^-\!\eta\;=\;\int_{S^*_yM}\operatorname{tr}\partial_{y\eta}\mathfrak{a}_l\,d\!\!\!{}^-\!\eta\,.\tag{4.1.13}$$

Substituting (4.1.12) in (4.1.11) and taking (4.1.13) into account we obtain (4.1.10).

2. The nonzero singularities. We shall need the following auxiliary lemmas.

LEMMA 4.1.7. *For each fixed* $T>0$ *the gradient of* $\varphi(T,y;y,\eta)$ *with respect to* (y,η) *vanishes at the point* (y_0,η_0) *if and only if* (y_0,η_0) *is* T*-periodic. Moreover, the point* (y_0,η_0) *is absolutely* T*-periodic if and only if the gradient* $\nabla_{y,\eta}\,(\varphi(T,y;y,\eta))$ *has an infinite order zero at* (y_0,η_0).

PROOF. In view of (2.4.4), (2.4.9) and (2.4.26) we have

$$\nabla_\eta\varphi(T,y;y,\eta)\;=\;F(T;y,\eta)\,(y-x^*(T;y,\eta))\,,\tag{4.1.14}$$

$$\nabla_y\,(\varphi(T,y;y,\eta))\;=\;(\xi^*(T;y,\eta)-\eta)+O(|y-x^*(T;y,\eta)|)\,,\tag{4.1.15}$$

where F is a smooth nonsingular matrix function. This implies the first statement of the lemma. The second statement is easily obtained by differentiating (4.1.14) and (4.1.15) with respect to y and η. □

LEMMA 4.1.8. *Assume that the phase function* φ *is real in a neighbourhood of the point* $(T,y_0;y_0,\eta_0)$, *where* $T>0$ *and* $y_0=x^*(T;y_0,\eta_0)$. *Let the conic neighbourhood* $O\subset T'\overset{\circ}{M}$ *of the point* (y_0,η_0) *be sufficiently small,* $(y,\eta)\in O$, *and* $\widetilde t_\star(y,\eta)$ *the* t*-solution of the equation* $\varphi(t,y;y,\eta)=0$ *such that* $\widetilde t_\star(y_0,\eta_0)=T$. *Then* $\nabla_{y,\eta}\widetilde t_\star(y,\eta)=0$ *if and only if the point* (y,η) *is* $\widetilde t_\star(y,\eta)$*-periodic, and* $\nabla_{y,\eta}\widetilde t_\star$ *has an infinite order zero at the point* (y,η) *if and only if* (y,η) *is absolutely* $\widetilde t_\star(y,\eta)$*-periodic.*

PROOF. Differentiating the identity $\varphi(\widetilde t_\star,y;y,\eta)\equiv0$ with respect to η and y we obtain

$$(\widetilde t_\star)_\eta\,\varphi_t(\widetilde t_\star,y;y,\eta)\;=\;-\varphi_\eta(\widetilde t_\star,y;y,\eta)\,,\tag{4.1.16}$$

$$(\widetilde t_\star)_y\,\varphi_t(\widetilde t_\star,y;y,\eta)\;=\;-\varphi_x(\widetilde t_\star,y;y,\eta)\;-\;\varphi_y(\widetilde t_\star,y;y,\eta)\,.\tag{4.1.17}$$

By (2.4.6) $\varphi_t(t,x,;y,\eta) \neq 0$ for $(t,x,;y,\eta)$ close to $(T,y_0;y_0,\eta_0)$. Therefore Lemma 4.1.7 and (4.1.16), (4.1.17) imply the first statement of the lemma. The second statement is easily proved by differentiating (4.1.16) and (4.1.17) with respect to y and η. □

Let $\Pi_{T,i} \subset O \cap S^*M$ and $\Pi^a_{T,i} \subset O \cap S^*M$ be the sets of T-periodic and absolutely T-periodic points of the canonical transformation corresponding to the ith leg of our Hamiltonian billiards. Lemma 4.1.8 implies

COROLLARY 4.1.9. *For all nonnegative T^-, T^+, $0 < T^- \leqslant T^+ \leqslant T_+$, the sets $\bigcup_{T^- \leqslant T \leqslant T^+} \Pi_{T,i}$ and $\bigcup_{T^- \leqslant T \leqslant T^+} \Pi^a_{T,i}$ are measurable, and*

$$\operatorname{meas}\left(\bigcup_{T^- \leqslant T \leqslant T^+} (\Pi_{T,i} \setminus \Pi^a_{T,i})\right) = 0\,.$$

PROOF. It is sufficient to prove the corollary assuming that O is a small conic neighbourhood of a T-periodic point $(y_0,\eta_0) \in T'\overset{\circ}{M}$, $T \in [T^-,T^+]$, and that the difference $T^+ - T^-$ is sufficiently small. Then, by Corollary 2.4.14, we can choose a phase function $\varphi \in \mathfrak{F}_i$, which is real for $t \in [T^-,T^+]$, $(y,\eta) \in O$ and x close to $x^*(T;y_0,\eta_0) = y$. Now Lemma 4.1.8 implies that the sets $\bigcup_{T^- \leqslant T \leqslant T^+} \Pi_{T,i}$ and $\bigcup_{T^- \leqslant T \leqslant T^+} \Pi^a_{T,i}$ are closed in $O \cap S^*M$, and consequently, measurable.

By Lemma 4.1.8 the set $\bigcup_{T^- \leqslant T \leqslant T^+} (\Pi_{T,i} \setminus \Pi^a_{T,i})$ consists of points (y,η) at which the function $\nabla_{y,\eta}\widetilde{t}_\star$ has zeros of finite (but not infinite) order. Clearly, if

$$\partial_{y_i}\partial_y^\alpha\partial_\eta^\beta\widetilde{t}_\star(y,\eta) \neq 0 \quad \text{or} \quad \partial_{\eta_i}\partial_y^\alpha\partial_\eta^\beta\widetilde{t}_\star(y,\eta) \neq 0,$$

then in a neighbourhood of (y,η) the function $\partial_\eta^\alpha\partial_\eta^\beta\widetilde{t}_\star$ can vanish only on a conic submanifold of codimension one. By induction we obtain that the set of points at which $\nabla_{y,\eta}\widetilde{t}_\star$ has finite order zeros is contained in a countable union of conic submanifolds of codimension one. Therefore its intersection with S^*M has measure zero. □

PROPOSITION 4.1.10. *Let* $\operatorname{cone\,supp}(\hat\rho\,\mathfrak{a})$ *lie in a sufficiently small conic neighbourhood of the point $(t_0;y_0,\eta_0)$, $t_0 \neq 0$ (in particular, $0 \notin \operatorname{supp}\hat\rho$). Then for any $\varphi \in \mathfrak{F}_i$*

$$\begin{aligned} &\int_M \mathcal{F}^{-1}_{t\to\lambda}\Big[\hat\rho(t)\,\mathcal{I}_{\varphi,a}(t,y,y)\Big]\,dy \\ (4.1.18)\qquad &= \lambda^{n+l-1}\int_{\Pi^a} e^{i\lambda T^\star + i\mathfrak{f}_c(T^\star;y,\widetilde\eta)}\,\hat\rho(T^\star)\,\mathfrak{a}_l(T^\star;y,\widetilde\eta)\,dy\,d\!\!\!{}^-\widetilde\eta \;+\; o(\lambda^{n+l-1}) \end{aligned}$$

as $\lambda \to +\infty$. Here Π^a is the set of absolutely periodic points of the corresponding canonical transformation, $T^\star = T^\star(y,\widetilde\eta)$ is the period function, and $\mathfrak{f}_c(T^\star;y,\widetilde\eta)$ is the phase shift generated by the passage of the closed trajectory $\big(x^*(t;y,\widetilde\eta),\xi^*(t;y,\widetilde\eta)\big)$, $0 \leqslant t \leqslant T^\star(y,\widetilde\eta)$, through caustics.

REMARK 4.1.11. We take in (4.1.18) the value of $T^\star$ for which $(T^\star(y,\eta);y,\widetilde\eta)$ lies in $\operatorname{cone\,supp}(\hat\rho\,\mathfrak{a}_l)$. If $\operatorname{supp}(\hat\rho\,\mathfrak{a}_l)$ is sufficiently small, then for each periodic point (y,η) this value is uniquely defined. It will be clear from the proof that the period function $T^\star$ coincides with the restriction of a smooth function to a closed set. Therefore the integral on the right-hand side of (4.1.18) is well defined.

PROOF OF PROPOSITION 4.1.10. We can assume without loss of generality that the phase function φ is real on $\operatorname{supp}(\hat\rho\,\varsigma\,\mathfrak{a}_l)$ (see Corollary 2.4.14). Let $\widetilde{t}_\star(y,\eta) := t_\star(y;y,\eta)$ and $\widetilde{r}_\star(y,\eta) := r_\star(y;y,\eta)$. Then, in view of (4.1.2), (4.1.6) and Remark 4.1.3 it is sufficient to prove that

$$\begin{aligned}(4.1.19)\qquad &\int_{S^*M} e^{i\lambda\widetilde{t}_\star}\,\hat\rho(\widetilde{t}_\star)\,\mathfrak{a}_l(\widetilde{t}_\star;y,\widetilde{\eta})\,\widetilde{r}_\star^{\,n+l}\,\left.d_\varphi\right|_{x=y}\,d\!\!{}^-\widetilde{\eta}\,dy\\ &\quad= \int_{\Pi^a} e^{i\lambda T^\star + i\mathfrak{f}_c(T^\star;y,\widetilde{\eta})}\,\hat\rho(T^\star)\,\mathfrak{a}_l(T^\star;y,\widetilde{\eta})\,dy\,d\!\!{}^-\widetilde{\eta} + o(1),\qquad \lambda\to+\infty.\end{aligned}$$

Let $f\in C_0^\infty(\mathbb{R})$, $f(s)=1$ for $|s|\leqslant 1$ and $f(s)=0$ for $|s|\geqslant 2$. We split the integral on the left-hand side of (4.1.19) into a sum of two integrals with the same integrand multiplied by $f(\varepsilon^{-1}|\nabla_{y,\widetilde{\eta}}\widetilde{t}_\star|^2)$ and $\left(1-f(\varepsilon^{-1}|\nabla_{y,\widetilde{\eta}}\widetilde{t}_\star|^2)\right)$, respectively. The function $\widetilde{t}_\star$ is positively homogeneous in η of degree 0, so Lemma 4.1.8 implies that the first integral uniformly (with respect to λ) converges to the integral over the set Π as $\varepsilon\to 0$. By Corollary 4.1.9 it coincides with the integral over Π^a.

In the second integral we replace $e^{i\lambda\widetilde{t}_\star}$ by

$$-i\,\lambda^{-1}\,|\nabla_{y,\widetilde{\eta}}\widetilde{t}_\star|^{-2}\,\left(\nabla_{y,\widetilde{\eta}}\widetilde{t}_\star\cdot\nabla_{y,\widetilde{\eta}}\,e^{i\lambda\widetilde{t}_\star}\right)$$

and then integrate by parts. This shows that the second integral is $O(\lambda^{-1})$ for each fixed ε. Thus, the left-hand side of (4.1.19) can be written as

$$\int_{\Pi^a} e^{i\lambda\widetilde{t}_\star}\,\hat\rho(\widetilde{t}_\star)\,\mathfrak{a}_l(\widetilde{t}_\star;y,\widetilde{\eta})\,\widetilde{r}_\star^{\,n+l}\,\left.d_\varphi\right|_{x=y}\,d\!\!{}^-\widetilde{\eta}\,dy\;+\;o(1),\qquad \lambda\to+\infty.$$

Since $(\widetilde{t}_\star(y,\eta),y;y,\eta)\in\mathfrak{C}_i$, we have $\left.\widetilde{t}_\star\right|_\Pi = T^\star$ and $\left.\widetilde{r}_\star\right|_\Pi\equiv 1$. By Definitions 1.5.9 and 1.7.2, $(\left.d_\varphi\right|_{x=y})|_{\Pi^a} = e^{-i\alpha\pi/2}$, where $\alpha=\alpha(y,\widetilde{y})$ is the Maslov index of the trajectory $\left(x^*(t;y,\widetilde{\eta}),\xi^*(t;y,\widetilde{\eta})\right)$, $0\leqslant t\leqslant T^\star(y,\widetilde{\eta})$, and $-\alpha\pi/2=\mathfrak{f}_c(T^\star;y,\widetilde{\eta})$. The proof is complete. □

PROPOSITION 4.1.12. *Let* $\operatorname{cone\,supp}(\hat\rho\,\mathfrak{a})$ *lie in a sufficiently small conic neighbourhood of the point* $(t_0;y_0,\eta_0)$, $t_0\neq 0$ *(in particular,* $0\notin\operatorname{supp}\hat\rho$*). If there are no* T*-periodic points* (y,η) *such that* $(T;y,\eta)\in\operatorname{cone\,supp}(\hat\rho\,\mathfrak{a}_l)$*, then*

$$\int_M \mathcal{F}^{-1}_{t\to\lambda}\left[\hat\rho(t)\,\mathcal{I}_{\varphi,a}(t,y,y)\right]dy = O(\lambda^{-\infty}),\qquad \lambda\to+\infty.$$

PROOF. In the same way as in the proof of Proposition 4.1.10 we obtain that the function $\widetilde{t}_\star$ has no stationary points on $\bigcup_j\operatorname{supp}(\hat\rho\,\mathfrak{a}_j)$. Therefore all the integrals on the right-hand side of (4.1.2) define rapidly decreasing functions of λ. This proves the proposition. □

Let $\Pi_{y,T,i}$, $\Pi^a_{y,T,i}$ be the sets defined in the same way as $\Pi_{y,T}$, $\Pi^a_{y,T}$ in Section 1.8, where the subscript i indicates that we assume $(x^*(T;y,\eta),\xi^*(T;y,\eta))$ to lie on the ith leg of our Hamiltonian billiards. The following two lemmas and corollary are proved in the same way as Lemmas 4.1.7, 4.1.8 and Corollary 4.1.9.

LEMMA 4.1.13. *For each fixed* $T > 0$ *and* $(x,y) \in \overset{\circ}{M} \times \overset{\circ}{M}$ *the gradient* $\nabla_\eta \varphi(T,y;y,\eta)$ *vanishes at the point* $\eta_0 \in S^*_y M$ *if and only if* $\eta_0 \in \Pi_{y,T}$. *Moreover, the gradient* $\nabla_\eta \varphi(T,y;y,\eta)$ *has an infinite order zero at* $\eta_0 \in S^*_y M$ *if and only if* $\eta_0 \in \Pi^a_{y,T}$.

LEMMA 4.1.14. *Assume that the phase function* φ *is real in a neighbourhood of the point* $(T,y;y,\eta_0)$, *where* $T > 0$ *and* $y = x^*(T;y,\eta_0)$. *Let* $\widetilde{t}_\star(y,\eta)$ *be a local* t*-solution of the equation* $\varphi(t,y;y,\eta) = 0$, *defined for* η *close to* η_0 *and such that* $t_\star(y,\eta_0) = T$. *Then* $\nabla_\eta \widetilde{t}_\star(y,\eta) = 0$ *if and only if* $\eta/h(y,\eta) \in \Pi_{y,T}$ *with* $T = \widetilde{t}_\star(y,\eta)$, *and* $\nabla_\eta \widetilde{t}_\star(y,\eta) = 0$ *has an infinite order zero at the point* η *if and only if* $\eta/h(y,\eta) \in \Pi^a_{y,T}$ *with* $T = \widetilde{t}_\star(y,\eta)$.

COROLLARY 4.1.15. *For all nonnegative* T^-, T^+, $0 < T^- \leqslant T^+ \leqslant T_+$, *the sets* $\bigcup_{T^- \leqslant T \leqslant T^+} \Pi_{y,T,i}$ *and* $\bigcup_{T^- \leqslant T \leqslant T^+} \Pi^a_{y,T,i}$ *are measurable, and*

$$\operatorname{meas}\left(\bigcup_{T^- \leqslant T \leqslant T^+} (\Pi_{y,T,i} \setminus \Pi^a_{y,T,i}) \right) = 0 .$$

PROPOSITION 4.1.16. *Let* cone supp $(\hat\rho\, \mathfrak{a})$ *lie in a sufficiently small conic neighbourhood of the point* $(t_0;y,\eta_0)$, $t_0 \neq 0$ *(in particular,* $0 \notin \operatorname{supp}\hat\rho$*). Then for any* $\varphi \in \mathfrak{F}_i$,

$$\begin{aligned} &\mathcal{F}^{-1}_{t\to\lambda}\Big[\hat\rho(t)\, \mathcal{I}_{\varphi,a}(t,y,y)\Big] \\ &\qquad = \lambda^{n+l-1} \int_{\Pi^a_y} e^{i\lambda T^\star_y + i\mathfrak{f}_c(T^\star_y;y,\widetilde\eta)}\, |\det \xi^*_\eta(T^\star_y;y,\widetilde\eta)|^{1/2} \\ &\qquad\qquad \times \hat\rho(T^\star_y)\, \mathfrak{a}_l(T^\star_y;y,\widetilde\eta)\, d\widetilde\eta \,+\, o(\lambda^{n+l-1}) \end{aligned} \tag{4.1.20}$$

as $\lambda \to +\infty$, *where* $T^\star_y = T^\star_y(\widetilde\eta)$ *is the value of* T *for which* $\widetilde\eta \in \Pi^a_{y,T}$, *and* $\mathfrak{f}_c(T^\star_y;y,\widetilde\eta)$ *is the phase shift generated by the passage of the trajectory*

$$\left(x^*(t;y,\widetilde\eta), \xi^*(t;y,\widetilde\eta)\right), \qquad 0 \leqslant t \leqslant T^\star(y,\widetilde\eta),$$

through caustics.

REMARK 4.1.17. Since supp$(\hat\rho\, \mathfrak{a}_l)$ is small, for each $\eta \in \Pi^a_y$ there exists at most one value of $T^\star_y$ such that $\hat\rho(T^\star_y)\, \mathfrak{a}_l(T^\star_y;y,\eta) \neq 0$. The function $T^\star_y$ is measurable as the restriction of a smooth function to a closed set.

PROOF OF PROPOSITION 4.1.16. The proof almost completely repeats that of Proposition 4.1.10. We assume that the phase function φ is real on supp$(\hat\rho\, \varsigma\, \mathfrak{a}_l)$, and then it is sufficient to prove that

$$\begin{aligned} &\int_{S^*_y M} e^{i\lambda \widetilde{t}_\star}\, \hat\rho(\widetilde{t}_\star)\, \mathfrak{a}_l(\widetilde{t}_\star;y,\widetilde\eta)\, \widetilde{r}_\star^{\,n+l}\; d_\varphi\big|_{x=y}\, d\widetilde\eta \\ &\quad = \int_{\Pi^a_y} e^{i\lambda T^\star_y + i\mathfrak{f}_c(T^\star_y;y,\widetilde\eta)}\, |\det \xi^*_\eta(T^\star_y;y,\widetilde\eta)|^{1/2}\, \hat\rho(T^\star_y)\, \mathfrak{a}_l(T^\star_y;y,\widetilde\eta)\, d\widetilde\eta \,+\, o(1) \end{aligned}$$

as $\lambda \to +\infty$. This equality is obtained in the same way as (4.1.19). The only difference is that we have to deal with the gradient $\nabla_{\widetilde\eta} \widetilde{t}_\star$ (instead of $\nabla_{y,\widetilde\eta} \widetilde{t}_\star$) and to apply Lemma 4.1.14 and Corollary 4.1.15 (instead of Lemma 4.1.8 and Corollary 4.1.9). □

We shall also need the following result which is proved in the same manner as Proposition 4.1.16.

PROPOSITION 4.1.18. *Assume that the conditions of Proposition* 4.1.16 *are fulfilled. Let* K *be a compact subset of* $\overset{\circ}{M}$, *and* $\widetilde{K}$ *the projection of* $\operatorname{supp} \mathfrak{a}_l$ *on* $\mathbb{R}_t \times K$. *If*

$$\operatorname{meas}_y\{\widetilde{\eta} \in S_y^* M \cap \operatorname{supp} \mathfrak{a}_l(t; y, \cdot) \,:\, |y - x^*(t; y, \widetilde{\eta})| < \varepsilon\} \to 0\,, \qquad \varepsilon \to 0\,,$$

uniformly with respect to $(t, y) \in \widetilde{K}$, *then*

$$\mathcal{F}^{-1}_{t\to\lambda}\Big[\hat{\rho}(t)\, \mathcal{I}_{\varphi,a}(t, y, y)\Big] \;=\; o(\lambda^{n+l-1})\,, \qquad \lambda \to +\infty\,,$$

uniformly with respect to $y \in K$.

4.2. Singularity of the wave group at $t = 0$

In the remainder of this chapter we will consider a Lagrangian distribution of a particular type: namely, the Schwartz kernel of the operator (4.0.1). This section is devoted to the calculation of the singularity of this Schwartz kernel at $t = 0$.

Let $\hat{\rho}_1$ be a function of the class $C_0^\infty(\mathbb{R})$ such that $\hat{\rho}_1|_{t=0} = 1$, $D_t\hat{\rho}_1|_{t=0} = 0$, and $\operatorname{supp} \hat{\rho}_1 \subset [-1, 1]$. Set $\hat{\rho}_T(t) := \hat{\rho}_1(t/T)$, where T is a positive parameter.

1. Pointwise case. In this subsection we prove

THEOREM 4.2.1. *Uniformly over* $y \in M$ *we have*

(4.2.1)

$$\begin{aligned}
\mathcal{F}^{-1}_{t\to\lambda}\Big[\hat{\rho}_{T_\delta}(t)\, \mathbf{u}_{P,Q}(t, y, y)\Big] \;&=\; \lambda^{n+\mathbf{l}+\mathbf{m}-1} \int_{S_y^* M} P_{\mathbf{l}}\, \overline{Q_{\mathbf{m}}}\, d\widetilde{\eta} \\
&+\; \lambda^{n+\mathbf{l}+\mathbf{m}-2} \int_{S_y^* M} \left(P_{\mathbf{l}} \overline{Q_{\mathrm{sub}}} + \overline{Q_{\mathbf{m}}} P_{\mathrm{sub}} + \frac{\{\overline{Q_{\mathbf{m}}}, P_{\mathbf{l}}\}}{2i} \right) d\widetilde{\eta} \\
&-\; (n + \mathbf{l} + \mathbf{m} - 1)\lambda^{n+\mathbf{l}+\mathbf{m}-2} \int_{S_y^* M} \left(\frac{A_{\mathrm{sub}}\, P_{\mathbf{l}} \overline{Q_{\mathbf{m}}}}{2m} - \frac{P_{\mathbf{l}}\{h, \overline{Q_{\mathbf{m}}}\} - \overline{Q_{\mathbf{m}}}\{h, P_{\mathbf{l}}\}}{2i} \right) d\widetilde{\eta} \\
&+\; O(\lambda^{n+\mathbf{l}+\mathbf{m}-3})\,, \qquad \lambda \to +\infty\,.
\end{aligned}$$

Here T_δ is the number defined in the introduction to the chapter.

PROOF OF THEOREM 4.2.1. Let us fix an arbitrary point $y_0 \in M$. Due to the compactness of M it is sufficient to prove Theorem 4.2.1 for y lying in a small neighbourhood of y_0. The case $y_0 \in \partial M$ is trivial because all the terms in (4.2.1) are zero. So further on we suppose that $y_0 \notin \partial M$.

Without loss of generality we shall also assume that the conic supports of the pseudodifferential operators P and Q lie in small conic neighbourhoods of $T'_{y_0} M$. This can always be achieved by multiplying P and Q by smooth spatial cut-offs from the right and left, respectively.

By the results of 3.4.6 $\mathbf{u}_{P,Q}(t, x, y)$ is a standard Lagrangian distribution of order $\mathbf{l} + \mathbf{m}$ with principal symbol

$$\begin{aligned}
\mathfrak{a}_{\mathbf{l}+\mathbf{m}}(t; y, \eta) \;&=\; \overline{Q_{\mathbf{m}}(x^*(t; y, \eta), \xi^*(t; y, \eta))} \\
&\times \exp\Big(-\frac{i}{2m}\, h^{1-2m}(y, \eta) \int_0^t A_{\mathrm{sub}}\big(x^*(\tau; y, \eta), \xi^*(\tau; y, \eta)\big)\, d\tau\Big)\, P_{\mathbf{l}}(y, \eta)\,.
\end{aligned} \tag{4.2.2}$$

Let $\mathfrak{a}(t;y,\eta) \sim \mathfrak{a}_{\mathbf{l}+\mathbf{m}}(t;y,\eta)+\mathfrak{a}_{\mathbf{l}+\mathbf{m}-1}(t;y,\eta)+\cdots$ be the full symbol of $\mathbf{u}_{P,Q}(t,x,y)$. The term $\mathfrak{a}_{\mathbf{l}+\mathbf{m}-1}$ depends, of course, on our choice of the phase function.

Without loss of generality we may assume that $\operatorname{supp}\hat\rho_{T_\delta}$ lies in an arbitrarily small neighbourhood of the origin. Indeed, if we replace our original $\hat\rho_{T_\delta}$ by a similar function with smaller support, then their difference $\hat\gamma$ will be a function of the class $C_0^\infty(\mathbb{R})$ such that $\hat\gamma|_{t=0} = D_t\hat\gamma|_{t=0} = 0$, and $\operatorname{supp}\hat\gamma \subset [-T_\delta, T_\delta]$. Then by Corollary 4.1.5 and Proposition 4.1.1 we have

$$\mathcal{F}^{-1}_{t\to\lambda}\Big[\hat\gamma(t)\,\mathbf{u}_{P,Q}(t,y,y)\Big] \;=\; O(\lambda^{n+\mathbf{l}+\mathbf{m}-3}),$$

so this term can be included into the remainder term of the required asymptotic formula (4.2.1). In the above argument we used the fact that we do not have loops of length $\leqslant T_\delta$; see the introduction to this chapter.

Let us fix an (arbitrary) local coordinate system, and use this coordinate system both for x and y. Let us choose a phase function $\varphi(t,x;y,\eta)$ such that in our local coordinates $\varphi = \langle x-x^*, \xi^*\rangle$ for (t,x,y) close to $(0,y_0,y_0)$; note that this choice of phase function is required by Corollary 4.1.5 which we intend to use. With our choice of the phase function and local coordinates the expression $\mathfrak{a}|_{t=0}$ coincides with the dual symbol of the pseudodifferential operator Q^*P; consequently (see 2.1.3)

(4.2.3)
$$\begin{aligned}\mathfrak{a}_{\mathbf{l}+\mathbf{m}-1}|_{t=0} &= (Q^*P)_{\mathrm{sub}} - (2i)^{-1}\operatorname{tr}\partial_{y\eta}(\overline{Q_{\mathbf{m}}}P_{\mathbf{l}}) \\ &= \overline{Q_{\mathbf{m}}}P_{\mathrm{sub}} + P_{\mathbf{l}}\overline{Q_{\mathrm{sub}}} + (2i)^{-1}\{\overline{Q_{\mathbf{m}}},P_{\mathbf{l}}\} - (2i)^{-1}\operatorname{tr}\partial_{y\eta}(\overline{Q_{\mathbf{m}}}P_{\mathbf{l}}).\end{aligned}$$

Substituting (4.2.2), (4.2.3) in (4.1.10), $l = \mathbf{l}+\mathbf{m}$, we obtain (4.2.1). □

REMARK 4.2.2. If P and Q are differential operators and the sum of their orders is even, then the second and third terms on the right-hand side of (4.2.1) vanish. This happens because in this case the integrands are odd functions with respect to η.

2. Integrated case. In order to proceed further we shall need the following

LEMMA 4.2.3. *Let $P_{\mathbf{l}}(y,\eta)$ and $Q_{\mathbf{m}}(y,\eta)$ be smooth functions positively homogeneous in η of degrees $\mathbf{l}$ and $\mathbf{m}$, respectively. Then*

$$\frac{n+\mathbf{l}+\mathbf{m}-1}{2}\int_{S^*M}(P_{\mathbf{l}}\{h,\overline{Q_{\mathbf{m}}}\} - \overline{Q_{\mathbf{m}}}\{h,P_{\mathbf{l}}\})\,dy\,d\!\!\!{}^-\eta \;=\; \int_{S^*M}\{P_{\mathbf{l}},\overline{Q_{\mathbf{m}}}\}\,dy\,d\!\!\!{}^-\eta.$$

PROOF OF LEMMA 4.2.3. Let us first note that it is sufficient to prove

$$(n+\mathbf{l}+\mathbf{m}-1)\int_{S^*M}P_{\mathbf{l}}\{h,\overline{Q_{\mathbf{m}}}\}\,dy\,d\!\!\!{}^-\eta \;=\; \int_{S^*M}\{P_{\mathbf{l}},\overline{Q_{\mathbf{m}}}\}\,dy\,d\!\!\!{}^-\eta \tag{4.2.4}$$

because an elementary symmetrization (interchange of $P_{\mathbf{l}}$ and $\overline{Q_{\mathbf{l}}}$, followed by subtraction from (4.2.4)) gives the required result.

Note also that it is sufficient to prove Lemma 4.2.3 under the assumption that the pseudodifferential operator P has a small conic support. The general situation is reduced to this one by performing a partition of unity, which is justified because the expressions on both sides of (4.2.4) are linear with respect to $P_{\mathbf{l}}$ and have an invariant nature.

The fact that P has small conic support implies that Q can be considered to have small conic support.

As the conic supports of P and Q are assumed to be small, we can fix an (arbitrary) local coordinate system, and use this coordinate system both for x and y. Fixation of local coordinates allows us to view the expressions on both sides of (4.2.4) as usual $(2n-1)$-dimensional surface integrals in a $2n$-dimensional Euclidean space.

Let us introduce some notation. For a (scalar) function $F(y,\eta)$ we introduce the following $2n$-dimensional "vector" functions:

$$\operatorname{grad} F := (F_y, F_\eta)\,, \qquad \operatorname{curl} F := (F_\eta, -F_y)\,.$$

For a $2n$-dimensional "vector" function $\vec{V}(y,\eta) = \big(V^{(1)}(y,\eta), \dots, V^{(2n)}(y,\eta)\big)$ we define the (scalar) divergence as

$$\operatorname{div} \vec{V} := \sum_{j=1}^{n} \big(V^{(j)}_{y_j} + V^{(n+j)}_{\eta_j}\big)\,.$$

Let us now prove (4.2.4) under the assumption $\mathbf{l} + \mathbf{m} > 1 - n$.

Let $\vec{V} \in C_0^\infty(T'M)$ be a $2n$-dimensional "vector" function with small conic support. Note that the "vector" (h_y, h_η) is normal to the $(2n-1)$-dimensional surface $S^*M \subset T'M$, and $(2\pi)^n\,|\operatorname{grad} h|\, dy\, đ\widetilde{\eta}$ is the Euclidean measure on S^*M. Consequently, applying the Gauss-Ostrogradskii formula we obtain

$$\begin{aligned}(4.2.5)\qquad &\int_{h(y,\eta)\leqslant 1} \big(\langle \operatorname{grad} P_{\mathbf{l}}, \vec{V}\rangle + P_{\mathbf{l}} \operatorname{div} \vec{V}\big)\, dy\, đ\eta \\ &\qquad = \int_{h(y,\eta)\leqslant 1} \operatorname{div}\big(P_{\mathbf{l}}\vec{V}\big)\, dy\, đ\eta \;=\; \int_{S^*M} P_{\mathbf{l}} \langle \operatorname{grad} h, \vec{V}\rangle\, dy\, đ\widetilde{\eta}\,.\end{aligned}$$

Obviously, formula (4.2.5) remains true if the components of $\vec{V}$ are smooth homogeneous (in η) functions, provided the degree of homogeneity is sufficiently high. Namely, for (4.2.5) to hold the first n components of $\vec{V}$ should have a degree of homogeneity greater than $-n-\mathbf{l}$, and the last n components, a degree greater than $1-n-\mathbf{l}$.

Substituting $\vec{V} = \operatorname{curl} \overline{Q_{\mathbf{m}}}$ in (4.2.5), and using the facts that $\operatorname{div}\operatorname{curl} f = 0$ and $\langle \operatorname{grad} f, \operatorname{curl} g\rangle = -\{f,g\}$ for any functions f, g, we obtain

$$\int_{h(y,\eta)\leqslant 1} \{P_{\mathbf{l}}, \overline{Q_{\mathbf{m}}}\}\, dy\, đ\eta \;=\; \int_{S^*M} P_{\mathbf{l}} \{h, \overline{Q_{\mathbf{m}}}\}\, dy\, đ\widetilde{\eta}\,.$$

By homogeneity this formula implies (4.2.4).

If $\mathbf{l} + \mathbf{m} < 1 - n$, then the lemma is proved by the same procedure, but with integration over the set $\{(y,\eta) : h(y,\eta) \geqslant 1\}$ instead of $\{(y,\eta) : h(y,\eta) \leqslant 1\}$. Finally, if $\mathbf{l} + \mathbf{m} = 1 - n$, then we apply the lemma to the function $h^\varepsilon P_{\mathbf{l}}$ (homogeneous of degree $\mathbf{l} + \varepsilon > 1 - n - \mathbf{m}$) instead of $P_{\mathbf{l}}$, and obtain the required formula by taking the limit as $\varepsilon \to 0$. □

An immediate consequence of Theorem 4.2.1 and Lemma 4.2.3 is

THEOREM 4.2.4. *We have*

$$\begin{aligned} \mathcal{F}^{-1}_{t\to\lambda}\Big[\hat\rho_{T_\delta}(t)\,\operatorname{Tr}\mathbf{U}_{P,Q}(t)\Big] &= \lambda^{n+\mathbf{l}+\mathbf{m}-1}\int_{S^*M} P_{\mathbf{l}}\,\overline{Q_{\mathbf{m}}}\,dy\,đ\eta \\ &+ \lambda^{n+\mathbf{l}+\mathbf{m}-2}\int_{S^*M}\left(P_{\mathbf{l}}\overline{Q_{\mathrm{sub}}}+\overline{Q_{\mathbf{m}}}P_{\mathrm{sub}}+\frac{\{P_{\mathbf{l}},\overline{Q_{\mathbf{m}}}\}}{2i}\right)dy\,đ\eta \\ &- (n+\mathbf{l}+\mathbf{m}-1)\,\lambda^{n+\mathbf{l}+\mathbf{m}-2}\int_{S^*M}\frac{A_{\mathrm{sub}}\,P_{\mathbf{l}}\overline{Q_{\mathbf{m}}}}{2m}\,dy\,đ\eta + O(\lambda^{n+\mathbf{l}+\mathbf{m}-3}) \end{aligned} \tag{4.2.6}$$

as $\lambda\to+\infty$.

REMARK 4.2.5. The integrand in the second term on the right-hand side of (4.2.6) is the subprincipal symbol of the pseudodifferential operator PQ^*. This is not surprising because

$$\operatorname{Tr}\mathbf{U}_{P,Q}(t) = \operatorname{Tr}\big(Q^*\exp(-it\mathcal{A}^{1/(2m)})P\big) = \operatorname{Tr}\big(\exp(-it\mathcal{A}^{1/(2m)})PQ^*\big).$$

REMARK 4.2.6. If PQ^* is a differential operator of even order, then the second and third terms on the right-hand side of (4.2.6) vanish. This happens because in this case the integrands are odd functions with respect to η (cf. Remark 4.2.2).

4.3. Singularities of the wave group at $t \neq 0$ for admissible pseudodifferential cut-offs

In this section we assume that $\operatorname{cone}\operatorname{supp} P$ is a T_+-admissible set (in the sense of Definition 1.3.26).

1. Integrated case. The following two theorems describe the nonzero singularities of $\operatorname{Tr}\mathbf{U}_{P,Q}(t)$. In the second theorem we assume simple reflection, though a similar result can be obtained in the same manner for the general case.

THEOREM 4.3.1. *Let* $\hat\gamma\in C_0^\infty(\mathbb{R})$, $\operatorname{supp}\hat\gamma\subset(0,T_+)$. *Assume that* $\operatorname{cone}\operatorname{supp} P$ *is a* T_+*-admissible set which contains no starting points of* T*-periodic billiard trajectories with periods* $T\in\operatorname{supp}\hat\gamma$. *Then*

$$\mathcal{F}^{-1}_{t\to\lambda}\Big[\hat\gamma(t)\,\operatorname{Tr}\mathbf{U}_{P,Q}(t)\Big] = O(\lambda^{-\infty}), \qquad \lambda\to+\infty.$$

PROOF. The results of 3.4.6 imply that $\mathbf{u}_{P,Q}(t;x,y)$ is a finite sum of standard oscillatory integrals associated with different legs of billiard trajectories originating from $\operatorname{cone}\operatorname{supp} P$. Now the theorem immediately follows from Proposition 4.1.12.

THEOREM 4.3.2. *Let* $\hat\gamma\in C_0^\infty(\mathbb{R})$, $\operatorname{supp}\hat\gamma\subset(0,T_+)$. *Assume that* $\operatorname{cone}\operatorname{supp} P$ *is a* T_+*-admissible set and that the simple reflection condition is fulfilled. Then*

$$\begin{aligned} &\mathcal{F}^{-1}_{t\to\lambda}\Big[\hat\gamma(t)\,\operatorname{Tr}\mathbf{U}_{P,Q}(t)\Big] \\ &\qquad = \lambda^{n+\mathbf{l}+\mathbf{m}-1}\int_{\Pi^a}\sum_{k\in\mathbb{N}} e^{ik(\lambda\mathbf{T}+\mathbf{q})}\,\hat\gamma(k\mathbf{T})\,P_{\mathbf{l}}(y,\tilde\eta)\,\overline{Q_{\mathbf{m}}(y,\tilde\eta)}\,dy\,đ\tilde\eta \\ &\qquad\quad + o(\lambda^{n+\mathbf{l}+\mathbf{m}-1}), \qquad \lambda\to+\infty, \end{aligned} \tag{4.3.1}$$

where $\mathbf{T} = \mathbf{T}(y, \widetilde{\eta})$ *is the minimal positive period and* $\mathbf{q} = \mathbf{q}(y, \widetilde{\eta})$ *is the total phase shift along the primitive closed trajectory originating from* $(y, \widetilde{\eta})$ *(see Definition 1.7.3).*

Note that the summation in (4.3.1) is in fact carried out over a finite set because $\mathbf{T}(y,\eta) \geqslant T_\delta > 0$ for all $(y,\eta) \in \operatorname{cone\,supp} P$. Therefore, in view of Lemmas 1.3.23, 1.3.24 and 1.7.5, the right-hand side of (4.3.1) is well defined.

PROOF. The results of 3.4.6 imply that $\mathbf{u}_{P,Q}(t;x,y)$ is a finite sum of standard oscillatory integrals associated with different legs of our Hamiltonian billiards. The corresponding principal symbols are

$$\overline{Q_{\mathbf{m}}\left(x^*(t;y,\eta), \xi^*(t;y,\eta)\right)}\, e^{i(\mathfrak{f}_r(t;y,\eta)+\mathfrak{f}_s(t;y,\eta))}\, P_{\mathrm{l}}(y,\eta)\,.$$

Here $\mathfrak{f}_r(t;y,\eta)$ and $\mathfrak{f}_s(t;y,\eta)$ are the phase shifts along the billiard trajectory $(x^*(s;y,\eta), \xi^*(s;y,\eta))$, $0 \leqslant s \leqslant t$, generated by the reflections and subprincipal symbol of A, respectively.

Let us assume (without loss of generality; see 3.4.2) that $\operatorname{cone\,supp} P$ is sufficiently small and choose a covering of $[0, T_+]$ by sufficiently small intervals (t_j, t_j'), $j = 0, 1, \dots, p$, such that

1. $t_j' - t_j < T_\delta/2$ for all j.
2. $t_{j-1} < t_j$, $t_{j-1}' < t_j'$, and $[t_{j-1}, t_{j-1}'] \cap [t_{j+1}, t_{j+1}'] = \varnothing$.
3. If $x^*(t;y,\eta) \notin \partial M$ for all $t \in [t_j, t_j']$ and $(y,\eta) \in \operatorname{cone\,supp} P$ then there exist associated phase function $\varphi_j(t,x;y,\eta)$ which is real for all $(y,\eta) \in \operatorname{cone\,supp} P$, $t \in [t_j, t_j']$, and x close to $x^*(t;y,\eta)$.

Then we can split the function $\hat{\gamma}$ into a finite sum of C_0^∞-functions $\hat{\gamma}_j$ with small supports such that $\operatorname{supp} \hat{\gamma}_j \subset (t_j, t_j')$ and either

$$(x^*(t;y,\eta), \xi^*(t;y,\eta)) \notin \operatorname{cone\,supp} Q \tag{4.3.2}$$

for all $(y,\eta) \in \operatorname{cone\,supp} P$, $t \in \operatorname{supp} \hat{\gamma}_j$, or

$$x^*(t;y,\eta) \notin \partial M\,, \qquad \forall (y,\eta) \in \operatorname{cone\,supp} P\,, \qquad \forall t \in \operatorname{supp} \hat{\gamma}_j\,. \tag{4.3.3}$$

If (4.3.2) is fulfilled, then $\hat{\gamma}_j\, \mathbf{u}_{P,Q}$ is a smooth function and its Fourier transform is rapidly decreasing. Otherwise (4.3.3) implies that $\hat{\gamma}_j\, \mathbf{u}_{P,Q}$ is represented by one standard oscillatory integral. Now

- if

$$\Pi_T \cap \operatorname{cone\,supp} P = \varnothing \quad \text{for all } T \in \operatorname{supp} \hat{\gamma}_j\,,$$

then by Theorem 4.3.1 $\mathcal{F}^{-1}_{t\to\lambda}\left[\hat{\gamma}_j \operatorname{Tr} \mathbf{U}_{P,Q}\right]$ is a rapidly decreasing function;

- if

$$\Pi_T \cap \operatorname{cone\,supp} P \neq \varnothing \quad \text{for some } T \in \operatorname{supp} \hat{\gamma}_j\,, \tag{4.3.4}$$

then, since $t_j' - t_j < T_\delta/2$, for each $(y,\eta) \in \Pi \cap \operatorname{cone\,supp} P$ there exists just one period $T_j^\star(y,\eta) \in \operatorname{supp} \hat{\gamma}_j$, and by Proposition 4.1.10

$$\begin{aligned} &\mathcal{F}^{-1}_{t\to\lambda}\left[\hat{\gamma}_j(t) \operatorname{Tr} \mathbf{U}_{P,Q}(t)\right] \\ &\quad = \lambda^{n+\mathrm{l}+\mathbf{m}-1} \int_{\Pi^a} e^{i\lambda T_j^\star + \mathfrak{f}(T_j^\star; y, \widetilde{\eta})}\, \hat{\gamma}_j(T_j^\star)\, P_{\mathrm{l}}(y,\widetilde{\eta})\, \overline{Q_{\mathbf{m}}(y,\widetilde{\eta})}\, dy\, đ\widetilde{\eta} \\ &\qquad + o(\lambda^{n+\mathrm{l}+\mathbf{m}-1})\,, \qquad \lambda \to +\infty\,. \end{aligned} \tag{4.3.5}$$

For each j satisfying (4.3.4) and $(y,\eta) \in \Pi_{T_j^\star(y,\eta)} \cap \operatorname{cone\,supp} P$ there exists a unique k such that

$$T_j^\star(y,\eta) \;=\; t_\star^{(j)} \;=\; k\,\mathbf{T}(y,\eta)\,,$$

and then $\mathfrak{f}\left(T_j^\star(y,\eta),y,\eta\right) = k\,\mathbf{q}(y,\eta)$. Since k is uniquely determined by j and (y,η), and $T_j^\star(y,\eta) \in \operatorname{supp}\hat\gamma_i \cap \operatorname{supp}\hat\gamma_j$ implies $T_i^\star(y,\eta) = T_j^\star(y,\eta) = k\,\mathbf{T}(y,\eta)$, we obtain

$$\sum_j e^{i\lambda T_j^\star + \mathfrak{f}(T_j^\star;y,\eta)}\,\hat\gamma_j(T_j^\star) \;=\; \sum_{k\in\mathbb{N}} e^{ik(\lambda\mathbf{T}+\mathbf{q})}\,\hat\gamma(k\mathbf{T})\,,$$

where the sum is taken over all the integers j satisfying (4.3.4). Summing up (4.3.5) over j we obtain (4.3.1). □

2. Pointwise case. Throughout this subsection we assume simple reflection and use the notation introduced in Section 1.8.

Let

$$\Pi_y^k \;=\; \{\tilde\eta \in \Pi_y^a \,:\, \Phi_y^j\tilde\eta \in \Pi_y^a\,,\; j=0,1,\dots,k-1\}\,.$$

Then $\Pi_y^a = \Pi_y^1 \supset \Pi_y^2 \supset \Pi_y^3 \supset \cdots$. If $\tilde\eta \in \Pi_y^k$, we define

$$T_y^{(k)}(\tilde\eta) \;=\; \sum_{j=0}^{k-1} \mathbf{T}_y(\Phi_y^j\tilde\eta)\,.$$

Clearly, $T_y^{(1)} = \mathbf{T}_y$ and $T_y^{(k)}$ is the value of t when the trajectory returns to the point y for the kth time.

LEMMA 4.3.3. *For any measurable set*

$$\Omega \;\subset\; \Pi_y^k \cap \{\eta \in S_y^*M \cap \operatorname{cone\,supp} P \,:\, T_y^{(k)} \leqslant T_+\}$$

and all $j = 1,\dots,k-1$ the sets $\Phi_y^j\Omega$ are also measurable and

$$\operatorname{meas}_y\left(\Phi_y^j\Omega\right) \;=\; \int_\Omega |\det \xi_\eta^*(T_y^{(j)}(\tilde\eta);y,\tilde\eta)\,|\,d\tilde\eta\,. \tag{4.3.6}$$

PROOF. Without loss of generality we can assume that

$$\Omega \;\subset\; \{\tilde\eta \in S_y^*M \,:\, |T - \mathbf{T}_y^{(j)}(\tilde\eta)| \leqslant \varepsilon\}$$

where $T \in (0,T_+]$ and ε is a small positive number, and that $\operatorname{cone\,supp} P$ lies in a small conic neighbourhood of a point $(y,\tilde\eta_0)$, $\tilde\eta_0 \in \Omega$.

Let us choose a coordinate system x in a neighbourhood of $x^*(T;y,\tilde\eta_0)$ such that $\det \xi_\eta^*(T;y,\tilde\eta_0) \neq 0$. Then by Lemma 2.4.13 there exists a nondegenerate phase function φ which is equal to $\langle x - x^*, \xi^*\rangle$ for (y,η) close to $\operatorname{cone\,supp} P$, $|t - T| \leqslant \varepsilon$, and x close to $x^*(T;y,\tilde\eta_0)$. Let $\tilde t_\star$ be the local t-solution of the equation $\langle x - x^*, \xi^*\rangle|_{x=y} = 0$. By Lemma 4.1.14, $\tilde t_\star$ coincides with $\mathbf{T}_y^{(j)}$ on Π_y^j.

Consider the map $S_y^*M \ni \eta \to \xi^*(\tilde t_\star;y,\eta)$ and denote by $J(\eta)$ its Jacobian. In view of Lemma 4.1.14, $J(\tilde\eta_0) = |\det \xi_\eta^*(\tilde t_\star;y,\tilde\eta_0)| \neq 0$, so this map is an isomorphism in a conical neighbourhood of $\tilde\eta_0$ and $d\xi = J(\eta)\,d\eta$. This implies (4.3.6) for our special choice of coordinates (if the coordinates y are the same as x). It remains to note that, since $x_\eta^*(t_\star;y,\tilde\eta_0)\big|_{\Pi_y^k} = 0$, the restriction of $|\det \xi_\eta^*(\tilde t_\star;y,\tilde\eta)|$ to Π_y^k does not depend on the choice of coordinates (assuming that the coordinates $\{x_j\}$ and $\{y_j\}$ are the same). □

Lemma 4.3.3 shows that the restriction of $|\det \xi^*_\eta(T_y^{(j)}; y, \widetilde{\eta})|$ to

$$\Pi_y^k \cap \{\eta \in S_y^* M \cap \operatorname{cone\,supp} P \,:\, T_y^{(k)} \leqslant T_+\} \tag{4.3.7}$$

is the Radon–Nikodym derivative of the measure $\operatorname{meas}_y(\Phi_y^j \cdot)$ with respect to meas_y. This fact and the obvious equalities

$$\xi^*(T_y^{(j)}; y, \widetilde{\eta}) = \xi^*(\mathbf{T}_y(\Phi_y^{j-1}\widetilde{\eta}); y, \Phi_y^{j-1}\widetilde{\eta}), \qquad i = 1, \dots, k,$$

imply that

$$|\det \xi^*_\eta(T_y^{(k)}; y, \widetilde{\eta})| = \prod_{j=0}^{k-1} |\det \xi^*_\eta\left(\mathbf{T}_y(\Phi_y^j\widetilde{\eta}); y, \Phi_y^j\widetilde{\eta}\right)| \tag{4.3.8}$$

almost everywhere on (4.3.7).

In view of (4.3.8), for an arbitrary function f defined on S_y^*M we have

$$e^{-i\mathfrak{f}(T_y^{(k)}; y, \widetilde{\eta})} |\det \xi^*_\eta(T_y^{(k)}; y, \widetilde{\eta})|^{1/2} f(x^*, \xi^*)\big|_{t=T_y^{(k)}} = U_y^k f \tag{4.3.9}$$

and

$$e^{-i\left(\lambda T_y^{(k)} + \mathfrak{f}(T_y^{(k)}; y, \widetilde{\eta})\right)} |\det \xi^*_\eta(T_y^{(k)}; y, \widetilde{\eta})|^{1/2} f(x^*, \xi^*)\big|_{t=T_y^{(k)}} = U_{y,\lambda}^k f \tag{4.3.10}$$

almost everywhere on (4.3.7), where $\mathfrak{f}(T_y^{(k)}; y, \widetilde{\eta})$ is the full phase shift.

THEOREM 4.3.4. *Under the conditions of Theorem 4.3.2, for any* $y \in \mathring{M}$,

$$\begin{aligned} &\mathcal{F}^{-1}_{t\to\lambda}\left[\hat{\gamma}(t)\, \mathbf{u}_{P,Q}(t, y, y)\right] \\ &\quad = \lambda^{n+\mathbf{l}+\mathbf{m}-1} \int_{\Pi_y^a} P_{\mathbf{l}} \sum_{k\in\mathbb{N}} \gamma(T_y^{(k)})\, \overline{(U_{y,\lambda}^k Q_{\mathbf{m}})}\, d\!\!{}^-\!\widetilde{\eta} \\ &\qquad + o(\lambda^{n+\mathbf{l}+\mathbf{m}-1}), \quad \lambda \to +\infty. \end{aligned} \tag{4.3.11}$$

PROOF. Applying the same procedure as in the proof of Theorem 4.3.2 we obtain that

$$\mathcal{F}^{-1}_{t\to\lambda}\left[\hat{\gamma}(t)\, \mathbf{u}_{P,Q}(t, y, y)\right] = \sum_j \mathcal{F}^{-1}_{t\to\lambda}\left[\hat{\gamma}_j(t)\, \mathbf{u}_{P,Q}(t, y, y)\right].$$

If $\Pi_{y,T} \cap \operatorname{cone\,supp} P = \varnothing$ for all $T \in \operatorname{supp} \hat{\gamma}_j$, then $\hat{\gamma}_j(t)\, \mathbf{u}_{P,Q}(t, y, y)$ is infinitely smooth in t, so the corresponding term on the right-hand side is rapidly decreasing. Otherwise, by Proposition 4.1.16,

$$\begin{aligned} &\mathcal{F}^{-1}_{t\to\lambda}\left[\hat{\gamma}_j(t)\, \mathbf{u}_{P,Q}(t, y, y)\right] \\ &\quad = \lambda^{n+\mathbf{l}+\mathbf{m}-1} \int_{\Pi_y^k} e^{i\lambda T_y^{(k)} + \mathfrak{f}(T_y^{(k)}; y, \widetilde{\eta})} P_{\mathbf{l}}(y, \widetilde{\eta}) |\det \xi^*_\eta(T_y^{(k)}; y, \widetilde{\eta})|^{1/2} \\ &\qquad \times \hat{\gamma}_j(T_y^{(k)})\, \overline{Q_{\mathbf{m}}(x^*, \xi^*)}\Big|_{t=T_y^{(k)}} d\!\!{}^-\!\widetilde{\eta} + o(\lambda^{n+\mathbf{l}+\mathbf{m}-1}), \end{aligned} \tag{4.3.12}$$

for some value of k which is determined by j. Summing up (4.3.12) over j we get

$$\mathcal{F}^{-1}_{t\to\lambda}\Big[\hat\gamma(t)\,\mathbf{u}_{P,Q}(t,y,y)\Big] = \lambda^{n+\mathfrak{l}+\mathbf{m}-1}\int_{\Pi^a_y} P_{\mathfrak{l}}(y,\widetilde\eta)$$

$$\times\sum_{k\in\mathbb{N}}\left(\hat\gamma(T^{(k)}_y)\,e^{i\lambda T^{(k)}_y+\mathfrak{f}(T^{(k)}_y;y,\widetilde\eta)}\,|\det\xi^*_\eta(T^{(k)}_y;y,\eta)|^{1/2}\,\overline{Q_{\mathbf{m}}(x^*,\xi^*)}\Big|_{t=T^{(k)}_y}\right)đ\widetilde\eta$$

modulo $o(\lambda^{n+\mathfrak{l}+\mathbf{m}-1})$. Now (4.3.11) follows from (4.3.10). □

Similarly, Proposition 4.1.18 implies the following

THEOREM 4.3.5. *Let the conditions of Theorem* 4.3.2 *be fulfilled, and let* K *be a compact subset of* $\overset{\circ}{M}$. *If*

$$\operatorname{meas}_y\{\widetilde\eta\in\operatorname{supp}P_{\mathfrak{l}}|_{S^*_yM}\;:\;\operatorname{dist}(y,x^*(t;y,\widetilde\eta))<\varepsilon\}\to 0\,,\qquad \varepsilon\to 0\,,$$

uniformly with respect to $t\in\operatorname{supp}\hat\gamma$ *and* $y\in K$, *then*

$$\mathcal{F}^{-1}_{t\to\lambda}\Big[\hat\gamma(t)\,\mathbf{u}_{P,Q}(t,y,y)\Big] = o(\lambda^{n+\mathfrak{l}+\mathbf{m}-1})\,,\qquad \lambda\to+\infty\,,$$

uniformly with respect to $y\in K$.

4.4. Singularities of the wave group at $t \neq 0$ for nonadmissible pseudodifferential cut-offs

In this section $\operatorname{cone\,supp}P$ is not necessarily T_+-admissible.

1. Integrated case. The aim of this subsection is to prove the following analogues of Theorems 4.3.1, 4.3.2.

THEOREM 4.4.1. *Let* $\hat\gamma\in C_0^\infty(\mathbb{R})$, $\operatorname{supp}\hat\gamma\subset(0,+\infty)$. *If the nonperiodicity and nonblocking conditions are fulfilled, then*

$$\mathcal{F}^{-1}_{t\to\lambda}\Big[\hat\gamma(t)\;\operatorname{Tr}\mathbf{U}_{P,P}(t)\Big] = o(\lambda^{n+2\mathfrak{l}-1})\,,\qquad \lambda\to+\infty\,. \tag{4.4.1}$$

THEOREM 4.4.2. *Let* $\hat\gamma\in C_0^\infty(\mathbb{R})$, $\operatorname{supp}\hat\gamma\subset(0,+\infty)$. *If the simple reflection condition is fulfilled, then*

$$\begin{aligned}\mathcal{F}^{-1}_{t\to\lambda}&\Big[\hat\gamma(t)\;\operatorname{Tr}\mathbf{U}_{P,P}(t)\Big]\\ &= \lambda^{n+2\mathfrak{l}-1}\int_{\Pi^a}\sum_{k\in\mathbb{N}} e^{ik(\lambda\mathbf{T}+\mathbf{q})}\,\hat\gamma(k\mathbf{T})\,|P_{\mathfrak{l}}(y,\widetilde\eta)|^2\,dy\,đ\widetilde\eta\\ &\quad+o(\lambda^{n+2\mathfrak{l}-1})\,,\qquad\lambda\to+\infty\,.\end{aligned}\tag{4.4.2}$$

Here $\mathbf{T}=\mathbf{T}(y,\widetilde\eta)$ *is the minimal positive period and* $\mathbf{q}=\mathbf{q}(y,\widetilde\eta)$ *is the total phase shift along the primitive closed trajectory starting at* $(y,\widetilde\eta)$ (*see* 1.7.2).

The proof of Theorems 4.4.1, 4.4.2 given below is split into a series of auxiliary lemmas. The pseudodifferential operator P and function $\hat\gamma$ are assumed to be fixed, as well as a number $T_+>0$ such that $\operatorname{supp}\hat\gamma\subset(0,T_+)$.

In the notation of Section 1.3 put

$$K_1 = (S^*\overset{\circ}{M} \cap \operatorname{cone\,supp} P) \setminus O_{T_+}\,, \tag{4.4.3}$$

$$K_2 = K_1 \cup \Big(\Big(\bigcup_{0<T\leqslant T_+} \Pi_T\Big) \cap \operatorname{cone\,supp} P\Big)\,. \tag{4.4.4}$$

In other words, K_1 is the set of points in $S^*\overset{\circ}{M} \cap \operatorname{cone\,supp} P$ which are not T_+-admissible, and K_2 is the set of points in $S^*\overset{\circ}{M} \cap \operatorname{cone\,supp} P$ which are not T_+-admissible or which are starting points of T-periodic billiard trajectories with $T \in (0, T_+]$.

LEMMA 4.4.3. *Under the conditions of Theorem* 4.4.2 *the set* K_1 *is compact and* $\operatorname{meas} K_1 = 0$.

PROOF. Lemma 4.4.3 is an immediate consequence of Lemmas 1.3.27, 1.3.28 and 1.3.31. □

LEMMA 4.4.4. *Under the conditions of Theorem* 4.4.1 *the set* K_2 *is compact and* $\operatorname{meas} K_2 = 0$.

PROOF. Lemma 4.4.4 follows from Lemmas 1.3.27–1.3.29. □

We shall often use the following simple lemma which we will give without proof.

LEMMA 4.4.5. *Let* Ω_f, Ω_g *be open subsets of* $S^*\widehat{M}$ *such that* $\Omega_f \cup \Omega_g = S^*\widehat{M}$. *Then there exist real-valued functions* $f \in C_0^\infty(\Omega_f)$, $g \in C_0^\infty(\Omega_g)$ *such that*

$$f^2 + g^2 \equiv 1 \tag{4.4.5}$$

on $S^*\widehat{M}$.

In the above lemma $\widehat{M}$ is the extended manifold; see 1.1.2.

In the remainder of this subsection $K \subset S^*\overset{\circ}{M}$ is a compact set of measure zero.

Let $\varkappa$ be an arbitrary positive number. There exists an open set $\Omega_g \subset S^*\widehat{M}$ such that $K \subset \Omega_g \subset \overline{O}_g \subset S^*\overset{\circ}{M}$, and

$$\operatorname{meas} \Omega_g < (2\pi)^{-n} \varkappa\,. \tag{4.4.6}$$

We can also choose an open set $\Omega_f \subset S^*\widehat{M}$ such that

$$\Omega_f \cap K = \varnothing \tag{4.4.7}$$

and $\Omega_f \cup \Omega_g = S^*\widehat{M}$. We apply Lemma 4.4.5 and use the functions f, g from this lemma further on. Note that formulae (4.4.5), (4.4.6) imply

$$\int_{S^*M} g^2(y,\widetilde{\eta})\, dy\, đ\widetilde{\eta} < \varkappa\,, \tag{4.4.8}$$

whereas (4.4.7) implies

$$\operatorname{supp} f \cap K = \varnothing\,. \tag{4.4.9}$$

Let us extend f, g from $S^*\widehat{M}$ to $T'\widehat{M}$ as functions positively homogeneous in η of degree 0. We retain for the extended functions the original notation f, g.

LEMMA 4.4.6. *There exist pseudodifferential operators* $F, G \in \Psi^0$ *acting on* $\widehat{M}$ *with principal symbols* f, g, *respectively, such that* $\operatorname{cone\,supp} F = \operatorname{supp} f$, $\operatorname{cone\,supp} G = \operatorname{supp} g$, *and*

$$I - FF^* - GG^* \in \Psi^{-\infty}. \tag{4.4.10}$$

PROOF OF LEMMA 4.4.6. Let us write down the Schwartz kernels of the pseudodifferential operators F, G, I in our global invariant form using some global phase function $\varphi \in \mathfrak{F}^0$; see 2.7.5. Let $\mathfrak{a}(y,\eta)$, $\mathfrak{b}(y,\eta)$, $\mathfrak{c}(y,\eta)$ be the respective symbols of these Schwartz kernels, and let

$$\begin{aligned} \mathfrak{a}(y,\eta) &\sim f(y,\eta) + \mathfrak{a}_{-1}(y,\eta) + \mathfrak{a}_{-2}(y,\eta) + \cdots, \\ \mathfrak{b}(y,\eta) &\sim g(y,\eta) + \mathfrak{b}_{-1}(y,\eta) + \mathfrak{b}_{-2}(y,\eta) + \cdots, \\ \mathfrak{c}(y,\eta) &\sim 1 + \mathfrak{c}_{-1}(y,\eta) + \mathfrak{c}_{-2}(y,\eta) + \cdots \end{aligned}$$

be the asymptotic expansions of these symbols into series of positively homogeneous terms. The functions f, g, $\mathfrak{c}_{-1}, \mathfrak{c}_{-2}, \dots$ are given, whereas the functions $\mathfrak{a}_{-1}, \mathfrak{a}_{-2}, \dots$, $\mathfrak{b}_{-1}, \mathfrak{b}_{-2}, \dots$ are to be found.

Writing down (4.4.10) in terms of symbols we arrive at a recurrent system of the following form:

$$\begin{aligned} &2f \operatorname{Re} \mathfrak{a}_{-k} + 2g \operatorname{Re} \mathfrak{b}_{-k} \\ &\qquad = \mathfrak{c}_{-k} + \Phi_k(f, \mathfrak{a}_{-1}, \mathfrak{a}_{-2}, \dots, \mathfrak{a}_{1-k}) + \Phi_k(g, \mathfrak{b}_{-1}, \mathfrak{b}_{-2}, \dots, \mathfrak{b}_{1-k}), \end{aligned} \tag{4.4.11}$$

$k = 1, 2, \dots$, where $\Phi_k(\,\cdot\,)$ are some combinations of functions and their derivatives. Note that if we solve (4.4.11) recurrently, then at each step the right-hand side of (4.4.11) will be real because it can be viewed as the principal symbol of a self-adjoint pseudodifferential operator of order $-k$. A solution of (4.4.11) is $\mathfrak{a}_{-k} = f\mathfrak{h}_{-k}$, $\mathfrak{b}_{-k} = g\mathfrak{h}_{-k}$, where

$$\mathfrak{h}_{-k} = (\mathfrak{c}_{-k} + \Phi_k(f, \mathfrak{a}_{-1}, \mathfrak{a}_{-2}, \dots, \mathfrak{a}_{1-k}) + \Phi_k(g, \mathfrak{b}_{-1}, \mathfrak{b}_{-2}, \dots, \mathfrak{b}_{1-k}))/2.$$

Having determined recurrently the homogeneous terms of $\mathfrak{a}$, $\mathfrak{b}$ it remains only to apply Lemma 2.1.2. □

Denote $\mathbf{c} = \max_{S^*M} |P_1(y,\widetilde{\eta})|^2$. We now return from $\widehat{M}$ to our original manifold M and state the following two lemmas.

LEMMA 4.4.7. *Under the conditions of Theorem* 4.4.1 *for any positive* $\varkappa$ *there exist pseudodifferential operators* $P^{(1)}, P^{(2)} \in \Psi^1_0$ *satisfying*

$$\int_{S^*M} \left|P_1^{(2)}(y,\widetilde{\eta})\right|^2 dy\, d\!\!{}^-\widetilde{\eta} < \mathbf{c}\varkappa, \tag{4.4.12}$$

$$PP^* - P^{(1)}\left(P^{(1)}\right)^* - P^{(2)}\left(P^{(2)}\right)^* \in \Psi_0^{-\infty}, \tag{4.4.13}$$

such that $\operatorname{cone\,supp} P^{(1)}$ *is a* T_+*-admissible set which contains no starting points of* T*-periodic billiard trajectories with periods* $T \in (0, T_+]$.

LEMMA 4.4.8. *Under the conditions of Theorem* 4.4.2 *for any positive* $\varkappa$ *there exist pseudodifferential operators* $P^{(1)}, P^{(2)} \in \Psi^1_0$ *satisfying* (4.4.12), (4.4.13), *such that* $\operatorname{cone\,supp} P^{(1)}$ *is a* T_+*-admissible set.*

PROOF OF LEMMAS 4.4.7 AND 4.4.8. In the case of Lemma 4.4.7 set $K = K_2$ (see (4.4.4)), and in the case of Lemma 4.4.8 set $K = K_1$ (see (4.4.3)); this is justified by virtue of Lemmas 4.4.3, 4.4.4. Let F, G be the pseudodifferential operators from Lemma 4.4.6 corresponding to this particular choice of the compact set K, and let $\chi \in C_0^\infty(\mathring{M})$ be a function such that $\chi(x) = 1$ for $x_n \geqslant \delta$ (cf. (4.0.4)). Take $P^{(1)} = PF\chi$, $P^{(2)} = PG\chi$. Then

1. (4.4.12) follows from (4.4.8).
2. (4.4.13) follows from (4.4.10).
3. The required properties of $\operatorname{cone\,supp} P^{(1)}$ follow from (4.4.3), (4.4.4), and (4.4.9).

□

We are now prepared to proceed to the

PROOF OF THEOREMS 4.4.1 AND 4.4.2. Let $P^{(1)}$, $P^{(2)}$ be the pseudodifferential operators from Lemma 4.4.7 or 4.4.8, depending on whether we are proving Theorem 4.4.1 or 4.4.2. Formula (4.4.13) implies

$$\begin{aligned}\mathcal{F}^{-1}_{t\to\lambda}\Big[\hat\gamma(t)\operatorname{Tr}\mathbf{U}_{P,P}(t)\Big] &= \mathcal{F}^{-1}_{t\to\lambda}\Big[\hat\gamma(t)\operatorname{Tr}\mathbf{U}_{P^{(1)},P^{(1)}}(t)\Big] \\ &\quad + \mathcal{F}^{-1}_{t\to\lambda}\Big[\hat\gamma(t)\operatorname{Tr}\mathbf{U}_{P^{(2)},P^{(2)}}(t)\Big] + O(\lambda^{-\infty})\end{aligned} \tag{4.4.14}$$

as $\lambda \to +\infty$. Applying Theorem 4.2.4 with $P = Q = P^{(2)}$ we get

$$\begin{aligned}&\mathcal{F}^{-1}_{t\to\lambda}\Big[\hat\rho_{T_\delta}(t)\operatorname{Tr}\mathbf{U}_{P^{(2)},P^{(2)}}(t)\Big] \\ &\qquad = \lambda^{n+2l-1}\int_{S^*M}\big|P_1^{(2)}(y,\tilde\eta)\big|^2\,dy\,d\tilde\eta + O(\lambda^{n+2l-2}), \qquad \lambda\to+\infty.\end{aligned} \tag{4.4.15}$$

Let us choose the function $\hat\rho_1$ so that it is even and $\rho_1(\lambda) > 0$, $\forall\lambda\in\mathbb{R}$. Then we can apply Theorem B.3.1 to the nondecreasing function

$$N_{P^{(2)}}(\lambda) \;=\; \int_0^\lambda \mathcal{F}^{-1}_{t\to\mu}\Big[\operatorname{Tr}\mathbf{U}_{P^{(2)},P^{(2)}}(t)\Big]\,d\mu \;=\; \operatorname{Tr}\big((P^{(2)})^* E_\lambda P^{(2)}\big)\,,$$

which with account of (4.4.15), (4.4.12) produces the estimate

$$\limsup_{\lambda\to+\infty}\frac{\Big|\mathcal{F}^{-1}_{t\to\lambda}\Big[\hat\gamma(t)\operatorname{Tr}\mathbf{U}_{P^{(2)},P^{(2)}}(t)\Big]\Big|}{\lambda^{n+2l-1}} \leqslant \mathbf{C}\varkappa, \tag{4.4.16}$$

where $\mathbf{C} = \mathbf{c}\,C_{\rho_{T_\delta},\gamma} \geqslant 0$ is a constant independent of $\varkappa$.

Suppose the conditions of Theorem 4.4.1 are fulfilled. Then by Theorem 4.3.1

$$\mathcal{F}^{-1}_{t\to\lambda}\Big[\hat\gamma(t)\operatorname{Tr}\mathbf{U}_{P^{(1)},P^{(1)}}(t)\Big] = O(\lambda^{-\infty}), \qquad \lambda\to+\infty. \tag{4.4.17}$$

Formulae (4.4.14), (4.4.16), (4.4.17) and the fact that $\varkappa$ can be chosen to be arbitrarily small imply (4.4.1). Theorem 4.4.1 has been proved.

Suppose the conditions of Theorem 4.4.2 are fulfilled. Then by Theorem 4.3.2

$$\begin{aligned} &\mathcal{F}^{-1}_{t\to\lambda}\Big[\hat\gamma(t)\ \operatorname{Tr}\mathbf{U}_{P^{(1)},P^{(1)}}(t)\Big] \\ &\quad = \lambda^{n+2\mathrm{l}-1}\int_{\Pi^a}\sum_{k\in\mathbb{N}} e^{ik(\lambda\mathbf{T}+\mathbf{q})}\,\hat\gamma(k\mathbf{T})\,|P_1^{(1)}(y,\widetilde\eta)|^2\,dy\,đ\widetilde\eta \\ &\qquad + o(\lambda^{n+2\mathrm{l}-1}), \qquad \lambda\to+\infty. \end{aligned} \tag{4.4.18}$$

For the sake of brevity let us denote the integrals on the right-hand sides of (4.4.2), (4.4.18) by $\psi(\lambda)$, $\psi_\varkappa(\lambda)$, respectively. In view of formulae (4.4.13), (4.4.12) and Lemma 1.7.4 we have

$$|\psi(\lambda)-\psi_\varkappa(\lambda)| \leqslant \mathbf{C}'\varkappa, \tag{4.4.19}$$

where

$$\mathbf{C}' = \frac{\mathbf{c}\,T_+\,\max_{\mathbb{R}}|\hat\gamma(t)|}{\inf_\Pi \mathbf{T}(y,\widetilde\eta)} \geqslant 0$$

is a constant independent of λ and $\varkappa$. Formula (4.4.19) means that the function $\psi_\varkappa(\lambda)$ converges to the function $\psi(\lambda)$ uniformly over $\lambda\in\mathbb{R}$ as $\varkappa\to+0$. This fact together with formulae (4.4.14), (4.4.16), (4.4.18) and the fact that $\varkappa$ can be chosen to be arbitrarily small imply (4.4.2). Theorem 4.4.2 has been proved. □

2. Pointwise case. When dealing with the pointwise case we have to overcome the same difficulty as with the integrated case, namely, to estimate the contribution to our asymptotics of certain "bad" points $(y,\widetilde\eta)$. In the previous subsection this was done by means of a pseudodifferential partition (4.4.13), with $\operatorname{cone\,supp}P^{(2)}$ being small and containing all the "bad" points. This approach does not work in the pointwise case because we are not able to make cyclic permutations of operators. So in this subsection we shall use a partition of the type

$$P = P^{(1)} + P^{(2)} \tag{4.4.20}$$

instead of (4.4.13). The technique described below can also be used in the integrated case.

In this subsection we prove the following analogues of Theorems 4.3.5, 4.3.4.

THEOREM 4.4.9. *Assume that the simple reflection condition is fulfilled. Let $K\subset\overset{\circ}{M}$ be a compact set all the points of which are regular and nonfocal. Let $P,Q\in\Psi^{\mathrm{l}}_0$, and let $\hat\gamma\in C_0^\infty(\mathbb{R})$, $\operatorname{supp}\hat\gamma\subset(0,+\infty)$. Then*

$$\mathcal{F}^{-1}_{t\to\lambda}\Big[\hat\gamma(t)\,\mathbf{u}_{P,Q}(t,y,y)\Big] = o(\lambda^{n+2\mathrm{l}-1}), \qquad \lambda\to+\infty, \tag{4.4.21}$$

uniformly on K.

THEOREM 4.4.10. *Assume that the simple reflection condition is fulfilled. Let $y\in\overset{\circ}{M}$ be a regular point, $P,Q\in\Psi^{\mathrm{l}}_0$, and let $\hat\gamma\in C_0^\infty(\mathbb{R})$, $\operatorname{supp}\hat\gamma\subset(0,+\infty)$. Then*

$$\begin{aligned} &\mathcal{F}^{-1}_{t\to\lambda}\Big[\hat\gamma(t)\,\mathbf{u}_{P,Q}(t,y,y)\Big] \\ &\quad = \lambda^{n+2\mathrm{l}-1}\int_{\Pi^a_y} P_1\sum_{k\in\mathbb{N}}\gamma(T_y^{(k)})\,\overline{(U^k_{y,\lambda}Q_1)}\,đ\widetilde\eta \\ &\qquad + o(\lambda^{n+2\mathrm{l}-1}), \qquad \lambda\to+\infty. \end{aligned} \tag{4.4.22}$$

Let us fix an arbitrary $T_+ > 0$ such that $\operatorname{supp}\widehat{\gamma} \subset (0, T_+)$, and let F be a continuous function on $S^*\overset{\circ}{M}$ such that $F(y, \widetilde{\eta}) = 0$ if and only if $(y, \widetilde{\eta})$ is not a T_+-admissible point. For an arbitrary compact set $K \subset \overset{\circ}{M}$ we define the function F_K on $S^*\overset{\circ}{M}\big|_K$ by

$$F_K(y, \widetilde{\eta}) = \begin{cases} F(y, \widetilde{\eta}) \inf_{C_K \leqslant t \leqslant T_+} \operatorname{dist}(y, x^*(t; y, \widetilde{\eta})), & \text{if } (y, \widetilde{\eta}) \text{ is } T_+\text{-admissible}, \\ 0, & \text{otherwise}, \end{cases}$$

where the distance is taken in an arbitrary smooth Riemannian metric and C_K is the positive constant introduced in Lemma 1.8.10. Obviously, the function F_K is continuous on $S^*M|_K$, and $F_K(y, \widetilde{\eta}) = 0$ if and only if the point $(y, \widetilde{\eta})$ is not T_+-admissible or $\widetilde{\eta} \in \Pi_T$ for some $T \in (0, T_+]$.

LEMMA 4.4.11. *If all the points $y \in K$ are regular and nonfocal, then*

$$\operatorname{meas}_y\{\widetilde{\eta} \in S_y^* M \,:\, F(y, \widetilde{\eta}) \leqslant \varepsilon\} \;\to\; 0 \quad \text{as } \varepsilon \to 0$$

uniformly on K.

PROOF. If the statement of the lemma is not true, then there exist sequences $y_{(k)} \to y_0 \in K$ and $\varepsilon_k \to 0$ such that

$$\operatorname{meas}_{y_{(k)}}\{\widetilde{\eta} \in S^*_{y_{(k)}} M : F(y_{(k)}, \widetilde{\eta}) \leqslant \varepsilon_k\} \geqslant \text{const} > 0. \tag{4.4.23}$$

Let us choose some coordinates in a neighbourhood of y_0 and define the sets

$$\Omega_j = \bigcup_{k \geqslant j} \{\widetilde{\eta} \in S^*_{y_0} M \,:\, F\left(y_{(k)}, \widetilde{\eta}/h(y_{(k)}, \widetilde{\eta})\right) \leqslant \varepsilon_k\}.$$

Since F is continuous, we have $\bigcap_{j \geqslant 1} \Omega_j \subset \{\widetilde{\eta} \in S^*_{y_0} M \,:\, F(y_0, \widetilde{\eta}) = 0\}$. On the other hand, (4.4.23) implies that $\operatorname{meas}_{y_0} \bigcap_{j \geqslant 1} \Omega_j > 0$. This contradicts the fact that y_0 is a regular point. □

We shall also need the following

LEMMA 4.4.12. *Let K be a compact subset of $\overset{\circ}{M}$. Then for any $\hat{\gamma} \in C_0^\infty(\mathbb{R})$ and any $\delta > 0$ there exists a constant $\mathbf{c}_{\delta,\gamma} \geqslant 0$ such that for any pseudodifferential operators $P, Q \in \Psi_0^1$ satisfying (4.0.4) and*

$$\max_{y \in K} \int_{S_y^* M} |P_1(y, \widetilde{\eta})|^2 \, d\!\!\!{}^-\widetilde{\eta} \;\leqslant\; \varkappa_1^2, \quad \max_{y \in K} \int_{S_y^* M} |Q_1(y, \widetilde{\eta})|^2 \, d\!\!\!{}^-\widetilde{\eta} \;\leqslant\; \varkappa_2^2, \tag{4.4.24}$$

we have

$$\begin{aligned} |\,\mathcal{F}^{-1}_{t \to \lambda}\left[\hat{\gamma}(t)\, \mathbf{u}_{P,Q}(t, y, y)\right]| \;&\leqslant\; 2\,\varkappa_1\,(\varkappa_2^2 + 1)\, \mathbf{c}_{\delta,\gamma}\, \lambda^{n+2l-1} \;+\; C\,\lambda^{n+2l-2}, \\ |\,\mathcal{F}^{-1}_{t \to \lambda}\left[\hat{\gamma}(t)\, \mathbf{u}_{P,Q}(t, y, y)\right]| \;&\leqslant\; 2\,\varkappa_2\,(\varkappa_1^2 + 1)\, \mathbf{c}_{\delta,\gamma}\, \lambda^{n+2l-1} \;+\; C\,\lambda^{n+2l-2} \end{aligned}$$

uniformly over $y \in K$ and $\lambda > 1$, with some constant $C > 0$ depending on $\hat{\gamma}$, P and Q.

PROOF. We shall prove only the first inequality; the second can be obtained in the same manner.

Since $\mathbf{u}_{P,Q} = \mathbf{u}_{\varkappa_1^{-1/2}P,\varkappa_1^{1/2}Q}$, applying polarization (see 1.8.1) we obtain

$$\mathbf{u}_{P,Q}(\lambda,y,y) = \frac{1}{4}\,\mathbf{u}_{R^{(1)},R^{(1)}}(\lambda,y,y) - \frac{1}{4}\,\mathbf{u}_{R^{(2)},R^{(2)}}(\lambda,y,y) + \frac{i}{4}\,\mathbf{u}_{R^{(3)},R^{(3)}}(\lambda,y,y) - \frac{i}{4}\,\mathbf{u}_{R^{(4)},R^{(4)}}(\lambda,y,y),$$

where $R^{(1)} = \varkappa_1^{-1/2}P+\varkappa_1^{1/2}Q$, $R^{(2)} = \varkappa_1^{-1/2}P-\varkappa_1^{1/2}Q$, $R^{(3)} = \varkappa_1^{-1/2}P+i\varkappa_1^{1/2}Q$ and $R^{(4)} = \varkappa_1^{-1/2}P - i\varkappa_1^{1/2}Q$. Theorem 4.2.1 and formula (4.4.24) imply that

$$|\,\mathcal{F}^{-1}_{t\to\lambda}\left[\hat\rho_{T_\delta}(t)\,\mathbf{u}_{R^{(j)},R^{(j)}}(t,y,y)\right]\,| \leqslant 2\,\varkappa_1\,(\varkappa_2^2+1)\,\lambda^{n+2l-1} + C\,\lambda^{n+2l-2},$$

where C is a constant depending on ρ_{T_δ}, P and Q. Applying Theorem B.3.1 to the nondecreasing functions

$$N_j(\lambda) = \int_0^\lambda \mathcal{F}^{-1}_{t\to\mu}[\mathbf{u}_{R^{(j)},R^{(j)}}(t,y,y)]\,d\mu = \sum_{\lambda_k<\lambda} \left|(R^{(j)}v_k)(y)\right|^2$$

(where v_k are the eigenfunctions of $\mathcal{A}$) we see that the required estimate is valid for each $\mathbf{u}_{R^{(j)},R^{(j)}}$, $j=1,2,3,4$. So it holds for $\mathbf{u}_{P,Q}$.

PROOF OF THEOREM 4.4.9. In view of Lemma 4.4.11, given $\varepsilon>0$ we can split P into the sum (4.4.20) of pseudodifferential operators $P^{(1)},P^{(2)}\in\Psi^l_0$ such that

1. $\operatorname{cone\,supp} P^{(1)} \cap S^*M|_K \subset \{(y,\widetilde\eta) : F_K(y,\widetilde\eta) > \varepsilon\}$.
2. $\operatorname{cone\,supp} P^{(2)} \cap S^*M|_K \subset \{(y,\widetilde\eta) : F_K(y,\widetilde\eta) < 2\varepsilon\}$.
3. $\sup_{y\in K}\int_{S^*_yM}\left|P_1^{(1)}(y,\widetilde\eta)\right|^2 d\widetilde\eta \leqslant C$ and $\sup_{y\in K}\int_{S^*_yM}\left|P_1^{(2)}(y,\widetilde\eta)\right|^2 d\widetilde\eta \leqslant C_\varepsilon$, where C is a positive constant independent of ε, and $C_\varepsilon\to 0$ as $\varepsilon\to 0$.

In view of Theorem 4.3.5, the contribution to (4.4.21) of $\mathbf{u}_{P^{(1)},Q}$ is $o(\lambda^{n+2l-1})$ uniformly with respect to y. By Lemma 4.4.12 the contribution of $\mathbf{u}_{P^{(2)},Q}$ is estimated uniformly with respect to y by

$$\widetilde C_\varepsilon\lambda^{n+2l-1} + \widetilde C'_\varepsilon\,\lambda^{n+2l-2},$$

where $\widetilde C_\varepsilon \to 0$ as $\varepsilon\to 0$. Since ε can be chosen arbitrarily small, this implies (4.4.22). □

PROOF OF THEOREM 4.4.10. The set of T_+-admissible "directions" $\widetilde\eta\in S^*_yM$ is open and, since y is a regular point, its complement is of measure zero. Therefore there exist pseudodifferential operators $P^{(1)},P^{(2)}\in\Psi^l_0$ such that (4.4.20) holds and

1. $\operatorname{cone\,supp} P^{(1)}$ is a T_+-admissible set.
2. $\sup_{y\in K}\int_{S^*_yM}\left|P_1^{(1)}(y,\widetilde\eta)\right|^2 d\widetilde\eta \leqslant C$ and $\sup_{y\in K}\int_{S^*_yM}\left|P_1^{(2)}(y,\widetilde\eta)\right|^2 d\widetilde\eta \leqslant C_\varepsilon$, where C is a positive constant independent of ε, and $C_\varepsilon\to 0$ as $\varepsilon\to 0$.

The rest of the proof repeats that of Theorem 4.4.9 (we take $K=\{y\}$ and apply Theorem 4.3.4 instead of Theorem 4.3.5).

4.5. Singularity of the wave group at $t=0$ when the source is close to the boundary

In this section we consider the case when the pseudodifferential operators P, Q are parameter-dependent of the type described in 2.10.6. Note that the situation considered below differs from the situation considered in Sections 4.2–4.4 because there the number $\delta>0$ appearing in the condition (4.0.4) was assumed to be *fixed*.

Let $\widehat{M}$, $\widehat{A}$, $\widehat{\mathcal{A}}$ be extensions of M, A, $\mathcal{A}$, respectively (see 1.1.2). Further on we use the notation

$$\widehat{\mathbf{U}}_{P,Q}(t) \;:=\; Q^*\exp(-it\widehat{\mathcal{A}}^{1/(2m)})\,P \;=\; \int_{M_y}\widehat{\mathfrak{u}}_{P,Q}(t,\widehat{x},\widehat{y})\,(\,\cdot\,)\,d\widehat{y}$$

(cf. (4.0.1)). When the operators P, Q depend on the parameter d we shall indicate this parameter as an additional independent variable of $\mathbf{U}_{P,Q}$, $\widehat{\mathbf{U}}_{P,Q}$. The Schwartz kernels of the pseudodifferential operators P, Q are assumed to be extended from $M\times M$ to $\widehat{M}\times\widehat{M}$ as identically zero, and we retain for the extended operators the original notation P, Q.

The constants T_{ε_1}, ε_1 below are the ones introduced in Section 2.10.

THEOREM 4.5.1. *If $P^\pm$ is a parameter-dependent pseudodifferential operator of the type $(1;1,\varepsilon_1)^\pm$, then*

$$\operatorname{Tr}\mathbf{U}_{P^\pm,P^\pm}(d,t)\;-\;\operatorname{Tr}\widehat{\mathbf{U}}_{P^\pm,P^\pm}(d,t)\;\in\;d^\infty((-T_{\varepsilon_1}d^{1-\varepsilon_1}/2,T_{\varepsilon_1}d^{1-\varepsilon_1}/2)). \tag{4.5.1}$$

The traces in formula (4.5.1) are viewed, of course, as distributions in t.

In formula (4.5.1) and below we understand the d^∞-notation as in Definition 2.10.3, but without dependence on x and y. Namely, $f(d,t)\in d^\infty(T_-(d),T_+(d))$ means that for any $r,s\in\mathbb{N}$ the function $d^{-r}D_t^s f$ is uniformly bounded over $d\in(0,d_0]$, $t\in(T_-(d),T_+(d))$. (Any given derivative of f with respect to t vanishes faster than any given power of d as $d\to0$.)

PROOF OF THEOREM 4.5.1. For definiteness let us prove the theorem in the $+$ case.

Theorem 3.5.3 (and its version for the case of the extended manifold) allows us to construct the Schwartz kernels of the operators

$$\exp(-it\mathcal{A}^{1/(2m)})\,P^+,\qquad \exp(-it\widehat{\mathcal{A}}^{1/(2m)})\,P^+ \tag{4.5.2}$$

in the form of parameter-dependent oscillatory integrals. However in the parameter-dependent situation we do not have an analogue of Lemma 2.7.16, that is, a result concerning the action of a parameter-dependent pseudodifferential operator on a parameter-dependent time-dependent Lagrangian distribution. (When we produced the list of properties of parameter-dependent time-dependent Lagrangian distributions in 2.10.5 we restricted it to the bare minimum.) Consequently, we cannot construct the operators (4.5.2) first, and then act on them with $(P^+)^*$. This technical hitch can be easily overcome by constructing the operators

$$\exp(-it\mathcal{A}^{1/(2m)})\,P^+\,(P^+)^*,\qquad \exp(-it\widehat{\mathcal{A}}^{1/(2m)})\,P^+\,(P^+)^* \tag{4.5.3}$$

instead, where we can use all the standard formulae for dealing with the composition of pseudodifferential operators $P^+\,(P^+)^*$ (see 2.10.6). The order of terms does not matter because we are interested in finding the trace; see also Remark 4.2.5.

Further on we denote for, brevity, $Q^+ := P^+(P^+)^*$. The operators (4.5.3) and their Schwartz kernels are denoted by $\mathbf{U}_{Q^+}(d,t)$, $\widehat{\mathbf{U}}_{Q^+}(d,t)$ and $\mathbf{u}_{Q^+}(d,t,x,y)$, $\widehat{\mathbf{u}}_{Q^+}(d,t,\widehat{x},\widehat{y})$, respectively.

According to Theorem 3.5.3 we have an effective (modulo d^∞) representation of the distribution $\mathbf{u}_{Q^+}(d,t,x,y)$. Unfortunately, this representation is valid only on the asymmetric time interval $(-T_{\varepsilon_1}d/2, T_{\varepsilon_1}d^{1-\varepsilon_1}/2)$, whereas in order to prove (4.5.1) we need information about $\mathbf{u}_{Q^+}(d,t,x,y)$ on the larger symmetric time interval $(-T_{\varepsilon_1}d^{1-\varepsilon_1}/2, T_{\varepsilon_1}d^{1-\varepsilon_1}/2)$. This is the basic difficulty with estimates near the boundary, because in the general case we cannot extend $\mathbf{u}_{Q^+}(d,t,x,y)$ to $t \simeq -d^{1-\varepsilon_1}$ without encountering reflections. Below we describe a way of overcoming this difficulty.

Let us start by constructing the distribution $\widehat{\mathbf{u}}_{Q^+}(d,t,\widehat{x},\widehat{y})$, which is simpler than constructing $\mathbf{u}_{Q^+}(d,t,x,y)$ as we have to work on a manifold without boundary. Repeating the arguments from Section 3.5 we obtain an effective (modulo d^∞) representation of the distribution $\widehat{\mathbf{u}}_{Q^+}(d,t,\widehat{x},\widehat{y})$ on the symmetric time interval $(-T_{\varepsilon_1}d^{1-\varepsilon_1}/2, T_{\varepsilon_1}d^{1-\varepsilon_1}/2)$. But for $t \in (-T_{\varepsilon_1}d/2, T_{\varepsilon_1}d^{1-\varepsilon_1}/2)$ the oscillatory integrals representing $\mathbf{u}_{Q^+}(d,t,x,y)$ and $\widehat{\mathbf{u}}_{Q^+}(d,t,\widehat{x},\widehat{y})$ are the same, so

$$\mathbf{u}_{Q^+}(d,t,x,y) - \widehat{\mathbf{u}}_{Q^+}(d,t,x,y) \in d^\infty((-T_{\varepsilon_1}d/2, T_{\varepsilon_1}d^{1-\varepsilon_1}/2) \times M_x \times M_y),$$

and consequently

$$\operatorname{Tr}\mathbf{U}_{Q^+}(d,t) - \operatorname{Tr}\widehat{\mathbf{U}}_{Q^+}(d,t) \in d^\infty((-T_{\varepsilon_1}d/2, T_{\varepsilon_1}d^{1-\varepsilon_1}/2)). \tag{4.5.4}$$

But as the operators $\mathcal{A}$, $\widehat{\mathcal{A}}$ and Q^+ are self-adjoint we have

$$\operatorname{Tr}\mathbf{U}_{Q^+}(d,-t) = \overline{\operatorname{Tr}\mathbf{U}_{Q^+}(d,t)}, \qquad \operatorname{Tr}\widehat{\mathbf{U}}_{Q^+}(d,-t) = \overline{\operatorname{Tr}\widehat{\mathbf{U}}_{Q^+}(d,t)}.$$

Combining the latter formulae with (4.5.4) we obtain (4.5.1). □

Set $\hat\rho(d,t) = \hat\rho_1(2tT_{\varepsilon_1}^{-1}d^{\varepsilon_1-1})$, where $\hat\rho_1$ is the function defined at the beginning of Section 4.2. Let us denote by $\widehat{\mathbf{u}}(t,\widehat{x},\widehat{y})$ the Schwartz kernel of the operator $\exp(-it\widehat{\mathcal{A}}^{1/(2m)})$.

THEOREM 4.5.2. *If the constant T_{ε_1} is sufficiently small, then, uniformly over $d \in (0,d_0]$, $\lambda \geqslant d^{\varepsilon_1-1}$ and $y \in M$, we have*

$$\mathcal{F}^{-1}_{t\to\lambda}\left[\hat\rho(d,t)\,\widehat{\mathbf{u}}(t,y,y)\right] = \lambda^{n-1}\int_{S_y^*M} d\widetilde{\eta} + O(d^{2\varepsilon_1-2}\lambda^{n-3}). \tag{4.5.5}$$

PROOF OF THEOREM 4.5.2. Theorem 4.5.2 is proved in the same way as Theorem 4.2.1 (see also Remark 4.2.2), that is, by applying the stationary phase formula. The only difference is that the factor $\hat\rho(d,t)$ now depends on the small parameter d, so we have to exercise control over this additional parameter. The required control is achieved by rescaling time and the spectral parameter according to the formulae $\widetilde{t} = d^{\varepsilon_1-1}t$, $\widetilde{\lambda} = d^{1-\varepsilon_1}\lambda$. Now if we write the left-hand side of (4.5.5) as one oscillatory integral $\int(\,\cdot\,)\,d\widetilde{\eta}\,d\widetilde{t}$, we notice that the differentiation of the amplitude of this oscillatory integral with respect to $\widetilde{t}$ does not produce negative powers of d; this was the aim of our rescaling. It remains to note that the resulting asymptotic expansion will be in powers of $\widetilde{\lambda}^{-1}$ instead of λ^{-1}. The latter explains the structure of the remainder term in (4.5.5). □

CHAPTER 5

Proof of Main Results

Our final aim (and this is the subject of Chapter 5) is to prove the main theorems stated in Chapter 1. Sections 5.1–5.4 of this chapter are devoted to the proof of Theorems 1.2.1, 1.6.1, 1.7.6, and Section 5.5 — to the proof of Theorems 1.8.5, 1.8.7, 1.8.14, 1.8.17.

Throughout this chapter $\hat\rho_1$ is a real-valued even function of the class $C_0^\infty(\mathbb{R})$ such that $\hat\rho_1(0) = 1$, $\operatorname{supp}\hat\rho_1 \subset [-1,1]$, and $\rho_1(\lambda) > 0$, $\forall\lambda \in \mathbb{R}$, where $\rho_1(\lambda) = \mathcal{F}^{-1}_{t\to\lambda}(\hat\rho_1(t))$; cf. Section B.1. We shall also denote $\hat\rho_{T_\delta}(t) = \hat\rho_1(t/T_\delta)$, $\rho_{T_\delta}(\lambda) = \mathcal{F}^{-1}_{t\to\lambda}(\hat\rho_{T_\delta}(t)) \equiv T_\delta\rho_1(T_\delta\lambda)$, where $T_\delta > 0$ is the constant defined in the introduction to Chapter 4. By F_+ we denote the class of nondecreasing functions introduced in Appendix B.

5.1. Partition of the manifold M into three zones

In this section we do some preliminary work concerned with the partition of the manifold M into three zones according to the distance from the boundary.

1. General idea. Let us try to solve the problem (1.1.1), (1.1.2) approximately for $\lambda \to +\infty$. Although an individual eigenfunction cannot be effectively constructed, the combination of a large number of eigenfunctions may be asymptotically described due to averaging effects. So it is natural to study the spectral projection E_λ and the spectral function $e(\lambda, x, y)$ of the problem (1.1.1), (1.1.2) (see 1.8.1), instead of individual eigenfunctions.

The counting function is related to the spectral projection and the spectral function by (1.8.1). We shall evaluate the full trace of the spectral projection, i.e., the integral (1.8.1), by dividing the manifold M into three zones:

$$\begin{aligned}
&\text{the interior zone } x_n \gtrsim \delta\,,\\
&\text{the intermediate zone } \lambda^{\varepsilon-1} \lesssim x_n \lesssim \delta\,,\\
&\text{the boundary zone } x_n \lesssim \lambda^{\varepsilon-1}\,.
\end{aligned}$$

Here δ is a small positive number (at the final step of the proof of Theorem 1.6.1 it will be allowed to tend to zero), ε is an arbitrary number from the interval

$$0 < \varepsilon < 1/3\,, \tag{5.1.1}$$

and x_n is our specified "normal" coordinate; see 1.1.2.

The partition of the manifold M into three zones is a forced measure, caused by our inability to construct the full function $N(\lambda)$ in one go. The basis for such a partition is the obvious observation that the eigenfunctions of the problem (1.1), (1.2) have different structure in the interior of the manifold M and near its boundary ∂M. Boundary effects are observed near ∂M up to a distance of the order of a wavelength, i.e., of the order of λ^{-1}. This fact explains the singling out of the boundary zone. The partition of the remaining part of the manifold M into two zones (the interior and intermediate ones) is of a more delicate nature. The necessity of such a partition is due to two impeding factors which will appear in the course of our subsequent (Sections 5.2–5.4) asymptotic analysis. First, global resonance-type effects may appear, which can spoil the polynomial character of the second asymptotic term (Theorem 1.7.6). Second, it is technically difficult to give a sufficiently accurate uniform remainder estimate with respect to two parameters, the spectral parameter λ and the distance to the boundary d. Partition of the complement of the boundary zone into two parts — the interior and intermediate zones — will allow us to overcome these two difficulties separately: we shall struggle only with global effects in the interior zone, and search only for an accurate uniform two-parameter remainder estimate in the intermediate zone.

Note that requirements concerning the relative accuracy of the asymptotics of $e(\lambda, x, x)$ will be different in different zones. In the interior zone we will have to evaluate $e(\lambda, x, x)$ with the greatest relative accuracy because this zone makes the main contribution to $N(\lambda)$. In the intermediate zone requirements to the relative accuracy will be slightly lower. Finally, in the boundary zone the relative accuracy of the asymptotics will be the lowest: this relative error does not make a large contribution to $N(\lambda)$ because of the small width of the boundary zone. This difference in the required relative accuracy will enable us to use different (most convenient in each particular situation) techniques for the construction of spectral asymptotics in each zone. In the interior zone we will use in full the wave equation method (taking reflections into account), in the intermediate zone the simplified wave equation method (without considering reflections), and in the boundary zone the resolvent method.

The second Weyl term $c_1\lambda^{n-1}$ of the asymptotics of $N(\lambda)$ will originate in the boundary zone as a result of boundary effects. The condition $\varepsilon > 0$ ensures that these boundary effects will be practically unnoticeable in the intermediate zone (as has been mentioned above, boundary effects influence a band of typical width $x_n \sim \lambda^{-1}$). At the same time, the condition $\varepsilon < 1/3$ will enable us to evaluate the asymptotics of $e(\lambda, x, x)$ in the boundary zone very roughly by freezing the coefficients in the differential operators A, $B^{(j)}$. Throughout this section the number ε will be fixed once and for all. Therefore, in principle, it is possible to simply indicate a concrete ε from the interval (5.1.1), for example $\varepsilon = 1/4$.

The nonclassical second term $\mathbf{Q}(\lambda)\lambda^{n-1}$ will originate in the interior zone. It is associated with the nonzero singularities of the wave group.

2. Technical realization. Let us now describe in detail how we divide M into three zones by means of smooth cut-off functions.

Let $\chi_+(x_n) \in C^\infty(\mathbb{R}_+)$ be a real-valued function such that $\chi_+(x_n) = 0$ when $0 \leqslant x_n \leqslant 1$, $\chi_+(x_n) = 1$ when $x_n \geqslant 2$, and $d\chi_+ / dx_n > 0$ when $1 < x_n < 2$. Set

$$\chi_0(x_n) := \sqrt{\chi_+^2(x_n) - \chi_+^2(x_n/2)}\,, \qquad \chi_-(x_n) := \sqrt{1 - \chi_+^2(x_n/2)}\,.$$

We shall assume that the function χ_+ is chosen in such a way that χ_0 and χ_- are infinitely smooth. Obviously,

$$\operatorname{supp}\chi_+ = [1, +\infty)\,, \qquad \operatorname{supp}\chi_0 = [1, 4]\,, \qquad \operatorname{supp}\chi_- = [0, 4]\,.$$

Let $p \equiv p(\lambda, \delta, \varepsilon)$ be the minimal natural number such that $\delta/2^{p+1} \leqslant \lambda^{\varepsilon-1}$. We set $\delta_l = \delta/2^l$, $l = 1, 2, \dots, p+1$.

The unit function on M can be partitioned as

$$1 = \chi_+^2(x_n/\delta) + \sum_{l=1}^{p} \chi_0^2(x_n/\delta_l) + \chi_-^2(x_n/\delta_{p+1}). \tag{5.1.2}$$

On the right-hand side of (5.1.2) the term with χ_+ corresponds to the interior zone, those with χ_0 to the intermediate zone, and that with χ_- to the boundary zone.

By (5.1.2), (1.8.1), (1.2.2) we have

$$N(\lambda) = \mathrm{Tr}(\chi_+^2(x_n/\delta)\, E_\lambda) + \sum_{l=1}^{p} \mathrm{Tr}(\chi_0^2(x_n/\delta_l)\, E_\lambda) + \mathrm{Tr}(\chi_-^2(x_n/\delta_{p+1})\, E_\lambda) \tag{5.1.3}$$

and

$$c_0 = c_0^+(\delta) + \sum_{l=1}^{p} c_0^0(\delta_l) + c_0^-(\delta_{p+1}), \tag{5.1.4}$$

where

$$c_0^v(d) = \int_{A_{2m} \leqslant 1} \chi_v^2(x_n/d)\, dx\, đ\xi \equiv \frac{1}{n} \int_{S^*M} \chi_v^2(x_n/d)\, dx\, đ\widetilde{\xi} \tag{5.1.5}$$

$(v = +, -, 0)$. Denote also

$$\mathbf{Q}_\delta(\lambda) = \int_{\Pi^a} \frac{[\pi - \mathbf{q} - \lambda \mathbf{T}]_{2\pi}}{\mathbf{T}}\, \chi_+^2(y_n/d)\, dy\, đ\widetilde{\eta} \tag{5.1.6}$$

(cf. (1.7.9)).

We now state Propositions 5.1.1–5.1.3 which establish the asymptotic behaviour of the terms on the right-hand side of (5.1.3).

PROPOSITION 5.1.1. *We have*

$$\mathrm{Tr}(\chi_+^2(x_n/\delta)\, E_\lambda) = c_0^+(\delta)\, \lambda^n + O(\lambda^{n-1}), \qquad \lambda \to +\infty \tag{5.1.7}$$

(*the O-term may depend on δ*). *If the nonperiodicity and nonblocking conditions are fulfilled, then*

$$\mathrm{Tr}(\chi_+^2(x_n/\delta)\, E_\lambda) = c_0^+(\delta)\, \lambda^n + o(\lambda^{n-1}), \qquad \lambda \to +\infty. \tag{5.1.8}$$

If the simple reflection condition is fulfilled, then

$$\begin{aligned} c_0^+(\delta)\, \lambda^n + \mathbf{Q}_\delta\,(\lambda - o(1))\, \lambda^{n-1} - o(\lambda^{n-1}) &\leqslant \mathrm{Tr}(\chi_+^2(x_n/\delta)\, E_\lambda) \\ &\leqslant c_0^+(\delta)\, \lambda^n + \mathbf{Q}_\delta\,(\lambda + o(1))\, \lambda^{n-1} + o(\lambda^{n-1}), \qquad \lambda \to +\infty, \end{aligned} \tag{5.1.9}$$

or equivalently,

$$\begin{aligned} c_0^+(\delta)\,(\lambda - \varkappa)^n + \mathbf{Q}_\delta(\lambda - \varkappa)\, \lambda^{n-1} - o(\lambda^{n-1}) &\leqslant \mathrm{Tr}(\chi_+^2(x_n/\delta)\, E_\lambda) \\ &\leqslant c_0^+(\delta)\,(\lambda + \varkappa)^n + \mathbf{Q}_\delta(\lambda + \varkappa)\, \lambda^{n-1} + o(\lambda^{n-1}), \qquad \lambda \to +\infty, \end{aligned} \tag{5.1.10}$$

for any $\varkappa > 0$.

PROPOSITION 5.1.2. *Uniformly over* d *from the interval*

$$\lambda^{\varepsilon-1} < d \leqslant d_0 \tag{5.1.11}$$

we have

$$\mathrm{Tr}(\chi_0^2(x_n/d)\, E_\lambda) = c_0^0(d)\, \lambda^n + O(d^{\varepsilon'} \lambda^{n-1}), \qquad \lambda \to +\infty, \tag{5.1.12}$$

where ε' *is some positive number (depending in general on* ε*).*

Uniformity in (5.1.12) means that there exist positive constants C, Λ independent of λ and d such that

$$\frac{\left|\mathrm{Tr}(\chi_0^2(x_n/d)\, E_\lambda) - c_0^0(d)\lambda^n\right|}{d^{\varepsilon'}\lambda^{n-1}} \leqslant C \tag{5.1.13}$$

for all $\lambda \geqslant \Lambda$ and d satisfying (5.1.11).

Note that the concrete value of ε' is unimportant for our purposes because we need only the fact that $\varepsilon' > 0$. In principle ε' can be evaluated from formulae (5.3.1), (5.3.3), (5.3.4); elementary analysis of these formulae shows that $\varepsilon' = \varepsilon/3$ is an acceptable value.

PROPOSITION 5.1.3. *Uniformly over* d *from the interval*

$$\lambda^{\varepsilon-1}/2 < d \leqslant \lambda^{\varepsilon-1} \tag{5.1.14}$$

we have

$$\mathrm{Tr}(\chi_-^2(x_n/d)\, E_\lambda) = c_0^-(d)\, \lambda^n + c_1\, \lambda^{n-1} + o(\lambda^{n-1}), \qquad \lambda \to +\infty, \tag{5.1.15}$$

where c_1 *is the coefficient defined in Section* 1.6.

Let us show now that Theorems 1.2.1, 1.6.1, 1.7.6 follow from Propositions 5.1.1–5.1.3.

If we set $d = \delta_l$, $l = 1, 2, \dots, p$, in (5.1.12) and sum over l, we obtain

$$\sum_{l=1}^{p} \mathrm{Tr}(\chi_0^2(x_n/\delta_l)\, E_\lambda) = \left(\sum_{l=1}^{p} c_0^0(\delta_l)\right) \lambda^n + \frac{1-(1/2)^{p\varepsilon'}}{2^{\varepsilon'}-1}\, O(\delta^{\varepsilon'}\lambda^{n-1}) \tag{5.1.16}$$

(in summing up the O-terms we have used the formula for the sum of a geometric progression). It is important to note that the O-term in (5.1.16) is uniform over all δ from the interval (5.1.11). Since the factor before the O-term in (5.1.16) is uniformly bounded over all natural p, (5.1.16) can be rewritten in the form

$$\sum_{l=1}^{p} \mathrm{Tr}(\chi_0^2(x_n/\delta_l)\, E_\lambda) = \left(\sum_{l=1}^{p} c_0^0(\delta_l)\right) \lambda^n + O(\delta^{\varepsilon'}\lambda^{n-1}). \tag{5.1.17}$$

Adding up (5.1.7) and (5.1.15) (with $d = \delta_{p+1}$) to (5.1.17) and taking (5.1.3), (5.1.4) into account, we obtain

$$N(\lambda) = c_0\, \lambda^n + c_1\, \lambda^{n-1} + O(\lambda^{n-1}) + O(\delta^{\varepsilon'}\lambda^{n-1}) + o(\lambda^{n-1}). \tag{5.1.18}$$

For a fixed δ, (5.1.18) turns into (1.2.1). This proves Theorem 1.2.1.

If in the summation we use (5.1.8) instead of (5.1.7), then in place of (5.1.18) we obtain

$$N(\lambda) = c_0\, \lambda^n + c_1\, \lambda^{n-1} + O(\delta^{\varepsilon'}\lambda^{n-1}) + o(\lambda^{n-1}). \tag{5.1.19}$$

Since δ can be taken arbitrarily small and the O-term is uniform over δ, formula (5.1.19) implies (1.6.1). This proves Theorem 1.6.1.

If in the summation we use (5.1.10) instead of (5.1.7), then in place of (5.1.18) we obtain

$$\begin{aligned} & c_0\,(\lambda-\varkappa)^n+\mathbf{Q}_\delta(\lambda-\varkappa)\,\lambda^{n-1}-O(\delta^{\varepsilon'}\lambda^{n-1})-o(\lambda^{n-1}) \\ & \leqslant \operatorname{Tr}(\chi_+^2(x_n/\delta)\,E_\lambda) \\ & \leqslant c_0\,(\lambda+\varkappa)^n+\mathbf{Q}_\delta(\lambda+\varkappa)\,\lambda^{n-1}+O(\delta^{\varepsilon'}\lambda^{n-1})+o(\lambda^{n-1}). \end{aligned} \tag{5.1.20}$$

The number δ can be taken arbitrarily small, the O-term is uniform over δ, and the function $\mathbf{Q}_\delta(\lambda)$ converges to the function $\mathbf{Q}(\lambda)$ uniformly over $\lambda\in\mathbb{R}$ as $\delta\to+0$. Therefore formula (5.1.20) implies (1.7.4). (Here we had to use the notation $\varkappa$ instead of ε because in this chapter ε is already being used for a different purpose.) This proves Theorem 1.7.6.

Thus the proof of Theorems 1.2.1, 1.6.1, 1.7.6 has been reduced to the proof of Propositions 5.1.1–5.1.3. We shall prove Propositions 5.1.1–5.1.3 in Sections 5.2–5.4, respectively.

REMARK 5.1.4. We partitioned the manifold M into three zones using smooth cut-off functions, though it might seem more natural to construct pointwise asymptotics of the spectral function $e(\lambda,x,x)$ and then integrate in x. This requires some explanation. In the interior zone it is absolutely essential to deal with some integral of $e(\lambda,x,x)$, not $e(\lambda,x,x)$ itself. Indeed, in obtaining pointwise asymptotics of the spectral function one has to impose much more restrictive geometrical conditions on our billiards, with loops instead of periodic trajectories; see Section 1.8 for details. In the intermediate and boundary zones it is possible in principle to deal with pointwise asymptotics of $e(\lambda,x,x)$, but such an approach is technically more difficult. The general fact is that integration in x with smooth cut-off functions enables us to use microlocal techniques with maximal success and to control well the remainder terms. In other words, integration gives an additional averaging effect, thus improving asymptotic estimates.

REMARK 5.1.5. An estimate of the remainder term of the form (5.1.12) in the case $m=1$ was obtained by Seeley [**Se1**], [**Se2**], and so we shall call (5.1.12) a *Seeley-type asymptotics*. The asymptotics (5.1.12) with $\varepsilon'=0$, i.e.,

$$\operatorname{Tr}(\chi_0^2(x_n/d)\,E_\lambda)=c_0^0(d)\,\lambda^n+O(\lambda^{n-1}), \tag{5.1.21}$$

is due to Brüning [**Br**] and will be called a *Brüning-type asymptotics*. Note that summing Brüning-type asymptotics (5.1.21) instead of (5.1.12) would result in the remainder $p\,O(\lambda^{n-1})=O(\lambda^{n-1}\ln\lambda)$ in (5.1.17) since $p=(1-\varepsilon+o(1))\log_2\lambda$ as $\lambda\to+\infty$. Thus, the Brüning-type asymptotics (5.1.21) is insufficient for proving Theorems 1.2.1, 1.6.1, 1.7.6 as it leads to an extra $\ln\lambda$ factor in the remainder term. It is important that in (5.1.12), $\varepsilon'>0$.

REMARK 5.1.6. In the case $m=1$, Seeley obtained (5.1.12) by considering one reflection from the boundary. In Seeley's method it is essential to consider at least one reflection, because without reflections the method produces only the rougher estimate (5.1.21). The reason for considering at least one reflection is fundamental: Tauberian arguments (see Theorem B.2.4) provide the necessary accuracy only if the length of the time interval on which we solve the wave equation is much greater that the distance to the boundary d. This seems to contradict the claim made in subsection 1 that we intend to prove (5.1.12) without considering reflections, and, moreover, do this for an arbitrary m. The explanation is that we use a method different from Seeley's: the increase of the time interval is achieved by splitting (with the help of pseudodifferential cut-offs) the phase space into two parts, namely, waves going from the boundary and to the boundary. These two types of waves are constructed on asymmetrical time intervals: waves going from the boundary are constructed further into positive time, whereas waves going to the boundary are constructed further into negative time. We have started developing this ideology in Sections 2.10, 3.5, 4.5, and will bring it to a conclusion in Section 5.3. We had to develop this technique because Seeley's method may be adapted to higher order operators only if the simple reflection condition is fulfilled. In the general case we may encounter a situation when we have an infinite number of reflections on a time interval of length $\sim d$.

5.2. Asymptotics of the trace of the spectral projection in the interior zone

1. Proof of the first part of Proposition 5.1.1. In this subsection we prove formula (5.1.7).

Set $P = Q = \chi_+(x_n/\delta)$, $\mathbf{l} = \mathbf{m} = 0$, and

$$N_+(\lambda) := \operatorname{Tr}(\chi_+^2(x_n/\delta)\, E_\lambda) \equiv \operatorname{Tr}(Q^*\, E_\lambda\, P)$$

(this is the function appearing on the left-hand side of (5.1.7)). Then in the notation (4.0.1) we have

$$\operatorname{Tr}(\mathbf{U}_{P,Q}(t)) = \mathcal{F}_{\lambda\to t}(N'_+(\lambda)), \tag{5.2.1}$$

where N'_+ is the derivative of the monotone function N_+. The left- and right-hand sides of formula (5.2.1) are understood, of course, as distributions in t. Apart from λ the quantity N_+ depends also on δ, but throughout this section δ is assumed to be fixed.

Applying Theorem 4.2.4 with $P = Q = \chi_+(x_n/\delta)$ we get

$$\mathcal{F}^{-1}_{t\to\lambda}\Big(\hat\rho_{T_\delta}(t)\ \operatorname{Tr}\mathbf{U}_{P,Q}(t)\Big) \;=\; \lambda^{n-1}\int_{S^*M}\chi_+^2(y_n/\delta)\,dy\,đ\eta \;+\; O(\lambda^{n-3}) \tag{5.2.2}$$

as $\lambda\to+\infty$. Note that there are no λ^{n-2}-terms in (5.2.2) in view of Remark 4.2.6.

With account of (5.2.1) and (5.1.5) formula (5.2.2) can be rewritten as

$$(N'_+ * \rho_{T_\delta})(\lambda) \;=\; n\,c_0^+(\delta)\,\lambda^{n-1} + O(\lambda^{n-3}), \qquad \lambda\to+\infty. \tag{5.2.3}$$

As our function N_+ is of the class F_+ we have

$$(N'_+ * \rho_{T_\delta})(\lambda) \;=\; O(|\lambda|^{-\infty}), \qquad \lambda\to-\infty. \tag{5.2.4}$$

Formulae (5.2.3), (5.2.4) imply

$$(N_+ * \rho_{T_\delta})(\lambda) \;=\; \int_{-\infty}^{\lambda}(N'_+ * \rho_{T_\delta})(\mu)\,d\mu \;=\; c_0^+(\delta)\,\lambda^n \;+\; O(\lambda^{n-2}\ln\lambda) \tag{5.2.5}$$

as $\lambda\to+\infty$; here the logarithm is written in order to cover the case $n=2$.

Formula (5.2.3) allows us to apply Corollary B.2.2, which gives

$$N_+(\lambda) \;=\; (N_+ * \rho_{T_\delta})(\lambda) \;+\; O(\lambda^{n-1}), \qquad \lambda\to+\infty. \tag{5.2.6}$$

Substituting (5.2.5) into (5.2.6) we obtain (5.1.7).

2. Proof of the second part of Proposition 5.1.1. In this subsection we prove formula (5.1.8) under the assumption that the nonperiodicity and nonblocking conditions are fulfilled.

Let $\gamma(\lambda)$ be an arbitrary function from the Schwartz class $\mathcal{S}(\mathbb{R})$ such that $\hat{\gamma}(t) = \mathcal{F}_{\lambda\to t}[\gamma] \in C_0^\infty(\mathbb{R})$ and $\operatorname{supp}\hat{\gamma} \subset (0,+\infty)$. Applying Theorem 4.4.1 with $P = \chi_+(x_n/\delta)$ we get

$$(N'_+ * \gamma)(\lambda) \;=\; \mathcal{F}^{-1}_{t\to\lambda}\Big(\hat{\gamma}(t)\,\operatorname{Tr}\mathbf{U}_{P,P}(t)\Big) \;=\; o(\lambda^{n-1})\,, \qquad \lambda\to+\infty\,. \tag{5.2.7}$$

Formulae (5.2.3), (5.2.7) allow us to apply Theorem B.5.1, which gives

$$N_+(\lambda) \;=\; (N_+ * \rho_{T_\delta})(\lambda) \,+\, o(\lambda^{n-1})\,, \qquad \lambda\to+\infty\,. \tag{5.2.8}$$

Substituting (5.2.5) into (5.2.8) we obtain (5.1.8).

3. Proof of the third part of Proposition 5.1.1. In this subsection we prove formula (5.1.9) assuming simple reflection.

It is easy to check that the function $\mathbf{Q}_\delta$ possesses properties similar to those of $\mathbf{Q}$; see 1.7.2. In particular, $\mathbf{Q}_\delta$ is uniformly bounded,

$$|\,\mathbf{Q}_\delta(\lambda)\,| \;\leqslant\; \pi\int_{\Pi^a}\mathbf{T}^{-1}\,\chi_+^2(y_n/\delta)\,dy\,d\!\!\!{}^{-}\!\eta\,,$$

and for all $\varkappa>0$ we have

$$\mathbf{Q}_\delta(\lambda+\varkappa)\,-\,\mathbf{Q}_\delta(\lambda) \;\geqslant\; -\varkappa\int_{\Pi^a}\chi_+^2(y_n/\delta)\,dy\,d\!\!\!{}^{-}\!\eta \;\geqslant\; -\varkappa\, n\, c_0^+(\delta)\,.$$

Let C be a sufficiently large positive constant. Then, in view of the above properties of the function $\mathbf{Q}_\delta$, the function

$$N_1(\lambda) \;=\; \begin{cases} c_0^+(\delta)\,\lambda^n \,+\, \mathbf{Q}_\delta(\lambda)\,\lambda^{n-1} \,+\, C\,\lambda^{n-1}\,, & \lambda>0\,,\\ 0\,, & \lambda\leqslant 0\,,\end{cases}$$

is nondecreasing. Let

$$N_2(\lambda) \;=\; \begin{cases} N_+(\lambda) \,+\, C\,\lambda^{n-1}\,, & \lambda>0\,,\\ 0\,, & \lambda\leqslant 0\,.\end{cases}$$

Then both functions N_1 and N_2 belong to the class F_+, so we can apply Theorem B.4.1. Formula (5.1.9) will follow from (B.4.3) if we establish (B.4.1) and (B.4.2) with $\nu=n-1$. Thus, the proof of the third part of Proposition 5.1.1 has been reduced to the proof of formulae (B.4.1), (B.4.2).

Let $\alpha(\lambda)\in\mathcal{S}(\mathbb{R})$, $\alpha'=d\alpha/d\lambda$, $\hat{\alpha}=\mathcal{F}_{\lambda\to t}(\alpha)$, $\hat{\alpha}'=d\hat{\alpha}/dt$. We have

$$\begin{aligned}
\int_0^\infty \alpha'(\lambda-\mu)\,\mu^n\,d\mu \;&=\; \int\alpha'(\lambda-\mu)\,\mu^n\,d\mu \,+\, O(\lambda^{-\infty})\\
&=\; n\int\alpha(\lambda-\mu)\,\mu^{n-1}\,d\mu \,+\, O(\lambda^{-\infty})\\
&=\; n\,\lambda^{n-1}\int\alpha(\mu)\,d\mu \,-\, n(n-1)\,\lambda^{n-2}\int\mu\,\alpha(\mu)\,d\mu \,+\, O(\lambda^{n-3})\\
&=\; n\,\hat{\alpha}(0)\,\lambda^{n-1} \,-\, i\,n(n-1)\,\hat{\alpha}'(0)\,\lambda^{n-2} \,+\, O(\lambda^{n-3})\,, \quad \lambda\to+\infty\,,
\end{aligned} \tag{5.2.9}$$

and, analogously,

$$\begin{aligned}\int_0^\infty \alpha'(\lambda-\mu)\,\mu^{n-1}\,d\mu &= \int \alpha'(\lambda-\mu)\,\mu^{n-1}\,d\mu \,+\, O(\lambda^{-\infty}) \\ &= (n-1)\,\hat\alpha(0)\,\lambda^{n-2} \,+\, O(\lambda^{n-3})\,, \quad \lambda\to+\infty\,.\end{aligned} \tag{5.2.10}$$

We also have

$$\begin{aligned}\int_0^\infty \alpha'(\lambda-\mu)\,\mathbf{Q}_\delta(\mu)\,\mu^{n-1}\,d\mu &= \int \alpha'(\lambda-\mu)\,\mathbf{Q}_\delta(\mu)\,\mu^{n-1}\,d\mu \,+\, O(\lambda^{-\infty}) \\ &= \int \alpha'(\mu)\,\mathbf{Q}_\delta(\lambda-\mu)\,(\lambda-\mu)^{n-1}\,d\mu \,+\, O(\lambda^{-\infty}) \\ &= a_\alpha(\lambda)\,\lambda^{n-1} \,+\, b_\alpha(\lambda)\,\lambda^{n-2} \,+\, O(\lambda^{n-3})\,,\end{aligned} \tag{5.2.11}$$

where

$$a_\alpha(\lambda) \;=\; \int \alpha'(\mu)\,\mathbf{Q}_\delta(\lambda-\mu)\,d\mu \;=\; \int \alpha(\mu)\,\mathbf{Q}'_\delta(\lambda-\mu)\,d\mu\,, \tag{5.2.12}$$

and

$$\begin{aligned}b_\alpha(\lambda) &= -\,(n-1)\int \mu\,\alpha'(\mu)\,\mathbf{Q}_\delta(\lambda-\mu)\,d\mu \\ &= (n-1)\int \alpha(\mu)\,(\mathbf{Q}_\delta(\lambda-\mu) - \mu\,\mathbf{Q}'_\delta(\lambda-\mu))\,d\mu\,.\end{aligned} \tag{5.2.13}$$

Note that the fact that the function $\mathbf{Q}_\delta(\lambda)$ is uniformly bounded implies that the functions $a_\alpha(\lambda)$, $b_\alpha(\lambda)$ are uniformly bounded.

According to (5.1.6), (1.7.10)

$$\mathbf{Q}_\delta(\lambda-\mu) = \,2\int_{\Pi^a} \mathbf{T}^{-1}\left(\sum_{k=1}^\infty k^{-1}\sin k((\lambda-\mu)\mathbf{T}+\mathbf{q})\right)\chi_+^2(y_n/\delta)\,dy\,d\!\!{}^-\tilde\eta\,,$$

$$\mathbf{Q}'_\delta(\lambda-\mu) = \,2\int_{\Pi^a} \left(\sum_{k=1}^\infty \cos k((\lambda-\mu)\mathbf{T}+\mathbf{q})\right)\chi_+^2(y_n/\delta)\,dy\,d\!\!{}^-\tilde\eta\,,$$

where the series converge in the sense of distributions. Substituting these formulae in (5.2.12), (5.2.13) after elementary calculations we arrive at

$$a_\alpha(\lambda) = \int_{\Pi^a} \sum_{k\in\mathbb{Z}\setminus\{0\}} e^{ik(\lambda\mathbf{T}+\mathbf{q})}\,\hat\alpha(k\mathbf{T})\,\chi_+^2(y_n/\delta)\,dy\,d\!\!{}^-\tilde\eta\,, \tag{5.2.14}$$

(5.2.15)

$$b_\alpha(\lambda) = -i(n-1)\int_{\Pi^a} \sum_{k\in\mathbb{Z}\setminus\{0\}} e^{ik(\lambda\mathbf{T}+\mathbf{q})}\left(\frac{\hat\alpha(k\mathbf{T})}{k\mathbf{T}} + \hat\alpha'(k\mathbf{T})\right)\chi_+^2(y_n/\delta)\,dy\,d\!\!{}^-\tilde\eta.$$

Formulae (5.2.9)–(5.2.11) imply

$$\begin{aligned}(N_1' * \alpha)(\lambda) \;&=\; (N_1 * \alpha')(\lambda) \\ &= \big(n\,c_0^+(\delta)\,\hat\alpha(0) \,+\, a_\alpha(\lambda)\big)\,\lambda^{n-1} \\ &\quad + \big(b_\alpha(\lambda) \,+\, (n-1)\,C\,\hat\alpha(0) \\ &\qquad - in(n-1)\,c_0^+(\delta)\,\hat\alpha'(0)\big)\lambda^{n-2} + O(\lambda^{n-3}),\end{aligned} \tag{5.2.16}$$

$$(5.2.17)\qquad \begin{aligned}(N_2' * \alpha)(\lambda) &= (N_2 * \alpha')(\lambda)\\ &= (N_+' * \alpha)(\lambda) + (n-1)\, C\, \hat{\alpha}(0)\, \lambda^{n-2} + O(\lambda^{n-3}).\end{aligned}$$

Set $\alpha = \rho_{T_\delta}$. Then $\hat{\rho}_{T_\delta}(0) = 1$, $\hat{\rho}'_{T_\delta}(0) = 0$. Moreover, because $\operatorname{supp} \hat{\rho}_{T_\delta}$ is sufficiently close to 0, formulae (5.2.14), (5.2.15) give $a_{\rho_{T_\delta}}(\lambda) \equiv b_{\rho_{T_\delta}}(\lambda) \equiv 0$. So, with account of (5.2.3), formulae (5.2.16), (5.2.17) in this case take the form

$$\begin{aligned}(N_1' * \rho_{T_\delta})(\lambda) &= n\, c_0^+(\delta)\, \lambda^{n-1} + (n-1)\, C\, \lambda^{n-2} + O(\lambda^{n-3}), \qquad \lambda \to +\infty,\\ (N_2' * \rho_{T_\delta})(\lambda) &= n\, c_0^+(\delta)\, \lambda^{n-1} + (n-1)\, C\, \lambda^{n-2} + O(\lambda^{n-3}), \qquad \lambda \to +\infty;\end{aligned}$$

consequently

$$(N_1' * \rho_{T_\delta})(\lambda) = (N_2' * \rho_{T_\delta})(\lambda) + O(\lambda^{n-3}), \qquad \lambda \to +\infty.$$

Integrating the latter formula we obtain

$$(N_1 * \rho_{T_\delta})(\lambda) = (N_2 * \rho_{T_\delta})(\lambda) + O(\lambda^{n-2} \ln \lambda)), \qquad \lambda \to +\infty$$

(cf. (5.2.5)). Thus, we have proved formula (B.4.1) with $\nu = n-1$ and $\rho = \rho_{T_\delta}$.

Now set $\alpha = \gamma$, where $\gamma(\lambda)$ is an arbitrary function from the Schwartz class $\mathcal{S}(\mathbb{R})$ such that $\hat{\gamma}(t) = \mathcal{F}_{\lambda \to t}[\gamma] \in C_0^\infty(\mathbb{R})$ and $\operatorname{supp} \hat{\gamma} \subset (0, +\infty)$. Then, with account of (5.2.14), formula (5.2.16) in this case takes the form

$$(N_1' * \gamma)(\lambda) = \lambda^{n-1} \int_{\Pi^a} \sum_{k \in \mathbb{N}} e^{ik(\lambda \mathbf{T} + \mathbf{q})}\, \hat{\gamma}(k\mathbf{T})\, \chi_+^2(y_n/\delta)\, dy\, đ\tilde{\eta} + O(\lambda^{n-2}).$$

On the other hand, by Theorem 4.4.2 formula (5.2.17) in this case takes the form

$$(N_2' * \gamma)(\lambda) = \lambda^{n-1} \int_{\Pi^a} \sum_{k \in \mathbb{N}} e^{ik(\lambda \mathbf{T} + \mathbf{q})}\, \hat{\gamma}(k\mathbf{T})\, \chi_+^2(y_n/\delta)\, dy\, đ\tilde{\eta} + o(\lambda^{n-1}).$$

Consequently

$$(N_1' * \gamma)(\lambda) = (N_2' * \gamma)(\lambda) + o(\lambda^{n-1}), \qquad \lambda \to +\infty.$$

Thus, we have proved formula (B.4.2) with $\nu = n-1$.

The proof of Proposition 5.1.1 is complete.

5.3. Asymptotics of the trace of the spectral projection in the intermediate zone

This section is devoted to the proof of Proposition 5.1.2. We shall, however, prove this proposition in an extended form. Namely, we shall prove

PROPOSITION 5.3.1. *Let ε_1, ε_4 be positive numbers satisfying the inequalities*

$$(5.3.1)\qquad \varepsilon_1 \leqslant 1/2, \qquad \varepsilon_1 < \varepsilon_4.$$

Then, uniformly over $d \in (0, d_0]$ and $\lambda \in \mathbb{R}_+$, we have

$$(5.3.2)\qquad \operatorname{Tr}(\chi_0^2(x_n/d)\, E_\lambda) = c_0^0(d)\, \lambda^n + O(d^{1-n(1+\varepsilon_4)}) + O(d^{\varepsilon_1} \lambda^{n-1}).$$

It is easy to see that Proposition 5.3.1 implies Proposition 5.1.2 with

$$(5.3.3)\qquad \frac{n\varepsilon_4}{n-1+n\varepsilon_4} < \varepsilon < 1,$$

$$(5.3.4)\qquad \varepsilon' = \min\Bigl(\varepsilon_1, \frac{(n-1)\varepsilon}{1-\varepsilon} - n\varepsilon_4\Bigr).$$

Thus, we have obtained an explicit estimate for the exponent ε' appearing in Proposition 5.1.2: given a positive $\varepsilon < 1$ one has to choose positive ε_1, ε_4 satisfying the inequalities (5.3.1), (5.3.3), and then ε' is determined from (5.3.4).

Before proving Proposition 5.3.1 we will have to prove several auxiliary lemmas. The constants T_{ε_1}, ε_1, ε_2 below are the ones introduced in Section 2.10.

LEMMA 5.3.2. *There exist parameter-dependent pseudodifferential operators* $P^{\pm}$ *of order* 0 *of the type* $(1;1,\varepsilon_1)^{\pm}$ *(see Definition* 2.10.13) *such that the Schwartz kernel of the operator*

$$\chi_0^2(x_n/d) \;-\; P^+(P^+)^* \;-\; P^-(P^-)^* \tag{5.3.5}$$

is of the class $d^{-\infty}(M_x \times M_y)$ *(see Definition* 2.10.15).

PROOF OF LEMMA 5.3.2. Let $\theta \in C^\infty(\mathbb{R})$ be a real-valued function such that $\operatorname{supp}\theta \subset [-1,+\infty)$, $\theta^2(z)+\theta^2(-z) \equiv 1$, and let χ_+ be as in 2.10.6. We construct the Schwartz kernels of the pseudodifferential operators $P^{\pm}$ in the form of parameter-dependent oscillatory integrals (2.1.20) with amplitudes

$$a^{\pm}(d,x;y,\eta) \sim \sum_{k=0}^{\infty} a^{\pm}_{-k}(d,x;y,\eta)\,, \tag{5.3.6}$$

where asymptotic convergence is understood as described in Section 2.10, and the $a^{\pm}_{-k}$ have the structure

$$a^{\pm}_{-k}(d,x;y,\eta) \;=\; \chi_0(x_n/d)\,\theta(\pm d^{-\varepsilon_1} h_{\xi_n}(y,\eta))\,\mathfrak{h}_{-k}(d,y,\eta)\,\chi_+(d^{1+\varepsilon_2}h(y,\eta))\,.$$

Here the $\mathfrak{h}_{-k}$ are unknown functions on $(0,d_0]\times T'M_y(d)$ positively homogeneous in η of degree $-k$ and satisfying

$$d^{2k\varepsilon_1}\,\left|\partial_y^\beta \partial_\eta^\gamma\,\mathfrak{h}_{-k}\right| \;\leqslant\; c_{\beta\gamma}\,d^{-(|\beta|+|\gamma|)\varepsilon_1}\,|\eta|^{-k-|\gamma|}$$

(cf. (2.10.17)). Note that the above conditions imply

$$d^{2k\varepsilon_1}\,a^{\pm}_{-k}(d,x;y,\eta) \;\in\; S^{-k}(1;1,\varepsilon_1)^{\pm}$$

which ensures the asymptotic convergence of the series (5.3.6) by the time-independent version of Lemma 2.10.2.

Now, disregarding the dependence on d we evaluate the symbol of the Schwartz kernel of the pseudodifferential operator

$$P^+(P^+)^* \;+\; P^-(P^-)^*\,, \tag{5.3.7}$$

we find its homogeneous terms (appearing in the asymptotic expansion as $|\eta| \to +\infty$), and equate them to the homogeneous terms of the symbol of the Lagrangian distribution

$$\chi_0(x_n/d)\,\delta(x-y)\,\chi_0(y_n/d)\,. \tag{5.3.8}$$

This leads to a recurrent system for the determination of the functions $\mathfrak{h}_{-k}$, $k = 0,1,2,\ldots$; see also the proof of Lemma 4.4.10. Solving it we obtain $\mathfrak{h}_0 \equiv 1$, and so on. This whole construction is possible because both the Schwartz kernel of the operator (5.3.7) and the distribution (5.3.8) have the same product structure $\chi_0(x_n/d)\,(\,\cdot\,)\,\chi_0(y_n/d)$; in fact, it is easy to see that the functions $\mathfrak{h}_{-k}$ do not depend on our choice of the function χ_0.

It remains only to note that under our construction the Schwartz kernel of the operator (5.3.5) is described by an absolutely convergent integral, the integrand of which is asymptotically (with respect to the two parameters $d \to +0$ and $|\eta| \to +\infty$) zero for $h(y,\eta) \geqslant 2d^{-1-\varepsilon_2}$. As in 2.10.7, this implies that the integral in question is of the class $d^{-\infty}(M_x \times M_y)$. □

As usual, let λ_k and v_k, $k = 1, 2, \dots$, be the eigenvalues and the orthonormalized eigenfunctions of the problem (1.1.1), (1.1.2), and $(\,\cdot\,,\,\cdot\,)$ be our scalar product (1.1.10).

LEMMA 5.3.3. *Let ε_4 be a constant satisfying the condition*

$$\varepsilon_4 > \varepsilon_2\,, \tag{5.3.9}$$

and let Q be a parameter-dependent operator the Schwartz kernel of which is of the class $d^{-\infty}(M_x \times M_y)$. Then

$$(Qv_k, v_k) = O(d^\infty k^{-\infty}) \tag{5.3.10}$$

uniformly over $d \in (0, d_0]$ and k such that $\lambda_k \geqslant d^{-1-\varepsilon_4}$.

More precisely, (5.3.10) means that the quantity $d^{-r}k^s(Qv_k, v_k)$ is uniformly bounded for any $r, s \in \mathbb{N}$.

PROOF OF LEMMA 5.3.3. For any $q \in \mathbb{N}$ we have $(Qv_k, v_k) = \lambda_k^{-q}(A^qQv_k, v_k)$, where A^qQ can be estimated using (2.10.20). As q can be taken to be arbitrarily large we arrive at (5.3.10). □

Set

$$\hat\rho(d,t) = \hat\rho_1(2tT_{\varepsilon_1}^{-1}d^{\varepsilon_1-1}), \quad \rho(d,\lambda) = \mathcal{F}^{-1}_{t\to\lambda}(\hat\rho(d,t)) \equiv \frac{T_{\varepsilon_1}d^{1-\varepsilon_1}}{2}\,\rho_1\!\left(\frac{T_{\varepsilon_1}d^{1-\varepsilon_1}\lambda}{2}\right).$$

Further on we denote

$$\mathbf{U}_Q(d,t) := \exp(-it\mathcal{A}^{1/(2m)})\,Q\,, \qquad \widehat{\mathbf{U}}_Q(d,t) := \exp(-it\widehat{\mathcal{A}}^{1/(2m)})\,Q\,,$$

where $\widehat{\mathcal{A}}$ is an extension of the operator $\mathcal{A}$; see Section 4.5.

LEMMA 5.3.4. *Let ε_4 and Q be as in Lemma 5.3.3. Then, uniformly over $d \in (0, d_0]$ and $\lambda \geqslant 2d^{-1-\varepsilon_4}$, we have*

$$\mathcal{F}^{-1}_{t\to\lambda}\Big(\hat\rho(d,t)\,\operatorname{Tr}\mathbf{U}_Q(d,t)\Big) = O(d^\infty\lambda^{-\infty})\,.$$

PROOF OF LEMMA 5.3.4. We have

$$\mathcal{F}^{-1}_{t\to\lambda}\Big(\hat\rho(d,t)\,\operatorname{Tr}\mathbf{U}_Q(d,t)\Big) = \sum_{\lambda_k < d^{-1-\varepsilon_4}} \rho(d,\lambda-\lambda_k)\,(Qv_k,v_k) + \sum_{\lambda_k \geqslant d^{-1-\varepsilon_4}} \rho(d,\lambda-\lambda_k)\,(Qv_k,v_k)\,.$$

The first sum on the right-hand side of this formula is small because $\rho(d,\mu)$ is small for $|\mu| \gg d^{\varepsilon_1-1}$, and the second is small by (5.3.10). □

REMARK 5.3.5. Lemma 5.3.4 remains true with $\widehat{\mathbf{U}}_Q(d,t)$ instead of $\mathbf{U}_Q(d,t)$. The proof is exactly the same.

Denote

$$\mathbf{U}_0(d,t) \;:=\; \exp(-it\mathcal{A}^{1/(2m)})\,\chi_0^2(y_n/d)\,.$$

Lemmas 5.3.2, 5.3.4, Remark 5.3.5, and Theorems 4.5.1, 4.5.2 immediately imply

LEMMA 5.3.6. *If the constant* T_{ε_1} *is sufficiently small, then, uniformly over* $d\in(0,d_0]$ *and* $\lambda\geqslant 2d^{-1-\varepsilon_4}$, *we have*

$$\mathcal{F}^{-1}_{t\to\lambda}\Big(\hat\rho(d,t)\;\mathrm{Tr}\,\mathbf{U}_0(d,t)\Big) \;=\; n\,c_0^0(d)\,\lambda^{n-1}+\,O(d^{2\varepsilon_1-1}\lambda^{n-3})\,. \tag{5.3.11}$$

We are now prepared to proceed to the

PROOF OF PROPOSITION 5.3.1. Set $N_0(d,\lambda):=\mathrm{Tr}(\chi_0^2(x_n/d)\,E_\lambda)$ (this is the function appearing on the left-hand side of (5.3.2)). Then formula (5.3.11) can be rewritten as

$$(N_0'*\rho)(d,\lambda) \;=\; n\,c_0^0(d)\,\lambda^{n-1}+\,O(d^{2\varepsilon_1-1}\lambda^{n-3}) \tag{5.3.12}$$

(cf. (5.2.3)).

On the basis of (5.3.12) we intend to apply Theorem B.2.4. We cannot, however, do it directly because Theorem B.2.4 requires information on $(N_0'*\rho)(d,\lambda)$ for all $\lambda\geqslant 2T_{\varepsilon_1}^{-1}d^{\varepsilon_1-1}$, whereas we can use (5.3.12) only for $\lambda\geqslant 2d^{-1-\varepsilon_4}$. In order to overcome this difficulty let us introduce the function

$$\widetilde N_0(d,\lambda):=\begin{cases}0 \quad\text{if}\quad \lambda\leqslant 3d^{-1-\varepsilon_4},\\ N_0(d,\lambda)-N_0(d,3d^{-1-\varepsilon_4})\quad\text{if}\quad \lambda>3d^{-1-\varepsilon_4}.\end{cases}$$

Then the fact that $\rho(d,\mu)$ is small for $|\mu|\gg d^{\,\varepsilon_1-1}$ leads to the estimates

$$(\widetilde N_0'*\rho)(d,\lambda) \;=\; n\,c_0^0(d)\,\lambda^{n-1}+\,O(d^{2\varepsilon_1-1}\lambda^{n-3}),\qquad \lambda\geqslant 4d^{-1-\varepsilon_4}\,, \tag{5.3.13}$$

$$(\widetilde N_0'*\rho)(d,\lambda) \;=\; O(d^\infty),\qquad |\lambda|\leqslant 2d^{-1-\varepsilon_4}\,, \tag{5.3.14}$$

$$(\widetilde N_0'*\rho)(d,\lambda) \;=\; O(d^\infty|\lambda|^{-\infty}),\qquad \lambda<-2d^{-1-\varepsilon_4}\,. \tag{5.3.15}$$

In view of the inequality $0\leqslant(\widetilde N_0'*\rho)(d,\lambda)\leqslant(N_0'*\rho)(d,\lambda)$ we also have

$$(\widetilde N_0'*\rho)(d,\lambda) \;=\; O(d\lambda^{n-1}),\qquad 2d^{-1-\varepsilon_4}<\lambda<4d^{-1-\varepsilon_4}\,. \tag{5.3.16}$$

Applying Theorem B.2.4 on the basis of the estimates (5.3.13), (5.3.14), (5.3.16), we obtain

$$\widetilde N_0(d,\lambda) \;=\; (\widetilde N_0*\rho)(d,\lambda)\,+\,O(d^{\,\varepsilon_1}\lambda^{n-1}),\qquad \lambda\geqslant 2T_{\varepsilon_1}^{-1}d^{\,\varepsilon_1-1}\,. \tag{5.3.17}$$

Further on we assume that

$$\lambda\geqslant 4d^{-1-\varepsilon_4}\,. \tag{5.3.18}$$

Formulae (5.3.13)–(5.3.16), (5.3.18) imply

$$\begin{aligned}(\widetilde N_0*\rho)(d,\lambda) &= \int_{-\infty}^{\lambda}(\widetilde N_0'*\rho)(d,\mu)\,d\mu\\ &= c_0^0(\delta)\,\lambda^n+\,O(d^{1-n(1+\varepsilon_4)})\,+\,O(d^{2\varepsilon_1-1}\lambda^{n-2}\ln\lambda)\,.\end{aligned} \tag{5.3.19}$$

Substituting (5.3.19) into (5.3.17) we get

$$\widetilde{N}_0(d,\lambda) = c_0^0(\delta)\,\lambda^n + O(d^{1-n(1+\varepsilon_4)}) + O(d^{\varepsilon_1}\lambda^{n-1}). \tag{5.3.20}$$

It is known [**Agm1**], [**Agm2**] that the spectral function admits the rough estimate

$$e(\lambda,x,x) = O(\lambda^n) \tag{5.3.21}$$

uniformly over $\lambda \in \mathbb{R}_+$ and $x \in M$. This implies

$$N_0(d, 3d^{-1-\varepsilon_4}) = O(d^{1-n(1+\varepsilon_4)}). \tag{5.3.22}$$

Recalling that $N_0(d,\lambda) = \widetilde{N}_0(d,\lambda) + N_0(d,3d^{-1-\varepsilon_4})$, we obtain from (5.3.20), (5.3.22) the required formula (5.3.2).

Thus, we have proved (5.3.2), but under the additional restrictions (5.3.9) and (5.3.18). Condition (5.3.9) can be replaced by the weaker condition $\varepsilon_4 > \varepsilon_1$ if one notes that we have freedom in choosing ε_2 which allows us to pick ε_2 arbitrarily close to ε_1 (see (2.10.2)). Condition (5.3.18) is removed if one notes that for $0 \leqslant \lambda < 4d^{-1-\varepsilon_4}$ formula (5.3.2) is a consequence of (5.3.21). □

5.4. Asymptotics of the trace of the spectral projection in the boundary zone

In this section we shall prove Proposition 5.1.3. We shall, however, prove it in the following extended form.

PROPOSITION 5.4.1. *Let* $\mathbf{d}_{\min}$, $\mathbf{d}_{\max}$, ε, ε_5 *be positive numbers satisfying the inequalities* $\mathbf{d}_{\min} \leqslant \mathbf{d}_{\max}$ *and*

$$\varepsilon_5 < \varepsilon < \frac{1}{3}, \qquad \varepsilon_5 < \frac{1-3\varepsilon}{n+3}. \tag{5.4.1}$$

Then, uniformly over d *from the interval*

$$\mathbf{d}_{\min}\lambda^{\varepsilon-1} \leqslant d \leqslant \mathbf{d}_{\max}\lambda^{\varepsilon-1}, \tag{5.4.2}$$

the following asymptotics holds as $\lambda \to +\infty$:

$$\operatorname{Tr}(\chi_-^2(x_n/d)\,E_\lambda) = c_0^-(d)\,\lambda^n + c_1\,\lambda^{n-1} + O(\lambda^{n-1-\varepsilon_5}), \tag{5.4.3}$$

where χ_- *is the function defined at the beginning of* 5.1.2, *and the coefficients* $c_0^-(d)$, c_1 *are defined in accordance with formulae* (5.1.5), (1.6.2), *respectively.*

Uniformity of the remainder term in the above proposition means that there exist positive constants C, Λ, independent of λ and d, such that

$$\frac{\left|\operatorname{Tr}(\chi_-^2(x_n/d)\,E_\lambda) - c_0^-(d)\lambda^n - c_1\lambda^{n-1}\right|}{\lambda^{n-1-\varepsilon_5}} \leqslant C$$

for all $\lambda \geqslant \Lambda$ and d satisfying (5.4.2).

Note that if Λ is sufficiently large, then d is sufficiently small (see (5.4.2)), and consequently in the strip $\{x_n \leqslant d\}$ adjoining ∂M we can use local coordinates of the form $x = (x', x_n)$ with prescribed x_n; see also 1.1.2. More precisely, this means that we are able to cover the compact $\{x_n \leqslant d\}$ by a finite number of coordinate patches in which local coordinates have the above form.

Throughout this section we assume, without loss of generality, that

$$2m > n \tag{5.4.4}$$

(recall that $2m$ is the order of our partial differential operator and n is the dimension of the manifold). The condition (5.4.4) can always be satisfied by taking a power of the original spectral problem.

The proof of Proposition 5.1.3 is split into a sequence of auxiliary results.

1. Further partition of the manifold (along the boundary). Set

$$c_0(x) := \int_{A_{2m}(x,\xi)\leqslant 1} đ\xi \equiv \frac{1}{n}\int_{S_x^* M} đ\widetilde{\xi}, \tag{5.4.5}$$

$$c_1(x') := \int_{T'_{x'}\partial M} \operatorname{shift}^+(1;x',\xi')\, đ\xi'. \tag{5.4.6}$$

It is easy to see that $c_0(x)$ and $c_1(x')$ are densities on M and ∂M, respectively. Comparing formulae (5.4.5), (5.4.6) with (5.1.5), (1.6.2) we get

$$c_0^-(d) = \int_M c_0(x)\, \chi_-^2(x_n/d)\, dx\,, \tag{5.4.7}$$

$$c_1 = \int_{\partial M} c_1(x')\, dx'\,. \tag{5.4.8}$$

Note that the possibility of splitting the Riemann integral (1.6.2) into iterated Riemann integral (5.4.8), (5.4.6) is not completely obvious because the integrand $\operatorname{shift}^+(1;x',\xi')$ is, in the general case, discontinuous. However the necessary justification is provided by an argument similar to the one given at the beginning of 1.6.4.

LEMMA 5.4.2. *The densities $c_0(x)$ and $c_1(x')$ are infinitely smooth.*

PROOF OF LEMMA 5.4.2. The smoothness of $c_0(x)$ is obvious.

Our proof of the smoothness of $c_1(x')$ will be essentially based on the observation that when we integrate $\operatorname{shift}^+(1;x',\xi')$ with respect to ξ' we obtain a spectral characteristic of a problem with constant coefficients in the half-space $\mathbb{R}^n_+$ (see (5.4.20), (5.4.21) below, with z' instead of x'). This problem in the half-space is homogeneous in the sense that the full symbols of the partial differential operator and of the operators describing boundary conditions are homogeneous functions of $\xi \in \mathbb{R}^n$. This homogeneity together with the fact that the coefficients are constant is a powerful property which leads to the smooth dependence of spectral characteristics under the perturbation of coefficients.

In order to prove the smoothness of $c_1(x')$ let us consider the function $\mathbf{f}^+(\mu;x',\xi')$ defined by formula (1.6.7); now we indicate explicitly the dependence of $\mathbf{f}^+$ on (x',ξ'). The function $\mathbf{f}^+$ is well defined for all $(\mu;x',\xi') \in \mathbb{C}\times T^*\partial M$ such that μ is not in the spectrum of (1.1.7), (1.1.8), and moreover, it is an infinitely smooth function (see the beginning of A.3.1). We also have

$$\mathbf{f}^+(\lambda^{2m}\mu;x',\lambda\xi') = \lambda^{-2m}\,\mathbf{f}^+(\mu;x',\xi')\,, \qquad \forall\lambda > 0 \tag{5.4.9}$$

(cf. (1.6.17)). Let us now fix a $\mu_0 \in \mathbb{C} \setminus \mathbb{R}_+$. The smoothness of $\mathbf{f}^+$ and the rescaling property (5.4.9) imply

$$|\partial_{x'}^{\alpha'} \mathbf{f}^+(\mu_0; x', \xi')| \leqslant \frac{\text{const}_{\alpha'}}{(1+|\xi'|^2)^m}, \qquad \text{const}_{\alpha'} > 0. \tag{5.4.10}$$

The estimate (5.4.10) holds in any local coordinates x for any nonnegative multi-index α' uniformly over T^*K', where K' is an arbitrary compact set in $\Omega \cap \partial M$ and $\Omega \subset M$ is the coordinate patch. The constants $\text{const}_{\alpha'}$ depend, of course, on the choice of μ_0, as well as on the choice of local coordinates and of the compact set. Note that for $\alpha' = 0$ the estimate (5.4.10) follows from Lemmas 1.6.7, 1.6.5 and formula (1.6.18).

Let us also introduce the density

$$\mathbf{f}^+(\mu_0; x') := \int_{T'_{x'}\partial M} \mathbf{f}^+(\mu_0; x', \xi')\, d\!\!{}^-\xi'. \tag{5.4.11}$$

Formulae (5.4.10), (5.4.4) imply that the integral (5.4.11) converges absolutely, and so does the integral of any partial derivative of the integrand with respect to x'; consequently

$$\mathbf{f}^+(\mu_0; x') \in C^\infty(\partial M) \tag{5.4.12}$$

(recall that μ_0 is fixed). Using Lemma 1.6.7 and the rescaling property (1.6.17) we can rewrite formula (5.4.11) as

$$\begin{aligned} \mathbf{f}^+(\mu_0; x') &= \int_{T'_{x'}\partial M} \left(\int_0^{+\infty} \frac{\text{shift}^+(\nu; x', \xi')}{(\nu-\mu_0)^2}\, d\nu \right) d\!\!{}^-\xi' \\ &= \int_{T'_{x'}\partial M} \left(\int_0^{+\infty} \frac{\text{shift}^+(1; x', \nu^{-1/(2m)}\xi')}{(\nu-\mu_0)^2}\, d\nu \right) d\!\!{}^-\xi' \\ &= \int_{T'_{x'}\partial M} \left(\int_0^{+\infty} \frac{\nu^{(n-1)/(2m)}\, \text{shift}^+(1; x', \xi')}{(\nu-\mu_0)^2}\, d\nu \right) d\!\!{}^-\xi' \\ &= \left(\int_0^{+\infty} \frac{\nu^{(n-1)/(2m)}}{(\nu-\mu_0)^2}\, d\nu \right) \left(\int_{T'_{x'}\partial M} \text{shift}^+(1; x', \xi')\, d\!\!{}^-\xi' \right). \end{aligned} \tag{5.4.13}$$

Comparing the latter expression with (5.4.6) we conclude that the density $\mathbf{f}^+(\mu_0; x')$ differs from the density $c_1(x')$ by a constant factor, so in view of (5.4.12) $c_1(x') \in C^\infty(\partial M)$. □

From now on let us fix some local coordinates $x = (x', x_n)$ and let $\Omega \subset M$ be the corresponding coordinate patch such that $\Omega' := \Omega \cap \partial M \neq \varnothing$. Without loss of generality we shall assume that the coordinate patch Ω is tubular, that is, $\Omega = \Omega' \times [0, \delta)$ where $\Omega' \subset \partial M$ and δ is some positive number. Thus, Ω' will be the coordinate patch corresponding to the local coordinates x'.

Let us fix an arbitrary function $\chi \in C^\infty(\partial M)$ such that $0 \leqslant \chi \leqslant 1$ and $\operatorname{supp} \chi \subset \Omega'$. Then in order to prove (5.4.3) it is sufficient to prove

$$\begin{aligned} \operatorname{Tr}(\chi^2(x')\, \chi_-^2(x_n/d)\, E_\lambda) &= \lambda^n \int_\Omega c_0(x)\, \chi^2(x')\, \chi_-^2(x_n/d)\, dx \\ &\quad + \lambda^{n-1} \int_{\Omega'} c_1(x')\, \chi^2(x')\, dx' + O(\lambda^{n-1-\varepsilon_5}). \end{aligned} \tag{5.4.14}$$

Indeed, (5.4.3) is reduced to (5.4.14) by a partition of unity in a neighbourhood of ∂M (see also formulae (5.4.7), (5.4.8)). So further on we deal with (5.4.14). The advantage of dealing with (5.4.14) is that now we have a fixed local coordinate system. Fixation of the coordinate system is necessary because eventually we intend to approximate our original spectral problem (1.1.1), (1.1.2) on the manifold M by a problem with constant coefficients in the half-space $\mathbb{R}^n_+$.

Let $Q'(z',d)$ be the closed $(n-1)$-dimensional cube with centre $z' \in \mathbb{R}^{n-1}$ of side $4d$, that is, $Q'(z',d) = \{x' : |x_k - z_k| \leqslant 2d,\ k = 1, \dots, n-1\}$. Let $q'(x';z',d)$ be the characteristic function of this cube, that is,

$$q'(x';z',d) = \begin{cases} 1 & \text{if } \ x' \in Q'(z',d)\,, \\ 0 & \text{if } \ x' \notin Q'(z',d)\,. \end{cases}$$

Similarly, let $Q(z',d)$ be the closed n-dimensional cube with centre $(z',2d) \in \mathbb{R}^n$ of side $4d$, that is, $Q(z',d) = \{x : |x_k - z_k| \leqslant 2d,\ k = 1, \dots, n-1,\ |x_n - 2d| \leqslant 2d\}$. Let $q(x;z',d)$ be the characteristic function of this cube, that is,

$$q(x;z',d) = \begin{cases} 1 & \text{if } \ x \in Q(z',d)\,, \\ 0 & \text{if } \ x \notin Q(z',d)\,. \end{cases}$$

Note that the identity $\operatorname{supp} \chi(x_n/d) = [0, 4d]$ (see 5.1.2) implies

$$\operatorname{supp}(q'(\,\cdot\,; z',d)\, \chi_-(\,\cdot\,/d)) = \operatorname{supp} q(\,\cdot\,; z',d) = Q(z',d)\,.$$

Then in order to prove (5.4.14) it is sufficient to prove

$$\begin{aligned} &\operatorname{Tr}(q'(x';z',d)\, \chi^2(x')\, \chi_-^2(x_n/d)\, E_\lambda) \\ &\quad = \lambda^n \int_{Q(z',d)} c_0(x)\ \chi^2(x')\, \chi_-^2(x_n/d)\, dx \\ &\qquad + \lambda^{n-1} \int_{Q'(z',d)} c_1(x')\, \chi^2(x')\, dx' + O(\lambda^{(n-1)\varepsilon - \varepsilon_5}) \end{aligned} \tag{5.4.15}$$

uniformly over all $z' \in K'$, where K' is an arbitrary compact set in Ω' such that $\operatorname{supp} \chi$ lies in the interior of K'. Indeed, (5.4.14) is reduced to (5.4.15) by partitioning a small neighbourhood of $\operatorname{supp} \chi$ into $(n-1)$-dimensional cubes of side $4d$. The number of cubes needed to carry out this partition is $O(d^{1-n}) \equiv O(\lambda^{(n-1)(1-\varepsilon)})$, and this explains the structure of the remainder term in (5.4.15).

Finally, in order to prove (5.4.15) it is sufficient to prove

$$\begin{aligned} &\operatorname{Tr}(q'(x';z',d)\, \chi^2(x')\, \chi_-^2(x_n/d)\, E_\lambda) \\ &\quad = \lambda^n\, c_0(z',0) \int_{Q(z',d)} \chi^2(x')\, \chi_-^2(x_n/d)\, dx \\ &\qquad + \lambda^{n-1}\, c_1(z') \int_{Q'(z',d)} \chi^2(x')\, dx' + O(\lambda^{(n-1)\varepsilon - \varepsilon_5})\,. \end{aligned} \tag{5.4.16}$$

Indeed, formula (5.4.16) differs from (5.4.15) only in that on the right-hand side we have "frozen" the coefficients in $c_0(x)$ and $c_1(x')$. The additional error induced by this freezing of coefficients is small because these functions are smooth (see Lemma 5.4.2) and because integration is carried out over cubes of side $\sim \lambda^{\varepsilon - 1}$. This additional error is included in $O(\lambda^{(n-1)\varepsilon - \varepsilon_5})$ because $\varepsilon_5 \leqslant 1 - 2\varepsilon$; see (5.4.1).

Denote for brevity

$$\mathbf{c}_0(z',d) := c_0(z',0)\int_{Q(z',d)} \chi^2(x')\,\chi_-^2(x_n/d)\,dx\,, \tag{5.4.17}$$

$$\mathbf{c}_1(z',d) := c_1(z')\int_{Q'(z',d)} \chi^2(x')\,dx' \tag{5.4.18}$$

(obviously, $\mathbf{c}_0(z',d)$, $\mathbf{c}_1(z',d)$ are of the order of d^n, d^{n-1}, respectively). Then (5.4.16) can be rewritten as

$$\begin{aligned}&\operatorname{Tr}(q'(x';z',d)\,\chi^2(x')\,\chi_-^2(x_n/d)\,E_\lambda)\\ &\qquad = \mathbf{c}_0(z',d)\,\lambda^n + \mathbf{c}_1(z',d)\,\lambda^{n-1} + O(\lambda^{(n-1)\varepsilon-\varepsilon_5})\,.\end{aligned} \tag{5.4.19}$$

Thus, we have reduced the proof of (5.4.3) to the proof of (5.4.19). The asymptotic formula (5.4.19) should be proved uniformly over d from the interval (5.4.2) and $z'\in K'$, where K' is an arbitrary compact in Ω' such that $\operatorname{supp}\chi\subset K'$.

2. Reduction to a problem for resolvents. Consider now the auxiliary spectral problem with constant coefficients

$$A_{2m}(z',0,D_x)v = \nu v\,, \tag{5.4.20}$$

$$\left(B_{m_j}^{(j)}(z',D_x)v\right)\Big|_{x_n=0} = 0\,, \qquad j=1,2,\dots,m, \tag{5.4.21}$$

in the half-space $\mathbb{R}^n_+ \equiv \{x_n \geqslant 0\}$. This problem depends on the point z' as a parameter. Let us denote by $\mathbf{A}^+(z')$ the self-adjoint operator in $L_2(\mathbb{R}^n_+)$ associated with (5.4.20), (5.4.21), and by $\mathbf{E}^+_\nu(z')$ and $\mathbf{R}^+_\mu(z')$ the respective spectral projection and resolvent.

Let us also consider the spectral problem (5.4.20) on the whole space $\mathbb{R}^n$ without boundary conditions, and let us denote by $\mathbf{A}(z')$ the corresponding self-adjoint operator in $L_2(\mathbb{R}^n)$. By $\mathbf{E}_\nu(z')$ and $\mathbf{R}_\mu(z')$ we shall denote the respective spectral projection and resolvent.

Obviously, $\sigma(\mathbf{A}^+(z')) = \sigma(\mathbf{A}(z')) = \mathbb{R}_+$.

Our auxiliary spectral problems in $\mathbb{R}^n_+$ and $\mathbb{R}^n$ admit a separation of variables if one takes the x'-dependence to be of the form $e^{i\langle x',\xi'\rangle}$. The resulting operators in $L_2(\mathbb{R}_+)$ and $L_2(\mathbb{R})$ will be distinguished (throughout this section) from the operators in $L_2(\mathbb{R}^n_+)$ and $L_2(\mathbb{R}^n)$ by the explicit indication of the parameters (z',ξ'); the same applies to spectral projections and resolvents.

The most straightforward way of proving (5.4.19) would be to replace the spectral projection E_λ of our original problem (1.1.1), (1.1.2) by the spectral projection $\mathbf{E}_{\lambda^{2m}}$ of the problem (5.4.20), (5.4.21), thus reducing our analysis to that of a simpler problem. However it is difficult to estimate directly the error induced by such a replacement. So we adopt a technique which reduces (with a controllable error) the comparison of spectral projections to the comparison of resolvents. The latter is much easier because on the whole the resolvent is a very stable mathematical object. In particular, the resolvent possesses the property of quasilocality (see Lemma 5.4.6 below for details) and this property will play a crucial role in our estimates.

Let $N(\nu)$ be a real-valued monotone nondecreasing function on $\mathbb{R}$ such that $N(\nu) = 0$ for $\nu \leqslant 0$ and there exists some constant $0 < p < 1$ such that

$\nu^{-p}N(\nu) \to 0$ as $\nu \to +\infty$. For $\mu \in \mathbb{C} \setminus \mathbb{R}_+$ set

$$M(\mu) := \int_{-0}^{+\infty} \frac{dN(\nu)}{\nu - \mu} \equiv \int_{0}^{+\infty} \frac{N(\nu)}{(\nu-\mu)^2}\, d\nu\,.$$

The holomorphic function $M(\mu)$ is the Stieltjes transform of the (positive) measure $dN(\nu)$. Let $\varkappa$ be a complex number with $\operatorname{Im}\varkappa > 0$, and let $L(\varkappa)$ be the oriented arc of the circumference $|\mu| = |\varkappa|$ in the complex μ-plane going counterclockwise from $\varkappa$ to $\overline{\varkappa}$. Then we have the following estimate due to Pleijel [**Pl**]:

$$\left| N(\nu) + \frac{1}{2\pi i} \int_{L(\varkappa)} M(\mu)\, d\mu \right| \leqslant (\operatorname{Im}\varkappa)\, |M(\varkappa)|\, \sqrt{1+\pi^{-2}}\,. \tag{5.4.22}$$

Pleijel's formula (5.4.22) can be viewed as an approximate inversion formula for the Stieltjes transform of a positive measure. A simple proof of (5.4.22) can be found in [**Agm2,** Sect. 2].

From now on we specify the choice of the complex number $\varkappa$ appearing in (5.4.22) in the following way:

$$\operatorname{Re}\varkappa = \lambda^{2m}, \qquad \operatorname{Im}\varkappa = \lambda^{2m-\varepsilon-\varepsilon_5}. \tag{5.4.23}$$

Applying Pleijel's formula (5.4.22) to the function

$$N(\nu) = \operatorname{Tr}(q'(x';z',d)\, \chi^2(x')\, \chi_-^2(x_n/d)\, E_{\nu^{1/(2m)}})$$

with $\varkappa = \varkappa(\lambda)$ given by (5.4.23), we conclude that in order to prove (5.4.19) it is sufficient to prove

$$\begin{aligned} -\frac{1}{2\pi i} \int_{L(\varkappa(\lambda))} & \operatorname{Tr}(q'(x';z',d)\, \chi^2(x')\, \chi_-^2(x_n/d)\, \mathcal{R}_\mu)\, d\mu \\ &= \mathbf{c}_0(z',d)\, \lambda^n \;+\; \mathbf{c}_1(z',d)\, \lambda^{n-1} \;+\; O(\lambda^{(n-1)\varepsilon-\varepsilon_5}) \end{aligned} \tag{5.4.24}$$

and

$$\operatorname{Tr}(q'(x';z',d)\, \chi^2(x')\, \chi_-^2(x_n/d)\, \mathcal{R}_{\varkappa(\lambda)}) \;=\; O(\lambda^{n\varepsilon-2m})\,. \tag{5.4.25}$$

Here $\mathcal{R}_\mu = (\mathcal{A} - \mu I)^{-1}$ is the resolvent of the operator $\mathcal{A}$ associated with our original spectral problem (1.1.1′), (1.1.2) (see also 1.1.7).

Let us introduce some additional notation. Let $\|\cdot\|_1$ denote the nuclear norm of an operator. Let $(-\mu)^{1/(2m)}$, $\mu \in \mathbb{C}\setminus\mathbb{R}_+$, be the branch of the $2m$th root of $-\mu$ specified by the condition $(-\mu)^{1/(2m)} > 0$ for $\mu < 0$. Finally, let us denote by $\varphi(z',d)$ the restriction operator acting from $L_2(M)$ to $L_2(Q(z',d))$, and by $\varphi^*(z',d)$ the extension operator acting from $L_2(Q(z',d))$ to $L_2(M)$, that is,

$$(\varphi^*(z',d)v)(x) \;=\; \begin{cases} 0\,, & x \in M \setminus Q(z',d)\,, \\ v(x)\,, & x \in Q(z',d)\,. \end{cases}$$

Similarly, let us denote by $\psi(z',d)$ the restriction operator acting from $L_2(\mathbb{R}^n_+)$ to $L_2(Q(z',d))$, and by $\psi^*(z',d)$ the extension operator acting from $L_2(Q(z',d))$ to $L_2(\mathbb{R}^n_+)$, that is,

$$(\psi^*(z',d)v)(x) \;=\; \begin{cases} 0\,, & x \in \mathbb{R}^n_+ \setminus Q(z',d)\,, \\ v(x)\,, & x \in Q(z',d)\,. \end{cases}$$

Suppose that we have proved

$$\|\varphi(z',d)\mathcal{R}_\mu\varphi^*(z',d)-\psi(z',d)\mathbf{R}_\mu^+(z')\psi^*(z',d)\|_1 = O\left(\frac{\lambda^{2m+(n-2)\varepsilon-2\varepsilon_5}}{\operatorname{dist}^2(\mu,\mathbb{R}_+)}\right) \tag{5.4.26}$$

and

$$\begin{aligned}&\operatorname{Tr}(q'(x';z',d)\,\chi^2(x')\,\chi_-^2(x_n/d)\,\mathbf{R}_\mu^+(z'))\\ &\qquad= \frac{\pi n\mathbf{c}_0(z',d)}{2m\sin\left(\frac{\pi n}{2m}\right)}(-\mu)^{(n-2m)/(2m)}\\ &\qquad+\frac{\pi(n-1)\mathbf{c}_1(z',d)}{2m\sin\left(\frac{\pi(n-1)}{2m}\right)}(-\mu)^{(n-1-2m)/(2m)}+O(\lambda^{-\infty})\end{aligned} \tag{5.4.27}$$

uniformly over $\mu\in L(\varkappa(\lambda))$. Then formulae (5.4.26), (5.4.27) would imply (5.4.24), (5.4.25).

Thus, we have reduced the proof of (5.4.3) to the proof of (5.4.26), (5.4.27). The asymptotic formulae (5.4.26), (5.4.27) should be proved uniformly over d from the interval (5.4.2), $z'\in K'$, and $\mu\in L(\varkappa(\lambda))$.

3. Proof of (5.4.27). Let $\mathbf{r}^+(\mu,x,y;z')$, $\mathbf{r}(\mu,x,y;z')$ be the integral kernels of $\mathbf{R}_\mu^+(z')$, $\mathbf{R}_\mu(z')$. With account of (5.4.5) we have

$$\begin{aligned}\mathbf{r}(\mu,x,x;z') &= \int_{T'_{(z',0)}M}\frac{d\!\!{}^-\xi}{A_{2m}(z',0,\xi)-\mu} = \left(\int_0^{+\infty}\frac{r^{n-1}\,dr}{r^{2m}-\mu}\right)\left(\int_{S^*_{(z',0)}M}\widetilde{d\!\!{}^-\xi}\right)\\ &= n\,c_0(z',0)\,(-\mu)^{(n-2m)/(2m)}\int_0^{+\infty}\frac{r^{n-1}\,dr}{r^{2m}+1} = \frac{\pi n c_0(z',0)}{2m\sin\left(\frac{\pi n}{2m}\right)}(-\mu)^{(n-2m)/(2m)};\end{aligned}$$

consequently (see (5.4.17))

$$\operatorname{Tr}(q'(x';z',d)\,\chi^2(x')\,\chi_-^2(x_n/d)\,\mathbf{R}_\mu(z')) = \frac{\pi n\mathbf{c}_0(z',d)}{2m\sin\left(\frac{\pi n}{2m}\right)}(-\mu)^{(n-2m)/(2m)}.$$

Subtracting the latter formula from (5.4.27) we conclude that proving (5.4.27) is equivalent to proving

$$\begin{aligned}&\operatorname{Tr}(q'(x';z',d)\,\chi^2(x')\,\chi_-^2(x_n/d)\,(\mathbf{R}_\mu^+(z')-\theta\mathbf{R}_\mu(z')\theta^*))\\ &\qquad=\frac{\pi(n-1)\mathbf{c}_1(z',d)}{2m\sin\left(\frac{\pi(n-1)}{2m}\right)}(-\mu)^{(n-1-2m)/(2m)}+O(\lambda^{-\infty}).\end{aligned} \tag{5.4.28}$$

Here by θ we denote the restriction operator acting from $L_2(\mathbb{R}^n)$ to $L_2(\mathbb{R}^n_+)$, and by θ^* the extension operator acting from $L_2(\mathbb{R}^n_+)$ to $L_2(\mathbb{R}^n)$, that is,

$$(\theta^*v)(x) = \begin{cases}0, & x_n<0,\\ v(x), & x_n\geqslant 0.\end{cases}$$

Formulae (5.4.13), (5.4.6) imply

$$\int_0^{+\infty} (\mathbf{r}^+(\mu,x,x;z') - \mathbf{r}(\mu,x,x;z'))\,dx_n \equiv \mathbf{f}^+(\mu;z')$$
$$= c_1(z')\,(-\mu)^{(n-1-2m)/(2m)} \int_0^{+\infty} \frac{\nu^{(n-1)/(2m)}}{(\nu+1)^2}\,d\nu$$
$$= \frac{\pi(n-1)c_1(z')}{2m\sin\left(\frac{\pi(n-1)}{2m}\right)}\,(-\mu)^{(n-1-2m)/(2m)};$$

consequently (see (5.4.18))

$$\operatorname{Tr}(q'(x';z',d)\,\chi^2(x')\,(\mathbf{R}^+_\mu(z') - \theta\mathbf{R}_\mu(z')\theta^*))$$
$$= \frac{\pi(n-1)\mathbf{c}_1(z',d)}{2m\sin\left(\frac{\pi(n-1)}{2m}\right)}\,(-\mu)^{(n-1-2m)/(2m)}\,.$$

Comparing the latter formula with (5.4.28) we conclude that proving (5.4.28) is equivalent to proving

$$\operatorname{Tr}(q'(x';z',d)\,\chi^2(x')\,\chi_+^2(x_n/(2d))\,(\mathbf{R}^+_\mu(z') - \theta\mathbf{R}_\mu(z')\theta^*)) = O(\lambda^{-\infty})\,, \tag{5.4.29}$$

where $\chi_+(x_n) := \sqrt{1 - \chi_-^2(2x_n)}$ (see the beginning of 5.1.2).

As $\mathbf{r}^+(\mu,x,x;z')$, $\mathbf{r}(\mu,x,x;z')$ do not depend on x', proving (5.4.29) is equivalent to proving

$$\int_0^{+\infty} \chi_+^2(x_n/(2d))\,(\mathbf{r}^+(\mu,z',x_n,z',x_n;z') - \mathbf{r}(\mu,z',x_n,z',x_n;z'))\,dx_n = O(\lambda^{-\infty})\,. \tag{5.4.30}$$

By separation of variables, in order to prove (5.4.30) it is sufficient to prove

$$\int_0^{+\infty} \chi_+^2(x_n/(2d))\,(\mathbf{r}^+(\mu,x_n,x_n;z',\xi') - \mathbf{r}(\mu,x_n,x_n;z',\xi'))\,dx_n = O((1+|\xi'|^2)^{-\infty}\lambda^{-\infty})\,, \tag{5.4.31}$$

where $\mathbf{r}^+(\mu,x_n,y_n;z',\xi')$, $\mathbf{r}(\mu,x_n,y_n;z',\xi')$ are the integral kernels of $\mathbf{R}^+_\mu(z',\xi')$, $\mathbf{R}_\mu(z',\xi')$. The estimate (5.4.31) should be proved uniformly over d from the interval (5.4.2), $(z',\xi')\in T^*K'$, and $\mu\in L(\varkappa(\lambda))$.

Let const >0 be a sufficiently large constant so that $\lambda^{2m}\notin\sigma(\mathbf{A}^+(z',\xi'))$ for $|\xi'|\geqslant \mathrm{const}\,\lambda$ (cf. (1.6.18)). Then for $|\xi'|\geqslant\mathrm{const}\,\lambda$ the required estimate (5.4.31) follows from Lemma A.4.2 and formula (5.4.2) by a rescaling argument. For $|\xi'|\leqslant\mathrm{const}\,\lambda$ the required estimate (5.4.31) follows from Lemma A.4.3, formulae (5.4.2), (5.4.23) and $\varepsilon_5<\varepsilon$ (see (5.4.1)) by a rescaling argument. Thus, we have proved (5.4.31), and consequently we have proved (5.4.27).

REMARK 5.4.3. Lemma A.4.3 used above is based on the following observation. Let $\mathbf{A}(\zeta) := a_0 + a_1\zeta + \cdots + a_k\zeta^k$, $a_k\neq 0$, $k\geqslant 2$, be a polynomial with real coefficients. Consider the algebraic equation $\mathbf{A}(\zeta) = i\delta$ where $\delta\in\mathbb{R}\setminus\{0\}$ is a parameter. Let $\zeta^+(\delta)$ and $\zeta^-(\delta)$ be roots of this algebraic equation with positive and negative imaginary part, respectively. Then there exists a positive constant

$\mathbf{C}$ such that uniformly over the choice of these roots and uniformly over small real perturbations of the coefficients $a_0, a_1, \ldots, a_k$ we have

$$|\zeta^+(\delta) - \zeta^-(\delta)| \geqslant \mathbf{C}\sqrt{|\delta|}\,.$$

The appearance of this square root is crucial for our arguments.

Before we proceed to the proof of the estimate (5.4.26) we will have to establish some general facts concerning the resolvents of the problems (1.1.1), (1.1.2) and (5.4.20), (5.4.21). This is done in the next two subsections.

4. General properties of the resolvent $\mathbf{R}^+_\mu(z')$. In this subsection we denote the spectral parameter by μ, and consider complex values of μ of the form $\mu = \Lambda^{2m} e^{\pm i\gamma}$, where $0 < \gamma \leqslant \pi$, $\Lambda > 0$. Our aim is to obtain estimates for the resolvent of the problem (5.4.20), (5.4.21) which would be uniform over γ, Λ, and z' (point at which we have frozen the coefficients).

The results of this subsection and the next one are similar to those of [**AgrVi**], the basic difference being that we have the additional parameter γ which gives a factor of γ^{-1} in all our estimates for the resolvents. We are able to control the influence of the parameter γ due to the self-adjointness of our problems.

We will be looking at the following nonhomogeneous version of the problem (5.4.20), (5.4.21):

$$A_{2m}(z', 0, D_x)v = \mu v + f\,, \tag{5.4.32}$$

$$\Lambda^{-m_j}\left(B^{(j)}_{m_j}(z', D_x)v\right)\Big|_{x_n=0} = g_j\,, \qquad j = 1, 2, \ldots, m, \tag{5.4.33}$$

where $f \in L_2(\mathbb{R}^n_+)$, $g_j \in H^{2m-m_j-1/2}(\mathbb{R}^{n-1})$ are given functions. The solution of the problem (5.4.32), (5.4.33) can be written down as

$$v = \mathbf{R}^+_\mu(z')\, f + \mathbf{S}_\mu(z')\, \vec{g}\,, \tag{5.4.34}$$

where $\vec{g} = \{g_1, \ldots, g_m\}$ and $\mathbf{S}_\mu(z')$ is a bounded linear operator acting from $\prod_{j=1}^m H^{2m-m_j-1/2}(\mathbb{R}^{n-1})$ to $H^{2m}(\mathbb{R}^n_+)$.

We denote

$$(v, w)_{L_2(\mathbb{R}^n_+)} = \int_{\mathbb{R}^n_+} v(x)\, \overline{w(x)}\, dx\,,$$

$$(v, w)_{L_2(\mathbb{R}^{n-1})} = \int_{\mathbb{R}^{n-1}} v(x')\, \overline{w(x')}\, dx'$$

(scalar products in $L_2(\mathbb{R}^n_+)$, $L_2(\mathbb{R}^{n-1})$), and

$$(v, w)_{H^s(\mathbb{R}^n_+)} = (v, w)_{L_2(\mathbb{R}^n_+)} + \sum_{|\alpha|=s} (\partial^\alpha_x v, \partial^\alpha_x w)_{L_2(\mathbb{R}^n_+)}\,, \qquad s = 0, 1, 2, \ldots,$$

$$(v, w)_{H^s(\mathbb{R}^{n-1})} = (v, w)_{L_2(\mathbb{R}^{n-1})} + \int_{\mathbb{R}^{n-1}} |\xi'|^{2s}\, \hat{v}(\xi')\, \overline{\hat{w}(\xi')}\, d\!\!{}^-\xi'\,, \qquad s \in [0, +\infty)$$

(scalar products in the Sobolev spaces $H^s(\mathbb{R}^n_+)$, $H^s(\mathbb{R}^{n-1})$); here $\hat{v}(\xi')$, $\hat{w}(\xi')$ are the Fourier transforms of $v(\xi')$, $w(\xi')$.

In studying the mapping properties of the operators $\mathbf{R}^+_\mu(z')$, $\mathbf{S}_\mu(z')$ it is, however, inconvenient to use the standard Sobolev spaces described above. It is more natural to deal with weighted Sobolev spaces $H^s(\mathbb{R}^n_+;\Lambda)$, $H^s(\mathbb{R}^{n-1};\Lambda)$ defined by the scalar products

$$(v,w)_{H^s(\mathbb{R}^n_+;\Lambda)} = (v,w)_{L_2(\mathbb{R}^n_+)} + \Lambda^{-2s}\sum_{|\alpha|=s}(\partial_x^\alpha v, \partial_x^\alpha w)_{L_2(\mathbb{R}^n_+)}, \qquad s=0,1,2,\dots,$$

$$(v,w)_{H^s(\mathbb{R}^{n-1};\Lambda)} = (v,w)_{L_2(\mathbb{R}^{n-1})} + \Lambda^{-2s}\int_{\mathbb{R}^{n-1}}|\xi'|^{2s}\hat{v}(\xi')\overline{\hat{w}(\xi')}\,d\xi', \quad s\in[0,+\infty),$$

respectively.

Let $\mathbf{F}_\Lambda$, $\mathbf{P}_\Lambda(z')$ be the linear operators mapping

$$v(x)\to\left\{(v)\big|_{x_n=0}, \Lambda^{-1}\left(D_{x_n}v\right)\big|_{x_n=0},\dots,\Lambda^{1-2m}\left(D_{x_n}^{2m-1}v\right)\big|_{x_n=0}\right\},$$

$$v(x)\to\left\{\Lambda^{-m_1}\left(B_{m_1}^{(1)}(z',D_x)v\right)\Big|_{x_n=0},\dots,\Lambda^{-m_m}\left(B_{m_m}^{(m)}(z',D_x)v\right)\Big|_{x_n=0}\right\},$$

respectively. We have

$$\|\mathbf{F}_\Lambda\|_{H^{2m}(\mathbb{R}^n_+;\Lambda)\to\prod_{j=1}^{2m}H^{2m-j+1/2}(\mathbb{R}^{n-1};\Lambda)} = O(\Lambda^{1/2}), \tag{5.4.35}$$

$$\|\mathbf{P}_\Lambda(z')\|_{H^{2m}(\mathbb{R}^n_+;\Lambda)\to\prod_{j=1}^{m}H^{2m-m_j-1/2}(\mathbb{R}^{n-1};\Lambda)} = O(\Lambda^{1/2}) \tag{5.4.36}$$

uniformly over $0<\gamma\leqslant\pi$, $\Lambda>0$, and $z'\in K'$, where K' is an arbitrary compact set in Ω'. Indeed, for $\Lambda=1$ the estimates (5.4.35), (5.4.36) follow from Theorem 8.3 of [**LioMag,** Chap. 1, Sect. 8.2], whereas the case of arbitrary $\Lambda>0$ is reduced to the case $\Lambda=1$ by rescaling (change of independent variable $\widetilde{x}=\Lambda x$).

Let $\mathbf{G}_\Lambda$ be a right inverse of $\mathbf{F}_\Lambda$, and let $\mathbf{Q}_\Lambda(z')$ be a right inverse of $\mathbf{P}_\Lambda(z')$; that is,

$$\mathbf{F}_\Lambda\,\mathbf{G}_\Lambda = I, \qquad \mathbf{P}_\Lambda(z')\,\mathbf{Q}_\Lambda(z') = I. \tag{5.4.37}$$

We shall choose $\mathbf{G}_\Lambda$, $\mathbf{Q}_\Lambda(z')$ in such a way that

$$\|\mathbf{G}_\Lambda\|_{\prod_{j=1}^{2m}H^{2m-j+1/2}(\mathbb{R}^{n-1};\Lambda)\to H^{2m}(\mathbb{R}^n_+;\Lambda)} = O(\Lambda^{-1/2}), \tag{5.4.38}$$

$$\|\mathbf{Q}_\Lambda(z')\|_{\prod_{j=1}^{m}H^{2m-m_j-1/2}(\mathbb{R}^{n-1};\Lambda)\to H^{2m}(\mathbb{R}^n_+;\Lambda)} = O(\Lambda^{-1/2}) \tag{5.4.39}$$

uniformly over $0<\gamma\leqslant\pi$, $\Lambda>0$, and $z'\in K'$. For $\Lambda=1$ the existence of operators $\mathbf{G}_\Lambda$, $\mathbf{Q}_\Lambda(z')$ satisfying (5.4.37)–(5.4.39) follows from Theorem 8.3 of [**LioMag,** Chap. 1, Sect. 8.2], whereas the case of arbitrary $\Lambda>0$ is reduced to the case $\Lambda=1$ by rescaling.

LEMMA 5.4.4. *We have*

$$\|\mathbf{R}_\mu^+(z')\|_{L_2(\mathbb{R}^n_+)\to H^{2m}(\mathbb{R}^n_+;\Lambda)} = O(\gamma^{-1}\Lambda^{-2m}), \tag{5.4.40}$$

$$\|\mathbf{S}_\mu(z')\|_{\prod_{j=1}^{m}H^{2m-m_j-1/2}(\mathbb{R}^{n-1};\Lambda)\to H^{2m}(\mathbb{R}^n_+;\Lambda)} = O(\gamma^{-1}\Lambda^{-1/2}), \tag{5.4.41}$$

uniformly over $0<\gamma\leqslant\pi$, $\Lambda>0$, *and* $z'\in K'$.

PROOF OF LEMMA 5.4.4. The case of arbitrary $\Lambda > 0$ is reduced to the case $\Lambda = 1$ by rescaling. So further on in this proof we assume that $\Lambda = 1$.

The problem (5.4.32), (5.4.33) is regular elliptic (see [**LioMag,** Chap. 2, Sect. 1.4], [**RoShSo,** Sect. 2.4]) and self-adjoint. The corresponding operator $\mathbf{A}^+(z')$ acts in $L_2(\mathbb{R}^n_+)$ and its domain of definition is

$$D(\mathbf{A}^+(z')) = \{v \in H^{2m}(\mathbb{R}^n_+) : \left.(B^{(j)}_{m_j}(z', D_x)v)\right|_{x_n=0} = 0, \quad j = 1, 2, \ldots, m\}.$$

It is easy to see that $\sigma(\mathbf{A}^+(z')) = \mathbb{R}_+$.

By Theorem 4.1 from [**LioMag,** Chap. 2, Sect. 4.4] the norm $\|\cdot\|_{H^{2m}(\mathbb{R}^n_+)}$ is equivalent on $D(\mathbf{A}^+(z'))$ to the norm $\|(A_{2m}(z', 0, D_x) + I)(\,\cdot\,)\|_{L_2(\mathbb{R}^n_+)}$, so proving (5.4.40) is equivalent to proving

$$\|(A_{2m}(z', 0, D_x) + I)\mathbf{R}^+_\mu(z')\|_{L_2(\mathbb{R}^n_+)\to L_2(\mathbb{R}^n_+)} = O(\gamma^{-1}). \tag{5.4.42}$$

But

$$(A_{2m}(z', 0, D_x) + I)\,\mathbf{R}^+_\mu(z') = I + (1 + \mu)\,\mathbf{R}^+_\mu(z'), \tag{5.4.43}$$

and because the operator $\mathbf{A}^+(z')$ is self-adjoint and nonnegative,

$$\|\mathbf{R}^+_\mu(z')\|_{L_2(\mathbb{R}^n_+)\to L_2(\mathbb{R}^n_+)} \leqslant \pi\gamma^{-1}. \tag{5.4.44}$$

Formulae (5.4.43), (5.4.44) imply (5.4.42).

We have

$$\mathbf{S}_\mu(z')\,\vec{g} = \mathbf{Q}_1(z')\,\vec{g} - \mathbf{R}^+_\mu(z')\,(A_{2m}(z', 0, D_x) - \mu I)\,\mathbf{Q}_1(z')\,\vec{g}.$$

The above formula and formulae (5.4.39), (5.4.40) imply (5.4.41). □

The following lemma will enable us to describe the mapping properties of differential operators which are not necessarily homogeneous with respect to D_x and whose coefficients are not necessarily constant.

LEMMA 5.4.5. *For any integer $p \geqslant 0$ there exists a constant $C_p > 0$ such that for any integers $0 \leqslant q \leqslant p$, $r \leqslant p - q$, and any differential operators*

$$\mathcal{L} = \sum_{|\alpha|\leqslant r} \ell_\alpha(x)\,D^\alpha_x, \qquad \ell_\alpha \in C^\infty(\mathbb{R}^n_+),$$

$$\mathcal{L}' = \sum_{|\alpha'|\leqslant r} \ell'_\alpha(x')\,D^{\alpha'}_{x'}, \qquad \ell'_{\alpha'} \in C^\infty(\mathbb{R}^{n-1}),$$

we have

$$\|\mathcal{L}\|_{H^p(\mathbb{R}^n_+;\Lambda)\to H^q(\mathbb{R}^n_+;\Lambda)} \leqslant C_p\,\Lambda^r \max_{\substack{|\alpha|\leqslant r\\ |\beta|\leqslant q}} \Big(\Lambda^{-|\beta|} \sup_{x\in\mathbb{R}^n_+} |\partial^\beta_x \ell_\alpha|\Big), \tag{5.4.45}$$

$$\begin{aligned} &\|\mathcal{L}'\|_{H^{p+1/2}(\mathbb{R}^{n-1};\Lambda)\to H^{q+1/2}(\mathbb{R}^{n-1};\Lambda)} \\ &\qquad \leqslant C_p\,\Lambda^r \max_{\substack{|\alpha'|\leqslant r\\ |\beta'|\leqslant q+1}} \Big(\Lambda^{-|\beta'|} \sup_{x'\in\mathbb{R}^{n-1}} |\partial^{\beta'}_{x'} \ell'_{\alpha'}|\Big). \end{aligned} \tag{5.4.46}$$

PROOF OF LEMMA 5.4.5. The case of arbitrary $\Lambda>0$ is reduced to the case $\Lambda=1$ by rescaling. So further on in this proof we assume that $\Lambda=1$.

The estimate (5.4.45) is established by an explicit computation of the Sobolev norm $\|\mathcal{L}v\|_{H^q(\mathbb{R}^n_+)}$.

With the estimate (5.4.46) we encounter the difficulty that the explicit computation of the Sobolev norm $\|\mathcal{L}'v'\|_{H^{q+1/2}(\mathbb{R}^{n-1})}$ requires the use of the Fourier transform. However, this difficulty is easily overcome by an extension–restriction argument: first we extend the function $v'\in H^{p+1/2}(\mathbb{R}^{n-1})$ to a function $v\in H^{p+1}(\mathbb{R}^n_+)$ in such a way that $v|_{x_n=0}=v'$ and $\|v\|_{H^{p+1}(\mathbb{R}^n_+)}\leqslant \text{const}\,\|v'\|_{H^{p+1/2}(\mathbb{R}^{n-1})}$, then we use the estimate (5.4.45) with $\mathcal{L}'$, $p+1$, $q+1$ in place of $\mathcal{L}$, p, q, respectively, and finally we restrict $\mathcal{L}'v$ to $x_n=0$. □

Let us now prove that the integral kernel of the resolvent $\mathbf{R}^+_\mu(z')$ rapidly vanishes outside the diagonal $x=y$. We prove an *integral* version of this quasilocality property, that is, we consider the composition of $\mathbf{R}^+_\mu(z')$ with two cut-offs (left cut-off and right cut-off) and evaluate the norm of the resulting operator when the supports of the cut-offs are being moved away from each other. Estimating integrated quantities is simpler than obtaining pointwise estimates, and dealing with integrated quantities is sufficient for our needs (see subsection 6).

In what follows we deal with some functions $X'(x')$, $x'\in\mathbb{R}^{n-1}$, and $X(x)$, $Y(x)$, $x\in\mathbb{R}^n$. We shall consider "shifted" versions of these functions, that is, $X'(x'-\mathbf{x}')$, $X(x-\mathbf{x})$, $Y(x-\mathbf{y})$, where $x'\in\mathbb{R}^{n-1}$, $x\in\mathbb{R}^n$ are the independent variables and $\mathbf{x}'\in\mathbb{R}^{n-1}$, $\mathbf{x},\mathbf{y}\in\mathbb{R}^n_+$ are parameters. Furthermore, we shall consider the restriction of the functions $X(x-\mathbf{x})$, $Y(x-\mathbf{y})$ to $\mathbb{R}^n_+$, that is, we will allow the independent variable x to take values only from $\mathbb{R}^n_+$; in order to simplify notation we shall continue denoting these restrictions by $X(x-\mathbf{x})$, $Y(x-\mathbf{y})$. Finally, for the sake of simplicity we shall denote the operators of multiplication by the functions $X'(x'-\mathbf{x}')$, $X(x-\mathbf{x})$, $Y(x-\mathbf{y})$ as the functions themselves. We shall also use the notation $|x'|:=\sqrt{x_1^2+\cdots+x_{n-1}^2}$, $|x|:=\sqrt{x_1^2+\cdots+x_n^2}$.

LEMMA 5.4.6 (quasilocality of the resolvent). *Let k be an arbitrary nonnegative integer, and let*

$$X'\in C^\infty(\mathbb{R}^{n-1}),\qquad X\in C^\infty(\mathbb{R}^n),\qquad Y\in L_\infty(\mathbb{R}^n)$$

be arbitrary functions satisfying

$$\operatorname{supp}X'\subset\{|x'|\leqslant\Lambda^{-1}\},\qquad \operatorname{supp}X\subset\{|x|\leqslant\Lambda^{-1}\},\qquad \operatorname{supp}Y\subset\{|x|\leqslant\Lambda^{-1}\},$$
$$|\partial^{\alpha'}_{x'}X'|\leqslant\Lambda^{|\alpha'|},\quad |\partial^\alpha_x X|\leqslant\Lambda^{|\alpha|},\quad |\partial^\alpha_x Y|\leqslant\Lambda^{|\alpha|}\quad \text{for}\quad |\alpha'|,|\alpha|\leqslant 2m,$$

and $\|Y\|_{L_\infty(\mathbb{R}^n)}\leqslant 1$. Then, uniformly over all such functions and over all

$$0<\gamma\leqslant\pi,\quad \Lambda>0,\quad z'\in K',\quad \mathbf{x},\mathbf{y}\in\mathbb{R}^n_+,\quad \mathbf{x}'\in\mathbb{R}^{n-1},\quad \mathbf{x}\neq\mathbf{y},\quad (\mathbf{x}',0)\neq\mathbf{y},$$

we have

(5.4.47)
$$\|X(x-\mathbf{x})\,\mathbf{R}^+_\mu(z')\,Y(x-\mathbf{y})\|_{L_2(\mathbb{R}^n_+)\to H^{2m}(\mathbb{R}^n_+;\Lambda)}=O(\gamma^{-k-1}\Lambda^{-k-2m}|\mathbf{x}-\mathbf{y}|^{-k}),$$

(5.4.48)
$$\begin{aligned}&\|X'(x'-\mathbf{x}')\,\mathbf{F}_\Lambda\,\mathbf{R}^+_\mu(z')\,Y(x-\mathbf{y})\|_{L_2(\mathbb{R}^n_+)\to\prod_{j=1}^{2m}H^{2m-j+1/2}(\mathbb{R}^{n-1};\Lambda)}\\&\quad=O(\gamma^{-k-1}\Lambda^{-k-2m+1/2}|(\mathbf{x}',0)-\mathbf{y}|^{-k}).\end{aligned}$$

Lemma 5.4.6 says in effect that the integral kernel $\mathbf{r}^+(\mu, x, y; z')$ of the resolvent $\mathbf{R}_\mu^+(z')$ vanishes faster than any negative power of $|x-y|$ as $|x-y| \to \infty$, and that this decay starts when $|x-y| \gg \gamma^{-1}\Lambda^{-1}$.

PROOF OF LEMMA 5.4.6. The case of arbitrary $\Lambda > 0$ is reduced to the case $\Lambda = 1$ by rescaling. So further on in this proof we assume that $\Lambda = 1$.

The estimate (5.4.48) is a consequence of (5.4.47), (5.4.35), (5.4.46). So below we prove (5.4.47).

For $k = 0$ or $|\mathbf{x}-\mathbf{y}| \leqslant 2^k + 1$ the estimate (5.4.47) follows from (5.4.40), (5.4.45); therefore we assume that $k > 0$ and

$$|\mathbf{x}-\mathbf{y}| > 2^k + 1 . \tag{5.4.49}$$

Let $\rho > 1$ be a parameter, and let $\upsilon \in C_0^\infty(\mathbb{R}_+)$ be a function such that $0 \leqslant \upsilon \leqslant 1$, $\operatorname{supp} \upsilon \subset [0,2]$, and $\upsilon = 1$ on $[0,1]$. Then, viewing x as the independent variable we can consider $\varsigma(x;\mathbf{y},\rho) := \upsilon(|x-\mathbf{y}|/\rho)$ as a function of the class $C_0^\infty(\mathbb{R}^n)$ (more precisely, as the restriction to $\mathbb{R}_+^n$ of a function from $C_0^\infty(\mathbb{R}^n)$). As usual, we shall denote the operator of multiplication by the function $\varsigma(x;\mathbf{y},\rho)$ as the function itself. Let us introduce the matrix operator

$$\mathbf{T}^\mu(\mathbf{y},\rho;z') = \begin{pmatrix} \mathbf{T}_\mu^{(1,1)}(\mathbf{y},\rho;z') & \mathbf{T}_\mu^{(1,2)}(\mathbf{y},\rho;z') \\ \mathbf{T}_\mu^{(2,1)}(\mathbf{y},\rho;z') & \mathbf{T}_\mu^{(2,2)}(\mathbf{y},\rho;z') \end{pmatrix},$$

where

$$\begin{aligned}
\mathbf{T}_\mu^{(1,1)}(\mathbf{y},\rho;z') &= \varsigma(x;\mathbf{y},\rho) - (A_{2m}(z',0,D_x)-\mu)\,\varsigma(x;\mathbf{y},\rho)\,\mathbf{R}_\mu^+(z'),\\
\mathbf{T}_\mu^{(2,1)}(\mathbf{y},\rho;z') &= -\mathbf{P}_1(z')\,\varsigma(x;\mathbf{y},\rho)\,\mathbf{R}_\mu^+(z'),\\
\mathbf{T}_\mu^{(1,2)}(\mathbf{y},\rho;z') &= -(A_{2m}(z',0,D_x)-\mu)\,\varsigma(x;\mathbf{y},\rho)\,\mathbf{S}_\mu^+(z'),\\
\mathbf{T}_\mu^{(2,2)}(\mathbf{y},\rho;z') &= \varsigma(x',0;\mathbf{y},\rho) - \mathbf{P}_1(z')\,\varsigma(x;\mathbf{y},\rho)\,\mathbf{S}_\mu^+(z').
\end{aligned}$$

Let us also denote, for the sake of brevity,

$$\mathcal{H} = L_2(\mathbb{R}_+^n) \times \prod_{j=1}^m H^{2m-m_j-1/2}(\mathbb{R}^{n-1}),$$

and view elements $\mathfrak{f}$ of this function space as columns with two components: from $L_2(\mathbb{R}_+^n)$ and $\prod_{j=1}^m H^{2m-m_j-1/2}(\mathbb{R}^{n-1})$, respectively. Lemmas 5.4.4, 5.4.5 and formula (5.4.35) imply

$$\|\mathbf{T}_\mu(\mathbf{y},\rho;z')\|_{\mathcal{H}\to\mathcal{H}} = O(\gamma^{-1}\rho^{-1}) \tag{5.4.50}$$

uniformly over $0 < \gamma \leqslant \pi$, $z' \in K'$, and $\rho > 1$. Here the negative powers of ρ appeared when we differentiated the product $\varsigma(x;\mathbf{y},\rho)\,\mathbf{R}_\mu^+(z')$ in accordance with Leibniz's rule.

We have the operator identity

$$\mathfrak{R}_\mu^+(\mathbf{y},\rho;z') = \widetilde{\mathfrak{R}}_\mu^+(\mathbf{y},\rho;z') + \mathfrak{R}_\mu^+(z') \times \mathbf{T}_\mu(\mathbf{y},\rho;z') \tag{5.4.51}$$

where the sign " $\times$ " stands for matrix multiplication, and

$$\begin{aligned}
\mathfrak{R}_\mu^+(\mathbf{y},\rho;z') &= \left(\mathbf{R}_\mu^+(z')\,\varsigma(x;\mathbf{y},\rho),\ \mathbf{S}_\mu^+(z')\,\varsigma(x',0;\mathbf{y},\rho)\right),\\
\widetilde{\mathfrak{R}}_\mu^+(\mathbf{y},\rho;z') &= \left(\varsigma(x;\mathbf{y},\rho)\,\mathbf{R}_\mu^+(z'),\ \varsigma(x;\mathbf{y},\rho)\,\mathbf{S}_\mu^+(z')\right),\\
\mathfrak{R}_\mu^+(z') &= \left(\mathbf{R}_\mu^+(z'),\ \mathbf{S}_\mu^+(z')\right)
\end{aligned}$$

are row operators.

But the matrix operator $\mathbf{T}_\mu(\mathbf{y},\rho;z')$ has been constructed in such a way that for any $\mathfrak{f}\in\mathcal{H}$ the support of $\mathbf{T}_\mu(\mathbf{y},\rho;z')\times\mathfrak{f}$ lies in a ball of radius 2ρ centred at $\mathbf{y}$, so

$$\mathbf{T}_\mu(\mathbf{y},\rho;z') = \varsigma(x;\mathbf{y},2\rho)\,\mathbf{T}_\mu(\mathbf{y},\rho;z').$$

Thus, (5.4.51) can be rewritten as

$$\mathfrak{R}_\mu^+(\mathbf{y},\rho;z') = \widetilde{\mathfrak{R}}_\mu^+(\mathbf{y},\rho;z') + \mathfrak{R}_\mu^+(\mathbf{y},2\rho;z')\times\mathbf{T}_\mu(\mathbf{y},\rho;z'). \tag{5.4.52}$$

Iterating this formula $k-1$ times we get

$$\begin{aligned}\mathfrak{R}_\mu^+(\mathbf{y},\rho;z') &= \mathfrak{S}_\mu^{(k-1)}(\mathbf{y},\rho;z') + \mathfrak{R}_\mu^+(\mathbf{y},2^k\rho;z')\times\mathbf{T}_\mu^{(k-1)}(\mathbf{y},\rho;z')\\ &\equiv \mathfrak{S}_\mu^{(k-1)}(\mathbf{y},\rho;z') + \mathfrak{R}_\mu^+(z')\times\mathbf{T}_\mu^{(k-1)}(\mathbf{y},\rho;z'),\end{aligned} \tag{5.4.53}$$

where the row operators $\mathfrak{S}_\mu^{(j)}(\mathbf{y},\rho;z')$ and the matrix operators $\mathbf{T}_\mu^{(j)}(\mathbf{y},\rho;z')$ are inductively defined as

$$\mathfrak{S}_\mu^{(0)}(\mathbf{y},\rho;z') \equiv \widetilde{\mathfrak{R}}_\mu^+(\mathbf{y},\rho;z'),\qquad \mathbf{T}_\mu^{(0)}(\mathbf{y},\rho;z') \equiv \mathbf{T}_\mu(\mathbf{y},\rho;z'),$$
$$\mathfrak{S}_\mu^{(j)}(\mathbf{y},\rho;z') = \mathfrak{S}_\mu^{(j-1)}(\mathbf{y},\rho;z') + \mathfrak{R}_\mu^+(\mathbf{y},2^j\rho;z')\times\mathbf{T}_\mu^{(j-1)}(\mathbf{y},\rho;z'),$$
$$\mathbf{T}_\mu^{(j)}(\mathbf{y},\rho;z') = \mathbf{T}_\mu(\mathbf{y},2^j\rho;z')\times\mathbf{T}_\mu^{(j-1)}(\mathbf{y},\rho;z').$$

The latter formula and formula (5.4.50) imply

$$\|\mathbf{T}_\mu^{(k-1)}(\mathbf{y},\rho;z')\|_{\mathcal{H}\to\mathcal{H}} = O(\gamma^{-k}\rho^{-k}). \tag{5.4.54}$$

The row operator $\mathfrak{S}_\mu^{(k-1)}(\mathbf{y},\rho;z')$ has been constructed in such a way that for any $\mathfrak{f}\in\mathcal{H}$ the support of $\mathfrak{S}_\mu^{(k)}(\mathbf{y},\rho;z')\times\mathfrak{f}$ lies in a ball of radius $2^k\rho$ centred at $\mathbf{y}$. Consequently, if

$$|\mathbf{x}-\mathbf{y}| > 2^k\rho+1, \tag{5.4.55}$$

then

$$X(x-\mathbf{x})\,\mathbf{R}_\mu^+(z')\,Y(x-\mathbf{y}) = X(x-\mathbf{x})\,\mathbf{R}_\mu^+(z')\,\mathbf{T}_\mu^{(k-1;1,1)}(\mathbf{y},\rho;z')\,Y(x-\mathbf{y}), \tag{5.4.56}$$

where $\mathbf{T}_\mu^{(k-1;1,1)}(\mathbf{y},\rho;z')$ is the upper left element of the matrix $\mathbf{T}_\mu^{(k-1)}(\mathbf{y},\rho;z')$.

Up till now we had freedom in our choice of the parameter ρ, the only restriction being the requirement $\rho>1$. Now we set

$$\rho\equiv\rho(\mathbf{x},\mathbf{y}) = \frac{|\mathbf{x}-\mathbf{y}|}{2^k+1}. \tag{5.4.57}$$

With such a choice of ρ we satisfy both the requirement $\rho>1$ (see (5.4.49)) and (5.4.55); consequently we can use (5.4.56). Formulae (5.4.56), (5.4.54), (5.4.57), (5.4.40), (5.4.45) imply (5.4.47). □

5. General properties of the resolvent $\mathcal{R}_\mu$. In this subsection we continue denoting the spectral parameter by μ, and consider complex values of μ of the form $\mu=\Lambda^{2m}e^{\pm i\gamma}$, where $0<\gamma\leqslant\pi$, $\Lambda\geqslant 1$. (Note the difference with the previous subsection where the parameter Λ was allowed to take any positive value, not necessarily $\geqslant 1$.) Our aim is to obtain estimates for the resolvent of the problem (1.1.1), (1.1.2) which are uniform over γ and Λ.

We will be looking at the following nonhomogeneous version of the problem (1.1.1), (1.1.2):

$$A(x, D_x)v = \mu v + f, \tag{5.4.58}$$

$$\Lambda^{-m_j}\left(B^{(j)}(x', D_x)v\right)\Big|_{\partial M} = g_j, \qquad j = 1, 2, \ldots, m, \tag{5.4.59}$$

where $f \in L_2(M)$, $g_j \in H^{2m-m_j-1/2}(\partial M)$ are given functions.

The problem (5.4.58), (5.4.59) is more complicated than (5.4.32), (5.4.33) because

1. We are working on the compact manifold M instead of the half-space $\mathbb{R}^n_+$.
2. The coefficients of the differential operators may be nonconstant.
3. The differential operators may contain lower order terms.

Consequently in our analysis of the problem (5.4.58), (5.4.59) we will not be able to use rescaling arguments in their pure form, as we did in the previous subsection. Nevertheless, we will show that all the main results proved for the problem (5.4.32), (5.4.33) remain true for the problem (5.4.58), (5.4.59).

The solution of the problem (5.4.58), (5.4.59) can be written down as

$$v = \mathcal{R}_\mu f + \mathcal{S}_\mu \vec{g}, \tag{5.4.60}$$

where $\vec{g} = \{g_1, \ldots, g_m\}$ and $\mathcal{S}_\mu$ is a bounded linear operator acting from $\prod_{j=1}^m H^{2m-m_j-1/2}(\partial M)$ to $H^{2m}(M)$.

Suppose that we have already defined (in some invariant manner) the Sobolev spaces $H^s(M)$, $H^s(\partial M)$ for half-densities on M, ∂M, respectively. We introduce weighted Sobolev spaces $H^s(M;\Lambda)$, $H^s(\partial M;\Lambda)$ defined by the scalar products

$$(v,w)_{H^s(M;\Lambda)} = (v,w)_{L_2(M)} + \Lambda^{-2s}(v,w)_{H^s(M)}, \qquad s = 0, 1, 2, \ldots,$$
$$(v,w)_{H^s(\partial M;\Lambda)} = (v,w)_{L_2(M)} + \Lambda^{-2s}(v,w)_{H^s(\partial M)}, \qquad s \in [0, +\infty),$$

respectively. Obviously, the norms $\|\cdot\|_{H^s(M;\Lambda)}$ and $\|\cdot\|_{H^s(\partial M;\Lambda)}$ are equivalent (with embedding constants independent of $\Lambda \geqslant 1$ and via a partition of unity) to the norms $\|\cdot\|_{H^s(\mathbb{R}^n_+;\Lambda)}$ and $\|\cdot\|_{H^s(\mathbb{R}^{n-1};\Lambda)}$, respectively.

Let $\mathcal{F}_\Lambda$, $\mathcal{P}_\Lambda$ be the linear operators mapping

$$v(x) \to \left\{(v)\big|_{x_n=0}, \Lambda^{-1}(D_{x_n}v)\big|_{x_n=0}, \ldots, \Lambda^{2m-1}(D_{x_n}^{2m-1}v)\big|_{x_n=0}\right\},$$
$$v(x) \to \left\{\Lambda^{-m_1}(B^{(1)}(x', D_x)v)\big|_{x_n=0}, \ldots, \Lambda^{-m_m}(B^{(m)}(x', D_x)v)\big|_{x_n=0}\right\},$$

respectively. From (5.4.35), (5.4.46) we obtain

$$\|\mathcal{F}_\Lambda\|_{H^{2m}(M;\Lambda)\to\prod_{j=1}^{2m} H^{2m-j+1/2}(\partial M;\Lambda)} = O(\Lambda^{1/2}), \tag{5.4.61}$$

$$\|\mathcal{P}_\Lambda\|_{H^{2m}(M;\Lambda)\to\prod_{j=1}^{m} H^{2m-m_j-1/2}(\partial M;\Lambda)} = O(\Lambda^{1/2}) \tag{5.4.62}$$

uniformly over $0 < \gamma \leqslant \pi$ and $\Lambda \geqslant 1$.

Let $\mathcal{G}_\Lambda$ be a right inverse of $\mathcal{F}_\Lambda$, and let $\mathcal{Q}_\Lambda$ be a right inverse of $\mathbf{P}_\Lambda$. We shall choose $\mathcal{G}_\Lambda$, $\mathcal{Q}_\Lambda$ in such a way that

$$\|\mathcal{G}_\Lambda\|_{\prod_{j=1}^{2m} H^{2m-j+1/2}(\partial M;\Lambda)\to H^{2m}(M;\Lambda)} = O(\Lambda^{-1/2}), \tag{5.4.63}$$

$$\|\mathcal{Q}_\Lambda\|_{\prod_{j=1}^{m} H^{2m-m_j-1/2}(\partial M;\Lambda)\to H^{2m}(M;\Lambda)} = O(\Lambda^{-1/2}) \tag{5.4.64}$$

uniformly over $0 < \gamma \leqslant \pi$ and $\Lambda \geqslant 1$. The existence of operators $\mathcal{G}_\Lambda$, $\mathcal{Q}_\Lambda$ with these properties can be established by writing them down explicitly (via a partition of unity) in terms of the operator $\mathbf{G}_\Lambda$ and using (5.4.38), (5.4.45), (5.4.46).

LEMMA 5.4.7. *We have*

$$\|\mathcal{R}_\mu\|_{L_2(M)\to H^{2m}(M;\Lambda)} = O(\gamma^{-1}\Lambda^{-2m}), \tag{5.4.65}$$

$$\|\mathcal{S}_\mu\|_{\prod_{j=1}^m H^{2m-m_j-1/2}(\partial M;\Lambda)\to H^{2m}(M;\Lambda)} = O(\gamma^{-1}\Lambda^{-1/2}), \tag{5.4.66}$$

uniformly over $0 < \gamma \leqslant \pi$ *and* $\Lambda \geqslant 1$.

PROOF OF LEMMA 5.4.7. In view of the definition of the weighted Sobolev space $H^{2m}(M;\Lambda)$ formula (5.4.65) is equivalent to the following two formulae:

$$\|\mathcal{R}_\mu\|_{L_2(M)\to L_2(M)} = O(\gamma^{-1}\Lambda^{-2m}),$$

$$\|\mathcal{R}_\mu\|_{L_2(M)\to H^{2m}(M)} = O(\gamma^{-1}).$$

The first of these two formulae follows from the self-adjointness and positivity of the operator $\mathcal{A}$, and the second from the fact that the norm $\|\cdot\|_{H^{2m}(M)}$ is equivalent on $D(\mathcal{A})$ to the norm $\|(A(x,D_x)(\cdot)\|_{L_2(M)}$ (see Theorem 4.1 from [**LioMag,** Chap. 2, Sect. 4.4], and also 1.1.7 for notation). The estimate (5.4.65) has been proved.

We have

$$\mathcal{S}_\mu\,\vec{g} = \mathcal{Q}_\Lambda\,\vec{g} - \mathcal{R}_\mu\,(A(x,D_x)-\mu I)\,\mathcal{Q}_\Lambda\,\vec{g}\,.$$

Using formula (5.4.45) we get

$$\begin{aligned}&\|A(x,D_x)-\mu I\|_{H^{2m}(M;\Lambda)\to L_2(M)}\\ &\qquad\leqslant \|A(x,D_x)\|_{H^{2m}(M;\Lambda)\to L_2(M)} + \mu\,\|I\|_{H^{2m}(M;\Lambda)\to L_2(M)} = O(\Lambda^{2m})\,.\end{aligned}$$

The above two formulae and (5.4.64), (5.4.65) imply (5.4.66). □

The resolvent $\mathcal{R}_\mu$ possesses a quasilocality property similar to that of the resolvent $\mathbf{R}_\mu^+(z')$, and in principle we could have proved for $\mathcal{R}_\mu$ an analogue of Lemma 5.4.6. We will, however, refrain from doing this because for our purposes it is sufficient to know only that one of the resolvents, $\mathbf{R}_\mu^+(z')$ or $\mathcal{R}_\mu$, has a quasilocality property. Naturally, we have chosen to deal with $\mathbf{R}_\mu^+(z')$ because it is simpler.

6. Proof of (5.4.26). Throughout this subsection we denote $\Lambda \equiv \Lambda(\lambda) = |\varkappa(\lambda)|^{1/(2m)}$, where $\varkappa$ is defined by (5.4.23). Of course, $\Lambda = \lambda(1+o(1))$ as $\lambda \to +\infty$. We consider complex μ of the form $\mu = \Lambda^{2m}e^{\pm i\gamma}$, where $0<\gamma\leqslant\pi$, $\lambda \geqslant c_1$, and c_1 is a sufficiently large positive constant. As in the two previous subsections, we denote the operators of multiplication by functions as the functions themselves. Moreover, in order to simplify notation we incorporate (implicitly) the restriction and extension operators (see the end of subsection 2) into our operators of multiplication by cut-off functions.

Our objective will be to prove the estimate

$$\|q(x;z',d)\,(\mathcal{R}_\mu - \mathbf{R}_\mu^+(z'))\,q(x;z',d)\|_1 = O\left(\frac{\lambda^{(n-2)\varepsilon-2\varepsilon_5-2m}}{\gamma^2}\right) \tag{5.4.67}$$

uniformly over

$$\gamma \geqslant c_2\,\lambda^{-\varepsilon-\varepsilon_5}\,, \tag{5.4.68}$$

d from the interval (5.4.2), and $z' \in K'$. Formula (5.4.67) is (5.4.26) rewritten in a more convenient form, and the condition (5.4.68) with a sufficiently small positive constant c_2 has been introduced in order to ensure that we cover all $\mu \in L(\varkappa(\lambda))$; see (5.4.23).

Let v be as in the proof of Lemma 5.4.6. Set

$$\varepsilon_6 := \frac{1-3\varepsilon-2\varepsilon_5}{n+1} > \varepsilon_5\,, \qquad \rho \equiv \rho(\lambda) := \lambda^{\varepsilon+\varepsilon_6-1}\,,$$
$$\varsigma(x;z',\rho) := v(|x-(z',0)|/\rho)\,.$$

Note that

$$\|q(x;z',d)\,(\mathcal{R}_\mu - \mathbf{R}^+_\mu(z'))\,q(x;z',d)\|_1 \leqslant \|\varsigma(x;z',\rho)\,(\mathcal{R}_\mu - \mathbf{R}^+_\mu(z'))\,q(x;z',d)\|_1$$

because for sufficiently large λ we have

$$\operatorname{supp} q(\,\cdot\,;z',d) \subset \{x \in \mathbb{R}^n_+ : \varsigma(x;z',\rho) = 1\}\,.$$

So in order to prove (5.4.67) it is sufficient to prove

$$\|\varsigma(x;z',\rho)\,(\mathcal{R}_\mu - \mathbf{R}^+_\mu(z'))\,q(x;z',d)\|_1 = O\left(\frac{\lambda^{(n-2)\varepsilon-2\varepsilon_5-2m}}{\gamma^2}\right). \tag{5.4.69}$$

Further on we deal with (5.4.69). The advantage of (5.4.69) is that the left cut-off is smooth.

By diam we will denote the diameter of a set in a Euclidean space.

We shall make use of the following interpolation inequality for the nuclear norm which is similar to Lemmas 2.1, 2.2 from [**Agm1**].

LEMMA 5.4.8. *For any $s \in \mathbb{N}$, $s > n$, there exists a positive constant C_s such that uniformly over all $r > 0$ and all operators*

$$U = \int_{\mathbb{R}^n_+} u(x,y)\,(\,\cdot\,)\,dy \;:\; L_2(\mathbb{R}^n_+) \to H^s(\mathbb{R}^n_+)$$

with $\operatorname{diam}(\operatorname{supp} u) \leqslant r$ *we have*

$$\|U\|_1 \leqslant C_s\, r^n \left(\|U\|_{L_2(\mathbb{R}^n_+)\to L_2(\mathbb{R}^n_+)}\right)^{1-n/s} \left(\|U\|_{L_2(\mathbb{R}^n_+)\to H^s(\mathbb{R}^n_+)}\right)^{n/s}.$$

The diameter of the support of the integral kernel of the operator

$$\varsigma(x;z',\rho)\,(\mathcal{R}_\mu - \mathbf{R}^+_\mu(z'))\,q(x;z',d)$$

does not exceed $4\rho \sim \lambda^{\varepsilon+\varepsilon_6-1}$, so applying Lemma 5.4.8 with $s = 2m$ we conclude that in order to prove (5.4.69) it is sufficient to prove

$$\|\varsigma(x;z',\rho)\,(\mathcal{R}_\mu - \mathbf{R}^+_\mu(z'))\,q(x;z',d)\|_{L_2(\mathbb{R}^n_+)\to L_2(\mathbb{R}^n_+)} = O(\gamma^{-2}\Lambda^{-2m}\rho)$$

and

$$\|\varsigma(x;z',\rho)\,(\mathcal{R}_\mu - \mathbf{R}^+_\mu(z'))\,q(x;z',d)\|_{L_2(\mathbb{R}^n_+)\to H^{2m}(\mathbb{R}^n_+)} = O(\gamma^{-2}\rho)\,.$$

But these two estimates are equivalent to one estimate

$$\|\varsigma(x;z',\rho)\,(\mathcal{R}_\mu - \mathbf{R}^+_\mu(z'))\,q(x;z',d)\|_{L_2(\mathbb{R}^n_+)\to H^{2m}(\mathbb{R}^n_+;\Lambda)} = O(\gamma^{-2}\Lambda^{-2m}\rho)\,. \tag{5.4.70}$$

Thus, we have reduced the proof of Proposition 5.4.1 to the proof of the estimate (5.4.70). This estimate should be proved uniformly over γ satisfying (5.4.68), d from the interval (5.4.2), and $z' \in K'$.

Set

$$\mathcal{T}^{(1)}_\mu(z',d,\rho) = q(x;z',d) - (A(x,D_x)-\mu)\,\varsigma(x;z',\rho)\,\mathbf{R}^+_\mu(z')\,q(x;z',d)\,,$$
$$\mathcal{T}^{(2)}_\mu(z',d,\rho) = -\mathcal{P}_\Lambda\,\varsigma(x;z',\rho)\,\mathbf{R}^+_\mu(z')\,q(x;z',d)\,.$$

We have the operator identity

$$\varsigma(x;z',\rho)\,(\mathcal{R}_\mu - \mathbf{R}^+_\mu(z'))\,q(x;z',d) = \varsigma(x;z',\rho)\,(\mathcal{R}_\mu\,\mathcal{T}^{(1)}_\mu(z',d,\rho) + \mathcal{S}_\mu\,\mathcal{T}^{(2)}_\mu(z',d,\rho))\,,$$

so using Lemma 5.4.7 and formula (5.4.45) we conclude that in order to prove (5.4.70) it is sufficient to prove

$$\|\mathcal{T}^{(1)}_\mu(z',d,\rho)\|_{L_2(\mathbb{R}^n_+)\to L_2(\mathbb{R}^n_+)} = O(\gamma^{-1}\rho) \tag{5.4.71}$$

and

$$\|\mathcal{T}^{(2)}_\mu(z',d,\rho)\|_{L_2(\mathbb{R}^n_+)\to\prod_{j=1}^m H^{2m-m_j-1/2}(\mathbb{R}^{n-1};\Lambda)} = O(\gamma^{-1}\Lambda^{-2m+1/2}\rho)\,. \tag{5.4.72}$$

Set

$$\widetilde{\mathcal{T}}^{(1)}_\mu(z',d,\rho) = q(x;z',d) - \varsigma(x;z',\rho)\,(A(x,D_x)-\mu)\,\mathbf{R}^+_\mu(z')\,q(x;z',d)\,,$$
$$\widetilde{\mathcal{T}}^{(2)}_\mu(z',d,\rho) = -\varsigma(x',0;z',\rho)\,\mathcal{P}_\Lambda\,\mathbf{R}^+_\mu(z')\,q(x;z',d)\,.$$

By Lemma 5.4.6 and a partition of unity we have

$$\|\mathcal{T}^{(1)}_\mu(z',d,\rho) - \widetilde{\mathcal{T}}^{(1)}_\mu(z',d,\rho)\|_{L_2(\mathbb{R}^n_+)\to L_2(\mathbb{R}^n_+)} = O(\Lambda^{-\infty})\,,$$

$$\|\mathcal{T}^{(2)}_\mu(z',d,\rho) - \widetilde{\mathcal{T}}^{(2)}_\mu(z',d,\rho)\|_{L_2(\mathbb{R}^n_+)\to\prod_{j=1}^m H^{2m-m_j-1/2}(\mathbb{R}^{n-1};\Lambda)} = O(\Lambda^{-\infty})$$

(note that in applying Lemma 5.4.6 here we made use of the condition (5.4.68)). Therefore, proving (5.4.71), (5.4.72) is equivalent to proving

$$\|\widetilde{\mathcal{T}}^{(1)}_\mu(z',d,\rho)\|_{L_2(\mathbb{R}^n_+)\to L_2(\mathbb{R}^n_+)} = O(\gamma^{-1}\rho) \tag{5.4.73}$$

and

$$\|\widetilde{\mathcal{T}}^{(2)}_\mu(z',d,\rho)\|_{L_2(\mathbb{R}^n_+)\to\prod_{j=1}^m H^{2m-m_j-1/2}(\mathbb{R}^{n-1};\Lambda)} = O(\gamma^{-1}\Lambda^{-2m+1/2}\rho)\,. \tag{5.4.74}$$

The operators $\widetilde{\mathcal{T}}^{(1)}_\mu(z',d,\rho)$, $\widetilde{\mathcal{T}}^{(2)}_\mu(z',d,\rho)$ can be rewritten in the form

$$\begin{aligned}\widetilde{\mathcal{T}}^{(1)}_\mu(z',d,\rho) &= \varsigma(x;z',\rho)\,(1 - (A(x,D_x)-\mu)\,\mathbf{R}^+_\mu(z'))\,q(x;z',d)\\ &= \varsigma(x;z',\rho)\,(A_{2m}(z',0,D_x) - A(x,D_x))\,\mathbf{R}^+_\mu(z'))\,q(x;z',d)\,,\\ \widetilde{\mathcal{T}}^{(2)}_\mu(z',d,\rho) &= \varsigma(x',0;z',\rho)\,(\mathbf{P}_\Lambda(z') - \mathcal{P}_\Lambda)\,\mathbf{R}^+_\mu(z')\,q(x;z',d)\,.\end{aligned}$$

Using the latter formulae as well as formulae $\|q(x;z',d)\|_{L_2(\mathbb{R}^n_+)\to L_2(\mathbb{R}^n_+)} = 1$ and (5.4.40), we conclude that in order to prove (5.4.73) and (5.4.74) it is sufficient to prove

$$\|\varsigma(x;z',\rho)\,(A_{2m}(z',0,D_x) - A(x,D_x))\|_{H^{2m}(\mathbb{R}^n_+;\Lambda)\to L_2(\mathbb{R}^n_+)} = O(\Lambda^{2m}\rho) \tag{5.4.75}$$

and

(5.4.76)
$$\|\varsigma(x',0;z',\rho)\,(\mathbf{P}_\Lambda(z') - \mathcal{P}_\Lambda)\|_{H^{2m}(\mathbb{R}^n_+;\Lambda)\to\prod_{j=1}^m H^{2m-m_j-1/2}(\mathbb{R}^{n-1};\Lambda)} = O(\Lambda^{1/2}\rho)\,.$$

Thus, we have reduced the proof of Proposition 5.4.1 to the proof of the estimates (5.4.75), (5.4.76).

By formula (5.4.45) we have

$$\|\varsigma(x;z',\rho)\,(A_{2m}(z',0,D_x) - A_{2m}(x,D_x))\|_{H^{2m}(\mathbb{R}^n_+;\Lambda)\to L_2(\mathbb{R}^n_+)} = O(\Lambda^{2m}\rho)\,,$$
$$\|\varsigma(x;z',\rho)\,(A_{2m}(x,D_x) - A(x,D_x))\|_{H^{2m}(\mathbb{R}^n_+;\Lambda)\to L_2(\mathbb{R}^n_+)} = O(\Lambda^{2m-1})\,.$$

The latter two formulae imply (5.4.75).

Further on for the sake of brevity we shall denote

$$\mathfrak{F}_k = H^{2m-k+1/2}(\mathbb{R}^{n-1};\Lambda)\,, \qquad \mathfrak{F} = \prod_{k=1}^{2m}\mathfrak{F}_k\,,$$
$$\mathfrak{P}_j = H^{2m-m_j-1/2}(\mathbb{R}^{n-1};\Lambda)\,, \qquad \mathfrak{P} = \prod_{j=1}^{m}\mathfrak{P}_j\,.$$

In order to prove (5.4.76) let us use (locally) the representations

$$\mathbf{P}_\Lambda(z') = \mathbf{V}(z',D_{x'})\,\mathbf{F}_\Lambda\,, \qquad \mathcal{P}_\Lambda = \mathcal{V}_\Lambda(x',D_{x'})\,\mathbf{F}_\Lambda\,,$$

where $\mathbf{V}(z',D_{x'})$, $\mathcal{V}_\Lambda(x',D_{x'})$ are $m\times(2m)$ matrix differential operators acting over $\Omega'\subset\mathbb{R}^{n-1}$; note that the operator $\mathbf{V}(z',D_{x'})$ is independent of Λ. Now, in view of (5.4.35), in order to prove (5.4.76) it is sufficient to prove

$$\|\varsigma(x',0;z',\rho)\,(\mathbf{V}(z',D_{x'}) - \mathcal{V}_\Lambda(x',D_{x'}))\|_{\mathfrak{F}\to\mathfrak{P}} = O(\rho)\,,$$

or, equivalently,

$$\|\varsigma(x',0;z',\rho)\,(\mathbf{V}^{(j,k)}(z',D_{x'}) - \mathcal{V}_\Lambda^{(j,k)}(x',D_{x'}))\|_{\mathfrak{F}_k\to\mathfrak{P}_j} = O(\rho) \tag{5.4.77}$$

for all matrix elements ($1\leqslant j\leqslant m$, $1\leqslant k\leqslant 2m$). By formula (5.4.46) we have

$$\|\varsigma(x',0;z',\rho)\,(\mathbf{V}^{(j,k)}(z',D_{x'}) - \mathbf{V}^{(j,k)}(x',D_{x'}))\|_{\mathfrak{F}_k\to\mathfrak{P}_j} = O(\rho)\,,$$
$$\|\varsigma(x',0;z',\rho)\,(\mathbf{V}^{(j,k)}(x',D_{x'}) - \mathcal{V}_\Lambda^{(j,k)}(x',D_{x'}))\|_{\mathfrak{F}_k\to\mathfrak{P}_j} = O(\Lambda^{-1})\,.$$

The latter two formulae imply (5.4.77).

The proof of Proposition 5.4.1 is complete.

5.5. Asymptotics of the spectral function

PROOF OF THEOREMS 1.8.5 AND 1.8.7. If $P = Q$, then Theorems 1.8.5 immediately follows from Theorems 4.2.1, 4.4.9 and Corollary B.2.2. Theorem 1.8.7 follows from Theorems 4.2.1, 4.4.9 and B.5.1. The general results (with $Q \neq P$) are easily obtained by polarization (see 1.8.1). □

PROOF OF LEMMAS 1.8.11 AND 1.8.12. In the notation of subsections 1.8.4 and 4.3.2, we have

$$(U^k_{y,\lambda} f)(\widetilde{\eta}) \;=\; e^{i\lambda T^{(k)}_y(\widetilde{\eta})}\,(U^k_y f)(\widetilde{\eta})\,, \qquad \forall f \in L_2(S^*_y M)\,. \tag{5.5.1}$$

In view of Lemma 1.8.10, the difference $T^{(k+1)}_y - T^{(k)}_y$ is bounded away from zero uniformly with respect to $\widetilde{\eta} \in \Pi^k_y$ and $k = 1, 2, \ldots$. This implies that the series

$$\sum_{k=1}^{\infty} (f, U^k_{y,\lambda} f)_{L_2(S^*_y M)} \;=\; \sum_{k=1}^{\infty} (f\,,\, e^{-i\lambda T^{(k)}_y} U^k_y\, f)_{L_2(S^*_y M)}$$

converges in the sense of distributions. Its sum is a distribution from the Schwartz class $\mathcal{S}(\mathbb{R})$ whose value on a test function γ is given by

$$\sum_{k=1}^{\infty} \int_{\Pi^k_y} \hat\gamma(-T^{(k)}_y)\, e^{i\mathfrak{f}(T^{(k)}_y;y,\widetilde{\eta})}\, f(\widetilde{\eta})\, \overline{(U^k_x f)(\widetilde{\eta})}\, đ\widetilde{\eta}\,. \tag{5.5.2}$$

If y lies outside the δ-neighbourhood of ∂M and $\operatorname{supp}\hat\gamma \in (-T_\delta, T_\delta)$, then all the terms in (5.5.2) are equal to zero. This proves Lemma 1.8.12.

It remains to prove that

$$\|f\|^2_{L_2(S^*_y M)} \;+\; \sum_{k=1}^{\infty} 2\operatorname{Re}\left(f, U^k_{y,\lambda}\, f\right)_{L_2(S^*_y M)} \tag{5.5.3}$$

is a positive measure. Clearly, (5.5.3) coincides with the $\mathcal{S}'$-limit of the functions

$$\Big(f\,,\, f + \sum_{k=1}^{\infty} \big(U^k_{y,\lambda+i\varkappa} + (U^k_{y,\lambda+i\varkappa})^*\big)\, f\Big)_{L_2(S^*_y M)} \tag{5.5.4}$$

as $\varkappa \to 0$, where $\varkappa > 0$ and $U_{y,\lambda+i\varkappa} = e^{i(\lambda+i\varkappa)\mathbf{T}^{(k)}_y}\, U_x$. Since the norm of the operators $U_{y,\lambda+i\varkappa}$ is strictly less than one, we have

$$\begin{aligned}
I \;+\; &\sum_{k=1}^{\infty} (U^k_{y,\lambda+i\varkappa} + (U^k_{y,\lambda+i\varkappa})^*) \\
&= (I - U_{y,\lambda+i\varkappa})^{-1}(I - U_{y,\lambda+i\varkappa})(I - U^*_{y,\lambda+i\varkappa})(I - U^*_{y,\lambda+i\varkappa})^{-1} \\
&\quad + \;(I - U_{y,\lambda+i\varkappa})^{-1} U_{y,\lambda+i\varkappa} \;+\; U^*_{y,\lambda+i\varkappa}(I - U^*_{y,\lambda+i\varkappa})^{-1} \\
&= (I - U_{y,\lambda+i\varkappa})^{-1}(I - U^*_{y,\lambda+i\varkappa} U_{y,\lambda+i\varkappa})(I - U^*_{y,\lambda+i\varkappa})^{-1} \;\geqslant\; 0\,.
\end{aligned}$$

Therefore for each fixed $\varkappa > 0$ the function (5.5.4) is nonnegative. This implies that the distribution (5.5.3) is nonnegative which can only be true if it is a positive Borel measure. □

PROOF OF LEMMA 1.8.13. Let us prove the uniform boundedness for nonnegative λ; the proof for negative λ is obtained in the same manner. Set

$$N_f(\lambda) \;=\; \begin{cases} \lambda + C \,+\, \|f\|^{-2}_{L_2(S^*_yM)}\,(f\,,\,\mathbf{Q}(y,\lambda)f)_{L_2(S^*_yM)} & \text{if } \lambda \geqslant 0\,, \\ 0 & \text{otherwise}\,, \end{cases}$$

where $f \in L_2(S^*_yM)$, $C = \|\mathbf{Q}(y,0)\|_{L_2(S^*_yM)}$. Since (5.5.3) is a positive measure, the function N_{f,λ_0} is nondecreasing for $\lambda \geqslant 0$ and belongs to the class F_+.

By the definition of T_δ (and Lemma 1.8.12) we have

$$(N_f * \rho_{T_\delta})(\lambda) = \int (\lambda + C - \mu)\,\rho_{T_\delta}(\mu)\,d\mu \,+\, O(\lambda^{-\infty}) = \lambda \,+\, C \,+\, O(\lambda^{-\infty})\,,$$

$$(N'_f * \rho_{T_\delta})(\lambda) = (N_f * \rho'_{T_\delta})(\lambda) = \int (\lambda + C - \mu)\,\rho'_{T_\delta}(\mu)\,d\mu \,+\, O(\lambda^{-\infty})$$

$$= \int \rho_{T_\delta}(\mu)\,d\mu \,+\, O(\lambda^{-\infty}) = 1 \,+\, O(\lambda^{-\infty})$$

as $\lambda \to +\infty$ uniformly in $f \in L_2(S^*_yM)$, $f \neq 0$. Applying Corollary B.2.2 we obtain

$$\frac{(\mathbf{Q}(y,\lambda)f\,,\,f)_{L_2(S^*_yM)}}{\|f\|^2_{L_2(S^*_yM)}} \;=\; O(1)\,, \qquad \lambda \to +\infty\,,$$

uniformly in f, which implies that the operator function $\mathbf{Q}(y,\lambda)$ is bounded uniformly with respect to $\lambda \in \mathbb{R}_+$.

Set $\widetilde{\mathbf{Q}}(y,\lambda) = \mathcal{F}^{-1}_{t\to\lambda}\left[t^{-1}\mathcal{F}_{\lambda\to t}[\mathbf{Q}(y,\lambda)]\right]$, which is well defined in view of Lemma 1.8.12. Then $\int_0^\lambda \mathbf{Q}(y,\mu)\,d\mu = \widetilde{\mathbf{Q}}(y,\lambda) + C_{\mathbf{Q}}$ with some (constant) self-adjoint operator $C_{\mathbf{Q}}$. Since the domain of $C_{\mathbf{Q}}$ is the whole space $L_2(S^*_yM)$, it is bounded.

If $C = \sup_{\mu\in\mathbb{R}} \|\mathbf{Q}(y,\mu)\|$, then the function

$$C\lambda \,+\, \|f\|^{-2}_{L_2(S^*_yM)}\left(f\,,\,\widetilde{\mathbf{Q}}(y,\lambda)f\right)_{L_2(S^*_yM)}$$

is nondecreasing. Now the boundedness of $\widetilde{\mathbf{Q}}(y,\lambda)$ is proved in the same way as that of $\mathbf{Q}(y,\lambda)$. $\square$

PROOF OF THEOREM 1.8.14. By (5.5.1)

$$\mathbf{Q}(y,\lambda) \;=\; \sum_{k=1}^{\infty}\left((iT^{(k)}_y)^{-1}\,e^{i\lambda T^{(k)}_y}\,(U^k_y)^* \;-\; (iT^{(k)}_y)^{-1}\,e^{-i\lambda T^{(k)}_y}\,U^k_y\right), \tag{5.5.5}$$

where the series converges in the weak operator topology and in the sense of distributions in λ.

Let C be a positive constant, and let

$$\begin{aligned} N_1(\lambda) &= c_{0;P}(y)\,\lambda^{n+2l} \,+\, (c_{1;P}(y) + C)\,\lambda^{n+2l-1} \,+\, \mathbf{Q}_P(y,\lambda)\,\lambda^{n+2l-1}\,, \\ N_2(\lambda) &= e_{P,P}(\lambda,y,y) \,+\, C\,\lambda^{n+2l-1} \end{aligned}$$

for $\lambda \geqslant 0$ and $N_1(\lambda) = N_2(\lambda) = 0$ for $\lambda \leqslant 0$. We assume C to be sufficiently large so that the function N_1 belong to the class F_+; this can always be achieved in view of Lemmas 1.8.11 and 1.8.13. By direct calculation (using the same arguments as in 5.2.3) we obtain that

$$(N_1 * \rho_{T_\delta})(\lambda) \;=\; c_{0;P}(y)\,\lambda^{n+2l} \,+\, (c_{1;P}(y) + C)\,\lambda^{n+2l-1} \,+\, o(\lambda^{n+2l-1})$$

and, in virtue of (5.5.1),

$$(N_1' * \gamma)(\lambda) \;=\; \lambda^{n+2l-1} \int_{\Pi_y^a} P_1 \sum_{k\geqslant 1} \hat\gamma(T_y^{(k)})\, \overline{(U_{y,\lambda}^k P_1)}\, đ\tilde\eta \;+\; o(\lambda^{n+2l-1})$$

for all $\gamma \in \mathcal{S}(\mathbb{R})$ such that $\operatorname{supp}\hat\gamma \in (0,+\infty)$. These formulae and Theorems 4.2.1, 4.4.10 imply that

$$(N_1 * \rho_{T_\delta})(\lambda) \;=\; (N_2 * \rho_{T_\delta})(\lambda)\,, \qquad (N_1' * \gamma)(\lambda) \;=\; (N_2' * \gamma)$$

modulo $o(\lambda^{n+2l-1})$. Now Theorem 1.8.14 follows from Theorem B.4.1. □

PROOF OF PROPOSITION 1.8.16. Let $\widetilde U_y$ be a unitary dilation of U_y acting in the extended Hilbert space $\widetilde H$ with the inner product $(\cdot,\cdot)_{\widetilde H}$. Then, by the spectral theorem, for all $f \in \widetilde H$ we have

$$(f, \widetilde U_y^k f)_{\widetilde H} \;=\; \int_{\mathbb{S}} e^{-iks}\, d\mu_f(s)\,, \qquad \forall k \in \mathbb{Z}\,,$$

$$\left(f\,,\, \{\pi + \arg \widetilde U_y - \lambda T\}_{2\pi} f\right)_{\widetilde H} \;=\; \int_{\mathbb{S}} \{\pi + s - \lambda T\}_{2\pi}\, d\mu_f(s)\,, \tag{5.5.6}$$

where μ_f is some positive Borel measure on the unit circumference $\mathbb{S}$. The Fourier expansion of the function $\{\pi + s - \lambda T\}_{2\pi}$ has the form

$$\{\pi + s - \lambda T\}_{2\pi} \;=\; 2\,T^{-1} \sum_{k=1}^{\infty} k^{-1} \sin k(\lambda T - s) \;=\; \sum_{k\in\mathbb{Z}\setminus\{0\}} (ikT)^{-1}\, e^{ik(\lambda T - s)}\,.$$

Clearly, the series on the right-hand side converges in $L_2(\mathbb{R})$ uniformly with respect to $s \in \mathbb{S}$. Therefore, by (5.5.6),

$$\left(f\,,\, \{\pi + \arg \widetilde U_y - \lambda T\}_{2\pi} f\right)_{\widetilde H} \;=\; \sum_{k\in\mathbb{Z}\setminus\{0\}} (ikT)^{-1}\, e^{ik\lambda T} \int_{\mathbb{S}} e^{-iks}\, d\mu_f(s)\,, \tag{5.5.7}$$

where the series converges in $L_2(\mathbb{R})$. It is easy to see that the Fourier transform of the function (5.5.7) vanishes for $t \in (-T,T)$ and that its derivative (understood in the sense of distributions in λ) coincides with

$$\sum_{k\in\mathbb{Z}\setminus\{0\}} e^{ik\lambda T} \int_{\mathbb{S}} e^{-iks}\, d\mu_f(s) \;=\; \sum_{k\in\mathbb{Z}\setminus\{0\}} e^{ik\lambda T}\, (f, \widetilde U_y^k f)_{\widetilde H}\,.$$

If $f \in L_2(S_y^* M)$, then the right-hand side takes the form

$$\left(f\,,\, \sum_{k=1}^{\infty} \left(U_{y,\lambda}^k + (U_{y,\lambda}^k)^*\right) f\right)_{L_2(S_y^* M)}\,.$$

This proves the theorem. □

In order to prove Theorem 1.8.17 we shall have to deal with a special dilation $\widetilde U_y$ of the operator U_y.

DEFINITION 5.5.1. A unitary dilation $\widetilde{U}$ is said to be *minimal* if the linear combinations of the elements $\widetilde{U}^k f$, $f \in H$, $k \in \mathbb{Z}$, form a dense subset of $\widetilde{H}$.

The minimal unitary dilation always exists [**Sz.-NaFoi**]. Moreover, we have the following two theorems (see [**Sz.-NaFoi**]).

THEOREM 5.5.2. *If a contraction operator U does not have a nontrivial invariant subspace H_0 such that $U|_{H_0}$ is a unitary operator, then the spectrum of its minimal unitary dilation $\widetilde{U}$ is absolutely continuous.*

THEOREM 5.5.3. *Let U be a contraction operator in the Hilbert space H. Then $H = H_1 \oplus H_2$, where $H_1, H_2 \subset H$ are invariant subspaces of the operator U such that $U|_{H_1}$ is a unitary operator and $U|_{H_2}$ satisfies the conditions of Theorem 5.5.2.*

PROOF OF THEOREM 1.8.17. Let $\widetilde{U}_y$ be a minimal unitary dilation of U_y acting in the extended Hilbert space $\widetilde{H}$, and H_1, H_2 be the invariant subspaces of U_y given by Theorem 5.5.3. Clearly, $\widetilde{H} = H_1 \oplus \widetilde{H}_2$, where $H_1, \widetilde{H}_2 \subset \widetilde{H}$ are invariant subspaces of the operator U such that $\widetilde{U}_y|_{H_1} = U_y|_{H_1}$ and $\widetilde{U}_y|_{\widetilde{H}_2}$ is a minimal unitary dilation of the operator $U_y|_{H_2}$. By Theorem 5.5.2 the spectrum of $\widetilde{U}_y|_{\widetilde{H}_2}$ is continuous. Therefore $f \in L_2(S_y^*M)$ is orthogonal to all the eigenfunctions of U_y corresponding to eigenvalues lying on $\mathbb{S}$ if and only if f is orthogonal to all the eigenfunctions of the operator $\widetilde{U}_y$. Furthermore, this is true if and only if the measure μ_f (introduced in the proof of Proposition 1.8.16) is continuous, i.e., $\mu_f(\{s\}) = 0$ for any point $s \in \mathbb{S}$ (see, for example, [**BirSo**]).

Let $\lambda > 0$, let k_λ be the integer part of $(2\pi)^{-1}\lambda T$, and let $\varkappa > 0$ be sufficiently small. Then

$$\{\pi + s - \lambda T\}_{2\pi} - \{\pi + s - (\lambda - \varkappa)T\}_{2\pi} = -\varkappa T + 2\pi\,\chi_\varkappa^-(s),$$
$$\{\pi + s - \lambda T\}_{2\pi} - \{\pi + s - (\lambda + \varkappa)T\}_{2\pi} = \varkappa T - 2\pi\,\chi_\varkappa^+(s),$$

where $\chi_\varkappa^-$, $\chi_\varkappa^+$ are the characteristic functions of the intervals

$$[(\lambda - \varkappa)T - 2\pi k_\lambda\,,\, \lambda T - 2\pi k_\lambda)\,, \qquad [\lambda T - 2\pi k_\lambda\,,\, (\lambda + \varkappa)T - 2\pi k_\lambda)\,,$$

respectively. Now by (5.5.6)

$$\Big(f\,,\, \{\pi + \arg\widetilde{U}_y - \lambda T\}_{2\pi} f\Big) - \Big(f\,,\, \{\pi + \arg\widetilde{U}_y - (\lambda - \varkappa)T\}_{2\pi} f\Big) \to 0\,,$$
$$\Big(f\,,\, \{\pi + \arg\widetilde{U}_y - \lambda T\}_{2\pi} f\Big) - \Big(f\,,\, \{\pi + \arg\widetilde{U}_y - (\lambda + \varkappa)T\}_{2\pi} f\Big) \to 2\pi\mu_f(\{s_\lambda\})$$

as $\varkappa \to 0$, where $s_\lambda = \lambda T - 2\pi k_\lambda$. The first formula means that the function (5.5.6) is left-continuous, and the second one implies that the function (5.5.6) is right-continuous if and only if the measure μ_f is continuous. The proof is complete. □

CHAPTER 6

Mechanical Applications

6.1. Membranes and acoustic resonators

In this section we consider the situation when M is a region in $\mathbb{R}^n$ and $A = -\Delta$, where $\Delta = \partial^2/\partial y_1^2 + \partial^2/\partial y_2^2 + \cdots + \partial^2/\partial y_n^2$ is the Laplacian in Cartesian coordinates. The boundary condition is either Dirichlet $v|_{\partial M} = 0$, or Neumann $(\partial v/\partial x_n)|_{\partial M} = 0$, where x_n is the distance to ∂M.

The cases $n = 2$ and $n = 3$ are, of course, the ones which have a physical meaning.

1. Membrane. In the case $n = 2$ the eigenvalue problem (1.1.1), (1.1.2) describes the vibrations of an isotropic membrane. Here $\lambda = \omega\sqrt{\frac{\rho}{T}}$, where ω is the vibration frequency, and ρ and T are the surface density and tension of the membrane, respectively. Throughout this chapter by "frequency" we mean "circular frequency"; consequently, the number of cycles per second is $\frac{\omega}{2\pi}$.

The function $v(y)$ is the deflection (normal displacement) of the membrane. The Dirichlet boundary condition describes a membrane fixed along its edge, and the Neumann boundary condition a membrane whose edge is free (i.e., there are no forces in the direction normal to the unperturbed surface of the membrane acting on its edge).

Let λ_k be an eigenvalue of (1.1.1), (1.1.2). Then the number $\omega_k = \lambda_k\sqrt{\frac{T}{\rho}}$ is called an *eigenfrequency* or *natural frequency*. An eigenfrequency corresponds to a vibration of the type $u(t,y) = e^{-it\omega_k}v_k(y)$ occurring without any external forces.

According to Examples 1.2.3, 1.6.14 and 1.6.15 we have

$$N(\lambda) = \frac{S}{4\pi}\lambda^2 \mp \frac{L}{4\pi}\lambda + o(\lambda), \qquad \lambda \to +\infty,$$

where S is the surface area of M, L is the length of ∂M, and the signs "minus" and "plus" correspond to the Dirichlet and Neumann boundary conditions, respectively.

2. Acoustic resonator. In the case $n = 3$ the eigenvalue problem (1.1.1), (1.1.2) describes the vibrations of an acoustic medium occupying a resonator (vessel). Here $\lambda = \frac{\omega}{c}$, where ω is the vibration frequency and c is the speed of sound in the medium.

The function $v(y)$ is the potential of displacements of the acoustic medium. The Dirichlet boundary condition describes a resonator with soft walls (zero pressure on ∂M), and the Neumann boundary condition a resonator with rigid walls (zero normal displacement on ∂M).

In this case according to Examples 1.2.3 and 1.6.16 we have

$$N(\lambda) = \frac{V}{6\pi^2}\lambda^3 \mp \frac{S}{16\pi}\lambda^2 + o(\lambda^2), \qquad \lambda \to +\infty,$$

where V is the volume of M, S is the surface area of ∂M, and the signs "minus" and "plus" correspond to the Dirichlet and Neumann boundary conditions, respectively.

6.2. Elastic plates

1. Statement of result. In this section we consider the situation when M is a region in $\mathbb{R}^2$ and $A = \Delta^2$, where $\Delta = \partial^2/\partial y_1^2 + \partial^2/\partial y_2^2$ is the Laplacian in Cartesian coordinates. The partial differential equation (1.1.1) in this case describes the flexural vibrations of an isotropic thin elastic plate. Here $\lambda = \left(\frac{12(1-\sigma^2)\rho\omega^2}{Eh^2}\right)^{1/4}$, where ω is the vibration frequency, h is the thickness of the plate, and ρ, E, and $0 < \sigma < \frac{1}{2}$ are the volume density, Young's modulus, and Poisson's ratio of the plate material, respectively. The function $v(y)$ is the deflection (normal displacement) of the middle surface of the plate.

In order to describe the boundary conditions we use the local coordinates $x = (x_1, x_2)$ from Example 1.6.14. We denote by $\mathbf{k}(x_1)$ the curvature of ∂M; the sign of the curvature is chosen in such a way that $\mathbf{k} > 0$ when the tangent to ∂M remains outside $\overset{\circ}{M}$ in a small neighbourhood of the point of tangency. By $\mathbf{k}'$ we denote the derivative of $\mathbf{k}$ with respect to x_1. Let us introduce the operators

$$F_0 = 1, \qquad F_1 = \frac{\partial}{\partial x_2}, \qquad F_2 = \frac{\partial^2}{\partial x_2^2} + \sigma\frac{\partial^2}{\partial x_1^2} - \sigma\mathbf{k}\frac{\partial}{\partial x_2},$$

$$F_3 = \frac{\partial^3}{\partial x_2^3} + (2-\sigma)\frac{\partial^3}{\partial x_1^2\partial x_2} + 3\mathbf{k}\frac{\partial^2}{\partial x_1^2} + (2-\sigma)\mathbf{k}'\frac{\partial}{\partial x_1} - (1+\sigma)\mathbf{k}^2\frac{\partial}{\partial x_2} - \mathbf{k}F_2.$$

We shall consider three types of boundary conditions, corresponding to the following choices of "boundary" operators in (1.1.2).

1. Clamped edge: $B^{(1)} = F_0$, $B^{(2)} = F_1$.
2. Hinge supported edge: $B^{(1)} = F_0$, $B^{(2)} = F_2$.
3. Free edge: $B^{(1)} = F_2$, $B^{(2)} = F_3$.

The clamped edge is what a pure mathematician would call the Dirichlet boundary conditions (see Example 1.1.10). The other two types of boundary conditions may appear to be exotic to a pure mathematician, and thus require an explanation. Consider the quadratic form

$$\mathcal{E}(v) := \iint_M \left(|\Delta v|^2 + 2(1-\sigma)(|v_{y_1y_2}|^2 - \operatorname{Re}(v_{y_1y_1}\overline{v_{y_2y_2}}))\right) dy_1\, dy_2\,.$$

The functional $\mathcal{E}$ is, up to a constant factor, the potential energy of the plate; see, e.g., [**GolLidTo,** p. 79, formula (1.9)]. Perturbing v by δv and integrating by parts we get

$$\delta\mathcal{E} = 2\operatorname{Re}\left(\iint_M (\Delta^2 v)\,\overline{\delta v}\, dy_1\, dy_2 + \int_{\partial M}\left((F_3 v)\,\overline{\delta v} - (F_2 v)\frac{\partial\overline{\delta v}}{\partial x_2}\right) dx_1\right).$$

Thus, when we vary $\mathcal{E}(v)$ without any constraints on v we obtain the free boundary conditions, and when we vary $\mathcal{E}(v)$ under the constraint $v|_{\partial M} = 0$ we obtain the

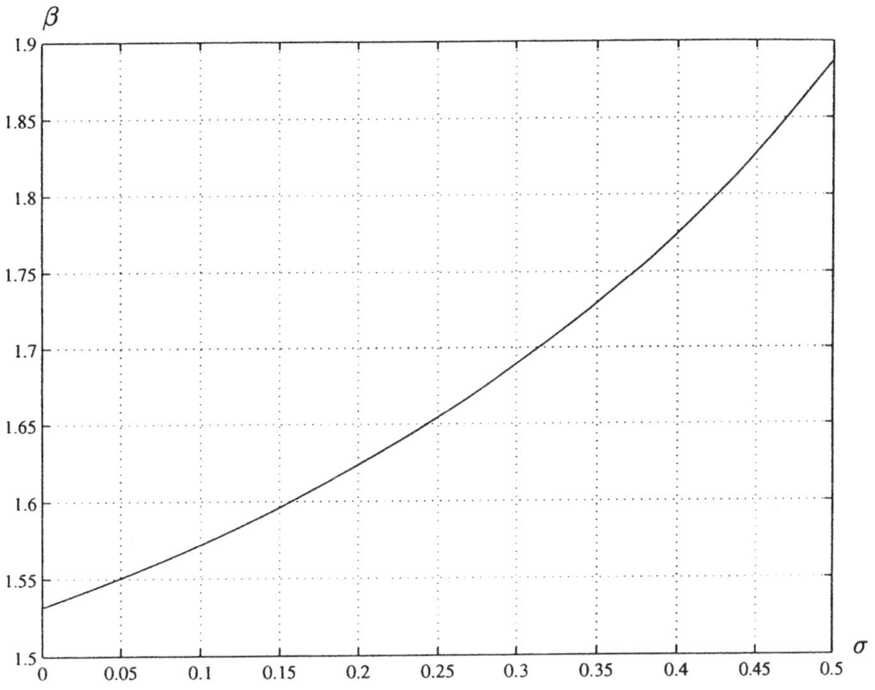

FIGURE 13. The coefficient β in the case of a free edge.

hinge supported boundary conditions. Conversely, for any function v satisfying any of the three types of boundary conditions described above we have $(\Delta^2 v, v) = \mathcal{E}(v)$.

Weyl's formula for the biharmonic operator has the form

$$N(\lambda) = \frac{S}{4\pi}\lambda^2 + \frac{\beta L}{4\pi}\lambda + o(\lambda), \qquad \lambda \to +\infty, \tag{6.2.1}$$

where S is the surface area of M, L is the length of ∂M, and β is a dimensionless coefficient the value of which is

(6.2.2)
$$\beta = -1 - \frac{\Gamma(3/4)}{\sqrt{\pi}\,\Gamma(5/4)} \approx -1.763,$$

(6.2.3)
$$\beta = -1,$$

(6.2.4)
$$\beta = 4\left(-1 + 4\sigma - 3\sigma^2 + 2(1-\sigma)\sqrt{1 - 2\sigma + 2\sigma^2}\right)^{-1/4} - 1 - \frac{4}{\pi}\int_0^1 \tan^{-1}\left[\left(\frac{1+(1-\sigma)\xi^2}{1-(1-\sigma)\xi^2}\right)^2 \sqrt{\frac{1-\xi^2}{1+\xi^2}}\right] d\xi$$

for the cases of clamped, hinge supported, and free edge, respectively; here Γ is the gamma function and $\tan^{-1}$ is the inverse $\tan$. For the case of a free edge the graph of β as a function of Poisson's ratio σ is shown in Figure 13.

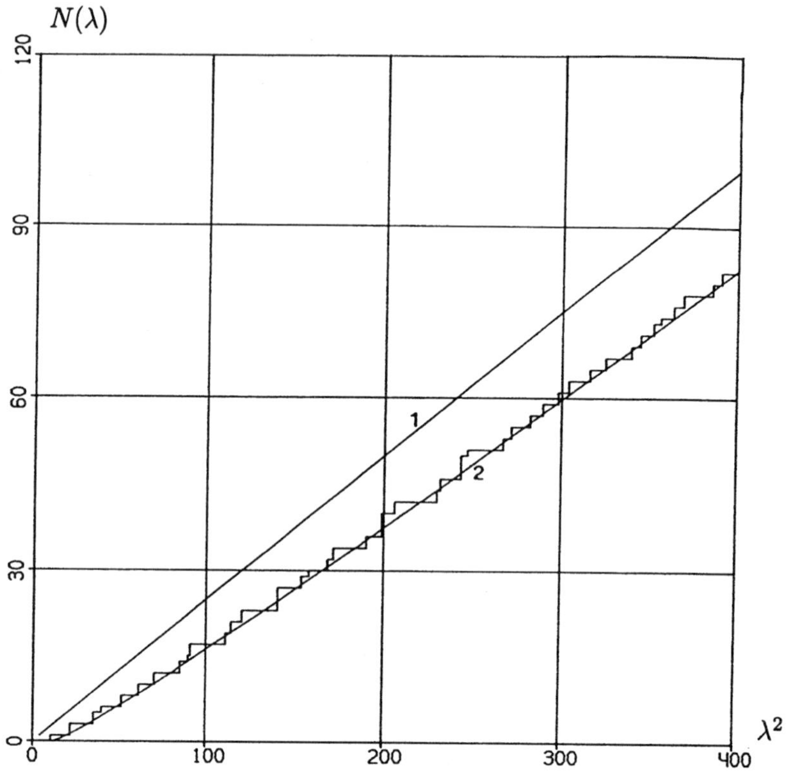

FIGURE 14. Eigenvalue distribution for a circular plate with a clamped edge.

The first asymptotic coefficient of (6.2.1) was evaluated in accordance with Example 1.2.3, whereas the second asymptotic coefficient will be evaluated in the next subsection. Note that the one-term asymptotic formula for the biharmonic operator was first obtained by Courant [**Cou**].

As an example let us consider a circular plate, i.e., $M = \{y \in \mathbb{R}^2 : y_1^2+y_2^2 \leqslant 1\}$. Figures 14–16 show numerical results for the cases of clamped, hinge supported, and free edge, respectively.

The independent variable, plotted along the horizontal axis, is λ^2; using λ^2 instead of λ is more natural from the mechanical point of view because λ^2 is proportional to the frequency ω. Each plot contains three lines: the stepwise line is the actual counting function $N(\lambda)$, and the two smooth lines are the graphs of the functions $\frac{S}{4\pi}\lambda^2$ (line 1) and $\frac{S}{4\pi}\lambda^2 + \frac{\beta L}{4\pi}\lambda$ (line 2). The actual $N(\lambda)$ was plotted using the numerical results from the handbook [**Gon**], and, for higher eigenvalues, the numerical results of the authors. Figures 14–16 demonstrate the remarkable effectiveness of the two-term asymptotic formula. On all three graphs Courant's one-term asymptotics $\frac{S}{4\pi}\lambda^2$ lies well away from the actual $N(\lambda)$, giving only a very rough approximation which does not feel the boundary conditions. On the other hand, the two-term asymptotics $\frac{S}{4\pi}\lambda^2 + \frac{\beta L}{4\pi}\lambda$ goes right through the actual $N(\lambda)$. The two-term asymptotics feels the boundary conditions through the coefficient β.

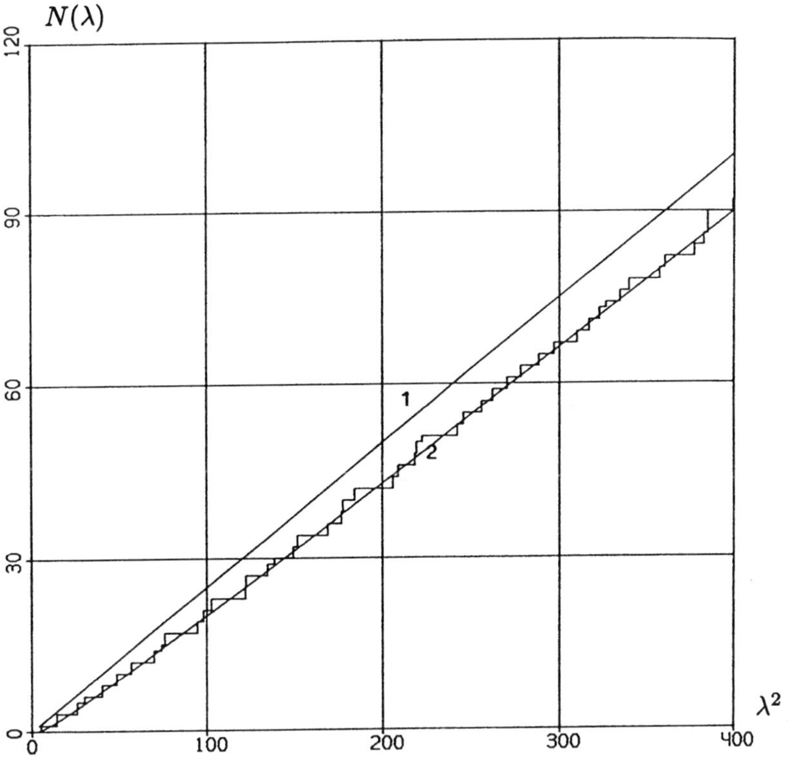

FIGURE 15. Eigenvalue distribution for a circular plate with hinge supported edge.

Note that the use of the two-term asymptotic formula (6.2.1) in the case of a circular plate is justified because M is convex and ∂M analytic; see 1.3.5.

2. Evaluation of the second asymptotic coefficient. In this subsection we demonstrate that for our problem (biharmonic operator) the second asymptotic coefficient is indeed

$$c_1 = \frac{\beta L}{4\pi}, \tag{6.2.5}$$

with β defined in accordance with (6.2.2)–(6.2.4). We do our calculations using the algorithm described in Section 1.6; the centrepiece of this algorithm is formula (1.6.23). The arguments go along the same lines as in Examples 1.6.14, 1.6.15. By producing explicit calculations for the case of the biharmonic operator we want to show that our algorithm for the calculation of the second asymptotic coefficient is a powerful tool in the analysis of concrete boundary value problems of mechanics and mathematical physics.

CLAMPED EDGE. The auxiliary one-dimensional spectral problem in this case is

$$d^4v/dx_2^4 - 2\xi_1^2 d^2v/dx_2^2 + \xi_1^4 v = \nu v, \tag{6.2.6}$$

$$v|_{x_2=0} = dv/dx_2|_{x_2=0} = 0, \tag{6.2.7}$$

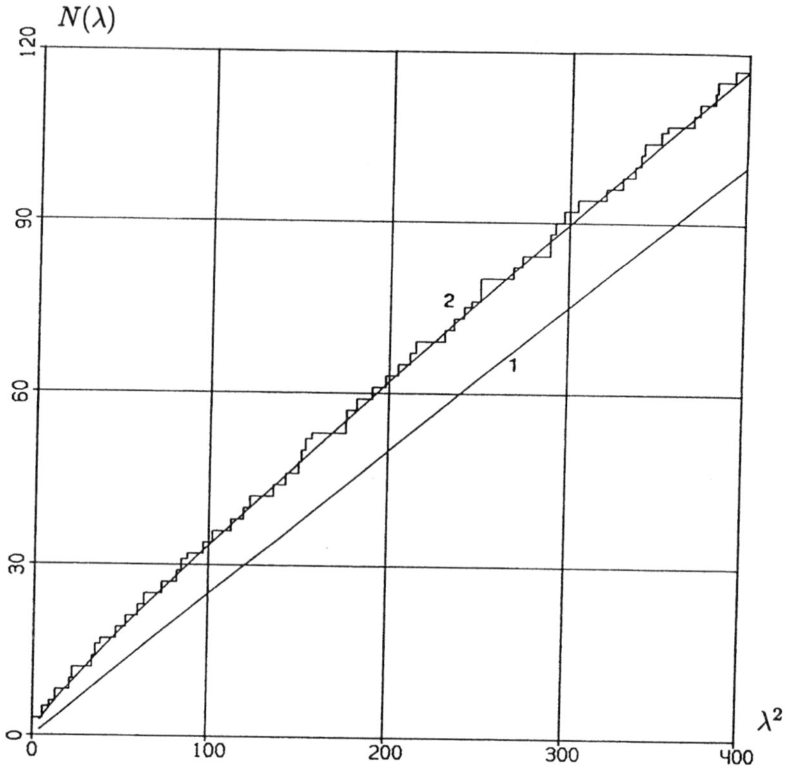

FIGURE 16. Eigenvalue distribution for a circular plate with a free edge.

with $v \equiv v(x_2)$, $x_2 \in \mathbb{R}_+$. As in Example 1.6.14, further on in this section we assume that $\xi_1 \neq 0$.

The problem (6.2.6), (6.2.7) has no eigenvalues, so

$$\mathbf{N}^+(\nu; x_1, \xi_1) \equiv 0\,. \tag{6.2.8}$$

The problem (6.2.6), (6.2.7) has only one threshold $\nu_1^{\mathrm{st}} = \xi_1^4$, and the continuous spectrum is the semi-infinite interval $[\xi_1^4, +\infty)$. The points $\nu > \xi_1^4$ of the continuous spectrum have multiplicity one, and the corresponding generalized eigenfunctions have the form

$$\begin{aligned} v(x_2) &= \sin\left(x_2\sqrt{\sqrt{\nu}-\xi_1^2}-\psi\right) + (\sin\psi)\, e^{-x_2\sqrt{\sqrt{\nu}+\xi_1^2}} \\ &= \frac{a_1^-\, e^{ix_2\zeta_1^-(\nu)}}{\sqrt{-2\pi A'(\zeta_1^-(\nu))}} + \frac{a_1^+\, e^{ix_2\zeta_1^+(\nu)}}{\sqrt{2\pi A'(\zeta_1^+(\nu))}} + \frac{a_2^+\, e^{ix_2\zeta_2^+(\nu)}}{\sqrt{2\pi A'(\zeta_2^+(\nu))}} \end{aligned}$$

(cf. (1.4.7)), where

$$\psi = \psi(\nu; \xi_1) = \tan^{-1}\left(\frac{\sqrt{\sqrt{\nu}-\xi_1^2}}{\sqrt{\sqrt{\nu}+\xi_1^2}}\right) \in (0, \pi/4)\,, \tag{6.2.9}$$

$\zeta_1^{\pm}(\nu) = \pm\sqrt{\sqrt{\nu} - \xi_1^2}$, $\zeta_2^{+}(\nu) = i\sqrt{\sqrt{\nu} + \xi_1^2}$, and

$$a_1^{\pm} = \mp i \sqrt{2\pi\sqrt{\nu}\sqrt{\sqrt{\nu} - \xi_1^2}\, e^{\mp i\psi}} \,. \tag{6.2.10}$$

As the strong simple reflection condition is satisfied, we can use formulae (1.6.15), (1.6.16). Substituting (6.2.10) in (1.6.15) we obtain for $\nu > \xi_1^4$

$$\arg_0 \det\big(i\, R(\nu; x_1, \xi_1)\big) = -\pi/2 - 2\psi(\nu; \xi_1) + 2\pi k \tag{6.2.11}$$

with an unknown integer k. Substituting (6.2.11), (6.2.9) in (1.6.16) we establish that $k = 0$. Thus,

$$\arg_0 \det\big(i\, R(\nu; x_1, \xi_1)\big) = \begin{cases} 0 & \text{if} \quad \nu \leqslant \xi_1^4\,, \\ -\pi/2 - 2\psi(\nu; \xi_1) & \text{if} \quad \nu > \xi_1^4\,. \end{cases} \tag{6.2.12}$$

According to (1.6.19)

$$c_1 = \int_0^L \int_{-\infty}^{+\infty} \left(\mathbf{N}^+(1; x_1, \xi_1) + \frac{\arg_0 \det\big(i\, R(1; x_1, \xi_1)\big)}{2\pi} \right) d\!\!{}^-\xi_1\, dx_1 \,. \tag{6.2.13}$$

Substituting (6.2.8), (6.2.12), and (6.2.9) in (6.2.13) and evaluating the integral we arrive at (6.2.5), (6.2.2).

HINGE SUPPORTED EDGE. The auxiliary one-dimensional spectral problem in this case is described by the equation (6.2.6) with boundary conditions

$$v|_{x_2=0} = d^2v/dx_2^2\big|_{x_2=0} = 0\,. \tag{6.2.14}$$

The operator associated with the spectral problem (6.2.6), (6.2.14) is the square of the operator associated with the spectral problem (1.6.22), (1.6.23). Consequently our coefficient c_1 is the same as in Example 1.6.14, which proves (6.2.5), (6.2.3).

FREE EDGE. The auxiliary one-dimensional spectral problem in this case is described by the equation (6.2.6) with boundary conditions

$$\left(d^2v/dx_2^2 - \sigma\xi_1^2 v\right)\big|_{x_2=0} = \left(d^3v/dx_2^3 - (2-\sigma)\xi_1^2 dv/dx_2\right)\big|_{x_2=0} = 0\,. \tag{6.2.15}$$

In the following analysis we assume that $\sigma \neq 0$. The fact that our formula (6.2.5), (6.2.4) remains valid for $\sigma = 0$ would follow by continuity; see Lemma A.4.1.

The problem (6.2.6), (6.2.15) has one eigenvalue

$$\nu_1 = \nu_1(\xi_1) = \left(-1 + 4\sigma - 3\sigma^2 + 2(1-\sigma)\sqrt{1 - 2\sigma + 2\sigma^2}\right)\xi_1^4\,. \tag{6.2.16}$$

We have $\nu_1 < \xi_1^4$, so this eigenvalue lies below the continuous spectrum. If we look at the dependence of the right-hand side of (6.2.16) on Poisson's ratio σ, we get Taylor's expansion

$$\nu_1 = (1 - \sigma^4/2 + O(\sigma^5))\,\xi_1^4 \quad \text{as} \quad \sigma \to 0\,. \tag{6.2.17}$$

It is remarkable that (6.2.17) contains no quadratic term. This suggests that for realistic values of σ the eigenvalue ν_1 is very close to the threshold ξ_1^4. Say, for $\sigma = \frac{1}{3}$ we get $\nu_1 = \sqrt{\frac{80}{81}}\,\xi_1^4$.

The existence of one eigenvalue (6.2.16) implies

$$\mathbf{N}^+(\nu; x_1, \xi_1) = \begin{cases} 0 & \text{for} \quad \nu \leqslant \nu_1(\xi_1), \\ 1 & \text{for} \quad \nu > \nu_1(\xi_1). \end{cases} \tag{6.2.18}$$

The generalized eigenfunctions now have the form

$$\begin{aligned} v(x_2) &= \sin\left(x_2\sqrt{\sqrt{\nu}-\xi_1^2} - \psi\right) + \frac{\sqrt{\nu}-(1-\sigma)\xi_1^2}{\sqrt{\nu}+(1-\sigma)\xi_1^2}(\sin\psi)\, e^{-x_2\sqrt{\sqrt{\nu}+\xi_1^2}} \\ &= \frac{a_1^-\, e^{ix_2\zeta_1^-(\nu)}}{\sqrt{-2\pi A'(\zeta_1^-(\nu))}} + \frac{a_1^+\, e^{ix_2\zeta_1^+(\nu)}}{\sqrt{2\pi A'(\zeta_1^+(\nu))}} + \frac{a_2^+\, e^{ix_2\zeta_2^+(\nu)}}{\sqrt{2\pi A'(\zeta_2^+(\nu))}}, \end{aligned}$$

where

$$\psi = \psi(\nu;\xi_1) = \tan^{-1}\left[\left(\frac{\sqrt{\nu}+(1-\sigma)\xi^2}{\sqrt{\nu}-(1-\sigma)\xi^2}\right)^2 \sqrt{\frac{\sqrt{\nu}-\xi^2}{\sqrt{\nu}+\xi^2}}\right] \in (0, \pi/2), \tag{6.2.19}$$

and the $a_1^\pm$ are given by formula (6.2.10) with our new ψ. Repeating the arguments used in the case of a clamped edge, we conclude that (6.2.12) holds with our new ψ. Substituting (6.2.18), (6.2.16), (6.2.12), and (6.2.19) in (6.2.13) we arrive at (6.2.5), (6.2.4).

6.3. Two- and three-dimensional elasticity

1. Statement of result. Let M be a region in $\mathbb{R}^n$, y Cartesian coordinates in $\mathbb{R}^n$, and v an n-component vector function. We consider the spectral problem for the system of equations

$$-c_t^2\,\Delta v - (c_l^2 - c_t^2)\,\operatorname{grad}\operatorname{div} v = \lambda^2 v \tag{6.3.1}$$

subject to the Dirichlet boundary conditions $v|_{\partial M} = 0$ (fixed boundary) or the conditions of free boundary. The latter are the variational boundary conditions generated by the quadratic functional

$$\mathcal{E}(v) := \int_M \left((c_l^2 - 2c_t^2)|\operatorname{div} v|^2 + \frac{c_t^2}{2}\sum_{i,j}|\partial_{y_j}v_i + \partial_{y_i}v_j|^2\right)dy\,.$$

The system (6.3.1) describes the vibrations of an isotropic elastic body; see [**LanLif,** Sect. 22]. Here $\lambda = \omega$ is the vibration frequency, and the constants c_l, c_t are the velocities of longitudinal and transverse waves, respectively. They are assumed to satisfy the inequality $c_l/c_t > \sqrt{2}$. Further we denote $\alpha = c_t^2\, c_l^{-2}$.

By considering (6.3.1) we are breaking the promise made in the Preface not to discuss systems in the book. However, we feel it necessary to outline the result for the system of elasticity because this system has played a special role in the development of the subject; see the next subsection.

The cases $n = 2$ and $n = 3$ are those of physical interest.

Note that problems in two-dimensional elasticity may arise in two ways: as a result of separation of variables in a infinite three-dimensional elastic cylinder, or in the study of tangential vibrations of a thin elastic plate. The expressions for the velocities c_l, c_t through Young's modulus, volume density and Poisson's ratio σ in these two physical models are different, but $\alpha = 1/2$ always corresponds to $\sigma = 0$.

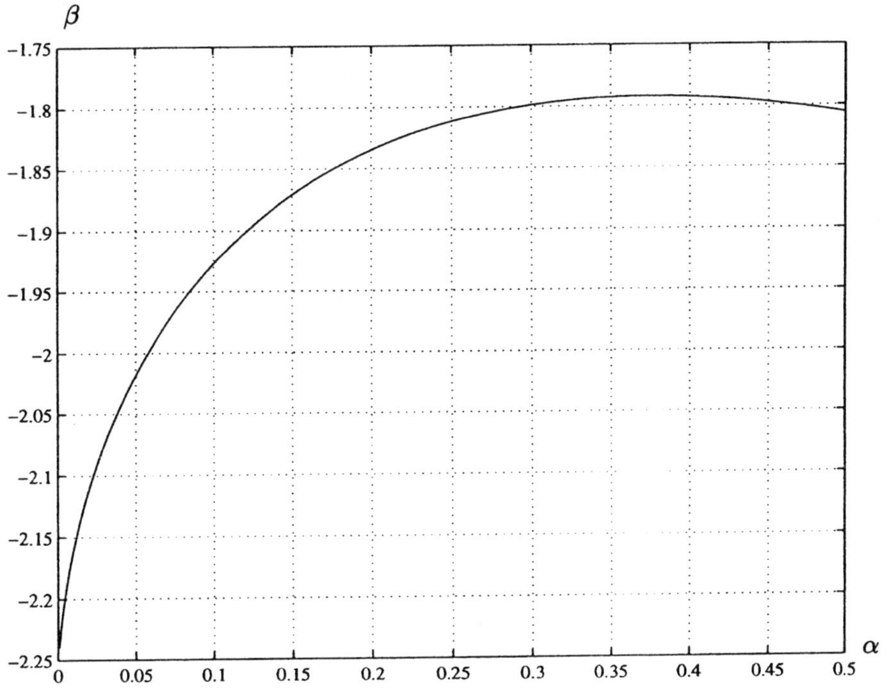

FIGURE 17. The coefficient β in the case of a fixed boundary.

Weyl's formula for two-dimensional elasticity has the form

$$N(\lambda) = \frac{(c_l^{-2}+c_t^{-2})S}{4\pi}\,\lambda^2 + \frac{\beta L}{4\pi c_t}\,\lambda + o(\lambda), \qquad \lambda \to +\infty,$$

where S is the surface area of M, L is the length of ∂M, and β is a dimensionless coefficient. The value of the coefficient β for a fixed boundary is

$$\beta = -1 - \sqrt{\alpha} - \frac{4}{\pi}\int_{\sqrt{\alpha}}^{1} \tan^{-1}\sqrt{(1-\alpha\xi^{-2})(\xi^{-2}-1)}\,d\xi.$$

For a free boundary,

$$\beta = 4\gamma^{-1} - 3 + \sqrt{\alpha} + \frac{4}{\pi}\int_{\sqrt{\alpha}}^{1} \tan^{-1}\frac{(2-\xi^{-2})^2}{4\sqrt{(1-\alpha\xi^{-2})(\xi^{-2}-1)}}\,d\xi,$$

where $0<\gamma<1$ is the root of the algebraic equation

$$\gamma^6 - 8\gamma^4 + 8(3-2\alpha)\gamma^2 - 16(1-\alpha) = 0.$$

The quantity $c_R = \gamma c_t$ has the physical meaning of the velocity of the Rayleigh wave; see [**LanLif**, Sect. 24]. The graph of β as a function of α for the cases of fixed and free boundary is shown in Figures 17 and 18, respectively.

Weyl's formula for three-dimensional elasticity has the form

$$N(\lambda) = \frac{(c_l^{-3}+2c_t^{-3})V}{6\pi^2}\,\lambda^3 + \frac{bS}{16\pi}\,\lambda^2 + o(\lambda^2), \qquad \lambda \to +\infty,$$

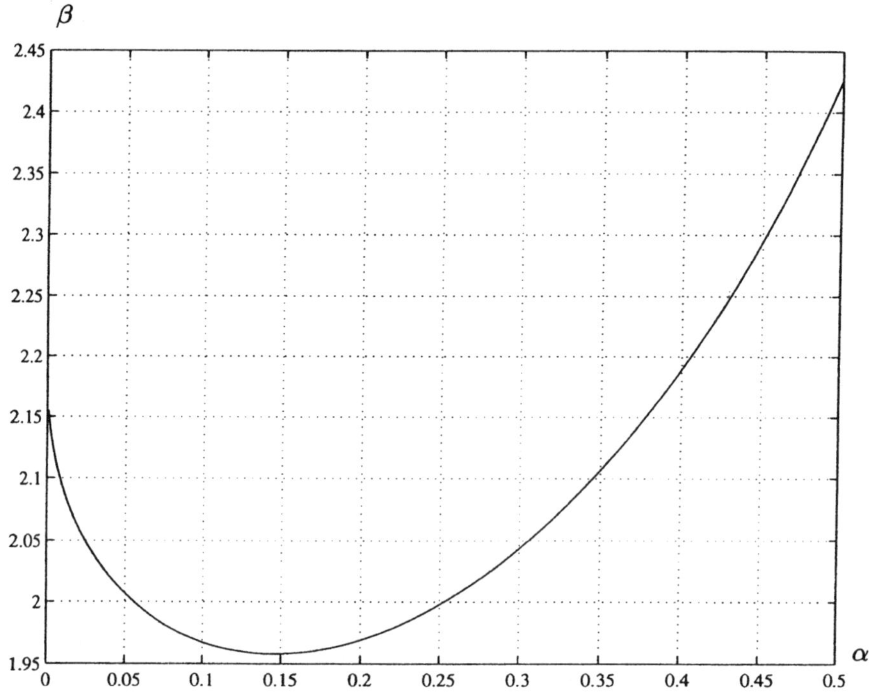

FIGURE 18. The coefficient β in the case of a free boundary.

where V is the volume of M, S is the surface area of ∂M, and

$$(6.3.2) \qquad b = -\frac{3c_l^4 + c_l^2 c_t^2 + 2c_t^4}{c_l^2 c_t^2 (c_l^2 + c_t^2)}, \qquad b = \frac{3c_l^4 - 3c_l^2 c_t^2 + 2c_t^4}{c_l^2 c_t^2 (c_l^2 - c_t^2)}$$

for the cases of fixed and free boundary, respectively.

Rigorous statements concerning classical two-term asymptotics for systems can be found in [**Va4,** Sect. 6], [**Va6**], [**SaVa1**]. These results deal with the case when the eigenvalues of the principal symbol (which is now a matrix function) have constant multiplicity. The latter is true for (6.3.1).

Without going into details let us note that the second asymptotic coefficient for systems can be evaluated by the algorithm from Section 1.6 with small modifications described in [**Va4,** Sect. 6]. Choosing normal coordinates and applying this algorithm to the elasticity system with free boundary it is easy to establish the existence of one eigenvalue $\nu_1(\xi') = c_R^2|\xi'|^2$ corresponding to the Rayleigh wave. In the three-dimensional case this eigenvalue does not appear explicitly in our final expression for the coefficient b only because a contour integration was carried out to simplify the result.

2. Historical background. The one-term asymptotic formula for $N(\lambda)$ in the case of the three-dimensional elasticity operator was obtained by Debye [**De**] in 1912. Debye arrived at this formula by considering the situation when M is a ball, and then extended the result to arbitrary shapes by physical arguments. Debye's analysis involved delicate manipulations with Bessel functions. Note that Debye could not use a cube for his calculations (as Rayleigh did for the Laplacian) because the elasticity problem in a cube does not allow separation of variables. The difficulty is with the boundary conditions: both the fixed and the free boundary conditions prevent one from separating variables. Later M. Born pointed out that Debye could have simplified matters by dealing with periodic boundary conditions, since in this case the problem in a cube allows separation of variables. One can guess that Debye did not adopt this approach because periodic boundary conditions do not have a physical meaning for a finite solid body.

Debye's work provided motivation for Weyl's research. Having started with the Laplacian, Weyl eventually produced (see [**We3**]) in 1915 a rigorous proof of the one-term asymptotics for elasticity.

Debye derived his one-term asymptotic formula for $N(\lambda)$ in order to evaluate the specific heat C of a (three-dimensional) body at low temperatures T:

$$C \approx \mathbf{c}_0 V T^3 \,, \tag{6.3.3}$$

where $\mathbf{c}_0$ is a constant expressed through the first asymptotic coefficient of $N(\lambda)$ and some physical constants, including Planck's constant. Formula (6.3.3) is known in theoretical physics as *Debye's law*.

Naturally, a two-term asymptotic formula for $N(\lambda)$ would lead to a correction in Debye's law:

$$C \approx \mathbf{c}_0 V T^3 + \mathbf{c}_1 S T^2 \,. \tag{6.3.4}$$

The correction term $\mathbf{c}_1 S T^2$ becomes noticeable in (6.3.4) if the temperature is sufficiently low and the surface area S is sufficiently large (say, if we are measuring the specific heat of a fine powder). This explains the interest of theoretical physicists in deriving the second asymptotic coefficient of $N(\lambda)$ for the elasticity operator. After a number of publications by different authors producing incorrect formulae for the second asymptotic coefficient, the correct formulae (6.3.2) were obtained by Dupuis, Mazo, and Onsager [**DupMazOn**] in 1960. The argumentation in [**DupMazOn**] is carried out on a physical level of rigour and the mathematical technique is different from that of Section 1.6.

6.4. Elastic shells

The notation in this and next sections is different from the rest of the book because we had to conform with traditions of shell theory. In particular h is not the Hamiltonian, but the shell thickness.

1. Statement of the problem. Let M be a smooth two-dimensional surface embedded in $\mathbb{R}^3$. Let $x = (x_1, x_2)$ be local coordinates on M, and let the surface be locally given by a three-component radius-vector $\mathbf{r}(x)$. We choose the coordinates x to be orthogonal, so that the first quadratic form of the surface is $d\mathbf{r}^2 = A_1^2 dx_1^2 + A_2^2 dx_2^2$, where $A_i = A_i(x) > 0$, $i = 1, 2$. We set $\mathbf{e}_i = \mathbf{e}_i(x) = A_i^{-1}\mathbf{r}_{x_i}$ (the subscript $_{x_i}$ indicates a partial derivative), and $\mathbf{n} = \mathbf{n}(x) = \mathbf{e}_1 \times \mathbf{e}_2$. Clearly, $\mathbf{e}_i$ is the unit vector in the direction of the coordinate line x_i, $\mathbf{n}$ is the unit normal to M, and the vectors $\{\mathbf{e}_1, \mathbf{e}_2, \mathbf{n}\}$ form a right triple. We shall assume for simplicity that our coordinate lines coincide with the curvature lines, so that the second quadratic form of the surface is $-\mathbf{n} \cdot d^2\mathbf{r} = R_1^{-1} A_1^2 dx_1^2 + R_2^{-1} A_2^2 dx_2^2$, where $d^2\mathbf{r} = \sum_{i,j} \mathbf{r}_{x_i x_j} dx_i dx_j$, and $R_i^{-1} = R_i^{-1}(x)$, $i = 1, 2$, are the principal curvatures. Our surface M may, of course, have a boundary ∂M which is a smooth one-dimensional curve.

A *shell* is an elastic body occupying the three-dimensional region

$$\{y \in \mathbb{R}^3 : \; y = \mathbf{r}(x) + x_3\mathbf{n}, \; x \in M, \; |x_3| \leqslant h/2\},$$

where $y = (y_1, y_2, y_3)$ are Cartesian coordinates, and $0 < h \ll 1$ is a parameter called the shell *thickness*. We shall assume the faces $\{x_3 = \pm h/2\}$ to be free, and the edge $\partial M \times [-h/2, h/2]$ to be fixed. The problem of free vibrations of such an elastic body is the one considered in the previous section. However, applying the results of the previous section does not make much sense when h is small: one would get asymptotic formulae which start working only at extremely high frequencies, namely, frequencies $\omega \gg h^{-1}$.

The proper way of telling whether the thickness is small or not is to introduce a characteristic length (or characteristic radius of curvature) R associated with the surface M, and to deal with the *relative* thickness h/R. In most technical applications the relative thickness is very small. It is quite usual to have $h/R \sim 10^{-3}$, whereas a shell with $h/R \sim 10^{-2}$ may be viewed as a rather thick one (this roughly corresponds to the hull of a submarine).

Therefore we shall fix our frequency range and study the behaviour of natural frequencies as $h \to 0$. In this case the elasticity equations (6.3.1) are reduced to the following system of three partial differential equations on M called *shell equations*:

$$\sum_{j=1}^{3} \mathcal{L}_{ij} u_j = \lambda u_i, \qquad i = 1, 2, 3. \tag{6.4.1}$$

The $\mathcal{L}_{ij}$ are the linear differential operators of shell theory which have the form

$$\mathcal{L}_{ij} = \frac{h^2}{12} n_{ij} + \ell_{ij}, \qquad i, j = 1, 2, 3. \tag{6.4.2}$$

Here n_{ij} and ℓ_{ij} are the moment and membrane operators, respectively.

We recall explicit expressions for ℓ_{ij}, n_{ij} from [**GolLidTo,** pp. 77, 78]:

$$\begin{aligned}
\ell_{ii} &= -\frac{1}{1-\sigma^2}\frac{1}{A_i}\frac{\partial}{\partial x_i}\frac{1}{A_iA_j}\frac{\partial}{\partial x_i}A_j - \frac{1}{2(1+\sigma)}\frac{1}{A_j}\frac{\partial}{\partial x_j}\frac{1}{A_iA_j}\frac{\partial}{\partial x_j}A_i \\
&\quad - \frac{1}{1+\sigma}{R_i}^{-1}{R_j}^{-1}, \\
\ell_{ij} &= -\frac{1}{1-\sigma^2}\frac{1}{A_i}\frac{\partial}{\partial x_i}\frac{1}{A_iA_j}\frac{\partial}{\partial x_j}A_i + \frac{1}{2(1+\sigma)}\frac{1}{A_j}\frac{\partial}{\partial x_j}\frac{1}{A_jA_i}\frac{\partial}{\partial x_i}A_j, \\
\ell_{i3} &= -\frac{1}{1-\sigma^2}\frac{1}{A_i}\frac{\partial}{\partial x_i}\left({R_i}^{-1} + {R_j}^{-1}\right) + \frac{1}{1+\sigma}\frac{1}{A_iR_j}\frac{\partial}{\partial x_i}, \\
\ell_{3i} &= \frac{1}{1-\sigma^2}\frac{1}{A_iA_j}\left({R_i}^{-1} + {R_j}^{-1}\right)\frac{\partial}{\partial x_i}A_j - \frac{1}{1+\sigma}\frac{1}{A_iA_j}\frac{\partial}{\partial x_i}\frac{A_j}{R_j}, \\
\ell_{33} &= \frac{1}{1-\sigma^2}\left({R_1}^{-2} + 2\sigma{R_1}^{-1}{R_2}^{-1} + {R_2}^{-2}\right)
\end{aligned}$$

(in the above formulae $i, j \leqslant 2$, $i \neq j$),

$$n_{ij} = 0 \qquad \text{for} \qquad i + j < 6,$$

$$n_{33} = \frac{1}{1-\sigma^2}\Delta_M^2 + \frac{1}{A_1A_2}\left(\frac{\partial}{\partial x_1}R_1^{-1}R_2^{-1}\frac{A_2}{A_1}\frac{\partial}{\partial x_1} + \frac{\partial}{\partial x_2}R_1^{-1}R_2^{-1}\frac{A_1}{A_2}\frac{\partial}{\partial x_2}\right),$$

where $0 < \sigma < \frac{1}{2}$ is Poisson's ratio and

$$\Delta_M = \frac{1}{A_1 A_2}\left(\frac{\partial}{\partial x_1}\frac{A_2}{A_1}\frac{\partial}{\partial x_1} + \frac{\partial}{\partial x_2}\frac{A_1}{A_2}\frac{\partial}{\partial x_2}\right)$$

is the surface Laplacian acting on functions. It can be checked that the matrix differential operators $(\ell_{ij})_{i,j=1}^{3}$ and $(n_{ij})_{i,j=1}^{3}$ are formally self-adjoint and nonnegative with respect to the standard $L_2(M)$-inner product on vector functions

$$(6.4.3) \qquad (u, v) = \iint_M (u_1\overline{v_1} + u_2\overline{v_2} + u_3\overline{v_3})\, dS\,,$$

$dS = A_1 A_2\, dx_1 dx_2$.

In (6.4.1) $\lambda = \frac{\rho_s \omega^2}{E}$, where ω is the vibration frequency, and ρ_s and E are the volume density and Young's modulus of the shell material, respectively. The vector function $u(x) = (u_1(x), u_2(x), u_3(x)$ is the displacement of the shell middle surface; its representation in the Cartesian coordinates in $\mathbb{R}^3$ is $\mathbf{u}(x) = u_1\mathbf{e}_1 + u_2\mathbf{e}_2 + u_3\mathbf{n}$. It is important to note that u_3 is the displacement in the *normal* direction.

The system of equations (6.4.1) has to be supplemented by the appropriate boundary conditions. As we assumed the original three-dimensional body to be fixed along the edge $\partial M \times [-h/2, h/2]$, these boundary conditions turn out to be

$$(6.4.4) \qquad u_1|_{\partial M} = u_2|_{\partial M} = u_3|_{\partial M} = \partial u_3/\partial x_1|_{\partial M} = 0\,.$$

Throughout this section we use near ∂M special local coordinates $x = (x_1, x_2)$ in which $\partial M = \{x_1 = 0\}$. This convention is contrary to the rest of the book, but is traditional for shell theory.

Clearly, apart from (6.4.4) there are many other meaningful boundary conditions for the shell equations (6.4.1), but we shall not discuss them for the sake of brevity.

In the special case when the surface M is flat (i.e., $R_1^{-1} \equiv R_2^{-1} \equiv 0$) the shell becomes a plate. It is easy to see that in this case the problem (6.4.1), (6.4.4) separates into two problems which were already considered in Sections 6.2, 6.3.

Further on we use the notation

$$(6.4.5) \quad |\xi|_x = \sqrt{A_1^{-2}\xi_1^2 + A_2^{-2}\xi_2^2}\,, \quad K(x,\xi) = |\xi|_x^{-2}(R_1^{-1}A_2^{-2}\xi_2^2 + R_2^{-1}A_1^{-2}\xi_1^2)\,,$$

where $\xi = (\xi_1, \xi_2) \in T'_x M$. Clearly, $|\xi|_x$ is the principal symbol of Δ_M, and $K(x,\xi)$ is the curvature of the normal section of the surface M at the point x in the direction $\xi_1 dx_1 + \xi_2 dx_2 = 0$.

2. Spectral properties of the shell operator. The shell equations (6.4.1) are simpler than the original elasticity equations (6.3.1) because they contain only two independent variables instead of three. On the other hand, their structure is nontrivial in that different equations of the system (6.4.1) have different orders: the orders are 2, 2, 4, respectively. Such systems should be treated in accordance with the theory developed by Agmon, Douglis, and Nirenberg [**AgmDoNir**]. It can be shown [**AsLid**], [**GolLidTo**] that the problem (6.4.1), (6.4.4) is indeed elliptic in the Agmon–Douglis–Nirenberg sense and generates a self-adjoint operator the spectrum of which is discrete and accumulates to $+\infty$. By $N(h,\lambda)$ we shall denote the counting function, that is, the number of eigenvalues of (6.4.1), (6.4.4) below a given λ. Here we wrote h as a variable to remind of the dependence of the operators (6.4.2) on the small parameter h.

We intend to fix a $\lambda > 0$ and study the asymptotic behaviour of the function $N(h,\lambda)$ as $h \to 0$. Let us stress that fixing h and letting λ tend to infinity would not make mechanical sense: one would get asymptotic formulae which start working for $\lambda \gg h^{-2}$, and this corresponds to frequencies at which one can no longer use the shell equations (6.4.1) and should switch to three-dimensional elasticity (6.3.1).

In studying the problem (6.4.1), (6.4.4) the first impulse is to put $h = 0$. This leads to the so-called membrane problem

$$\sum_{j=1}^{3} \ell_{ij} u_j = \lambda\, u_i\,, \qquad i = 1,2,3\,, \tag{6.4.6}$$

$$u_1|_{\partial M} = u_2|_{\partial M} = 0 \tag{6.4.7}$$

(note that the number of boundary conditions is different compared with (6.4.4)). The spectral problem (6.4.6), (6.4.7) is associated with a self-adjoint operator which can be viewed as the limit of the operator associated with (6.4.1), (6.4.4). Namely, let us denote by $E_\lambda(h)$ and E_λ the spectral projections of (6.4.1), (6.4.4) and (6.4.6), (6.4.7), respectively, and let $g = (g_1(x), g_2(x), g_3(x))$ be an arbitrary vector function from $L_2(M)$. Then, as shown in [**AsLid**], if $\lambda \in \mathbb{R}$ is not an eigenvalue of (6.4.6), (6.4.7), we have $\lim_{h\to 0} \|(E_\lambda(h) - E_\lambda)g\|_{L_2(M)} = 0$; see also [**S.-PaVa**].

Despite the convergence of spectral projections the spectra of (6.4.1), (6.4.4) and (6.4.6), (6.4.7) are completely different. In particular, the membrane problem (6.4.6), (6.4.7) always has an essential spectrum. This essential spectrum is a union of two sets: the interval

$$\left[\min_{T'M} K^2(x,\xi)\,,\ \max_{T'M} K^2(x,\xi)\right]$$

which is the set of λ where the ellipticity of $\left(\ell_{pq}\right)_{i,j=1}^{3}$ is violated, and the set of λ where the Shapiro–Lopatinskii condition is violated (this set can also be described explicitly; see [**AsLid**], [**GolLidTo**]). Here the ellipticity of ℓ_{ij} and the Shapiro–Lopatinskii condition are understood in the Agmon–Douglis–Nirenberg sense.

3. Spectral asymptotics for the shell operator. Let us fix some $\lambda_{\max} > 0$. Then uniformly over $\lambda \in [0, \lambda_{\max}]$ we have

$$N(h,\lambda) = a(\lambda)\, h^{-1} + O(h^{-9/10})\,, \qquad h \to 0\,, \tag{6.4.8}$$

where

$$a(\lambda) = \varkappa^2 \int_{\xi^4 + K^2(x,\xi) < \lambda} dx\, đ\xi \tag{6.4.9}$$

and $\varkappa = (12(1-\sigma^2))^{1/4}$. The asymptotics $N(h,\lambda) \sim a(\lambda)\, h^{-1}$ with $a(\lambda)$ given by (6.4.9) was initially conjectured by A. L. Gol'denveizer in 1970. The mathematical proof of (6.4.8), (6.4.9) based on a version of Courant's Dirichlet–Neumann bracketing technique is due to Aslanyan and Lidskii [**AsLid**]. Their work was followed by publications by other mathematicians who applied different methods and obtained improved remainder estimates under various assumptions on the level surfaces of $K^2(x,\xi) = \lambda$.

The following result is from [**Va9**]. Let us fix a $\lambda > 0$ such that this λ is not in the essential spectrum of the membrane problem (6.4.6), (6.4.7) and satisfies $\lambda > \max K^2(x,\xi)$. Then we have

$$N(h,\lambda) = a(\lambda)\,h^{-1} + O(h^{-1/2}), \qquad h \to 0. \tag{6.4.10}$$

If, in addition, the billiard system generated by the Hamiltonian

$$H(\lambda;x,\xi) = \frac{(A_1^{-2}\xi_1^2 + A_2^{-2}\xi_2^2)^{1/2}}{(\lambda - K^2(x,\xi))^{1/4}} \tag{6.4.11}$$

satisfies the nonblocking and nonperiodicity conditions, then

$$N(h,\lambda) = a(\lambda)\,h^{-1} + b(\lambda)\,h^{-1/2} + o(h^{-1/2}), \qquad h \to 0. \tag{6.4.12}$$

For a clamped edge (6.4.4) the coefficient $b(\lambda)$ is defined as follows. Assume for simplicity that ∂M coincides with the curvature lines, so $R_1^{-1}(x_2) \equiv R_1^{-1}(0,x_2)$ is the curvature of M in the cross section normal to ∂M, and $R_2^{-1}(x_2) \equiv R_2^{-1}(0,x_2)$ is that in the cross section normal to M and tangent to ∂M. Assume also that on ∂M, we have $R_1^{-1}R_2^{-1} \geqslant 0$ and $|R_2^{-1}| > |R_1^{-1}|$. Then

$$b(\lambda) = -\frac{\varkappa}{4\pi}\int_{\partial M}\left((\lambda - R_1^{-2}(x_2))^{1/4} + \frac{4}{\pi}\int_0^{(\lambda - R_1^{-2}(x_2))^{1/4}} \operatorname{Arg}\Delta(\lambda;x_2,\xi_2)d\xi_2\right)dx_2,$$

where $0 \leqslant \operatorname{Arg} < 2\pi$,

$$\Delta = \frac{i\,\det\big(a_{pq}\big)_{p,q=1}^{4}}{(\zeta_3-\zeta_2)(\zeta_4-\zeta_3)(\zeta_2-\zeta_4)},$$

ζ_q, $q = 1,2,3,4$, are the roots of the algebraic equation

$$(\zeta^2+\xi_2^2)^4 + (R_2^{-1}\zeta^2 + R_1^{-1}\xi_2^2)^2 = \lambda(\zeta^2+\xi_2^2)^2$$

specified by the conditions $\operatorname{Im}\zeta_1 = 0$, $\operatorname{Re}\zeta_1 > 0$, and $\operatorname{Im}\zeta_q > 0$, $q = 2,3,4$, and the a_{pq}, $p,q = 1,2,3,4$, are defined as

$$\begin{pmatrix} a_{1q} \\ a_{2q}\end{pmatrix} = \begin{pmatrix} \frac{\zeta_q^2}{1-\sigma^2} + \frac{\xi_2^2}{2(1+\sigma)} & \frac{i\zeta_q\xi_2}{2(1-\sigma)} \\ -\frac{i\zeta_q\xi_2}{2(1-\sigma)} & \frac{\zeta_q^2}{2(1+\sigma)} + \frac{\xi_2^2}{1-\sigma^2}\end{pmatrix}^{-1} \cdot \begin{pmatrix} -\frac{i(R_1^{-1}+\sigma R_2^{-1})\zeta_q}{1-\sigma^2} \\ -\frac{(R_2^{-1}+\sigma R_1^{-1})\xi_2}{1-\sigma^2}\end{pmatrix},$$

$a_{3q} = 1$, $a_{4q} = i\zeta_q$, $q = 1,2,3,4$.

4. Sketch of proof. One cannot obtain the sharp one-term asymptotic formula (6.4.10) or the two-term asymptotic formula (6.4.12) by applying the wave equation method directly to the problem (6.4.1), (6.4.4). The proof is based on the following ideas.

Let us denote the value of λ at which we are proving (6.4.12) by λ_{fix}.

First, one may assume without loss of generality that the eigenvalues of the operator $\big(\ell_{pq}\big)_{i,j=1}^{2}$ (upper left block of our full shell operator $\big(\mathcal{L}_{pq}\big)_{i,j=1}^{3}$) subject to the boundary conditions (6.4.7) are greater than λ_{fix}. Moreover, one may assume that our λ_{fix} is not an eigenvalue of the membrane problem (6.4.6), (6.4.7). Both these conditions can be satisfied by a finite rank perturbation independent of h, which may change the counting function only by $O(1)$.

Second, let us "freeze" the spectral parameter λ in the first two equations (6.4.1), that is, set in the first two equations $\lambda = \lambda_{fix}$ and view these two equations with the boundary conditions (6.4.7) as a differential constraint which determines the vector function (u_1, u_2) given a function u_3. Then we have the "real" spectral parameter λ (which is allowed to vary) only in the third equation (6.4.1). Denote the counting function of this new problem by $N(h, \lambda_{fix}, \lambda)$. It can be shown by variational arguments that $N(h, \lambda_{fix}) = N(h, \lambda_{fix}, \lambda_{fix})$.

Third, resolving the first two equations (6.4.1) with respect to (u_1, u_2) and substituting the result into the third equation we arrive at a scalar spectral problem

$$\frac{h^2}{12} n_{33} u_3 + V_{\lambda_{fix}} u_3 = \lambda u_3 \,, \tag{6.4.13}$$

$$u_3|_{\partial M} = \partial u_3/\partial x_1|_{\partial M} = 0 \,. \tag{6.4.14}$$

Here $V_{\lambda_{fix}}$ is a bounded scalar operator in $L_2(M)$. This operator is, in fact, a pseudodifferential operator of order 0 with principal symbol $K^2(x, \xi)$, and the subprincipal symbol of $(A_1 A_2)^{1/2} V_{\lambda_{fix}} (A_1 A_2)^{-1/2}$ is zero.

Fourth, let us fix the remaining spectral parameter λ in (6.4.13) and declare $\nu = 12h^{-2}$ to be the new spectral parameter. In other words, consider the spectral problem

$$A u_3 = \nu B_{\lambda_{fix}} u_3 \,, \tag{6.4.15}$$

($A = n_{33}$, $B_{\lambda_{fix}} = \lambda_{fix} I - V_{\lambda_{fix}}$) subject to the boundary conditions (6.4.14). Denote by $N_+(\lambda_{fix}, \nu)$ the number of *positive* eigenvalues of (6.4.15), (6.4.14) less than a given ν. Here we had to stress the word "positive" because the operator $B_{\lambda_{fix}}$ is not necessarily semibounded from below. It can be shown by variational arguments that $N(h, \lambda_{fix}, \lambda_{fix}) = N_+(\lambda_{fix}, 12h^{-2})$.

Thus, we have reduced the original spectral problem (6.4.1), (6.4.4) to the spectral problem (6.4.15), (6.4.14). The latter is a scalar problem without a small parameter, and we should be looking for asymptotic formulae for $N_+(\lambda_{fix}, \nu)$ as $\nu \to +\infty$. The only difference between (1.1.1′), (1.1.2) and (6.4.15), (6.4.14) is that in the latter problem we have a pseudodifferential weight $B_{\lambda_{fix}}$.

The problem (6.4.15), (6.4.14) can be handled along the same lines as in Chapters 2–5. In particular, in the interior zone one should construct the wave group by dealing with the "wave" operator $D_t^4 B_{\lambda_{fix}} - A(x, D_x)$.

Note that $B_{\lambda_{fix}}$ is an operator of the Boutet de Monvel type [**BdM**] and dealing with such an operator directly is inconvenient. This hitch can be overcome if in the process of technical realization of our standard scheme (Chapters 2-5) applied to (6.4.15), (6.4.14) one reintroduces the first two differential equations (6.4.1) (with $\lambda = \lambda_{fix}$) and writes $B_{\lambda_{fix}} u_3$ in terms of (u_1, u_2, u_3). In this way one has to deal only with differential operators.

More details can be found in [**Va8**], [**Va9**].

5. Discussion. Let us consider two examples. The first example is a cylindrical shell of radius 1, length 2 and thickness $h = 0.004$. The second is a truncated conical shell with meridian $r = z$, $\frac{1}{\sqrt{2}} \leqslant z \leqslant \sqrt{2}$ (here we use cylindrical coordinates; cf. Example 1.3.9) and thickness h=0.01. In both cases $\sigma = 0.3$ and the edges are clamped. Figures 19 and 20 show numerical results for these two examples.

Each plot contains three lines: the stepwise line is the actual counting function $N(h, \lambda)$, and the two smooth lines are the graphs of the functions $a(\lambda)h^{-1}$ (line

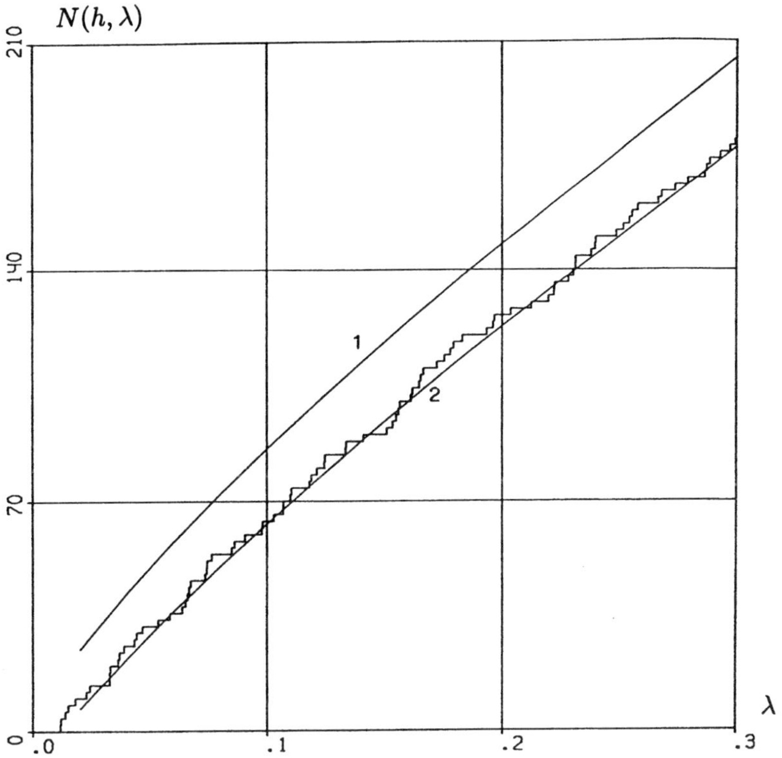

FIGURE 19. Eigenvalue distribution for a cylindrical shell.

1) and $a(\lambda)h^{-1} + b(\lambda)h^{-1/2}$ (line 2). The actual $N(h, \lambda)$ was plotted using the author's own numerical results (Figure 19) and numerical results from [**AsLid,** p. 149] (Figure 20). As usual, the two-term asymptotic formula shows itself to be very effective.

Let us comment on whether the use of the two-term asymptotic formula is justified in these two examples (see precise conditions in subsection 3). As in both these examples M is a surface of revolution, it is not too difficult to analyze the billiard trajectories and establish that we have nonblocking and nonperiodicity. However, we are also supposed to check that λ is not in the essential spectrum of the membrane problem (6.4.6), (6.4.7) and satisfies $\lambda > \max K^2(x,\xi)$. In both examples $\max K^2(x,\xi) = 1$ and the essential spectrum of the membrane problem (6.4.6), (6.4.7) is the union of the interval $[0,1]$ and the point 1.0012 (the latter is the value of λ at which the Shapiro–Lopatinskii condition for (6.4.6), (6.4.7) fails). This means that the whole graph in Figure 19 and two thirds of the graph in Figure 20 lie in the zone where we are not supposed to use (6.4.12). One can only conclude that our conditions on λ are probably too restrictive, and that it may be possible to give a mathematical proof of (6.4.12) for λ lying on the essential spectrum of the membrane problem, as long as the level surfaces $K^2(x,\xi) = \lambda$ are not too bad.

Shell theory is interesting in that it provides natural examples of periodic Hamiltonian and billiard flows. Indeed, shells in the form of a full sphere or a sufficiently large spherical cap provide such examples because by formula (6.4.11)

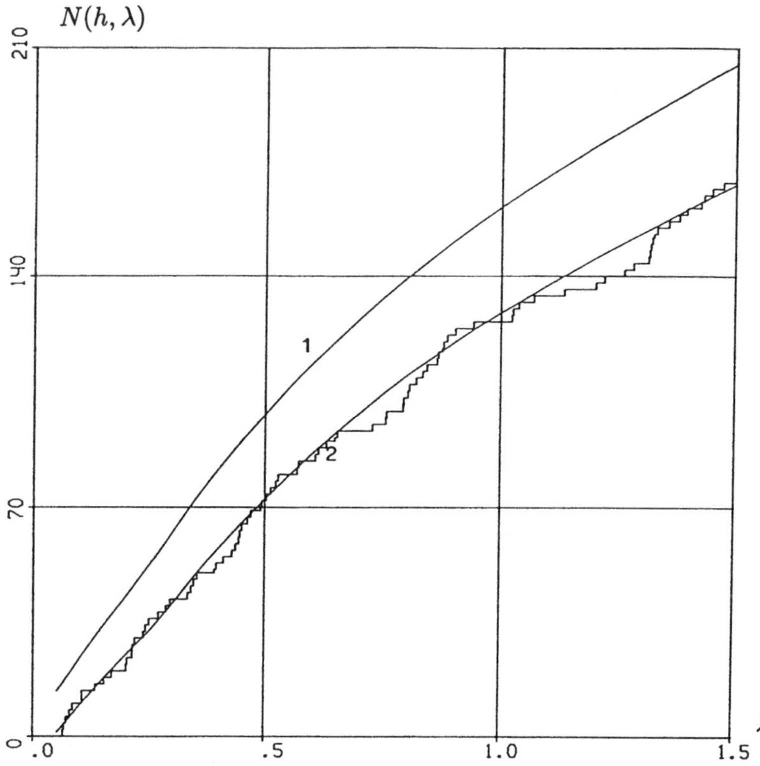

FIGURE 20. Eigenvalue distribution for a truncated conical shell.

our Hamiltonian trajectories in this case are geodesics and we have already looked at this situation; see the discussion after Lemma 1.3.19. Thus, for a full spherical shell or spherical cap which is greater than or equal to a hemisphere, one expects to observe clusters of eigenvalues. Numerical results of Niordson [**Nio**] show that this is exactly the case.

In the case of a hemispherical shell of radius R it is interesting to compare the numerical results [**Nio**] with the asymptotic formula from Example 1.7.13. The problems considered in [**Nio**] and in Example 1.7.13 are different, but if one replaces the pseudodifferential operator $V_{\lambda_{fix}}$ in (6.4.13) by the operator of multiplication by R^{-2} (its principal symbol), then the differential equations coincide up to a renormalization of the spectral parameter. Elementary calculations show that the asymptotic formula from Example 1.7.13 correctly predicts the wide gaps and the very sharp clusters in the spectrum of the hemispherical shell. The positions of these clusters are, however, slightly shifted as compared with the positions predicted in Example 1.7.13. This shift is explained by two factors:

1. The boundary of the hemisphere in [**Nio**] is assumed to be free, not clamped as in Example 1.7.13.
2. $V_{\lambda_{fix}}$ is not exactly the operator of multiplication by a constant.

Finally, let us elaborate on the analogy between shell equations and the Schrödinger equation. This analogy becomes evident if one reduces shell equations to the scalar problem (6.4.13), (6.4.14). Both (6.4.13), (6.4.14) and the spectral problem for the Schrödinger operator have a small parameter at the higher derivative and both have a potential V. The differences between the two problems are:

1. The potential in (6.4.13) is pseudodifferential.
2. The equation (6.4.13) is a fourth order equation.
3. Shells are usually compact and require boundary conditions, whereas in the Schrödinger case the problem is normally stated in $\mathbb{R}^n$.

The first difference is probably the most crucial one. The fact that the potential is pseudodifferential leads to an astonishing variety of absolutely different situations, which can be realized by choosing curvatures of different signs, different absolute values, and with different dependence on x. Many of these situations are analyzed in [**AsLid**], [**GolLidTo**].

6.5. Hydroelasticity

In this section we examine free vibrations of a closed ($\partial M = \varnothing$) shell filled with an ideal compressible fluid.

The surface M divides $\mathbb{R}^3$ into two parts: a bounded domain G_i (interior of the shell) and an unbounded domain G_e (exterior of the shell). The spectral problem being considered is

$$\sum_{j=1}^{3} \mathcal{L}_{ij} u_j = \lambda\, u_i\,, \qquad i = 1, 2, \tag{6.5.1}$$

$$\sum_{j=1}^{3} \mathcal{L}_{3j} u_j = \lambda\, u_3 + \frac{\rho_f \lambda}{\rho_s h}\, \psi|_M\,, \tag{6.5.2}$$

$$-\frac{c_f^2}{c_s^2} \Delta\psi = \lambda\, \psi \quad \text{in } G_i\,, \tag{6.5.3}$$

$$\left.\frac{\partial \psi}{\partial n}\right|_M = u_3\,. \tag{6.5.4}$$

Here

1. $\mathcal{L}_{ij}$, u_j, λ, ρ_s, E, σ and h are as in the previous section.
2. $\psi = \psi(y)$ is the potential of displacements of the fluid, i.e., $\operatorname{grad}\psi$ is the vector function of fluid displacements.
3. $y = (y_1, y_2, y_3)$ are Cartesian coordinates in $\mathbb{R}^3$.
4. $\Delta = \partial^2/\partial y_1^2 + \partial^2/\partial y_2^2 + \partial^2/\partial y_3^2$.
5. ρ_f and c_f are the fluid density and the speed of sound in the fluid, respectively.
6. $c_s := \sqrt{E/\rho_s}$ (characteristic speed of sound in the shell material).
7. $\partial/\partial n$ is the derivative along the outward normal to M.

The term $\frac{\rho_f \lambda}{\rho_s h}\, \psi|_M$ in (6.5.2) describes the dynamical pressure of the fluid acting on the shell, and (6.5.4) is the so-called nonpenetration condition (the fluid cannot go through the shell).

The first difficulty with (6.5.1)–(6.5.4) is that this system contains an "extra" occurrence of the spectral parameter λ, that is, the one in the term $\frac{\rho_f \lambda}{\rho_s h}\, \psi|_M$. So it is not *a priori* clear whether (6.5.1)–(6.5.4) is a spectral problem for a self-adjoint operator in some Hilbert space. In order to overcome this difficulty let us rewrite

(6.5.2) in equivalent form

$$(6.5.2') \qquad \sum_{j=1}^{3} \mathcal{L}_{3j} u_j + \frac{\rho_f c_f^2}{\rho_s c_s^2 h}\,(\Delta\psi)|_M = \lambda\, u_3$$

(here we used (6.5.3)). The system (6.5.1), (6.5.2′), (6.5.3) can be written as

$$(6.5.5) \qquad Af = \lambda f\,,$$

where A is the 4×4 matrix operator

$$A = \begin{pmatrix} \mathcal{L}_{11} & \mathcal{L}_{12} & \mathcal{L}_{13} & 0 \\ \mathcal{L}_{21} & \mathcal{L}_{22} & \mathcal{L}_{23} & 0 \\ \mathcal{L}_{31} & \mathcal{L}_{32} & \mathcal{L}_{33} & \frac{\rho_f c_f^2}{\rho_s c_s^2 h}\,(\Delta(\,\cdot\,))|_M \\ 0 & 0 & 0 & -\frac{c_f^2}{c_s^2}\Delta \end{pmatrix}$$

acting on quadruples $f = \big(u_1(x_1,x_2), u_2(x_1,x_2), u_3(x_1,x_2), \psi(y_1,y_2,y_3)\big)$. It is easy to see that on quadruples satisfying the nonpenetration condition (6.5.4) the spectral problem (6.5.5) is formally self-adjoint and nonnegative with respect to the inner product

$$(6.5.6) \qquad \iint_M (u_1\overline{v_1} + u_2\overline{v_2} + u_3\overline{v_3})\,dS + \frac{\rho_f}{\rho_s h}\iiint_{G_i} \operatorname{grad}\psi\cdot\overline{\operatorname{grad}\varphi}\,dV$$

(cf. (6.4.3)), $dV = dy_1dy_2dy_3$. It can be shown [**AsLidVa**] that the problem (6.5.5), (6.5.4) generates a self-adjoint operator in the Hilbert space with inner product (6.5.6), and that the spectrum of this operator is discrete and accumulates to $+\infty$. By $N(h,\lambda)$ we shall denote the counting function, that is, the number of eigenvalues of (6.5.1)–(6.5.4) below a given λ.

The following result is from [**AsLidVa**], [**Va2**]. Let is fix some $0 < \lambda_{\min} \leqslant \lambda_{\max}$. Then, uniformly over $\lambda \in [\lambda_{\min}, \lambda_{\max}]$, we have

$$(6.5.7) \qquad N(h,\lambda) = a(\lambda)\,h^{-6/5} + b(\lambda)\,h^{-4/5} + O(h^{-3/5})\,, \qquad h \to 0$$

(cf. (6.4.10)). If, in addition, the geodesic flow on M satisfies the nonperiodicity condition, then, uniformly over $\lambda \in [\lambda_{\min}, \lambda_{\max}]$,

$$(6.5.8) \quad N(h,\lambda) = a(\lambda)\,h^{-6/5} + b(\lambda)\,h^{-4/5} + c(\lambda)\,h^{-3/5} + o(h^{-3/5})\,, \qquad h \to 0$$

(cf. (6.4.12)). Here

$$a(\lambda) = \frac{S}{4\pi}\left(\frac{12(1-\sigma^2)\rho_f\lambda}{\rho_s}\right)^{2/5},$$

$$b(\lambda) = \frac{6(1-\sigma^2)^{3/5}}{5\pi}\left(\frac{12\rho_f\lambda}{\rho_s}\right)^{-2/5}\left(2\pi + \lambda S - \frac{3}{2}\iint_M \left(\frac{R_1^{-1}+R_2^{-1}}{2}\right)^2 dS\right),$$

$$c(\lambda) = \frac{1}{20\pi}\left(\frac{12(1-\sigma^2)\rho_f\lambda}{\rho_s}\right)^{1/5}\iint_M \frac{R_1^{-1}+R_2^{-1}}{2}\,dS\,,$$

with S being the surface area of M.

The proof of the asymptotic formulae (6.5.7), (6.5.8) is based on the reduction of the original spectral problem (6.5.1)–(6.5.4) to the scalar pseudodifferential spectral problem

$$\frac{h^2}{12} n_{33} u_3 + V_{\lambda_{fix}} u_3 = \lambda u_3 + \frac{\rho_f \lambda_{fix}}{\rho_s h} F_{\lambda_{fix}} u_3 \tag{6.5.9}$$

on M (cf. (6.4.13)), where $F_{\lambda_{fix}} : \partial\psi/\partial n|_M \to \psi|_M$ is the Neumann–Dirichlet operator for the Helmholtz equation (6.5.3). The latter is a pseudodifferential operator of order -1 with principal symbol $|\xi|_x^{-1}$, and the subprincipal symbol of $(A_1 A_2)^{1/2} F_{\lambda_{fix}} (A_1 A_2)^{-1/2}$ is $|\xi|_x^{-2} K(x,\xi)/2$ (see (6.4.5) for notation). The equation (6.5.9) is simpler than (6.4.13) because now the manifold has no boundary, but at the same time it is more complicated because the small parameter h comes into the equation twice and there is no obvious way of excluding it. Note that the main (according to their contribution to $N(h,\lambda)$) terms in (6.5.9) are $\frac{h^2}{12} n_{33} u_3$ and $\frac{\rho_f \lambda_{fix}}{\rho_s h} F_{\lambda_{fix}} u_3$, so we are dealing with the unusual situation when the lower order term cannot be disregarded. This happens because this lower order term contains a negative power of the small parameter h.

A special method for dealing with pseudodifferential operators with parameters was developed in [**Va1**], [**Va5**]. The application of this general result to (6.5.9) gives (6.5.7), (6.5.8). See [**AsLidVa**], [**Va2**] for more details.

It may seem strange that we are able to obtain up to three asymptotic terms for $N(h,\lambda)$. The explanation is that the small parameter h comes into our problem in a complicated way, and the term $b(\lambda)\, h^{-4/5}$ should really be viewed as a correction to the first asymptotic term. The second term proper is $c(\lambda)\, h^{-3/5}$. Its appearance is caused by the fact that the Neumann–Dirichlet operator has a nontrivial subprincipal symbol.

Suppose now that the fluid is not inside, but outside the shell. In other words, suppose that (6.5.2), (6.5.3) have been replaced by

$$\sum_{j=1}^{3} \mathcal{L}_{3j} u_j = \lambda\, u_3 - \frac{\rho_f \lambda}{\rho_s h} \psi|_M , \tag{6.5.10}$$

$$-\frac{c_f^2}{c_s^2} \Delta\psi = \lambda\,\psi \quad \text{in} \quad G_e . \tag{6.5.11}$$

In this case one can also associate a self-adjoint operator with the system (6.5.1), (6.5.10), (6.5.11), (6.5.4), and the spectrum of this self-adjoint operator is $\mathbb{R}_+$ (consequence of the unboundedness of G_e). The fact that the spectrum of the exterior problem fills the nonnegative half-line (or, in terms of frequency ω, the whole real line) is not very informative from the mechanical point of view. This inconvenience can be overcome by modifying our choice of function spaces and extending the resolvent by analyticity through the (continuous) spectrum onto the whole complex ω-plane; see [**LtinVa3**] and [**S.-HuS.-Pa,** Chap. 9]. The extended resolvent is meromorphic as a function of ω with poles at some complex ω_k which are called *resonances*. It turns out that a massive group of resonances lies close (in terms of the small parameter h) to the real line, and the real parts of the corresponding λ_k are distributed in accordance with (6.5.7), (6.5.8). The only difference is that in the case of the exterior problem the coefficient $c(\lambda)$ changes sign (this happens because the fluid is on the other side of the shell).

A detailed mathematical analysis of both the interior and exterior problems was carried out in [**Va2**], [**Va8**]. See also the review paper [**LtinVa3**].

REMARK 6.5.1. Besides the relative thickness h/R the hydroelasticity problem contains two other dimensionless parameters, namely, ρ_f/ρ_s and c_f/c_s. In the derivation of (6.5.7), (6.5.8) we assumed these two parameters to be fixed. In mechanical terms this means that we assumed these parameters to be of the order of 1. However, in reality this is not exactly the case. Say, for the pair water–duralumin we have $\rho_f/\rho_s \approx 0.357$ and $c_f/c_s \approx 0.288$, and for water–steel $\rho_f/\rho_s \approx 0.127$ and $c_f/c_s \approx 0.286$. The influence of these two additional parameters is such that they make the asymptotic convergence in (6.5.7), (6.5.8) not so good as one would have hoped. Basically, this implies that formula (6.5.8) is sufficiently accurate only in a relatively narrow frequency range. Things can be improved if one rewrites (6.5.8) in the form

$$N(h,\lambda) = \int dx\, d\!\!{}^{-}\!\xi + o(h^{-3/5}), \qquad h \to 0, \tag{6.5.12}$$

where integration is carried out over all $(x,\xi) \in T'M$ such that

$$\frac{h^2|\xi|_x^4}{12(1-\sigma^2)} + K^2(x,\xi) < \lambda + \frac{\rho_f \lambda}{\rho_s h |\xi|_x}\left(1 \pm \frac{K(x,\xi)}{2|\xi|_x}\right), \tag{6.5.13}$$

with sign "plus" for the interior problem and "minus" for exterior; note that (6.5.13) is obtained from (6.5.9) if one replaces the equality by the inequality and the operators by their principal and subprincipal symbols. From the purely mathematical point of view (6.5.12), (6.5.13) is equivalent to (6.5.8). However, in reality (6.5.12), (6.5.13) works in a much wider frequency range because we have avoided additional errors caused by expanding (6.5.13) in powers of h.

Appendix A

Spectral Problem on the Half-line

Anders Holst

This appendix is devoted to the analysis of a self-adjoint ordinary differential operator with constant coefficients on the half-line $\mathbb{R}_+ = [0, +\infty)$, with appropriate boundary conditions at the origin. The difference with existing works is that in our case the operator has an arbitrary (even) order and is not necessarily a power of a one-dimensional Laplacian.

The notation in this appendix sometimes differs from that in the rest of the book.

A.1. Basic facts

1. The operator $\mathbf{A}^+$. Let $A(\xi) = \sum_{j=0}^{2m} a_j \xi^j$ be a polynomial of degree $2m$ ($a_{2m} \neq 0$) with real coefficients, such that $A(\xi) \geqslant 0$ when $\xi \in \mathbb{R}$ and $\inf_{\xi \in \mathbb{R}} A(\xi) = 0$. We shall assume that we are given polynomials $B_1, \dots, B_m$ of degrees $0 \leqslant m_1 < \dots < m_m \leqslant 2m-1$ such that the operator $\mathbf{A}^+$ on $L_2(\mathbb{R}_+)$ with domain

$$D(\mathbf{A}^+) = \{u \in H^{2m}(\mathbb{R}_+) : (B_1(D)u)(+0) = \dots = (B_m(D)u)(+0) = 0\}$$

associated with the differential expression $A(D)$ is self-adjoint. (Later on we shall let $A, B_1, \dots, B_m$ depend on some real parameters but this is irrelevant at the moment.)

By the definition of self-adjointness we have

$$(A(D)u, v)_+ = (u, A(D)v)_+, \quad u, v \in D(\mathbf{A}^+),$$

where $(\cdot, \cdot)_+$ denotes the scalar product on $L_2(\mathbb{R}_+)$. On the other hand, if $u, v \in H^{2m}(\mathbb{R}_+)$, then it follows through integration by parts that

$$(A(D)u, v)_+ - (u, A(D)v)_+ = i\mathcal{A}^+(u, v), \tag{A.1.1}$$

where

$$\mathcal{A}^+(u, v) = \sum_{r,s=0}^{2m-1} a_{r+s+1} (D^r u)(+0) \overline{(D^s v)(+0)}$$

is called the boundary form (see e.g. [**CodLson**] or [**LtanSar**]); here and further on we use the convention $a_j = 0$ for $j > 2m$. Obviously the boundary form extends naturally to all functions $u, v \in C^{2m-1}(\mathbb{R}_+)$. We shall sometimes refer to (A.1.1) as Green's formula.

Supported by EPSRC grant GR/J74640.

DEFINITION A.1.1. We call $u \in C^{2m}(\mathbb{R}_+)\setminus L_2(\mathbb{R}_+)$ a *generalized eigenfunction of* $\mathbf{A}^+$ *corresponding to* $\nu \in \mathbb{R}$ if

$$(A(D) - \nu)u = 0 \quad \text{on} \quad \mathbb{R}_+,$$
$$(B_j(D)u)(+0) = 0, \quad j = 1, \dots, m,$$

and if $|u(x)|$ does not grow faster than a power of x as $x \to +\infty$.

Let $\xi_1, \dots, \xi_r$, $r \leqslant 2m$, be the roots of $A(\xi) = \nu$, and let $k_1, \dots, k_r$ be their multiplicities. Clearly, any eigenfunction of $\mathbf{A}^+$ corresponding to the eigenvalue ν has the form

$$u(x) = \sum_{\operatorname{Im}\xi_r > 0} \sum_{j=0}^{k_r - 1} c_{rj} x^j e^{ix\xi_r}, \tag{A.1.2}$$

with at least one $c_{jr} \neq 0$, and any generalized eigenfunction has the form

$$u(x) = \sum_{\operatorname{Im}\xi_r \geqslant 0} \sum_{j=0}^{k_r - 1} c_{rj} x^j e^{ix\xi_r},$$

with at least one $c_{rj} \neq 0$ for some r such that $\operatorname{Im}\xi_r = 0$. This motivates the following subsection.

2. The equation $A(\xi) = z$**.** Let Λ_0 denote the set of $z \in \mathbb{C}$ for which $A(\xi) = z$ has a multiple root. Clearly

$$\Lambda_0 = \{z : \ z = A(\xi) \text{ for some } \xi \in \mathbb{C} \text{ such that } A'(\xi) = 0\},$$

so there are at most $2m - 1$ elements in Λ_0. Note in particular that the point 0 belongs to Λ_0. If $A'(\xi_0) \neq 0$ and $A(\xi_0) = z_0$, it follows from the implicit function theorem that in a neighbourhood of z_0 there is an analytic function $\xi(z)$ such that $A(\xi(z)) = z$, $\xi(z_0) = \xi_0$ and

$$\xi'(z) = \frac{1}{A'(\xi(z))}. \tag{A.1.3}$$

We can continue $\xi(z)$ uniquely along any curve which does intersect Λ_0, and we get a uniquely defined analytic function $\xi(z)$ on any simply connected domain D in $\mathbb{C}\setminus\Lambda_0$ that contains z_0. If $z_0 < 0$, then the equation $A(\xi) = z_0$ has m roots $\xi^+_{1,0}, \dots, \xi^+_{m,0}$ with positive imaginary part and m roots $\xi^-_{1,0}, \dots, \xi^-_{m,0}$ with negative imaginary part. If, in addition, D is chosen in such a way that $D \cap \mathbb{R}_+ = \varnothing$, then there are $2m$ analytic functions $\xi^\pm_j(z)$ in D such that $A(\xi^\pm_j(z)) = z$ and $\xi^\pm_j(z_0) = \xi^\pm_{j,0}$. These functions satisfy $\pm\operatorname{Im}\xi^\pm_j(z) > 0$ everywhere in D, since $A(\xi) \in \mathbb{R}_+$ for real ξ. Clearly, we can choose D in such a way that the strips $\{0 < \pm\operatorname{Im} z < \varepsilon\}$ are contained in D for some $\varepsilon > 0$.

We shall be particularly interested in the limits when z approaches $\mathbb{R}_+$ from below or above. Let $\lambda \in \mathbb{R}_+$. The limits $\xi^\pm_j(\lambda + i0)$ are roots of the equation $A(\xi) = \lambda$. If $\lambda \notin \Lambda_0$, these are all distinct and $2q(\lambda)$ of them are real, where the nonnegative integer $q(\lambda)$ is constant on each connected component of $\mathbb{R}\setminus\Lambda_0$. Assume that $\xi^\pm_j(\lambda + i0) = \zeta^\pm$ is real. From the equation (A.1.3) we see that

$$\xi^\pm_j(\lambda + i\varepsilon) - \zeta^\pm = \frac{i\varepsilon}{A'(\zeta^\pm)} + o(\varepsilon),$$

and thus $\pm A'(\zeta^{\pm}) > 0$. On each connected component of $\mathbb{R}_+ \setminus \Lambda_0$, we can number the real roots in increasing order

$$\zeta_1^- < \zeta_1^+ < \cdots < \zeta_{q(\lambda)}^- < \zeta_{q(\lambda)}^+ .$$

3. Some results involving the boundary form.

LEMMA A.1.2. *Let $u(x) = x^j e^{ix\xi}$ and $v(x) = x^k e^{ix\eta}$, where $j,k \in \mathbb{N}$ and $\xi, \eta \in \mathbb{C}$. Then*

$$\mathcal{A}^+(u,v) = (-1)^k D_a^j D_b^k \left\{ \frac{A(a) - A(b)}{a-b} \right\}_{a=\xi, b=\overline{\eta}} . \tag{A.1.4}$$

PROOF. For fixed j,k the left-hand side of (A.1.4) is a polynomial in $\xi, \overline{\eta}$ and the same is true for the right-hand side, since

$$\frac{A(a) - A(b)}{a-b} = \sum_{j=1}^{2m} A^{(j)}(b) \frac{(a-b)^{j-1}}{j!}$$

is a polynomial in a, b. Thus it suffices to prove (A.1.4) for $\operatorname{Im} \xi > 0$, $\operatorname{Im} \eta > 0$. Then $u, v \in H^{2m}(\mathbb{R}_+)$ and we have (A.1.1). For $\operatorname{Im} a > 0$, $\operatorname{Im} b < 0$ we have

$$\begin{aligned}
\int_0^\infty &(A(D) x^j e^{ixa}) x^k e^{-ixb}\, dx - \int_0^\infty x^j e^{ixa} \left(A(D) x^k e^{-ibx}\right) dx \\
&= D_a^j (-D_b^k) \left(\int_0^\infty A(a) e^{ix(a-b)}\, dx - \int_0^\infty A(b) e^{ix(a-b)}\, dx \right) \\
&= D_a^j (-D_b)^k \frac{A(a) - A(b)}{-i(a-b)} = i D_a^j (-D_b)^k \left\{ \frac{A(a) - A(b)}{a-b} \right\}.
\end{aligned}$$

Setting $a = \xi$, $b = \overline{\eta}$ we obtain the required statement. □

COROLLARY A.1.3. *Assume that $\xi, \eta \in \mathbb{C}$ are zeros of $A(\xi) - z$ $(z \in \mathbb{C})$ of orders j_0 and k_0. Set*

$$\begin{aligned}
u(x) &= x^j e^{ix\xi}, \quad j < j_0, \\
v(x) &= x^k e^{ix\eta}, \quad k < k_0.
\end{aligned}$$

Then

$$\mathcal{A}^+(u,v) = \begin{cases} 0, & \xi \neq \overline{\eta}, \\ (-i)^j i^k \frac{j!k!}{(j+k+1)!} A^{(1+j+k)}(\xi), & \xi = \overline{\eta}. \end{cases}$$

PROOF. In the case $\xi \neq \overline{\eta}$ the required result follows immediately from Lemma A.1.2.

Let us treat the remaining case $\xi = \overline{\eta}$ with the help of Lemma A.1.2. Since $\mathcal{A}^+(u,v)$ is a continuous function of $\xi, \eta \in \mathbb{C}$, we have

$$\mathcal{A}^+(u,v) = \lim_{\substack{a \to \xi \\ b \to \xi}} D_a^j (-D_b)^k \left\{ \frac{A(a) - A(b)}{a-b} \right\}.$$

Note that

$$\frac{A(a) - A(b)}{a-b} = \int_0^1 A'(at + (1-t)b)\, dt$$

and consequently

$$D_a^j(-D_b)^k\left\{\frac{A(a)-A(b)}{a-b}\right\}=(-i)^j i^k\int_0^1 t^j(1-t)^k A^{(1+j+k)}(at+(1-t)b)\,dt.$$

When $a,b\longrightarrow\xi$ the right-hand side tends to

$$i^k(-i)^j A^{(j+k+1)}(\xi)\int_0^1 t^j(1-t)^k\,dt.$$

This completes the proof since

$$\int_0^1 t^j(1-t)^k\,dt=\frac{j!k!}{(j+k)!}\int_0^1 t^{j+k}\,dt=\frac{j!k!}{(j+k+1)!}.$$

□

COROLLARY A.1.4. *Assume that* $\nu\in\mathbb{R}_+\setminus\Lambda_0$. *If*

$$u(x)=\sum_{k=1}^{q(\nu)}c_k^+\frac{e^{ix\zeta_k^+}}{\sqrt{A'(\zeta_k^+)}}+\sum_{k=1}^{q(\nu)}c_k^-\frac{e^{ix\zeta_k^-}}{\sqrt{-A'(\zeta_k^-)}}+\sum_{\operatorname{Im}\xi_j^+>0}d_je^{ix\xi_j^+}\tag{A.1.5}$$

is a generalized eigenfunction for $\mathbf{A}^+$, *then*

$$0=\mathcal{A}^+(u,u)=\sum_{k=1}^{q(\nu)}|c_k^+|^2-\sum_{k=1}^{q(\nu)}|c_k^-|^2.\tag{A.1.6}$$

PROOF. It follows from Corollary A.1.3 that

$$\mathcal{A}^+(u,u)=\sum_{k=1}^{q(\nu)}|c_k^+|^2-\sum_{k=1}^{q(\nu)}|c_k^-|^2.$$

But if u is a generalized eigenfunction, then $\mathcal{A}^+(u,u)=0$, since in this case $\phi u\in D(\mathbf{A}^+)$ for any $\phi\in C_0^\infty(\mathbb{R})$ which is identically equal to one in a neighbourhood of the origin. □

4. The operator A and its resolvent. Our eventual objective is to compare the operator $\mathbf{A}^+$ with the corresponding operator "without boundary conditions". It is natural to define the latter on the full real line. We define $\mathbf{A}$ as the self-adjoint operator on $L_2(\mathbb{R})$ associated with the differential expression $A(D)$ acting on the domain $D(\mathbf{A})=H^{2m}(\mathbb{R})$.

REMARK A.1.5. At this stage the reader may become worried by the fact that $\mathbf{A}$ is not defined on the same Hilbert space (i.e., $L_2(\mathbb{R}_+)$) as $\mathbf{A}^+$. However, we can easily overcome this difficulty by introducing the operator $\mathbf{A}^{+-}$ on $L_2(\mathbb{R})$ which is assembled from "two copies" of $\mathbf{A}^+$. Namely, let us first introduce the operator $\mathbf{A}^-$ on $L_2(\mathbb{R}_-)$ with domain

$$D(\mathbf{A}^-)=\{u\in H^{2m}(\mathbb{R}_-):(\overline{B}_1(D)u)(-0)=\cdots=(\overline{B}_m(D)u)(-0)=0\},$$

associated with the differential expression $A(D)$. It is easy to see that $\mathbf{A}^-$ is self-adjoint. Moreover, there is a simple relationship between the operators $\mathbf{A}^-$ and $\mathbf{A}^+$. Indeed, if we introduce the (antilinear) operator $U:L_2(\mathbb{R}_-)\to L_2(\mathbb{R}_+)$ defined by $(U(v))(x) = \overline{v(-x)}$, then $D(\mathbf{A}^+) = U(D(\mathbf{A}^-))$ and $U(\mathbf{A}^-(\,\cdot\,)) = \mathbf{A}^+(U(\cdot))$. The latter leads to an obvious relationship between the spectral projections of $\mathbf{A}^-$ and $\mathbf{A}^+$; in particular, the spectra of $\mathbf{A}^-$ and $\mathbf{A}^+$ are the same and their resolvent kernels satisfy $\mathbf{r}_z^-(x,y) = \overline{\mathbf{r}_{\bar{z}}^+(-x,-y)}$. Now, let $D(\mathbf{A}^{+-})$ be the linear subspace of $L_2(\mathbb{R})$ consisting of all functions the restrictions of which to $(-\infty,0)$ and $(0,+\infty)$ belong to $H^{2m}(-\infty,0)$ and $H^{2m}(0,+\infty)$, respectively, and which satisfy the boundary conditions

$$\begin{aligned} B_j(D)u(+0) &= 0, \quad 1 \leqslant j \leqslant m, \\ \overline{B_j}(D)u(-0) &= 0, \quad 1 \leqslant j \leqslant m. \end{aligned}$$

We define the operator $\mathbf{A}^{+-}$ to be the self-adjoint operator on $L_2(\mathbb{R})$ associated with the differential expression $A(D)$ acting on the domain $D(\mathbf{A}^{+-})$. Obviously, the boundary conditions at the origin decouple the two axes and we have $\mathbf{A}^{+-} = \mathbf{A}^+ \oplus \mathbf{A}^-$. Consequently, the spectrum of $\mathbf{A}^{+-}$ is the same as that of $\mathbf{A}^+$ (but with double multiplicity), and if $\mathbf{r}_z^{+-}(x,y)$, $\mathbf{r}_z^+(x,y)$ denote the resolvent kernels of $\mathbf{A}^{+-}$, $\mathbf{A}^+$, respectively,

$$\mathbf{r}_z^{+-}(x,y) = \begin{cases} \mathbf{r}_z^+(x,y) & \text{if} \quad x,y>0, \\ \overline{\mathbf{r}_{\bar{z}}^+(-x,-y)} & \text{if} \quad x,y<0, \\ 0 & \text{if} \quad xy<0. \end{cases}$$

Thus, studying $\mathbf{A}^+$ is in fact equivalent to studying $\mathbf{A}^{+-}$.

The resolvent of $\mathbf{A}$ is easily calculated using the Fourier transform

$$\mathbf{R}_z f(x) = \frac{1}{2\pi}\int_{-\infty}^{\infty} e^{ix\xi}\frac{\hat{f}(\xi)}{(A(\xi)-z)}\,d\xi\,,$$

$$\mathbf{r}_z(x,y) = \begin{cases} i\sum_{\operatorname{Im} w>0}\operatorname{Res}_{\zeta=w}\left((A(\zeta)-z)^{-1}e^{i(x-y)\zeta}\right), & x-y>0, \\ -i\sum_{\operatorname{Im} w<0}\operatorname{Res}_{\zeta=w}\left((A(\zeta)-z)^{-1}e^{i(x-y)\zeta}\right), & x-y<0. \end{cases}$$

We see that the kernel of the resolvent depends only on $x-y$, which reflects the translational invariance of $\mathbf{A}$. Thus we have $\mathbf{r}_z(x,y) = \mathbf{r}_z^0(x-y)$, where $\mathbf{r}_z^0$ is the inverse Fourier transform of $(A(\xi)-z)^{-1}$.

In particular, in a neighbourhood of $z_0 \notin \mathbb{R}_+ \cup \Lambda_0$ we have, for $x \neq y$,

$$\mathbf{r}(z,x,y) = iY_+(x-y)\sum_{k=1}^{m}\frac{e^{i(x-y)\xi_k^+(z)}}{A'(\xi_k^+(z))} - iY_-(x-y)\sum_{k=1}^{m}\frac{e^{i(x-y)\xi_k^-(z)}}{A'(\xi_k^-(z))},$$

where Y_+ (Y_-) denotes the characteristic function of $\mathbb{R}_+$ ($\mathbb{R}_-$).

We shall abuse notation in so far as that we will use the notation $\mathbf{R}_z$ also for the operator of convolution with $\mathbf{r}_z^0$, allowing it to act on functions which are not necessarily in $L_2(\mathbb{R})$. Thus, for example, $\mathbf{r}_z^0(x) = (\mathbf{r}_z^0 * \delta_0)(x) = (\mathbf{R}_z\delta_0)(x)$, where δ_0 is the delta function.

When

$$\{\operatorname{Im} z>0\} \ni z \to \nu \in \mathbb{R}_+ \setminus \Lambda_0,$$

the kernel of the resolvent converges in $C_{loc}^\infty(\mathbb{R}^2 \setminus \{(x,x) : x \in \mathbb{R}\})$ to

$$\mathbf{r}_0(\nu+i0;x-y) = iY_+(x-y)\sum_{k=1}^{m}\frac{e^{i(x-y)\xi_k^+(\nu+i0)}}{A'(\xi_k^+(\nu+i0))} - iY_-(x-y)\sum_{k=1}^{m}\frac{e^{i(x-y)\xi_k^-(\nu+i0)}}{A'(\xi_k^-(\nu+i0))}.$$

This is no longer the kernel of a bounded operator on $L_2(\mathbb{R})$, since some of the numbers $\xi_k^{\pm}(\nu + i0)$ are real. However, it does represent a bounded operator from $L^1(\mathbb{R})$ to $L^\infty(\mathbb{R})$, and $\mathbf{r}_0(\nu + i0; \cdot)$ is a fundamental solution of $A(D) - \nu$. Recall that a fundamental solution of $P(D)$ is a distribution E such that $P(D)E = \delta_0$. If $u \in \mathcal{E}'$ we have

$$P(D)(E * u) = u = E * P(D)u\,.$$

If E is a fundamental solution, then so is $E+w$, where w is any solution of $P(D)w = 0$. When $z \notin \mathbb{R}_+$ the distribution $\mathbf{r}_z^0$ is the unique fundamental solution of $A(D) - z$ which vanishes at infinity.

Clearly, we can also let $z \in \{\operatorname{Im} z < 0\}$ tend to ν. We then obtain another operator $\mathbf{r}(\nu - i0)$ with similar properties. We have

$$\text{(A.1.7)} \qquad \mathbf{r}_{\nu+i0}(x,y) - \mathbf{r}_{\nu-i0}(x,y) = i\sum_{k=1}^{q(\nu)} \frac{e^{i(x-y)\zeta_k^+(\nu)}}{A'(\zeta_k^+(\nu))} - i\sum_{k=1}^{q(\nu)} \frac{e^{i(x-y)\zeta_k^-(\nu)}}{A'(\zeta_k^-(\nu))}\,,$$

since the contributions from the nonreal roots cancel out. Note in particular that this expression vanishes when $\nu < 0$.

Functions of the form $\mathbf{R}_{\nu+i0}f$ are called ν-outgoing; similarly, functions of the form $\mathbf{R}_{\nu-i0}f$ are called ν-incoming. The reason for this terminology is that if $f \in C_0^\infty$, then

$$(\mathbf{R}_{\nu+i0}f)(x) = \pm i\sum_{k=1}^{q(\nu)} \frac{e^{ix\zeta_k^{\pm}(\nu)}}{A'(\zeta_k^{\pm}(\nu))}\hat{f}(\zeta_k^{\pm}) + o(1) \quad \text{as } x \to \pm\infty.$$

Thus, the main contributions to $\mathbf{R}_{\nu+i0}f$ for large positive x are determined by the values of $\hat{f}$ at ζ_k^+. But a classical particle governed by the Hamiltonian $H(x,\xi) = A(\xi)$, with momentum ζ_k^+ and energy $A(\zeta_k^+) = \nu$, has velocity $\frac{dx}{dt} = A'(\zeta_k^+) > 0$. Similarly the main contributions for large negative x correspond to particles moving to the left. One can give a more convincing motivation by introducing in the appropriate way a time variable and considering the evolution of a wave packet, in which case A' turns out to be the group velocity. More generally we call a function on $\mathbb{R}_\pm$ of the form

$$\sum_{k=1}^{q(\nu)} b_k e^{ix\zeta_k^{\pm}(\nu)} + g(x),$$

where $g \in L^1(\mathbb{R}_\pm)$, outgoing and similarly for incoming functions.

5. The resolvent of $\mathbf{A}^+$. We now study the resolvent $\mathbf{R}_z^+ = (\mathbf{A}^+ - zI)^{-1}$ for $\operatorname{Im} z \neq 0$. If $f \in L_2(\mathbb{R}_+)$, then $u = \mathbf{R}_z^+ f$ is the unique solution in $L_2(\mathbb{R}_+)$ of

$$\begin{aligned} (A(D) - z)u &= f \quad \text{on } \mathbb{R}_+, \\ (B_j(D)u)(+0) &= 0, \quad 1 \leqslant j \leqslant m. \end{aligned}$$

Let $\theta : L_2(\mathbb{R}) \to L_2(\mathbb{R}_+)$ be the natural restriction operator and let $\theta^* : L_2(\mathbb{R}_+) \to L_2(\mathbb{R})$ be its adjoint, given by

$$(\theta^* f)(x) = \begin{cases} 0, & x < 0 \\ f(x), & x \geqslant 0. \end{cases}$$

If we set $w = u - \theta\mathbf{R}_z\theta^* f$, then

(A.1.8)
$$\begin{cases} (A(D) - z)w = 0 \quad \text{on} \quad \mathbb{R}_+, \\ (B_j(D)w)(+0) = -(B_j(D)\theta\mathbf{R}_z\theta^* f)(+0). \end{cases}$$

When $z \notin \mathbb{R}\cup\Lambda_0$, the equation $(A(D)-z)w = 0$ has precisely m linearly independent solutions $e^{ix\xi_k^+(z)}$, $k = 1, \dots, m$, in $L_2(\mathbb{R}_+)$. Setting

$$w(x) = \sum_{k=1}^{m} d_k e^{ix\xi_k^+(z)}$$

we conclude that (A.1.8) is equivalent to the linear system

(A.1.9)
$$\sum_{k=1}^{m} Q_{jk}^+(z) d_k = -(B_j(D)\theta\mathbf{R}_z\theta^* f)(+0), \quad 1 \leqslant j \leqslant m,$$

where $Q_{jk}^+(z) = B_j(\xi_k^+(z))$. Throughout this appendix when dealing with elements of matrices the first lower index stands for the number of the row, and the second for the number of the column.

For $z \in \mathbb{C} \setminus (\mathbb{R} \cup \Lambda_0)$ this matrix is invertible. Indeed, if $d \in \mathbb{C}^m$, $\|d\| \neq 0$, and $Q^+(z)d = 0$, then $w(x)$ is an eigenfunction of $\mathbf{A}^+$ corresponding to the nonreal eigenvalue z. This contradicts the assumption that $\mathbf{A}^+$ is self-adjoint. Thus the system (A.1.9) has a unique solution which determines $\mathbf{R}_z^+$.

As a function of z the matrix $Q^+(z)$ clearly inherits the properties of the $\xi_k^+(z)$. Thus it is an analytic function of $z \in D$, and when $\nu \in \mathbb{R}_+$ the limits $Q^+(\nu \pm i0)$ exist but are in general not equal. The functions $\nu \mapsto Q^+(\nu \pm i0)$ are real analytic on each connected component of $\mathbb{R}_+ \setminus \Lambda_0$.

PROPOSITION A.1.6. *Let $z \in \mathbb{C} \setminus (\mathbb{R} \cup \Lambda_0)$. Then*

$$\mathbf{r}_z^+(x, y) - \mathbf{r}_z(x, y) = i \sum_{k,l=1}^{m} G_{kl}(z) e^{ix\xi_k^+(z)} e^{-iy\xi_l^-(z)},$$

where

(A.1.10)
$$G_{kl}(z) = \sum_{j=1}^{m} \{Q^+(z)^{-1}\}_{kj} B_j(\xi_l^-(z))/A'(\xi_l^-(z)).$$

Before proceeding to the proof of Proposition A.1.6 let us make a few remarks.

Let $\nu \in \mathbb{R} \setminus \Lambda_0$. We shall see in the next section that $\det Q^+(\nu \pm i0) = 0$ if and only if ν is an eigenvalue of $\mathbf{A}^+$. In this case $(z-\nu)\{Q^+(z)\}^{-1}$ is uniformly bounded for small $z - \nu \in \{\operatorname{Im} z \neq 0\}$; in other words, we have a first order singularity at ν.

PROOF OF PROPOSITION A.1.6. For $y > 0$ we have

$$-(B_j(D)\mathbf{R}_z\delta_y)(+0) = -\frac{1}{2\pi} \int \frac{e^{i(0-y)\xi} B_j(\xi)}{A(\xi) - z}\, d\xi = i \sum_{l=1}^{m} \frac{B_j(\xi_l^-(z))}{A'(\xi_l^-(z))} e^{-iy\xi_l^-(z)}.$$

Thus, formula (A.1.10) follows immediately from (A.1.9) with $f = \delta_y$. □

For $x, y > 0$ we have $\mathbf{r}_z^{+-}(x,y) - \mathbf{r}_z(x,y) = \mathbf{r}_z^{+}(x,y) - \mathbf{r}_z(x,y)$ and for $x < 0 < y$ we have

$$\mathbf{r}_z^{+-}(x,y) - \mathbf{r}_z(x,y) = -\mathbf{r}_z(x,y) = i\sum_{k=1}^{m} \frac{e^{ix\xi_k^-(z)}\, e^{-iy\xi_k^-(z)}}{A'(\xi_k^-(z))}.$$

Since clearly similar arguments can be carried out for $y < 0$, it follows that $\mathbf{R}_z - \mathbf{R}_z^{+-}$ is a finite rank operator for any $z \in \mathbb{C} \setminus (\mathbb{R} \cup \Lambda_0)$. Thus Weyl's Theorem [RS3, Theorem XIII.14] immediately implies that

$$\sigma_{ess}(\mathbf{A}^+) = \sigma_{ess}(\mathbf{A}^{+-}) = \sigma_{ess}(\mathbf{A}).$$

The function $w = \theta^*(\mathbf{R}_z^+ f - \theta\mathbf{R}_z\theta^* f)$ satisfies $(A(D)-z)w = 0$ on $\mathbb{R}\setminus\{0\}$ and its restriction to $\mathbb{R}_+$ belongs to $H^{2m}(\mathbb{R}_+)$. We conclude that $(A(D)-z)w = P_{z,f}^+(D)\delta_0$ for some polynomial $P_{z,f}^+$ of degree $\leqslant 2m-1$, and that $w = \mathbf{R}_z P_{z,f}^+(D)\delta_0$. This can also be seen in terms of Green's formula:

LEMMA A.1.7. *Let $z \in \mathbb{C} \setminus \mathbb{R}$ and $f, g \in L_2(\mathbb{R}_+)$. We have*

$$\begin{aligned}(\mathbf{R}_z^+ f, g)_+ &= (\mathbf{R}_z\theta^* f, g)_+ - i\mathcal{A}^+(\mathbf{R}_z^+ f, \mathbf{R}_{\bar z}\theta^* g)\\ &= (\mathbf{R}_z\theta^* f, g)_+ + i\mathcal{A}^+(\mathbf{R}_z\theta^* f, \mathbf{R}_{\bar z}^+ g).\end{aligned}$$

The first equality implies that

$$\mathbf{R}_z^+ f = \theta\mathbf{R}_z\theta^* f - i\theta\mathbf{R}_z P_{z,f}^+(D)\delta_0,$$

where

$$P_{z,f}^+(D) = \sum_{r,s=0}^{2m-1} a_{r+s+1}(D^r\mathbf{R}_z^+ f)(+0)D^s.$$

PROOF OF LEMMA A.1.7. It follows from (A.1.1) that

$$((A(D)-z)u, v)_+ - (u, (A(D)-\bar z)v)_+ = i\mathcal{A}^+(u,v), \quad u, v \in H^{2m}(\mathbb{R}_+).$$

The first equality of Lemma A.1.11 follows from the above formula if we choose $u = \mathbf{R}_z^+ f$ and $v = \mathbf{R}_{\bar z}\theta^* g$, and the second if we choose $u = \mathbf{R}_z\theta^* f$ and $v = \mathbf{R}_{\bar z}^+ g$. □

6. The eigenvalues of $\mathbf{A}^+$.

THEOREM A.1.8. *Let $\nu \in \mathbb{R} \setminus \Lambda_0$. Then ν is an eigenvalue of $\mathbf{A}^+$ if and only if* $\det Q^+(\nu + i0) = 0$. *This condition is equivalent to* $\det Q^+(\nu - i0) = 0$, *and*

$$\dim\ker Q^+(\nu + i0) = \dim\ker Q^+(\nu - i0) = \dim\ker(\mathbf{A}^+ - \nu),$$

where $\dim\ker(\mathbf{A}^+ - \nu)$ *is the multiplicity of ν as an eigenvalue.*

Note that the conditions $\det Q^+(\nu \pm i0) = 0$ are independent of the numbering of the functions $\xi_k^+(z)$ although the function $\det Q^+(z)$ is only determined up to a factor ± 1.

PROOF. We may assume that the functions $\xi_k^+(z)$ are numbered so that

$$\xi_k^+(\nu + i0) = \zeta_k^+(\nu) \in \mathbb{R} \quad \text{for } 1 \leqslant k \leqslant q(\nu)$$

(see the end of subsection 2).

Assume at first that ν is an eigenvalue. A corresponding eigenfunction has the form

$$u(x) = \sum_{k=1}^{m} c_k e^{ix\xi_k^+(\nu+i0)},$$

with $c_1 = c_2 = \cdots = c_{q(\nu)} = 0$ and $|c_{q+1}| + \cdots + |c_m| > 0$. Applying the boundary conditions we see that

$$\sum_{k=1}^{m} B_j(\xi_k^+(\nu + i0))c_k = 0, \quad 1 \leqslant j \leqslant m,$$

so $Q^+(\nu + i0)$ has nonempty kernel. It is obvious that linearly independent eigenfunctions correspond to linearly independent elements in the kernel, i.e.,

$$\dim\ker Q^+(\nu + i0) \geqslant \dim\ker(\mathbf{A}^+ - \nu).$$

Next assume that $0 \neq d \in \ker Q^+(\nu + i0)$. This implies that

$$\widetilde{u}(x) = \sum_{k=1}^{q(\nu)} d_k e^{ix\zeta_k^+(\nu)} + \sum_{k=q+1}^{m} d_k e^{ix\xi_k^+(\nu+i0)}$$

is either an eigenfunction or a generalized eigenfunction. If the latter is true, i.e., if $|c_1| + |c_2| + \cdots + |c_{q(\nu)}| > 0$, then an application of Corollary A.1.3 yields the contradiction

$$0 = \mathcal{A}^+(\widetilde{u}, \widetilde{u}) = \sum_{k=1}^{q(\nu)} A'(\zeta_k^+(\nu))|d_k|^2.$$

Thus $\widetilde{u}$ is an eigenfunction. Again, it is easy to see that linearly independent elements in $\ker Q^+(\nu + i0)$ correspond to linearly independent eigenfunctions, i.e., $\dim\ker Q^+(\nu + i0) \leqslant \dim\ker(\mathbf{A}^+ - \nu)$.

The same argument works for $Q^+(\nu - i0)$. □

DEFINITION A.1.9. We let Λ_1 denote the set of eigenvalues in $\mathbb{C} \setminus \Lambda_0$. We set $\Lambda = \Lambda_0 \cup \Lambda_1$.

In the case $m = 1$ it is well known that there are no eigenvalues in the continuous spectrum, i.e., $\sigma_p(\mathbf{A}) \cap \sigma_c(\mathbf{A}) = \varnothing$. This is no longer true when $m > 1$ as the following proposition shows.

PROPOSITION A.1.10. *The multiplicity of an eigenvalue $\nu \notin \Lambda_0$ of the operator $\mathbf{A}^+$ does not exceed $m - q(\nu)$. Moreover, given a differential expression $A(D)$ of the type described in subsection 1 and a $\nu \in \mathbb{R} \setminus \Lambda_0$ such that $q(\nu) < m$, there is a choice of boundary operators $B_j(D)$ such that $\mathbf{A}^+$ is self-adjoint and ν is an eigenvalue of multiplicity $m - q(\nu)$ for $\mathbf{A}^+$. (Here $q(\nu) = 0$ when $\nu < 0$.)*

The first statement follows from the fact that there are at most $m - q(\nu)$ linearly independent exponentially decaying solutions of $(A(D) - \nu)u(x) = 0$, $x > 0$. The converse statement is also basically a question of counting dimensions; we omit the proof for the sake of brevity.

We shall need the following rough result for estimating later the number of eigenvalues.

LEMMA A.1.11. *Let $\nu_0 \in \mathbb{R}$ be an eigenvalue of $\mathbf{A}^+$ of multiplicity l. Then the function* $\det Q^+(\nu + i0)$ *has a zero of order at least l at ν_0, that is, $(\nu - \nu_0)^{-l} \times$* $\det Q^+(\nu + i0)$ *is bounded in some neighbourhood of ν_0.*

PROOF. Let N be a small complex neighbourhood of ν_0, and let $M(z)$ be a bounded square matrix-valued function in N for which there are l linearly independent vectors $c^{(j)}$, $1 \leqslant j \leqslant l$, such that $M(z)c^{(j)}$ are analytic and vanish at $z = \nu_0$. It is easy to see that in this case $(z - \nu_0)^{-l} \det M(z)$ is bounded in N.

If $A(\xi) = \nu_0$ has no multiple nonreal ξ-roots we obtain the required result by applying the above argument directly to our matrix $Q^+(z)$, since each eigenfunction $\sum c_k e^{ix\xi_k^+(\nu_0+i0)}$ gives an element in the kernel of $Q^+(\nu_0 + i0)$.

Suppose now that $A(\xi) = \nu_0$ has multiple nonreal ξ-roots, and let $m - q$ be the total number of ξ-roots with positive imaginary part. As usual let us enumerate the roots $\xi_k^+(z)$ of the equation $A(\xi) = z$, $\operatorname{Im} z > 0$, in such a way that $\operatorname{Im} \xi_k^+(\nu_0+i0) = 0$ for $1 \leqslant k \leqslant q$. Then not all of the functions $g_k(z, x) = e^{ix\xi_k^+(z)}$, $m - q + 1 \leqslant k \leqslant m$, are analytic at $z = \nu_0$. However, if C is a simple closed curve in $\{\operatorname{Im} z > 0\}$ enclosing all the $m - q$ roots $\xi_k^+(\nu_0 + i0)$, $m - q + 1 \leqslant k \leqslant m$, then the functions

$$h_s(z, x) = \frac{1}{2\pi i} \oint_C \frac{\xi^{s-q-1} e^{ix\xi}}{A(\xi) - z}\, d\xi, \quad s = q + 1, \dots, m,$$

are analytic functions of z in a neighbourhood of ν_0. It is also easy to see that they are linearly independent as functions of x. We can now apply the argument of the beginning of the proof to the matrix

$$M_{js}(z) = (B_j(D_x)h_s)(z, +0)$$

with $h_s(z, x) = e^{ix\xi_s^+(z)}$ for $1 \leqslant s \leqslant q$. It is easy to see that each eigenfunction $\sum_{s=q+1}^{m} c_s h_s(\nu_0, x)$ gives an element in the kernel of $M(\nu_0 + i0)$, and consequently $\det M(z)$ vanishes to order at least l at $z = \nu_0$.

On the other hand, for $z \in N \cap \{\operatorname{Im} z > 0\}$ we have

$$M(z) = Q^+(z) \begin{pmatrix} I_q & 0 \\ 0 & V_{m-q}(z) \end{pmatrix},$$

where

$$\{V_{m-q}(z)\}_{k's'} = \frac{(\xi_{k'+q}^+(z))^{s'-1}}{A'(\xi_{k'+q}^+(z))}, \quad 1 \leqslant k', s' \leqslant m - q.$$

Thus

$$\det Q^+(z) = \det M(z) / \det V_{m-q}(z) = \det M(z) \cdot \frac{\prod_{k=m-q+1}^{m} A'(\xi_k^+(z))}{\prod_{\substack{k,r \geqslant m-q+1 \\ k \neq r}} (\xi_k^+(z) - \xi_r^+(z))}.$$

A direct calculation shows that the last factor is bounded in a neighbourhood of ν_0; in fact it is of the order of $(z - \nu_0)^\sigma$ where σ is a sum of positive contributions from the nonreal multiple roots, each root of multiplicity p contributing $(p - 1)/2$. □

Let us fix a point $\nu \in \mathbb{R} \setminus \Lambda_0$. Obviously, there exists a (small) neighbourhood $N_\nu \subset \mathbb{C}$ of ν and functions $Q_{\nu+i0}^+(z)$, $Q_{\nu-i0}^+(z)$ analytic in N_ν which coincide with

$Q^+(z)$ in $N_\nu \cap \{\operatorname{Im} z > 0\}$, $N_\nu \cap \{\operatorname{Im} z < 0\}$, respectively. The following proposition is a refinement of Lemma A.1.11 in the case when we are outside Λ_0.

PROPOSITION A.1.12. *Let $\nu \in \mathbb{R} \setminus \Lambda_0$ be an eigenvalue of $\mathbf{A}^+$ of multiplicity l. Then at the point $z = \nu$ the functions* $\det Q^+_{\nu \pm i0}(z)$ *have zeros of order l (precisely). Moreover, the matrix functions $(Q^+_{\nu\pm i0}(z))^{-1}$ have simple poles at this point, and the residues are matrices of rank l.*

PROOF. It follows from Theorem A.1.8 that $\{Q^+_{\nu+i0}(z)\}^{-1}$ has a pole at $z = \nu$. We shall prove that this pole is simple. This will imply that $\det Q^+_{\nu+i0}(z)$ has a zero of order at most $\dim\ker Q^+_{\nu+i0}(\nu) = l$ at ν, and the opposite inequality will follow from Lemma A.1.11.

Suppose that $Q^{-1}_{\nu+i0}$ has a pole of order $p > 1$. We can then find a $v \in \mathbb{C}^m$ such that

$$\{Q^+_{\nu+i0}(\nu + i\varepsilon)\}^{-1} v = \frac{a}{\varepsilon^p} + O(\varepsilon^{1-p}), \quad \mathbb{C}^m \ni a \neq 0,$$

as $\varepsilon \to +0$. Consider now the family of solutions $u_\varepsilon \in L_2(\mathbb{R}_+)$ of the boundary value problems

$$\begin{aligned} (A(D) - \nu - i\varepsilon)u_\varepsilon &= 0 \quad \text{on} \quad \mathbb{R}_+, \\ (B_j(D)u_\varepsilon)(+0) &= v_j, \quad j = 1, \dots, m. \end{aligned}$$

Naturally, $u_\varepsilon(x) = \sum_{j,k=1}^m \{Q^+_{\nu+i0}(\nu + i\varepsilon)\}^{-1}_{kj} v_j e^{ix\xi_k^+(\nu+i\varepsilon)}$.

The self-adjointness of $\mathbf{A}^+$ implies that there are polynomials $C_j(\xi), D_j(\xi)$, $j = 1, \dots, m$, such that the boundary form may be written as

$$i\mathcal{A}^+(u, v) = \sum_{j=1}^m (B_j(D)u)(+0)\,\overline{(C_j(D)v)(+0)} + \sum_{j=1}^m (D_j(D)u)(+0)\,\overline{(B_j(D)v)(+0)}.$$

This implies that

$$\begin{aligned} 0 &= ((A(D) - \nu - i\varepsilon)u_\varepsilon, u_\varepsilon)_+ - (u_\varepsilon, (A(D) - \nu - i\varepsilon)u_\varepsilon)_+ \\ &= -2i\varepsilon \| u_\varepsilon \|_+^2 + i\mathcal{A}(u_\varepsilon, u_\varepsilon) = -2i\varepsilon \| u_\varepsilon \|_+^2 + O(\varepsilon^{-p}). \end{aligned} \tag{A.1.11}$$

We have, however,

$$\sup_{0 \leqslant x \leqslant 1} \left| u_\varepsilon(x) - \frac{1}{\varepsilon^p} \sum_{k=1}^m a_k e^{ix\xi_k^+(\nu+i0)} \right| = O(\varepsilon^{1-p}),$$

and consequently

$$\varepsilon \| u_\varepsilon \|_+^2 \geqslant \varepsilon^{1-2p} \int_0^1 \left| \sum_{k=1}^m a_k e^{ix\xi_k^+(\nu+i0)} \right|^2 dx + O(\varepsilon^{2-2p}).$$

Since $1 - 2p < -p$, this contradicts (A.1.11). □

PROPOSITION A.1.13. *Let N denote the total number of eigenvalues of $\mathbf{A}^+$, counting multiplicities. Then*

$$N \leqslant (m_1 + m_2 + \cdots + m_m) \frac{(2m-1)!}{(m!)^2}, \tag{A.1.12}$$

where the m_j are the degrees of the polynomials $B_j(\xi) = \sum_{s=0}^{m_j} b_{js}\xi^s$ appearing in the boundary conditions.

For $m = 1$ the estimate (A.1.12) is sharp. For $m \geqslant 2$ the estimate (A.1.12) can, in principle, be improved. However, for our purposes it is sufficient to establish that there is an upper bound for the number of eigenvalues which depends only on m. The latter follows from (A.1.12) since $m_j \leqslant m - 1 + j$.

The estimate (A.1.12) improves by a factor of 2 the estimate from [**LtinSu**] and [**Va7**].

PROOF. For $\varepsilon > 0$ sufficiently small, the strip $D = \{0 < \operatorname{Im} z < \varepsilon\} \subset \mathbb{C}$ is a simply connected domain which does not contain points from Λ_0. Consider the function $\det^2 Q^+(z)$ on D (see subsection 5). Here we prefer to deal with $\det^2 Q^+(z)$ instead of $\det Q^+(z)$ since the former is independent of the enumeration of the $\xi_k^+(z)$. The function $\det^2 Q^+(z)$ is a (single-valued) analytic function on D, and it has a continuous boundary value $\det^2 Q^+(\nu + i0)$, $\nu \in \mathbb{R}$. It follows from Lemma A.1.11 that the number of eigenvalues, N, is less than or equal to half the total multiplicity of the zeros of $\det^2 Q^+(\nu + i0)$ on $\mathbb{R}$.

When we continue $\det^2 Q^+(z)$ analytically from D along various curves in $\mathbb{C}\setminus\Lambda_0$ we get various branches of the form

$$f_\sigma(z) = \left(\det\left(\{B_j(\xi_{\sigma(k)}(z))\}_{j,k=1}^m\right)\right)^2, \quad z \in \mathbb{C}\setminus\Lambda_0,$$

where σ is an element in the set S_m^{2m} of all $\binom{2m}{m}$ choices of m elements out of $2m$ without respect to their order, and $\xi_1(z), \dots, \xi_{2m}(z)$ are the solutions of $A(\xi) = z$. Let S denote the subset of S_m^{2m} consisting of the σ which can be actually realized in our analytic continuation process. Consider the function

$$\mathbb{C}\setminus\Lambda_0 \ni z \mapsto f(z) = \prod_{\sigma\in S} f_\sigma(z).$$

By construction $f(z)$ is single-valued and analytic in $\mathbb{C}\setminus\Lambda_0$. Obviously, the function which we are interested in, $\det^2 Q^+(\nu+i0)$, is a factor in $f(z)$, so the number of zeros of $f(z)$ gives an upper bound for $2N$. Furthermore, f does not vanish identically since it is a product of analytic continuations of $\det^2 Q^+(\nu + i0)$. The fact that the latter is not identically zero follows from the invertibility of the matrix $Q^+(z)$ in the original strip D; see subsection 5.

The function $f(z)$ is bounded as $z \to z_0 \in \Lambda_0$, so it is analytic in the whole complex plane. Moreover, since $f(z)$ is polynomially bounded as $|z| \to \infty$, it is a polynomial. It remains to count the degree (number of zeros) of f. The degree of f with respect to the ξ_k is at most $2\cdot\binom{2m}{m}\cdot(m_1 + \cdots + m_m)$, and $|\xi_k(z)| \sim |z|^{1/(2m)}$ as $|z| \to \infty$. Thus, $f(z)$ is a polynomial of degree $\leqslant \frac{1}{m}\binom{2m}{m}(m_1 + \cdots + m_m)$. □

REMARK A.1.14. The set S may be a strict subset of S_m^{2m}. This can be seen from the example $A(\xi) = \xi^{2m}$. In this case we can choose $D = \mathbb{C}\setminus\mathbb{R}_+$ as our original simply connected domain, and setting $z = w^{2m}$ we get

$$\xi_k(w^{2m}) = e^{\frac{i\pi(k-1)}{m}}w, \quad 0 \leqslant \arg w < \frac{\pi}{m}, \quad 1 \leqslant k \leqslant 2m,$$

(the first m of these are the ξ_k^+). The set S in this case consists of $2m$ cyclic permutations; hence the polynomial $f(z)$ has degree $2\cdot 2m\cdot(m_1 + \cdots + m_m)$. We conclude that any corresponding $\mathbf{A}^+$ has at most $m_1 + \cdots + m_m$ eigenvalues. This is also obvious from the fact that the above parametrization turns the equation $\det Q^+(w^{2m}) = 0$ into an algebraic equation of degree $m_1 + \cdots + m_m$ for w.

A.2. The reflection matrix

1. Spectral representation for A. According to the spectral theorem we have, for $g \in C_0(\mathbb{R})$ and $u, v \in C_0^\infty(\mathbb{R})$,

$$(g(\mathbf{A})u, v) = (2\pi i)^{-1} \lim_{\varepsilon \to +0} \int g(\nu)((\mathbf{R}_{\nu+i\varepsilon} - \mathbf{R}_{\nu-i\varepsilon})u, v)\, d\nu\,.$$

Suppose now that we have the additional condition $\operatorname{supp} g \cap \Lambda_{00} = \varnothing$, where Λ_{00} denotes the set of thresholds, i.e. $\nu \in \Lambda_0$ such that $A(\xi) = \nu$ has multiple real ξ-roots. Then, in view of the uniform convergence of the integral kernels of the resolvents as $\varepsilon \to +0$ (see A.1.4), we can rewrite the above formula as

$$(g(\mathbf{A})u, v) = (2\pi i)^{-1} \int g(\nu)((\mathbf{R}_{\nu+i0} - \mathbf{R}_{\nu-i0})u, v)\, d\nu\,.$$

From formula (A.1.7) we see that

$$(2\pi i)^{-1}((\mathbf{R}_{\nu+i0} - \mathbf{R}_{\nu-i0})u, u) = (\Gamma(\nu)u, \Gamma(\nu)u)_q,$$

where $(\cdot,\cdot)_q$ denotes the scalar product in $\mathbb{C}^{2q(\nu)}$ and the vector $\Gamma(\nu)u \in \mathbb{C}^{2q(\nu)}$ is given by

$$(\Gamma(\nu)u)_j = \frac{\hat{u}(\zeta_j^+(\nu))}{(2\pi A'(\zeta_j^+(\nu)))^{1/2}}, \quad 1 \leqslant j \leqslant q(\nu)\,,$$

$$(\Gamma(\nu)u)_j = \frac{\hat{u}(\zeta_{j-q}^-(\nu))}{(-2\pi A'(\zeta_{j-q}^-(\nu)))^{1/2}}, \quad q(\nu) + 1 \leqslant j \leqslant 2q(\nu)\,.$$

Consequently

$$(g(\mathbf{A})u, v) = \int_0^\infty g(\nu)(\Gamma(\nu)u, \Gamma(\nu)v)_q\, d\nu\,. \tag{A.2.1}$$

Formula (A.2.1) also follows easily from Parseval's identity.

For fixed $\nu \in \mathbb{R} \setminus \Lambda_{00}$ we may consider $\Gamma(\nu)$ as a linear map from $C_0^\infty(\mathbb{R})$ to $\mathbb{C}^{2q(\nu)}$. It has the adjoint

$$\Gamma(\nu)^* : \mathbb{C}^{2q} \to \mathcal{D}'(\mathbb{R}),$$

given by

$$\Gamma(\nu)^* a = \sum_{j=1}^{q} \left(a_l^+ \frac{e^{ix\zeta_l^+(\nu)}}{(2\pi A'(\zeta_l^+(\nu)))^{1/2}} + a_l^- \frac{e^{ix\zeta_l^-(\nu)}}{(-2\pi A'(\zeta_l^-(\nu)))^{1/2}} \right),$$

where $a = (a_1^+, \dots, a_q^+, a_1^-, \dots, a_q^-)$. We see that actually the image of $\Gamma(\nu)^*$ is contained in C_B^∞, the set of functions that are bounded together with all their derivatives.

For any $a \in \mathbb{C}^{2q}$ we have

$$(A(D) - \nu)\Gamma(\nu)^* a = 0,$$

so $\Gamma(\nu)^* a$ is a generalized eigenfunction for $\mathbf{A}$, and formula (A.2.1) may be seen as a generalized eigenfunction expansion :

$$(g(\mathbf{A})u, v) = \int_0^\infty g(\nu) \sum_{j=1}^{2q(\nu)} (u, \Gamma(\nu)^* e^{(j)}) \, (\Gamma(\nu)^* e^{(j)}, v) \, d\nu \, .$$

Here $e^{(j)} \in \mathbb{C}^{2q(\nu)}$ is a vector with entry 1 in the jth position, and zeros everywhere else.

Note that $\Gamma(\nu)$ has a natural extension to $\mathcal{E}'$, the space of compactly supported distributions, and that by definition

$$\text{(A.2.2)} \qquad \frac{1}{2\pi i}((\mathbf{R}_{\nu+i0} - \mathbf{R}_{\nu-i0})u, v) = (\Gamma(\nu)^*\Gamma(\nu)u, v) = (\Gamma(\nu)u, \Gamma(\nu)v)_q,$$

for any $u, v \in \mathcal{E}'$. Here we use the fact that $\mathbf{R}_{\nu+i0} - \mathbf{R}_{\nu-i0}$ is an operator of convolution with a C^∞-function.

Let us introduce the space $C_0(\mathbb{R}_+ \setminus \Lambda_{00}, \mathbb{C}^{2q(\cdot)})$ of continuous functions, ϕ, with compact support in $\mathbb{R}_+ \setminus \Lambda_{00}$, such that $\phi(\nu) \in \mathbb{C}^{2q(\nu)}$. We let $L^2(\mathbb{R}_+, \mathbb{C}^{2q(\cdot)})$ denote the closure of $C_0(\mathbb{R}_+ \setminus \Lambda_{00}, \mathbb{C}^{2q(\cdot)})$ with respect to the norm

$$|| \, \phi \, || = \left(\int_0^\infty || \, \phi(\nu) \, ||^2_{\mathbb{C}^{2q(\nu)}} \, d\nu \right)^{1/2} .$$

This is a Hilbert space.

Let $\Omega \subset \mathbb{R}$ be measurable and denote its characteristic function by χ_Ω. Then

$$(\chi_\Omega(\mathbf{A})u, u) = \frac{1}{2\pi} \int_{A(\xi)\in\Omega} |\hat{u}(\xi)|^2 \, d\xi, \quad u \in L_2(\mathbb{R}) \, ,$$

which implies that $\mathbf{A}$ has purely absolutely continuous spectrum. Consequently, if $\{g_j\}$ is an increasing sequence of nonnegative functions in $C_0(\mathbb{R} \setminus \Lambda_{00})$ which converge pointwise to $1 - \chi_{\Lambda_{00}}$, then, for $u \in C_0^\infty$,

$$\begin{aligned}(u, u) &= \lim_{j\to\infty} (g_j(\mathbf{A})u, u) = \lim_{j\to\infty} \int g_j(\nu)(\Gamma(\nu)u, \Gamma(\nu)u)_q \, d\nu \\ &= \int_0^\infty (\Gamma(\nu)u, \Gamma(\nu)u)_q \, d\nu.\end{aligned}$$

From this it is easy to see that the mapping

$$\Gamma : C_0^\infty \to L^2(\mathbb{R}_+, \mathbb{C}^{2q(\cdot)})$$

given by $(\Gamma u)_j(\nu) = (\Gamma(\nu)u)_j$ may be extended to a unitary mapping $L_2(\mathbb{R}) \to L^2(\mathbb{R}_+, \mathbb{C}^{2q(\cdot)})$. We note that if g is, for instance, continuous, then

$$(\Gamma g(\mathbf{A})u)(\nu) = g(\nu)(\Gamma u)(\nu), \quad \nu \in \mathbb{R} \setminus \Lambda_{00}.$$

2. The reflection matrix. As usual, when dealing with a point $\nu \in \mathbb{R} \setminus \Lambda_0$, we shall assume that the functions $\xi_k^\pm(z)$ are numbered so that

$$\begin{aligned}\xi_k^\pm(\nu + i0) &= \zeta_k^\pm(\nu), \quad 1 \leqslant k \leqslant q(\nu), \\ \xi_k^+(\nu + i0) &= \overline{\xi_k^-(\nu + i0)}, \quad q(\nu) < k \leqslant m.\end{aligned}$$

Recall also that by Λ we denote the union of Λ_0 with the set of eigenvalues of $\mathbf{A}^+$.

PROPOSITION A.2.1. *Let $\nu \in \mathbb{R}_+ \setminus \Lambda$. Then for each $a^- \in \mathbb{C}^{q(\nu)}$ there is a unique generalized eigenfunction $\Gamma^+_{\text{in}}(\nu)^* a^-$ of $\mathbf{A}^+$ of the form*

$$\Gamma^+_{\text{in}}(\nu)^* a^- = \sum_{k=1}^{q(\nu)} a_k^- \frac{e^{ix\zeta_k^-(\nu)}}{\sqrt{-2\pi A'(\zeta_k^-(\nu))}} + \sum_{k=1}^{q(\nu)} a_k^+ \frac{e^{ix\zeta_k^+(\nu)}}{\sqrt{2\pi A'(\zeta_k^+(\nu))}} + \sum_{k=q(\nu)+1}^{m} d_k^+ e^{ix\xi_k^+(\nu+i0)}.$$

Moreover, there is a unitary matrix $R(\nu)$ (the reflection matrix) such that

$$a_j^+ = \sum_{k=1}^{q(\nu)} R_{jk}(\nu) a_k^-.$$

It is given by the formula

$$R_{jk}(\nu) = G_{jk}(\nu + i0)\sqrt{-A'(\zeta_k^-(\nu))A'(\zeta_j^+(\nu))}, \quad 1 \leqslant j, k \leqslant q(\nu), \tag{A.2.3}$$

with $G_{jk}(z)$ as in (A.1.10). Finally, the function $\nu \mapsto R(\nu)$ can be extended to $\mathbb{R}_+ \setminus \Lambda_{00}$ and the extension is real-analytic on each connected component of this set.

Thus, we have a linear mapping $\Gamma^+_{\text{in}}(\nu)^* : \mathbb{C}^q \to C_B^\infty(\mathbb{R}_+)$. Since $R(\nu)$ is unitary, we might as well specify $a^+ = R(\nu)a^-$. In this way we obtain another map

$$\Gamma^+_{\text{out}}(\nu)^* : \mathbb{C}^q \ni a^+ \mapsto \Gamma^+_{\text{out}}(\nu)^* a^+ = \Gamma^+_{\text{in}}(\nu)^* R^*(\nu) a^+ \in C_B^\infty(\mathbb{R}_+).$$

The reason for the subscripts in/out is that $\Gamma^+_{\text{in}}(\nu)^* a^-$ and $\Gamma^+_{\text{out}}(\nu)^* a^+$ are generalized eigenfunctions with specified incoming and outgoing parts, respectively.

PROOF OF PROPOSITION A.2.1. Set $d_k^+ = \frac{a_k^+}{\sqrt{2\pi A'(\zeta_k^+)}}$ for $k = 1, \ldots, q(\nu)$. An application of the boundary conditions to the function $\Gamma^+_{\text{in}}(\nu)^* a^-$ gives

$$(Q^+(\nu + i0)d^+)_j = -\sum_{r=1}^{q(\nu)} \frac{B_j(\zeta_r^-(\nu))}{\sqrt{-2\pi A'(\zeta_r^-(\nu))}} a_r^-. \tag{A.2.4}$$

Since $Q^+(\nu + i0)$ is invertible when $\nu \notin \Lambda$, this determines d^+, and in particular the a_k^+, uniquely. This proves the existence of the matrix $R(\nu)$ when $\nu \notin \Lambda$, and it follows from Corollary A.1.4 that $R(\nu)$ is unitary.

If $\nu_0 \notin \Lambda_0$ is an eigenvalue, then the right-hand side in (A.2.3) has a meromorphic extension to a neighbourhood of ν_0. The unitarity of $R(\nu)$ for real $\nu \neq \nu_0$ implies that $\lim_{\nu \to \nu_0} R(\nu)$ is bounded. Thus $\nu = \nu_0$ is a removable singularity.

Finally, let $\nu_0 \in (\Lambda_0 \setminus \Lambda_{00}) \cap \mathbb{R}$. For real ν close to ν_0, $\nu \neq \nu_0$, we can write

$$\Gamma^+_{\text{in}}(\nu)^* a^- = \sum_{k=1}^{q(\nu)} a_k^- \frac{e^{ix\zeta_k^-(\nu)}}{\sqrt{-2\pi A'(\zeta_k^-(\nu))}} + \sum_{k=1}^{q(\nu)} a_k^+ \frac{e^{ix\zeta_k^+(\nu)}}{\sqrt{2\pi A'(\zeta_k^+(\nu))}} + \sum_{s=q(\nu)+1}^{m} \tilde{d}_s^+(\nu) h_s(\nu, x),$$

where $h_s(z, x)$ are the functions introduced in the proof of Lemma A.1.11. Applying the boundary conditions, we conclude that

$$R_{jk}(\nu) = -\{M(\nu)\}_{jk}^{-1} B_k(\zeta_k^-(\nu)) \sqrt{-\frac{A'(\zeta_j^+(\nu))}{A'(\zeta_k^-(\nu))}}, \quad 1 \leqslant j, k \leqslant q(\nu),$$

where $\{M(z)\}_{jk} = (B_j(D_x)h_k)(z, +0)$. Here the right-hand side is meromorphic in a neighbourhood of ν_0, and we can proceed as in the previous case. $\square$

It is natural to regard $\Gamma^+_{\rm in}(\nu)^*$ and $\Gamma^+_{\rm out}(\nu)^*$ as linear maps $\mathbb{C}^{q(\nu)} \to C^\infty_B(\mathbb{R}_+)$, and to introduce their adjoints $\Gamma^+_{\rm in}(\nu)$, $\Gamma^+_{\rm out}(\nu)$ as maps $C_0^\infty(0,+\infty) \to \mathbb{C}^{q(\nu)}$. We shall see that they play the same rôle for the absolutely continuous part of $\mathbf{A}^+$ as $\Gamma(\nu)^*$ and $\Gamma(\nu)$ do for $\mathbf{A}$. For brevity we shall essentially only discuss $\Gamma^+_{\rm out}(\nu)$ since the arguments for $\Gamma^+_{\rm in}(\nu)$ are parallel.

By $\{\nu_j\}_{j=1}^N$ and $\{u_j\}_{j=1}^N$ we shall denote the eigenvalues and eigenfunctions of $\mathbf{A}^+$.

THEOREM A.2.2. *Let $u, v \in C_0^\infty(0,+\infty)$ and $g \in C_0(\mathbb{R})$. Then*

$$(g(\mathbf{A}^+)u, v)_+ = \int_{\mathbb{R}_+\setminus\Lambda} g(\nu)\,(\Gamma^+_{\rm out}(\nu)u, \Gamma^+_{\rm out}(\nu)v)_q\, d\nu + \sum_{j=1}^N g(\nu_j)\,(u,u_j)_+(u_j,v)_+\,.$$

REMARK A.2.3. Here as in the rest of this subsection we adopt the following convention which simplifies formulae involving both $\Gamma(\nu)$ and $\Gamma^+_{\rm out}(\nu)$: we identify $\Gamma^+_{\rm out}(\nu)u$ and $\Gamma^+_{\rm in}(\nu)u$ with the vectors $(\Gamma^+_{\rm out}(\nu)u, 0)$ and $(0, \Gamma^+_{\rm in}(\nu)u)$ in $\mathbb{C}^{2q(\nu)}$. Similarly, for $a = (a^+, a^-) \in \mathbb{C}^{2q(\nu)}$ we agree that

$$\Gamma^+_{\rm out}(\nu)^*a = \Gamma^+_{\rm out}(\nu)^*a^+$$
$$\Gamma^+_{\rm in}(\nu)^*a = \Gamma^+_{\rm in}(\nu)^*a^-.$$

The proof of Theorem A.2.2 is preceded by several auxiliary results.

In analogy with the previous subsection we start with the formula

$$(g(\mathbf{A}^+)u, u)_+ = \lim_{\varepsilon\to+0} \frac{1}{2\pi i}\int g(\nu)((\mathbf{R}^+_{\nu+i\varepsilon} - \mathbf{R}^+_{\nu-i\varepsilon})u, u)_+\, d\nu.$$

When $\nu \in \mathbb{R}\setminus\Lambda$ it follows easily from A.1.4 and Proposition A.1.6 that

$$\mathbf{r}^+_{\nu\pm i\varepsilon}(x,y) \to \mathbf{r}^+_{\nu\pm i0}(x,y) \equiv \mathbf{r}_{\nu\pm i0}(x,y) + i\sum_{k,l=1}^m G_{kl}(\nu\pm i0)e^{ix\xi_k^+(\nu\pm i0)}e^{-iy\xi_k^-(\nu\pm i0)},$$

as $\varepsilon \to +0$, and it is easy to see that the convergence is uniform over (ν, x, y) belonging to a fixed compact set in $(\mathbb{R}\setminus\Lambda)\times\mathbb{R}^2_+$; see formula (A.1.10) for the definition of the G_{kl}. It follows that for $g \in C_0(\mathbb{R}\setminus\Lambda)$ and $u \in C_0^\infty(0,+\infty)$ we have

$$(g(\mathbf{A}^+)u, u)_+ = \frac{1}{2\pi i}\int g(\nu)((\mathbf{R}^+_{\nu+i0} - \mathbf{R}^+_{\nu-i0})u, u)_+\, d\nu, \tag{A.2.5}$$

where $\mathbf{R}^+_{\nu\pm i0}$ are the operators with kernels $\mathbf{r}^+_{\nu\pm i0}(x,y)$. Repeating the arguments from the proof of Lemma A.1.7 we conclude that, for $u, v \in C_0^\infty(0,+\infty)$,

$$\begin{aligned}(\mathbf{R}^+_{\nu+i0}u, v)_+ &- (\mathbf{R}_{\nu+i0}u, v)_+ \\ &= -\,i\mathcal{A}^+(\mathbf{R}^+_{\nu+i0}u, \mathbf{R}_{\nu-i0}v) = ((\mathfrak{A}\mathbf{R}^+_{\nu+i0}u), \mathbf{R}_{\nu-i0}v)\,,\end{aligned} \tag{A.2.6}$$

where $(\mathfrak{A}\mathbf{R}^+_{\nu+i0}u)$ is the distribution

$$-i\sum_{r,s=0}^{2m-1} a_{r+s+1}(D^r\mathbf{R}^+_{\nu+i0}u)(+0)D^s\delta_0.$$

LEMMA A.2.4. *Let $g \in L^1(\mathbb{R})$, and let $\chi_\delta(x) = \chi(\delta x)$, where $\chi \in C_0^\infty(\mathbb{R})$ is a function which is identically one in a neighbourhood of the origin. Then*

$$\lim_{\delta \to +0} (g, \chi'_\delta)_\pm = 0.$$

PROOF OF LEMMA A.2.4. The result follows immediately from the inequality

$$|(g, \chi'_\delta)_\pm| \leqslant \delta \| g \|_1 \| \chi' \|_\infty.$$

□

LEMMA A.2.5. *Let $a, b \in \{\operatorname{Im} z \geqslant 0\}$ and let $\chi_\delta(x) = \chi(\delta x)$ be as in Lemma A.2.4. Then*

$$\lim_{\delta \to +0} (e^{\pm iax}, e^{\pm ibx} \chi'_\delta)_\pm = \begin{cases} \mp 1, & a = b \in \mathbb{R}, \\ 0, & \textit{otherwise}. \end{cases}$$

PROOF OF LEMMA A.2.5. For $a = b \in \mathbb{R}$ the result is obvious, and otherwise we can integrate by parts in

$$(e^{\pm iax}, e^{\pm ibx} \chi'_\delta)_\pm = \int_{\mathbb{R}_\pm} e^{\pm i(a - \bar{b})x} \delta \chi'(\delta x) \, dx$$

to conclude that the right-hand side vanishes faster than any power of δ. □

LEMMA A.2.6. *Let $\nu \in \mathbb{R}_+ \setminus \Lambda$ and $u \in C_0^\infty(0, +\infty)$. Then*

$$\Gamma^+_{\text{out}}(\nu) u - \Gamma(\nu) u = \Gamma(\nu)(\mathfrak{A} \mathbf{R}^+_{\nu + i0} u).$$

Here we are using the convention from Remark A.2.3.

PROOF OF LEMMA A.2.6. Let $\chi_\delta(x)$ be as in Lemma A.2.4. For any $a \in \mathbb{C}^{2q}$ we have

$$\begin{aligned} (u, \Gamma^+_{\text{out}}(\nu)^* a - \Gamma(\nu)^* a)_+ &= ((A(D) - \nu)\mathbf{R}^+_{\nu+i0} u, \Gamma^+_{\text{out}}(\nu)^* a - \Gamma(\nu)^* a)_+ \\ &= \lim_{\delta \to +0} ((A(D) - \nu)\mathbf{R}^+_{\nu+i0} u, \chi_\delta(\Gamma^+_{\text{out}}(\nu)^* - \Gamma_+(\nu)^*)a)_+. \end{aligned}$$

We can apply Green's formula to the expression under the limit sign to conclude that

$$\begin{aligned} &(u, \Gamma^+_{\text{out}}(\nu)^* a - \Gamma(\nu)^* a)_+ \\ &\qquad = i\mathcal{A}^+(\mathbf{R}^+_{\nu+i0} u, (\Gamma^+_{\text{out}}(\nu)^* - \Gamma(\nu)^*)a) \\ &\qquad \quad + \lim_{\delta \to +0} (\mathbf{R}^+_{\nu+i0} u, [A(D), \chi_\delta](\Gamma^+_{\text{out}}(\nu)^* - \Gamma(\nu)^*)a)_+. \end{aligned} \tag{A.2.7}$$

Note that so far we might as well have worked with $\mathbf{R}^+_{\nu - i0}$ instead, but by using $\mathbf{R}^+_{\nu+i0}$ we make the limit on the right-hand side equal to zero. Indeed, by construction $\Gamma^+_{\text{out}}(\nu)^* a - \Gamma(\nu)^* a$ is an incoming function on $\mathbb{R}_+$, whereas $\mathbf{R}^+_{\nu+i0} u$ is outgoing. A direct calculation shows that

$$\begin{aligned} &(\mathbf{R}^+_{\nu+i0} u, [A(D), \chi_\delta](\Gamma^+_{\text{out}}(\nu)^* a - \Gamma(\nu)^* a))_+ \\ &\qquad = -i(\mathbf{R}^+_{\nu+i0} u, \chi'_\delta A'(D)(\Gamma^+_{\text{out}}(\nu)^* - \Gamma(\nu)^*)a)_+ + O(\delta). \end{aligned}$$

It now follows from Lemmas A.2.4, A.2.5 that the second term on the right-hand side of (A.2.7) vanishes. Since furthermore both $\mathbf{R}^+_{\nu+i0} u$ and $\Gamma^+_{\text{out}}(\nu)^* a$ satisfy the boundary conditions corresponding to $\mathbf{A}^+$, it follows that $\mathcal{A}^+(\mathbf{R}^+_{\nu+i0}u, \Gamma^+_{\text{out}}(\nu)^*a) = 0$, so that the right-hand side in (A.2.7) equals

$$-i\mathcal{A}^+(\mathbf{R}^+_{\nu+i0}u, \Gamma(\nu)^*a) = ((\mathfrak{A}\mathbf{R}^+_{\nu+i0}u), \Gamma(\nu)^*a).$$

□

LEMMA A.2.7. *Let* $\nu \in \mathbb{R} \setminus \Lambda$ *and* $u \in C_0^\infty(0, +\infty)$. *Then*

$$\frac{1}{2\pi i}((\mathbf{R}^+_{\nu+i0} - \mathbf{R}^+_{\nu-i0})u, u)_+ = (\Gamma^+_{\text{out}}(\nu)u, \Gamma^+_{\text{out}}(\nu)u)_q. \tag{A.2.8}$$

PROOF OF LEMMA A.2.7. Since $\mathbf{A}^+$ is self-adjoint, we have for $\operatorname{Im} z \neq 0$

$$(\mathbf{R}^+_{\bar{z}}u, u)_+ = (u, \mathbf{R}^+_z u)_+ = \overline{(\mathbf{R}^+_z u, u)_+}.$$

Hence the left-hand side of (A.2.8) may be written as $\pi^{-1}\operatorname{Im}(\mathbf{R}^+_{\nu+i0}u, u)_+$.

By the first equality of formula (A.2.6) we have

$$\begin{aligned}(\mathbf{R}^+_{\nu+i0}u, u)_+ &= (\mathbf{R}_{\nu+i0}u, u)_+ - i\mathcal{A}^+(\mathbf{R}^+_{\nu+i0}u, \mathbf{R}_{\nu-i0}u)\\ &= (\mathbf{R}_{\nu+i0}u, u)_+ + i\mathcal{A}^+(\mathbf{R}^+_{\nu+i0}u, (\mathbf{R}_{\nu+i0} - \mathbf{R}_{\nu-i0})u)\\ &\quad - i\mathcal{A}^+(\mathbf{R}^+_{\nu+i0}u, \mathbf{R}_{\nu+i0}u).\end{aligned} \tag{A.2.9}$$

We shall now rewrite the second and third terms on the right.

By the definition of $(\mathfrak{A}\mathbf{R}^+_{\nu+i0}u)$ the second term may be rewritten as

$$-((\mathfrak{A}\mathbf{R}^+_{\nu+i0}u), (\mathbf{R}_{\nu+i0} - \mathbf{R}_{\nu-i0})u),$$

and in view of (A.2.2) and Lemma A.2.6 we have

$$\begin{aligned}-((\mathfrak{A}\mathbf{R}^+_{\nu+i0}u), (\mathbf{R}_{\nu+i0} - \mathbf{R}_{\nu-i0})u) &= 2\pi i\,(\Gamma(\nu)(\mathfrak{A}\mathbf{R}^+_{\nu+i0}u), \Gamma(\nu)u)_q\\ &= 2\pi i\,(\Gamma^+_{\text{out}}(\nu)u - \Gamma(\nu)u,\, \Gamma(\nu)u)_q\,.\end{aligned}$$

For the third term on the right of (A.2.9) we have

$$-i\mathcal{A}^+(\mathbf{R}^+_{\nu+i0}u,\, \mathbf{R}_{\nu+i0}) = i\mathcal{A}^+(\mathbf{R}^+_{\nu+i0}u,\, (\mathbf{R}^+_{\nu+i0} - \mathbf{R}_{\nu+i0})u)\,, \tag{A.2.10}$$

since $\mathbf{R}^+_{\nu+i0}u$ satisfies the boundary conditions corresponding to $\mathbf{A}^+$. In view of formula (A.2.6) the right-hand side of (A.2.10) equals

$$\begin{aligned}i\mathcal{A}^+(\mathbf{R}^+_{\nu+i0}u,\, \mathbf{R}_{\nu+i0}(\mathfrak{A}\mathbf{R}^+_{\nu+i0}u)) &= -\,((\mathfrak{A}\mathbf{R}^+_{\nu+i0}u),\, \mathbf{R}_{\nu+i0}(\mathfrak{A}\mathbf{R}^+_{\nu+i0}u))\\ &= -\,(\mathbf{R}_{\nu-i0}(\mathfrak{A}\mathbf{R}^+_{\nu+i0}u),\, (\mathfrak{A}\mathbf{R}^+_{\nu+i0}u))\,.\end{aligned}$$

We have shown that

$$\begin{aligned}(\mathbf{R}^+_{\nu+i0}u, u)_+ = (\mathbf{R}_{\nu+i0}u, u) &+ 2\pi i\,(\Gamma^+_{\text{out}}(\nu)u - \Gamma(\nu)u, \Gamma(\nu)u)_q\\ &- (\mathbf{R}_{\nu-i0}(\mathfrak{A}\mathbf{R}^+_{\nu+i0}u),\, (\mathfrak{A}\mathbf{R}^+_{\nu+i0}u))\,.\end{aligned}$$

The required result now follows if we take the imaginary part and use the fact that

$$\begin{aligned}&\frac{1}{2\pi i}((\mathbf{R}_{\nu+i0} - \mathbf{R}_{\nu-i0})(\mathfrak{A}\mathbf{R}^+_{\nu+i0}u), (\mathfrak{A}\mathbf{R}^+_{\nu+i0}u))\\ &\qquad = ||\,\Gamma(\nu)\mathfrak{A}\mathbf{R}^+_{\nu+i0}u\,||_q^2 = ||\,\Gamma^+_{\text{out}}(\nu)u - \Gamma(\nu)u\,||_q^2.\end{aligned}$$

□

We are now prepared to proceed to the

PROOF OF THEOREM A.2.2. Without loss of generality let us assume that $g \geqslant 0$ and that $u = v$. Combining Lemma A.2.7 with formula (A.2.5) we see that the theorem has been proved for all $g \in C_0(\mathbb{R} \setminus \Lambda)$.

Now let $g \in C_0(\mathbb{R})$. Take a monotone sequence $g_j \in C_0(\mathbb{R}\setminus\Lambda)$ such that $g_j \nearrow g$ pointwise on $\mathbb{R} \setminus \Lambda$. Then

$$\begin{aligned}(g_j(\mathbf{A}^+)u, u)_+ \\ = \int_0^\infty g_j(\nu)(\Gamma^+_{\text{out}}(\nu)u, \Gamma^+_{\text{out}}(\nu)u)_q \, d\nu \nearrow \int_{\mathbb{R}_+\setminus\Lambda} g(\nu)(\Gamma^+_{\text{out}}(\nu)u, \Gamma^+_{\text{out}}(\nu)u)_q \, d\nu \,,\end{aligned}$$

and

$$(g_j(\mathbf{A}^+)u, u)_+ \nearrow (g(\mathbf{A}^+)u, u)_+ - \sum_{\nu\in\Lambda} g(\nu)((\mathbf{E}^+_{\nu+0}u, u)_+ - (\mathbf{E}^+_\nu u, u)_+)$$

as $j \to \infty$; here $\mathbf{E}^+_\nu$ is the spectral family for $\mathbf{A}^+$. It remains to note that $\mathbf{E}^+_{\nu+0} - \mathbf{E}^+_\nu$ is the projection on the eigenspace corresponding to the eigenvalue ν (in particular, it is zero if ν is not an eigenvalue). □

REMARK A.2.8. If g is compactly supported in $\mathbb{R} \setminus \Lambda$, we have good control over the function $g(\nu)((\mathbf{R}_{\nu+i0} - \mathbf{R}_{\nu-i0})u, u)$. Using this and formula (A.2.5) it is easy to see that $\mathbf{A}^+$ is absolutely continuous on $\mathbb{R} \setminus \Lambda$. Since the complement of this set consists of a finite number of points, there can be no singular continuous spectrum, and we have

$$(P_{\text{ac}}(\mathbf{A}^+)u, u) = \int_{\mathbb{R}_+\setminus\Lambda} (\Gamma^+_{\text{out}}(\nu)u, \Gamma^+_{\text{out}}(\nu)u)_q \, d\nu \,, \tag{A.2.11}$$

where $P_{\text{ac}}(\mathbf{A}^+)$ is the orthogonal projection onto the absolutely continuous subspace of the operator $\mathbf{A}^+$.

In the previous subsection we defined the operator Γ starting with $\Gamma(\nu)$. Similarly, we introduce now the operator

$$\Gamma^+_{\text{out}} : \quad C_0^\infty(0, +\infty) \ni u \mapsto \Gamma^+_{\text{out}} u \in L^2(\mathbb{R}_+; \mathbb{C}^{2q(\cdot)}),$$

where $\Gamma^+_{\text{out}} u$ is the L^2-function defined by the formula $(\Gamma^+_{\text{out}} u)(\nu) = \Gamma^+_{\text{out}}(\nu)u$ for $\nu \in \mathbb{R}_+ \setminus \Lambda$. It follows from (A.2.11) that Γ^+_{out} can be extended to an isometry

$$P_{\text{ac}}(\mathbf{A}^+)L_2(\mathbb{R}_+) \to L^2(\mathbb{R}_+, \mathbb{C}^{2q(\cdot)}).$$

We conclude this subsection by defining the corresponding notions for $\mathbf{A}^{+-}$. First we briefly consider $\mathbf{A}^-$. For any nonzero vector $a^+ \in \mathbb{C}^{q(\nu)}$ there is a unique generalized eigenfunction of the form

$$\begin{aligned}\Gamma^-_{\text{in}}(\nu)^* a^+ &= \sum_{k=1}^{q(\nu)} a_k^+ \frac{e^{ix\zeta_k^+(\nu)}}{\sqrt{2\pi A'(\zeta_k^+(\nu))}} + \sum_{k=1}^{q(\nu)} a_k^- \frac{e^{ix\zeta_k^-(\nu)}}{\sqrt{-2\pi A'(\zeta_k^-(\nu))}} \\ &\quad + \sum_{k=q(\nu)+1}^{m} d_k^+ e^{ix\overline{\xi_k^+(\nu+i0)}} \,.\end{aligned} \tag{A.2.12}$$

Indeed, if such a function exists, then $\overline{\Gamma_{\rm in}^-(\nu)^*a^+(-x)}$ is a generalized eigenfunction of $\mathbf{A}^+$, and a simple calculation shows that it equals

$$\Gamma_{\rm out}^+(\nu)^*\overline{a^+} = \Gamma_{\rm in}^+(\nu)^*R(\nu)^*\overline{a^+} = \Gamma_{\rm in}^+(\nu)^*\overline{a^-} \tag{A.2.13}$$

(here $R(\nu)$ is the reflection matrix for $\mathbf{A}^+$ introduced in Proposition A.2.1). Conversely, $\overline{(\Gamma_{\rm out}^+(\nu)^*\overline{a^+})(-x)}$ has the required properties. There is a unitary $q(\nu)\times q(\nu)$ matrix $\widetilde{R}(\nu)$ such that $a_k^- = \sum_{j=1}^{q(\nu)} \widetilde{R}_{kj}a_j^+$ in (A.2.12). From the second equality in (A.2.13) we see that $\overline{a^-} = R(\nu)^*\overline{a^+}$, or

$$\widetilde{R}(\nu)a^+ = a^- = \overline{R(\nu)^*}a^+ = R(\nu)^T a^+.$$

Thus $\widetilde{R}(\nu) = R(\nu)^T$. We define the map $\Gamma_{\rm out}^-(\nu)^*$ by the equality $\Gamma_{\rm out}^-(\nu)^*a^- = \Gamma_{\rm in}^-(\nu)^*\overline{R(\nu)}a^-$.

Similarly to the case of the operator $\mathbf{A}^+$, we can introduce the adjoint maps

$$\Gamma_{\rm in}^-(\nu), \Gamma_{\rm out}^-(\nu) : C_0^\infty(-\infty,0) \to \mathbb{C}^{q(\nu)},$$

and they satisfy

$$(g(\mathbf{A}^-)u,u)_- = \int_0^\infty g(\nu)(\Gamma_{\rm out}^-(\nu)u, \Gamma_{\rm out}^-(\nu)u)_q\, d\nu,$$

for $g \in C_0(\mathbb{R}\setminus\Lambda)$ and $u \in C_0^\infty(-\infty,0)$.

Definition A.2.9. *Let $\nu \in \mathbb{R}_+\setminus\Lambda$, and let $a = (a^+,a^-)$ and $b = (b^+,b^-)$ be vectors in $\mathbb{C}^{2q(\nu)}$. Then*

$$\begin{aligned}\Gamma_{\rm in}^{+-}(\nu)^*a &= \theta_+^*\,\Gamma_{\rm in}^+(\nu)^*a^- + \theta_-^*\,\Gamma_{\rm in}^-(\nu)^*a^+,\\ \Gamma_{\rm out}^{+-}(\nu)^*b &= \theta_+^*\,\Gamma_{\rm out}^+(\nu)^*b^+ + \theta_-^*\,\Gamma_{\rm out}^-(\nu)^*b^-,\end{aligned}$$

where $\theta_\pm^*$ are the extension operators from $L^2(\mathbb{R}_\pm)$ to $L_2(\mathbb{R})$.

Thus $\Gamma_{\rm in}^{+-}(\nu)^*a$ and $\Gamma_{\rm out}^{+-}(\nu)^*b$ are generalized eigenfunctions of $\mathbf{A}^{+-}$ with specified incoming and outgoing parts, respectively.

Similarly to the case of the operator $\mathbf{A}^+$, we can introduce their adjoints

$$\Gamma_{\rm in}^{+-}(\nu), \Gamma_{\rm out}^{+-}(\nu) : C_0^\infty(\mathbb{R}\setminus\{0\}) \to \mathbb{C}^{2q(\nu)}.$$

Let $u_\pm$ denote the restrictions of $u \in C_0^\infty(\mathbb{R}\setminus\{0\})$ to $\mathbb{R}_\pm$. Then our maps satisfy

$$\begin{aligned}(\Gamma_{\rm in}^{+-}(\nu)u)_j &= (\Gamma_{\rm in}^-(\nu)u_-)_j, & 1\leqslant j\leqslant q(\nu),\\ (\Gamma_{\rm in}^{+-}(\nu)u)_{j+q(\nu)} &= (\Gamma_{\rm in}^+(\nu)u_+)_j, & 1\leqslant j\leqslant q(\nu),\\ (\Gamma_{\rm out}^{+-}(\nu)u)_j &= (\Gamma_{\rm out}^+(\nu)u_+)_j, & 1\leqslant j\leqslant q(\nu),\\ (\Gamma_{\rm out}^{+-}(\nu)u)_{j+q} &= (\Gamma_{\rm in}^-(\nu)u_-)_j, & 1\leqslant j\leqslant q(\nu).\end{aligned}$$

Lemma A.2.10. *For $\nu \in \mathbb{R}_+\setminus\Lambda$ we have*

$$\Gamma_{\rm in}^{+-}(\nu)^* = \Gamma_{\rm out}^{+-}(\nu)^*S(\nu), \tag{A.2.14}$$

where $S(\nu)$ is the unitary $2q(\nu)\times 2q(\nu)$-matrix

$$\begin{pmatrix} 0 & R(\nu)\\ R^T(\nu) & 0\end{pmatrix}.$$

Note that (A.2.14) is equivalent to $S^*(\nu)\Gamma^{+-}_{\text{out}}(\nu) = \Gamma^{+-}_{\text{in}}(\nu)$ or, since $S(\nu)$ is unitary,

$$\Gamma^{+-}_{\text{out}}(\nu) = S(\nu)\,\Gamma^{+-}_{\text{in}}(\nu)\,. \tag{A.2.15}$$

As $S(\nu)$ maps the coefficients at the incoming terms of a solution to those at the outgoing terms, it is perhaps natural to call it the *scattering matrix* for the pair $\mathbf{A}^{+-}, \mathbf{A}$. In the next subsection we shall see that this agrees with the notion of scattering matrix in the sense of (time dependent) scattering theory.

PROOF OF LEMMA A.2.10. The equality (A.2.14) follows from Definition A.2.9 and the formulae

$$\Gamma^{+}_{\text{out}}(\nu)^* R(\nu) a^- = \Gamma^{+}_{\text{in}}(\nu)^* a^-,$$
$$\Gamma^{-}_{\text{out}}(\nu)^* R(\nu)^T a^+ = \Gamma^{+}_{\text{in}}(\nu)^* a^+.$$

The unitarity of $S(\nu)$ is an immediate consequence of that of $R(\nu)$. □

Clearly we can use $\Gamma^{+-}_{\text{in}}(\nu)$, $\Gamma^{+-}_{\text{out}}(\nu)$ to introduce isometric maps Γ^{+-}_{in}, Γ^{+-}_{out} taking $P_{\text{ac}}(\mathbf{A}^{+-})L^2(\mathbb{R})$ into $L^2(\mathbb{R}_+, \mathbb{C}^{2q(\cdot)})$ in the same manner as we obtained Γ^{+}_{in}, Γ^{+}_{out}.

3. The classical scattering matrix for the pair $\mathbf{A}, \mathbf{A}^{+-}$. We shall now introduce the wave operators $W_\pm$ and see that they are closely related to the operators Γ^{+-}_{out}, Γ^{+-}_{in} (actually, $\Gamma^{+-}_{\text{out}} W_+ = \Gamma^{+-}_{\text{in}} W_- = \Gamma$).

DEFINITION A.2.11. *Let A, B be self-adjoint operators on a Hilbert space $\mathcal{H}$ and let $P_{\text{ac}}(A)$ denote the orthogonal projection on the absolutely continuous subspace of A. The wave operators for the pair A, B are the operators*

$$W_\pm = \operatorname*{s\text{-}lim}_{t\to\pm\infty} e^{itB} e^{-itA} P_{\text{ac}}(A)$$

(if these limits exist).

Here s-lim denotes the strong limit, that is $W_+ u = \lim_{t\to\infty} e^{itB} e^{-itA} P_{\text{ac}}(A) u$ for each $u \in \mathcal{H}$. When W_+ exists, it is usually not true that

$$\lim_{t\to\infty} \| W_+ - e^{itB} e^{-itA} P_{\text{ac}}(A) \|_{\mathcal{H}} = 0.$$

It is known (see, e.g., [**Ya**] or volume 3, section XI.3, of [**ReSim**]) that if the wave operators exist, then they are isometric and their ranges are contained in the absolutely continuous subspace of B. We say that the wave operators are *complete* if their ranges, $\operatorname{Ran} W_\pm$, equal the absolutely continuous subspace of B, and that they are *asymptotically complete* if, in addition, B has no singular continuous spectrum. Note also that we have

$$\operatorname*{s\text{-}lim}_{t\to\pm\infty} e^{i(t+t_0)B} e^{-iAt} = \operatorname*{s\text{-}lim}_{s\to\pm\infty} e^{isB} e^{-i(s-t_0)A}.$$

Consequently the wave operators have the intertwining property

$$e^{it_0 B} W_\pm = W_\pm e^{it_0 A}.$$

Taking the Laplace transform with respect to t we see that

$$(B - zI)^{-1} W_\pm = W_\pm (A - zI)^{-1}$$

($\operatorname{Im} z \neq 0$), which implies that $W_{\pm} D(A) \subset D(B)$ and that

$$W_{\pm} A = B W_{\pm}. \tag{A.2.16}$$

PROPOSITION A.2.12. *The wave operators for the pair* $\mathbf{A}, \mathbf{A}^{+-}$ *exist.*

Before the (sketch of the) proof we illustrate the meaning of the wave operators. Let $\phi \in L_2(\mathbb{R})$ and set $\phi_{\pm} = W_{\pm}\phi$. Then

$$\| e^{-it\mathbf{A}}\phi - e^{-it\mathbf{A}^{+-}}\phi_{\pm} \| = \| e^{it\mathbf{A}^{+-}} e^{-it\mathbf{A}}\phi - \phi_{\pm} \| \to \| W_{\pm}\phi - \phi_{\pm} \| = 0$$

as $t \to \pm\infty$. Thus, $\phi_{\pm}$ are initial states for the equation

$$\partial_t \Phi(t,x) = -i\mathbf{A}^{+-}\Phi(t,x)$$

for which the solutions look like the solution of

$$\begin{cases} \partial_t \Phi(t,x) = -i\mathbf{A}\Phi(t,x), \\ \Phi(0,x) = \phi(x) \end{cases}$$

in the limits $t \to \pm\infty$. (In the "$-$" case we are solving the equation backwards in time.)

SKETCH OF PROOF. Since $\| e^{it\mathbf{A}^{+-}} e^{-it\mathbf{A}} \| = 1$ for all t, it is sufficient to prove that the limits $\lim_{t\to\pm\infty} e^{it\mathbf{A}^{+-}} e^{-it\mathbf{A}} u$ exist for a dense set of u's. We can, for instance, assume that $\hat{u} \in C_0^\infty(\mathbb{R})$ vanishes in a neighbourhood of the (real) zeros of $A'(\xi)$.

If $\chi \in C_0^\infty$, then a stationary phase argument shows that

$$\| e^{it\mathbf{A}^{+-}} \chi(x) e^{-it\mathbf{A}} u \| = \| \chi(x) e^{-it\mathbf{A}} u \| \to 0$$

as $t \to \pm\infty$. Indeed, if $f(x,\xi) \in C_0^\infty(K_1 \times K_2)$ and $A'(\xi) \neq 0$ on K_2, then for any integer N there is a constant C_N such that

$$\left| \int f(x,\xi) e^{i(x\xi - tA(\xi))}\, d\xi \right| \leqslant C_N (1 + |t|)^{-N},$$

since we may integrate by parts when $|t|$ is sufficiently large. Consequently, it suffices to verify that

$$\lim_{t\to\pm\infty} e^{it\mathbf{A}^{+-}} (1 - \chi_0) e^{-it\mathbf{A}} u$$

exists, where $\chi_0 \in C_0^\infty$ is a function which equals 1 in a neighbourhood of the origin. Since $(1 - \chi_0)$ vanishes in a neighbourhood of the origin, we have $(1 - \chi_0) D(\mathbf{A}) \subset D(\mathbf{A}^{+-})$. In particular, if

$$W(t) = e^{it\mathbf{A}^{+-}} (1 - \chi_0) e^{-it\mathbf{A}},$$

then

$$\begin{aligned} \frac{d}{dt}(W(t)u) &= i e^{it\mathbf{A}^{+-}} (\mathbf{A}^{+-}(1 - \chi_0) - (1 - \chi_0)\mathbf{A}) e^{-it\mathbf{A}} u \\ &= i e^{it\mathbf{A}^{+-}} [\chi_0, A(D)] e^{-it\mathbf{A}} u. \end{aligned}$$

Since

$$W(T)u = u + \int_0^T \frac{d}{dt}(W(t)u)\, dt,$$

the existence of the two limits will follow if we prove that

$$\int_{-\infty}^{\infty} \left\| \frac{d}{dt} W(t)u \right\| \, dt < \infty. \tag{A.2.17}$$

The operator $e^{it\mathbf{A}^{+-}}$ is unitary, so

$$\| e^{it\mathbf{A}^{+-}} [\chi_0, A(D)] e^{-it\mathbf{A}} u \| = \| [\chi_0, A(D)] e^{-it\mathbf{A}} u \|.$$

This means that to prove (A.2.17) it is sufficient to show that there is a constant C_u, independent of t, such that

$$\| [\chi_0, A(D)] e^{-it\mathbf{A}} u \| \leqslant C_u (1 + t^2)^{-1}. \tag{A.2.18}$$

The inequality (A.2.18) follows from another stationary phase argument. □

REMARK A.2.13. In the main text of this book we deal with the unitary groups $e^{-it\mathbf{A}^{1/(2m)}}$ and $e^{-it(\mathbf{A}^{+-})^{1/(2m)}}$ (more precisely, their analogues for partial differential operators). For this reason it is perhaps more natural to consider the wave operators for the pair $\mathbf{A}^{1/(2m)}$ and $(\mathbf{A}^{+-})^{1/(2m)}$. It is a consequence of the so-called invariance principle for the wave operators that these agree with the ones considered above. Indeed, arguing as in the proof given above, one sees that

$$\begin{aligned} \| (W_+ - I) e^{-is\mathbf{A}^{1/(2m)}} u \| &\leqslant \int_0^\infty \| [\chi_0, A(D)] e^{-i(t\mathbf{A} + s\mathbf{A}^{1/(2m)})} u \| \, dt \\ &\leqslant \widetilde{C}_u \int_0^\infty (1 + t + s)^{-2} \, dt. \end{aligned}$$

This implies that

$$\begin{aligned} 0 &= \lim_{s \to \infty} e^{is(\mathbf{A}^{+-})^{1/(2m)}} (W_+ - I) e^{-is\mathbf{A}^{1/(2m)}} u \\ &= \lim_{s \to \infty} \big(W_+ - e^{is(\mathbf{A}^{+-})^{1/(2m)}} e^{-is\mathbf{A}^{1/(2m)}} \big) u \end{aligned}$$

in view of (A.2.16).

THEOREM A.2.14. *The wave operators $W_\pm$ for our pair $\mathbf{A}, \mathbf{A}^{+-}$ are unitary mappings*

$$L_2(\mathbb{R}) \to P_{\mathrm{ac}}(\mathbf{A}^{+-}) L_2(\mathbb{R}),$$

and the maps $\Gamma_{\mathrm{in}}^{+-}$, $\Gamma_{\mathrm{out}}^{+-}$ are unitary mappings

$$P_{\mathrm{ac}}(\mathbf{A}^{+-}) L_2(\mathbb{R}) \to L^2(\mathbb{R}_+, \mathbb{C}^{2q(\cdot)}).$$

In particular, the wave operators are asymptotically complete and $W_\pm^* \mathbf{A}_{\mathrm{c}}^{+-} W_\pm = \mathbf{A}$. The operators $\Gamma_{\mathrm{in}}^{+-}$, $\Gamma_{\mathrm{out}}^{+-}$ are related to $W_\pm$ by the formula

$$\Gamma_{\mathrm{out}}^{+-} W_+ = \Gamma = \Gamma_{\mathrm{in}}^{+-} W_- . \tag{A.2.19}$$

For the proof of Theorem A.2.14 we shall need the following lemma, the proof of which is very close to that of Lemma A.2.6.

LEMMA A.2.15. *Let $\chi_R(x) = \chi(Rx)$ where $\chi \in C_0^\infty(\mathbb{R})$ is a function which is identically one in a neighbourhood of the origin, and let K be a compact set in $\mathbb{R}_+ \setminus \Lambda$. Then for any $v \in C_0^\infty(0, +\infty)$*

$$-\Gamma(\nu)[\,A(D), \chi_R\,]\mathbf{R}^+_{\nu+i0}v \to \Gamma^+_{\mathrm{out}}(\nu)v - \Gamma(\nu)v, \quad R \to +\infty,$$

uniformly over $\nu \in K$.

PROOF OF LEMMA A.2.15. It is no restriction to assume that K is connected, so that $q(\nu) \equiv q$ is constant on K. Let $e^{(j)}$, $1 \leqslant j \leqslant 2q$, be an orthonormal basis in $\mathbb{C}^{2q}$. It is obviously enough to prove that each of the $2q$ functions

$$\nu \mapsto (\Gamma(\nu)[\,A(D), \chi_R]\mathbf{R}^+_{\nu+i0}v, e^{(j)})_q \equiv ([\,A(D), \chi_R\,]\mathbf{R}^+_{\nu+i0}v, \Gamma(\nu)^* e^{(j)})_+\,,$$

$j = 1, \dots, 2q$, converges uniformly to the corresponding component of $\Gamma^+_{\mathrm{out}}(\nu)v - \Gamma(\nu)v$. By Green's formula we have

$$\begin{aligned}([\,A(D), \chi_R\,]\mathbf{R}^+_{\nu+i0}v, \Gamma(\nu)^* e^{(j)})_+ &= i\mathcal{A}^+(\mathbf{R}^+_{\nu+i0}v, \Gamma(\nu)^* e^{(j)}) - (\chi_R v, \Gamma(\nu)^* e^{(j)})_+ \\ &= -(\Gamma(\nu)(\mathfrak{A}\mathbf{R}^+_{\nu+i0}v), e^{(j)})_q - (\chi_R v, \Gamma(\nu)^* e^{(j)}).\end{aligned}$$

When $\nu \in K$ the modulus of the second term on the right may be estimated by a constant times

$$R^{-1} \sup_{\nu \in K} |\Gamma(\nu)^* e^{(j)}|,$$

and consequently the required statement follows from Lemma A.2.6. □

PROOF OF THEOREM A.2.14. We first assume that (A.2.19) is true. We already know that $\Gamma_\pm$ and $W_\pm$ are isometric, and it will follow from (A.2.19) that they are surjective. This will imply the first two statements.

It remains to prove (A.2.19). Since all the operators involved are bounded, formula (A.2.19) will follow if we prove that it is true on functions u such that $\Gamma u \in C_0^\infty(\mathbb{R}_+ \setminus \Lambda, \mathbb{C}^{2q(\cdot)})$. We shall study $(W_+ u, v)$ for such u with $v \in C_0^\infty(\mathbb{R} \setminus \{0\})$. To simplify the notation further we shall even assume $v \in C_0^\infty(0, +\infty)$, so that $e^{it\mathbf{A}^{+-}}v = \theta_+^* e^{it\mathbf{A}^+}v$. Let χ_R be as in the previous lemma. As in the proof of Proposition A.2.12 we get

$$\begin{aligned}(W_+ u, v) - (u, v) &= i\int_0^\infty (e^{it\mathbf{A}^{+-}}[A(D), \chi_R]e^{-it\mathbf{A}}u, v)_+\,dt \\ &= -i\int_0^\infty (e^{-it\mathbf{A}}u, [A(D), \chi_R]e^{-it\mathbf{A}^+}v)_+\,dt,\end{aligned}$$

where the integral is absolutely convergent. Using our spectral representation of $\mathbf{A}$ we may rewrite this as

$$\begin{aligned}(W_+ u, v) - (u, v) &= \lim_{\varepsilon \to +0} -i\int_0^\infty e^{-\varepsilon t}(e^{-it\mathbf{A}}u, [A(D), \chi_R]e^{-i\mathbf{A}^+ t}v)_+\,dt \\ &= \lim_{\varepsilon \to +0} -i\int_0^\infty \int_0^\infty (\Gamma(\nu)u, \Gamma(\nu)[A(D), \chi_R]e^{-it(\mathbf{A}^+ - \nu - i\varepsilon)}v)_q\,d\nu dt.\end{aligned}$$

Since $\Gamma(\nu)u \in C_0^\infty(\mathbb{R}_+ \setminus \Lambda)$, the ν-integral is taken over a finite interval, and we may change the order of integration

$$\begin{aligned}(W_+u, v) - (u, v) &= -\lim_{\varepsilon\to+0}\int_0^\infty (\Gamma(\nu)u, \Gamma(\nu)[A(D), \chi_R]\mathbf{R}^+_{\nu+i\varepsilon}v)_q\, d\nu \\ &= -\int_0^\infty (\Gamma(\nu)u, \Gamma(\nu)[A(D), \chi_R]\mathbf{R}^+_{\nu+i0}v)_q\, d\nu\,.\end{aligned}$$

In the last equality we used the fact that the integrand is continuous in ν, ε when $\nu \in \operatorname{supp}(\Gamma(\nu)u)$ and $\varepsilon \geqslant 0$. In view of Lemma A.2.15, it follows that

$$(W_+u, v) - (u, v) = \int_0^\infty (\Gamma(\nu)u, (\Gamma^{+-}_{\text{out}}(\nu) - \Gamma(\nu))u)_q\, d\nu\,.$$

We conclude that

$$(W_+u, v) = (\Gamma u, \Gamma^{+-}_{\text{out}} v)_0\,, \tag{A.2.20}$$

where $(\cdot,\cdot)_0$ denotes the scalar product in $L^2(\mathbb{R}_+, \mathbb{C}^{2q(\cdot)})$. This obviously extends by continuity to general $u, v \in L_2(\mathbb{R})$.

By the isometry of the wave operators, and formula (A.2.20) applied to $v = W_+w$, we have

$$(u, w) = (W_+u, W_+w) = (\Gamma u, \Gamma^{+-}_{\text{out}} W_+w)_0\,.$$

Since we also have $(u, w) = (\Gamma u, \Gamma w)_0$, the first equality (A.2.19) follows. The other equality is proved in the same manner. □

COROLLARY A.2.16. *We have* $W_+ = (\Gamma^{+-}_{\text{out}})^*\Gamma$ *and* $W_- = (\Gamma^{+-}_{\text{in}})^*\Gamma$.

DEFINITION A.2.17. *The scattering operator is the operator* $S = W_+^* W_-$.

Since $\operatorname{Ran} W_+ = \operatorname{Ran} W_-$ and the $W_\pm$ are isometric, it follows that S is unitary. From the intertwining property (A.2.16) it follows that

$$e^{it\mathbf{A}}S = Se^{it\mathbf{A}}. \tag{A.2.21}$$

To get an interpretation of S, we note that $W_+Su = W_-u$, so that if

$$\|\, e^{-it\mathbf{A}^{+-}}v - e^{-it\mathbf{A}}u \,\| \to 0 \qquad \text{as} \qquad t \to -\infty,$$

then

$$\|\, e^{-it\mathbf{A}^{+-}}v - e^{-it\mathbf{A}}Su \,\| \to 0 \qquad \text{as} \qquad t \to +\infty.$$

The asymptotic completeness means that any $w \in L_2(\mathbb{R})$ may be written as an orthogonal sum

$$w = P_{\text{ac}}(\mathbf{A}^{+-})w + \sum_{j=1}^N (w, u_j)u_j\,,$$

where ν_j, u_j, $1 \leqslant j \leqslant N$, are the eigenvalues and the orthonormal eigenfunctions of $\mathbf{A}^{+-}$, and

$$P_{\text{ac}}(\mathbf{A}^{+-})w \in \operatorname{Ran} W_+ = \operatorname{Ran} W_-\,.$$

Consequently,

$$e^{-it\mathbf{A}^{+-}}w = e^{-it\mathbf{A}^{+-}}P_{\text{ac}}w + \sum_{j=1}^N e^{-it\nu_j}(w, u_j)u_j\,,$$

and there is a $u_0 \in L_2(\mathbb{R})$ such that

$$\lim_{t\to-\infty} || e^{-it\mathbf{A}^{+-}} P_{\mathrm{ac}} w - e^{-it\mathbf{A}} u_0 || = 0,$$

$$\lim_{t\to+\infty} || e^{-it\mathbf{A}^{+-}} P_{\mathrm{ac}} w - e^{-it\mathbf{A}} S u_0 || = 0.$$

The commutation property (A.2.21) can be used to show that there exists a family of $2q(\nu) \times 2q(\nu)$ matrices $\widetilde{S}(\nu)$, $\nu > 0$, such that

$$(Su, v) = \int_0^\infty (\widetilde{S}(\nu)\Gamma(\nu)u, \Gamma(\nu)v)_q \, d\nu$$

for all $u, v \in C_0^\infty$. The matrix-valued function $\widetilde{S}(\nu)$ is called the scattering matrix. In our situation we can prove the existence of such a map directly.

PROPOSITION A.2.18. *Let* $S(\nu)$ *be the matrix in Lemma* A.2.10. *Then*

$$(\Gamma S u)(\nu) \;=\; S(\nu)\,\Gamma(\nu) u \tag{A.2.22}$$

for almost all $\nu > 0$ *if* $u \in C_0^\infty(\mathbb{R})$. *Thus,* $S(\nu)$ *is the scattering matrix for the pair* $\mathbf{A}^{+-}$, $\mathbf{A}$.

PROOF OF PROPOSITION A.2.18. We have to prove that

$$(Su, v) \;=\; \int_0^\infty (S(\nu)\Gamma(\nu)u,\, \Gamma(\nu)v)_q \, d\nu, \quad u, v \in C_0^\infty(\mathbb{R}). \tag{A.2.23}$$

Using Corollary A.2.16 we see that

$$(Su, v) \;=\; (\Gamma^{+-}_{\mathrm{out}} (\Gamma^{+-}_{\mathrm{in}})^* \Gamma u,\, \Gamma v)_0 \,.$$

Since Γ^{+-} and Γ^* are surjective, we can find $\widetilde{u} = (\Gamma^{+-}_{\mathrm{in}})^* \Gamma u$ such that $u = \Gamma^* \Gamma^{+-}_{\mathrm{in}} \widetilde{u}$, and hence

$$(\Gamma S u,\, \Gamma v)_0 \;=\; (Su,\, v) \;=\; (\Gamma^{+-}_{\mathrm{out}} \widetilde{u},\, \Gamma v)_0$$

for all $v \in C_0^\infty(\mathbb{R})$. This means that $\Gamma\, Su = \Gamma^{+-}_{\mathrm{out}}\, \widetilde{u}$. In view of (A.2.15) we have

$$(\Gamma^{+-}_{\mathrm{out}} \widetilde{u})(\nu) \;=\; S(\nu)\, (\Gamma^{+-}_{\mathrm{in}} \widetilde{u})(\nu)$$

for almost all $\nu \in \mathbb{R}_+$, since this is true when $\widetilde{u} \in C_0^\infty(\mathbb{R})$. It follows that

$$(\Gamma S u)(\nu) \;=\; (\Gamma^{+-}_{\mathrm{out}} \widetilde{u})(\nu) \;=\; S(\nu)\, (\Gamma^{+-}_{\mathrm{in}} \widetilde{u})(\nu) \;=\; S(\nu)\, (\Gamma u)(\nu)$$

for almost all $\nu \in \mathbb{R}_+$, which implies (A.2.23). □

A.3. Trace formulae

1. The regularized trace of $\mathbf{E}^+_\nu - \theta \mathbf{E}_\nu \theta^*$. Set $T(z) = \mathbf{R}^+_z - \theta \mathbf{R}_z \theta^*$. We know that its integral kernel $T(z; x, y)$ is analytic with respect to z on $\mathbb{C} \setminus (\sigma(\mathbf{A}^+) \cup \Lambda)$ and that

$$T(z; x, y) = i \sum_{k,l=1}^{m} e^{i(x\xi_k^+(z) - y\xi_l^-(z))} G_{kl}(z) \tag{A.3.1}$$

where $G(z)$ was defined in Proposition A.1.6. Locally, on any simply connected domain in $\mathbb{C} \setminus (\sigma(\mathbf{A}^+) \cup \Lambda)$, this follows from Proposition A.1.6. But $T(z; x, y)$ cannot be multivalued on $\mathbb{C} \setminus (\sigma(\mathbf{A}^+) \cup \Lambda)$ by the abstract properties of the resolvent, so the analyticity of $T(z; x, y)$ is, in fact, global on $\mathbb{C} \setminus (\sigma(\mathbf{A}^+) \cup \Lambda)$. Formula (A.3.1) also shows that $T(z;x,x)$ is uniformly exponentially decaying as $x \to \infty$ when z belongs to a compact set that does not intersect $\sigma(\mathbf{A}^+) \cup \Lambda$.

Formula (A.3.1) does not work at points $z_0 \in \Lambda \backslash \mathbb{R}_+$. However, using techniques from the proof of Lemma A.1.11 it is easy to see that all the statements from the previous paragraph remain true near such points if they are not eigenvalues of $\mathbf{A}^+$. Namely, $T(z;x,y)$ is analytic with respect to z on $\mathbb{C} \setminus \sigma(\mathbf{A}^+)$, and $T(z;x,x)$ is uniformly exponentially decaying as $x \to \infty$ when z belongs to a compact set that does not intersect $\sigma(\mathbf{A}^+)$.

Finally, at negative eigenvalues of $\mathbf{A}^+$ our integral kernel $T(z;x,y)$ has simple poles with residues given by minus the orthogonal projections on the corresponding eigenspaces.

DEFINITION A.3.1. *For* $z \in \mathbb{C} \setminus \sigma(\mathbf{A}^+)$ *we set* $\mathbf{f}^+(z) = \operatorname{Tr}(\mathbf{R}_z^+ - \theta \mathbf{R}_z \theta^*)$.

Clearly $\mathbf{f}^+(z)$ is analytic, and the formula

$$\mathbf{f}^+(z) = -\sum_{k,l=1}^{m} \frac{1}{\xi_k^+(z) - \xi_l^-(z)} G_{kl}(z) \tag{A.3.2}$$

shows immediately that the limits $\mathbf{f}^+(\nu \pm i0)$ exist and are real-analytic on each connected component of $\mathbb{R}_+ \setminus \Lambda$.

LEMMA A.3.2. *Let* K *be a compact subset of* $\mathbb{R}_+ \setminus \Lambda$ *and let* $\chi \in C_0^\infty(\mathbb{R})$ *be a function which is identically equal to one in a neighbourhood of the origin. Then*

$$\int_0^\infty T(z;x,x)\chi(\delta x)\,dx \to \mathbf{f}^+(z)$$

as $\delta \to +0$, *uniformly over* $\{z \in \mathbb{C} : \operatorname{Re} z \in K,\ \operatorname{Im} z \neq 0\}$. *In particular,*

$$\mathbf{f}^+(\nu \pm i0) = \lim_{\delta \to +0} \int_0^\infty T(\nu \pm i0; x,x)\chi(\delta x)\,dx.$$

PROOF OF LEMMA A.3.2. Without loss of generality we may assume that K is a (closed) interval. Furthermore, since $x \to T(z;x,x)$ is uniformly exponentially decaying on $\{|\operatorname{Im} z| \geqslant \varepsilon\}$ for any $\varepsilon > 0$, it is sufficient to prove the statement for

$$z \in M_\varepsilon^\pm = \{z \in \mathbb{C} : \operatorname{Re} z \in K,\ 0 < \pm \operatorname{Im} z < \varepsilon\}$$

with $\varepsilon > 0$ so small that the representation (A.3.1) is valid on $M_\varepsilon^\pm$. By self-adjointness we have $T(z;x,x) = \overline{T(\overline{z};x,x)}$, so it is enough to consider M_ε^+. We then have

$$\int_0^\infty T(z;x,x)\chi(\delta x)\,dx = \sum_{k,l=1}^{m} G_{kl}(z)\, i \int_0^\infty e^{ix(\xi_k^+(z) - \xi_l^-(z))}\chi(\delta x)\,dx,$$

so in view of (A.3.2) it is sufficient to prove that if $\inf_{z \in M_\varepsilon^+} |\alpha(z)| > 0$ and $\operatorname{Im}\alpha(z) \geqslant 0$, then

$$\sup_{z \in M_\varepsilon^+} \left| i \int_0^\infty e^{ix\alpha(z)}\chi(\delta x)\,dx + \frac{1}{\alpha(z)} \right| = O(\delta). \tag{A.3.3}$$

Integrating by parts twice we obtain

$$i \int_0^\infty e^{ix\alpha(z)}\chi(\delta x)\,dx = -\frac{1}{\alpha(z)} - i\frac{\delta}{(\alpha(z))^2} \int_0^\infty e^{ix\alpha(z)}\delta\chi''(\delta x)\,dx,$$

which implies (A.3.3). □

We shall denote the spectral families of $\mathbf{A}$ and $\mathbf{A}^+$ by $\mathbf{E}_\nu$ and $\mathbf{E}_\nu^+$, respectively, and the integral kernels of the latter by $\mathbf{e}_\nu^+(x,y)$ and $\mathbf{e}_\nu(x,y)$. By L we shall denote a sufficiently large positive number, namely, a number satisfying $-L < \min(\Lambda \cap \mathbb{R})$.

LEMMA A.3.3. *For fixed ν the integral kernels $\mathbf{e}_\nu^+(x,y)$ and $\mathbf{e}_\nu(x,y)$ are smooth functions of $(x,y) \in \mathbb{R}_+ \times \mathbb{R}_+$. Moreover, these functions and all their derivatives are uniformly bounded.*

PROOF OF LEMMA A.3.3. Since $\mathbf{E}_\nu$ is the operator of convolution with the inverse Fourier transform of the characteristic function of the compact set $\{\xi \in \mathbb{R} : A(\xi) \leqslant \nu\}$, the statement about $\mathbf{e}_\nu$ is obvious.

We have

$$\mathbf{E}_\nu^+ = \mathbf{R}_{-L}^+(\mathbf{A}^+ + L)\mathbf{E}_\nu^+(\mathbf{A}^+ + L)\mathbf{R}_{-L}^+ . \tag{A.3.4}$$

The operator $(\mathbf{A}^+ + L)\mathbf{E}_\nu^+(\mathbf{A}^+ + L)$ is self-adjoint and maps $L_2(\mathbb{R}_+)$ continuously into $D(\mathbf{A}^+) \subset H^{2m}(\mathbb{R}_+)$. Hence, it follows from Theorem 3.1 of [**Agm1**] that $(\mathbf{A}^+ + L)\mathbf{E}_\nu^+(\mathbf{A}^+ + L)$ has an integral kernel $\mathbf{e}_{\nu,L}^+(x,y)$ which is continuous and uniformly bounded on $(0,+\infty)\times(0,+\infty)$. Consequently, (A.3.4) implies that

$$\mathbf{e}_\nu^+(x,y) = \int_{\mathbb{R}_+\times\mathbb{R}_+} \mathbf{r}_{-L}^+(x,z_1)\overline{\mathbf{r}_{-L}^+(y,z_2)}\mathbf{e}_{\nu,L}^+(z_1,z_2)\,dz_1dz_2. \tag{A.3.5}$$

But the integral kernel $\mathbf{r}_{-L}^+(x,y) \equiv \overline{\mathbf{r}_{-L}^+(y,x)}$ can be written down explicitly in terms of exponential functions (see A.1.5), and it is easy to see that it is $2m-2$ times continuously differentiable on $\mathbb{R}_+\times\mathbb{R}_+$; moreover, these derivatives are uniformly bounded and decay uniformly exponentially as $|x-y|\to\infty$. Consequently, (A.3.5) implies that $\mathbf{e}_\nu^+(x,y)$ has bounded continuous derivatives up to order $2m-2$ on $\mathbb{R}_+\times\mathbb{R}_+$.

It remains to note that in the above argument we can replace the original operator $\mathbf{A}^+$ by $(\mathbf{A}^+)^N$ for any natural odd N, with an obvious relationship between the spectral families of $\mathbf{A}^+$ and $(\mathbf{A}^+)^N$. This observation allows us to conclude that $\mathbf{e}_\nu^+(x,y)$ has bounded continuous derivatives up to order $2mN-2$ on $\mathbb{R}_+\times\mathbb{R}_+$. As N can be taken arbitrarily large, the required result follows. □

For $\nu > -L$, we shall take γ_ν to be some C^∞ curve, of finite length, from $\nu + i0$ to $\nu - i0$ in $\mathbb{C}\setminus[-L,+\infty)$ subject to the following restrictions: the curve γ_ν is symmetric with respect to the real axis and it is transversal to the real axis at $\nu \pm i0$. We shall parametrize γ_ν by $\varphi \in (0,2\pi)$, and for $\theta \in (0,\pi]$ we shall denote by γ_ν^θ the part of γ_ν corresponding to $\varphi\in[\theta,2\pi-\theta]$. The curve γ_ν^θ is obviously a compact set in $\mathbb{C}\setminus\sigma(\mathbf{A}^+)$.

LEMMA A.3.4. *Let $\nu \in \mathbb{R}_+\setminus\Lambda$ and let γ_ν be as above. Then*

$$\mathbf{e}_\nu^+(x,x) - \mathbf{e}_\nu(x,x) = -\frac{1}{2\pi i}\int_{\gamma_\nu} T(z;x,x)\,dz\,. \tag{A.3.6}$$

Furthermore,

$$\mathbf{e}_\nu^+(x,x) - \mathbf{e}_\nu(x,x) = \frac{1}{\pi x}\sum_{k,l=1}^{q(\nu)} C_{kl}(\nu)\,\mathrm{Im}\big(R_{kl}(\nu)\,e^{ix(\zeta_k^+(\nu)-\zeta_l^-(\nu))}\big) + O(1/x^2)$$

as $x \to +\infty$. *Here*

$$C_{kl}(\nu) = \frac{\sqrt{-A'(\zeta_k^+(\nu))A'(\zeta_l^-(\nu))}}{A'(\zeta_k^+(\nu)) - A'(\zeta_l^-(\nu))} > 0,$$

and $R_{kl}(\nu)$ *are the elements of the reflection matrix.*

PROOF OF LEMMA A.3.4. By Stone's formula (Theorem VII.13 of [**ReSim**]) and the fact that ν is not an eigenvalue we have

$$\mathbf{E}_\nu^+ = \frac{1}{2}\{\mathbf{E}_{\nu+0}^+ + \mathbf{E}_\nu^+\} = -\frac{1}{2\pi i}\lim_{\theta\to+0}\int_{\gamma_\nu^\theta}\mathbf{R}_z^+\,dz\,.$$

The same calculations are obviously valid if we replace $\mathbf{A}^+$ by $\mathbf{A}$. Consequently,

$$\mathbf{E}_\nu^+ - \theta\mathbf{E}_\nu\theta^* = -\frac{1}{2\pi i}\lim_{\theta\to+0}\int_{\gamma_\nu^\theta}\{\mathbf{R}_z^+ - \theta\mathbf{R}_z\theta^*\}\,dz.$$

The left-hand side is an integral operator with continuous kernel and the same is true for the integrand on the right-hand side. Moreover, the integral kernel of the integrand on the right-hand side is uniformly continuous with respect to z and has limits as $z \to \nu \pm i0$ (see the beginning of this subsection). Thus

$$\mathbf{e}_\nu^+(x,x) - \mathbf{e}_\nu(x,x) = -\frac{1}{2\pi i}\int_{\gamma_\nu} T(z;x,x)\,dz.$$

This proves (A.3.6).

By the self-adjointness of $\mathbf{A}$, $\mathbf{A}^+$ we have $T(\overline{z};x,x) = \overline{T(z;x,x)}$. A simple calculation shows that this implies

$$\text{(A.3.7)} \qquad \mathbf{e}_\nu^+(x,x) - \mathbf{e}_\nu(x,x) = -\frac{1}{\pi}\,\mathrm{Im}\int_{\varphi=0}^{\pi} T(z(\varphi);x,x)\,d(z(\varphi))\,.$$

If we set $h_{kl}(z) = \xi_k^+(z) - \xi_l^-(z)$ the integral on the right-hand side may be written as

$$I(x) \equiv i\sum_{k,l=1}^{m}\int_{\varphi=0}^{\pi} G_{kl}(z(\varphi))\,e^{ixh_{kl}(z(\varphi))}\,d(z(\varphi))\,.$$

The terms with $\mathrm{Im}\,h_{kl}(\nu+i0) > 0$ are uniformly exponentially decaying in x on the interval of integration, and consequently

$$I(x) \equiv i\sum_{k,l=1}^{q(\nu)}\int_{\varphi=0}^{\pi} G_{kl}(z(\varphi))\,e^{ixh_{kl}(z(\varphi))}\,d(z(\varphi)) + O(x^{-\infty})\,.$$

The functions $h_{kl}(z)$, $1 \leqslant k,l \leqslant q(\nu)$, can be analytically extended to a neighbourhood of $\{z(\varphi) : \varphi \in [0,\pi]\}$ and we have

$$\text{(A.3.8)} \qquad h'_{kl}(\nu+i0) = \frac{1}{A'(\zeta_k^+(\nu))} - \frac{1}{A'(\zeta_l^-(\nu))} = \frac{A'(\zeta_k^+(\nu)) - A'(\zeta_l^-(\nu))}{-A'(\zeta_l^-(\nu))A'(\zeta_k^+(\nu))} > 0$$

by (A.1.3). Thus, it is no restriction to assume that γ_ν is chosen such that $h'_{kl}(z(\varphi)) \neq 0$ for $0 \leqslant \varphi \leqslant \pi$, $1 \leqslant k,l \leqslant q(\nu)$ (there is only a finite number of points we should avoid in charting our curve γ_ν). Consequently, we can integrate by parts twice to obtain

$$I(x) = \sum_{k,l=1}^{q(\nu)} \left[\frac{G_{kl}(z)}{x h'_{kl}(z)} \, e^{ixh_{kl}(z)} \right]_{\nu+i0}^{z(\pi)} + O\left(\frac{1}{x^2}\right).$$

Since $\operatorname{Im} h_{kl}(z(\pi)) > 0$, the contributions from the terms emanating from the point $z(\pi)$ to the sum on the right-hand side are exponentially decaying in x. We conclude that

$$I(x) = -\sum_{k,l=1}^{q(\nu)} G_{kl}(\nu + i0) \, \frac{e^{ix(\zeta_k^+(\nu) - \zeta_l^-(\nu))}}{x h'_{kl}(\nu + i0)} + O\left(\frac{1}{x^2}\right).$$

Using (A.3.8) and (A.2.3), we conclude that our asymptotic formula for $I(x)$ can be rewritten as

$$I(x) = -\frac{1}{x} \sum_{k,l=1}^{q(\nu)} C_{kl}(\nu) \, R_{kl}(\nu) e^{ix(\zeta_k^+(\nu) - \zeta_l^-(\nu))} + O\left(\frac{1}{x^2}\right).$$

Finally, inserting this into (A.3.7) we obtain the required statement. □

DEFINITION A.3.5. We define

(A.3.9) $$\operatorname{shift}^+(\nu) = \int_0^\infty (\mathbf{e}_\nu^+(x,x) - \mathbf{e}_\nu(x,x)) \, dx \,, \qquad \nu \in \mathbb{R} \setminus \Lambda.$$

For $\nu < 0$ the above integral converges absolutely and $\operatorname{shift}^+(\nu) = N^+(\nu)$, the number of eigenvalues of $\mathbf{A}^+$ less than ν. For $\nu > 0$ convergence follows from Lemma A.3.4 and is not, generally speaking, absolute.

Denote by θ_X the characteristic function of the interval $[0, X]$. Then formula (A.3.9) is equivalent to

(A.3.10) $$\operatorname{shift}^+(\nu) = \lim_{X \to +\infty} \operatorname{Tr}(\theta_X (\mathbf{E}_\nu^+ - \theta \mathbf{E}_\nu \theta^*) \theta_X) \,, \qquad \nu \in \mathbb{R} \setminus \Lambda.$$

Note that the (self-adjoint) operator $\theta_X (\mathbf{E}_\nu^+ - \theta \mathbf{E}_\nu \theta^*) \theta_X$ is of trace class because it can be identified with an operator with smooth integral kernel acting over the finite interval $[0, X]$.

Formula (A.3.10) allows us to interpret $\operatorname{shift}^+(\nu)$ as the regularized trace of $\mathbf{E}_\nu^+$. Regularization is achieved in two steps: first, by subtracting the spectral projection of the "unperturbed" operator, and second, by introducing a spatial cut-off.

For $\nu < 0$ this regularization is really unnecessary because $\mathbf{E}_\nu = 0$, $\mathbf{E}_\nu^+$ is of trace class and $\operatorname{shift}^+(\nu) = \operatorname{Tr} \mathbf{E}_\nu^+ = N^+(\nu)$.

For $\nu > 0$ our two-step regularization is unavoidable. Since $\sigma_{\text{ess}}(\mathbf{A}^+) = \mathbb{R}_+$, our spectral projection $\mathbf{E}_\nu^+$ is a noncompact operator and, consequently, not of trace class. After the first step of regularization we obtain the operator $\mathbf{E}_\nu^+ - \theta \mathbf{E}_\nu \theta^*$ which is compact, so the situation is improved. Let us explain now why we still have to perform the second step, that is, why for $\nu \in \mathbb{R}_+ \setminus \Lambda$ the operator $\mathbf{E}_\nu^+ - \theta \mathbf{E}_\nu \theta^*$ is not, generally speaking, of trace class. A necessary condition for $\mathbf{E}_\nu^+ - \theta \mathbf{E}_\nu \theta^*$ to be of trace class is that the leading terms in the asymptotic formula from Lemma A.3.4 vanish, so that $s(x) = \mathbf{e}_\nu^+(x,x) - \mathbf{e}_\nu(x,x)$ is absolutely integrable. Indeed, if a self-adjoint operator G is of trace class, then the expression $\sum_{j=1}^\infty |(G\varphi_j, \varphi_j)|$

is finite and uniformly bounded over all orthonormal systems $\{\varphi_j\}_{j=1}^\infty$. However, it is easy to see, by choosing φ_j^h as $h^{-1/2}$ times the characteristic function of the interval $[(j-1)h, jh)$, that for any $X > 0$ and any $\varepsilon > 0$, we have

$$\sum_{j=1}^{\infty} |((\mathbf{E}_\nu^+ - \theta\mathbf{E}_\nu\theta^*)\varphi_j^h, \varphi_j^h)| > \int_0^X |s(x)|\,dx - \varepsilon\,,$$

for sufficiently small $h > 0$. Thus, if $\int_0^\infty |s(x)|\,dx = \infty$, the left-hand side in the above formula can be made arbitrarily large.

Of course, spatial regularization can be performed in many ways. The following lemma illustrates one possibility of using a smooth cut-off.

LEMMA A.3.6. *Let $\chi \in C_0^\infty(\mathbb{R})$ be a function as in Lemma* A.3.2. *Then*

$$\mathrm{shift}^+(\nu) = \lim_{\delta\to+0} \int_0^\infty (\mathbf{e}_\nu^+(x,x) - \mathbf{e}_\nu(x,x))\,\chi(\delta x)\,dx\,, \qquad \nu \in \mathbb{R}_+ \setminus \Lambda.$$

One can, of course, take χ^2 instead of χ, and then the above formula can be rewritten as

$$\mathrm{shift}^+(\nu) = \lim_{\delta\to 0} \mathrm{Tr}\left(\chi(\delta x)(\mathbf{E}_\nu^+ - \theta\mathbf{E}_\nu\theta^*)\chi(\delta x)\right).$$

PROOF OF LEMMA A.3.6. Set $S_\nu(x) = \int_0^x (\mathbf{e}^+(\nu,t,t) - \mathbf{e}(\nu,t,t))\,dt$. Integrating by parts and using the fact that $\mathrm{shift}^+(\nu) = \lim_{y\to\infty} S_\nu(y)$ we obtain

$$\int_0^\infty (\mathbf{e}_\nu^+(x,x) - \mathbf{e}_\nu(x,x))\,\chi(\delta x)\,dx = -\int_0^\infty S_\nu(x)\,\delta\chi'(\delta x)\,dx$$
$$= -\int_0^\infty S_\nu(y/\delta)\,\chi'(y)\,dy \to -\mathrm{shift}^+(\nu)\int_0^\infty \chi'(y)\,dy = \mathrm{shift}^+(\nu)\,.$$

□

PROPOSITION A.3.7. *Let $\nu \in \mathbb{R}_+ \setminus \Lambda$. Then*

$$\mathrm{shift}^+(\nu) = -\frac{1}{2\pi i}\oint_{\gamma_\nu} \mathbf{f}^+(z)\,dz.$$

PROOF OF PROPOSITION A.3.7. By Lemmas A.3.6, A.3.4 we have

$$\begin{aligned}\mathrm{shift}^+(\nu) &= \lim_{\delta\to+0}\int_0^\infty (\mathbf{e}^+(\nu,x,x) - \mathbf{e}(\nu,x,x))\,\chi(\delta x)\,dx \\ &= -\frac{1}{2\pi i}\lim_{\delta\to+0}\int_0^\infty \left(\oint_{\gamma_\nu} T(z;x,x)\,dz\right)\chi(\delta x)\,dx.\end{aligned}$$

For $\delta > 0$ we can change the order of integration in the integral on the right. Furthermore, the functions $F_\delta(z) = \int_0^\infty T(z;x,x)\,\chi(\delta x)\,dx$, $\delta > 0$, converge uniformly in $z \in \gamma_\nu$ to $\mathbf{f}^+(z)$ when $\delta \to +0$. Indeed, when $z \in \gamma_\nu$ is close to ν this follows from Lemma A.3.2, and on the remaining part of γ_ν the modulus of the integrand is uniformly exponentially decaying. Thus, by uniform convergence, we conclude that

$$\mathrm{shift}^+(\nu) = -\frac{1}{2\pi i}\lim_{\delta\to+0}\oint_{\gamma_\nu} F_\delta(z)dz = -\frac{1}{2\pi i}\int_{\gamma_\nu} \mathbf{f}^+(z)\,dz.$$

□

PROPOSITION A.3.8. *Let L be a positive number such that $-L$ is below the spectrum of the operator $\mathbf{A}^+$, and let $s_+(L)$ and $s_-(L)$ be the numbers of positive and negative eigenvalues of $\mathbf{R}^{+-}_{-L} - \mathbf{R}_{-L}$, respectively. Then*

$$-s_-(L) \leqslant 2\,\mathrm{shift}^+(\nu) \leqslant s_+(L)\,. \tag{A.3.11}$$

We postpone proving the inequality (A.3.11) until the next subsection for the following reason. If one compares two operators $\widetilde{\mathbf{A}}$, $\widetilde{\mathbf{A}}^+$ with discrete spectra an analogue of (A.3.11) can be derived by simple variational arguments, and then our case can, in principle, be handled by approximating $\mathbf{A}$, $\mathbf{A}^+$ by appropriate $\widetilde{\mathbf{A}}$, $\widetilde{\mathbf{A}}^+$ and performing a limiting process. This "discrete" approach was used in Section 4.1 of [**Va7**]. The "discrete" approach is natural, but the limiting process requires substantial technical justification. So in this book we use a different, "continuous", approach going back to M. G. Krein, and we devote to it the next subsection. There it will be shown that $-2\,\mathrm{shift}^+(\nu)$ equals (a.e. on $\mathbb{R}_+$) Krein's spectral shift function for the pair $\mathbf{A}, \mathbf{A}^{+-}$, and this will lead to a short proof of (A.3.11).

As $\mathrm{rank}(\mathbf{R}^{+-}_{-L} - \mathbf{R}_{-L}) = s_+(L) + s_-(L) \leqslant 2m$, Proposition A.3.8 immediately implies the following two corollaries.

COROLLARY A.3.9. *For any $\nu \in \mathbb{R} \setminus \Lambda$*

$$|\,\mathrm{shift}^+(\nu)| \leqslant m\,. \tag{A.3.12}$$

COROLLARY A.3.10. *For any $\mu, \nu \in \mathbb{R} \setminus \Lambda$*

$$|\,\mathrm{shift}^+(\mu) - \mathrm{shift}^+(\nu)| \leqslant m\,. \tag{A.3.13}$$

If we note that, for sufficiently small $\varepsilon > 0$, $\mathrm{shift}^+(-\varepsilon)$ is the number of negative eigenvalues of $\mathbf{A}^+$, we obtain

COROLLARY A.3.11. *The operator $\mathbf{A}^+$ has at most m negative eigenvalues.*

The estimate in Corollary A.3.11 can be improved: one can replace m by the number of B_j with degree higher than $m-1$; see [**LtinSu**].

PROPOSITION A.3.12. *The function $\nu \mapsto \mathrm{shift}^+(\nu)$ is real-analytic on $\mathbb{R} \setminus \Lambda$. In particular,*

$$(\mathrm{shift}^+)'\,(\nu) \;=\; \frac{1}{2\pi i}\,(\mathbf{f}^+(\nu + i0) \;-\; \mathbf{f}^+(\nu - i0)). \tag{A.3.14}$$

If $\nu_0 \in \Lambda$, then the limits

$$\mathrm{shift}^+(\nu_0 \pm 0) = \lim_{\varepsilon \to +0} \mathrm{shift}^+(\nu_0 \pm \varepsilon)$$

exist.

PROOF OF PROPOSITION A.3.12. Consider an interval $I = [\nu - h, \nu + h] \subset \mathbb{R}_+ \setminus \Lambda$. The functions $\mathbf{f}^+(\lambda \pm i0)$ are real-analytic on I. Using this, the analyticity of $\mathbf{f}^+(z)$ and Proposition A.3.7, we see that

$$\frac{\mathrm{shift}^+(\nu + s) - \mathrm{shift}^+(\nu)}{s} = \frac{1}{2\pi i s} \int_\nu^{\nu+s} (\mathbf{f}^+(\lambda + i0) - \mathbf{f}^+(\lambda - i0))\, d\lambda$$

when $0 < |s| \leqslant h$. Taking the limit as $s \to 0$ we obtain (A.3.14), and consequently, the real-analyticity of shift^+.

Consider now a point $\nu_0 \in \Lambda$. Obviously, in small one-sided neighbourhoods of ν_0 the function $\mathrm{shift}^+(\nu)$ admits power series expansions in fractional powers of $|\nu - \nu_0|$, and the occurring powers cannot be negative in view of Corollary A.3.9. This proves the existence of the one-sided limits. □

Proposition A.3.12 shows that Definition 1.6.4 in the main part of this book makes sense, i.e., we can extend $\mathrm{shift}^+(\nu)$ to $\nu \in \Lambda$ by setting $\mathrm{shift}^+(\nu) = \mathrm{shift}(\nu - 0)$. Of course, with this convention the inequalities (A.3.11)–(A.3.13) extend to all $\nu \in \mathbb{R}$.

2. The function $\mathrm{shift}^+(\nu)$ **and the spectral shift function of Krein.** We first recall some facts about the spectral shift function of Krein (see Chapter 8 in [**Ya**], in particular Sections 2 and 7).

Let H and H_0 be self-adjoint operators, bounded from below, and such that $(H-z)^{-1} - (H_0 - z)^{-1}$ is of trace class when $z \notin \mathbb{R}$. (It follows from the resolvent equation that it suffices to verify this condition for one $z \notin \mathbb{R}$.) Then $f(H) - f(H_0)$ is of trace class for all $f \in C^2(\mathbb{R})$ such that for some positive ε

$$(\lambda^2 f'(\lambda))' = O(\lambda^{-1-\varepsilon}) \qquad \text{as} \qquad \lambda \to +\infty.$$

There is a real-valued function $\xi \in L^1(\mathbb{R}, \frac{d\nu}{1+\nu^2})$, called the spectral shift function, such that

$$\mathrm{Tr}\,(f(H) - f(H_0)) = \int f'(s)\,\xi(s)\,ds\,. \tag{A.3.15}$$

This formula determines ξ up to a constant, and moreover we see that $\xi(s)$ is constant in each (connected) component of $\mathbb{R} \setminus (\sigma(H) \cup \sigma(H_0))$. Thus the constant can be specified by requiring that $\xi(s) = 0$ when $s < \min\{\sigma(H) \cup \sigma(H_0)\}$. The spectral shift function satisfies the relation

$$\det S(\nu) = \exp\,(-2\pi i \xi(\nu)) \quad \text{a.a.} \quad \nu \in \sigma_{\mathrm{ac}}(H_0), \tag{A.3.16}$$

where $S(\nu)$ is the scattering matrix for the pair H_0, H.

We shall now apply this approach with $H = \mathbf{A}^{+-}$ and $H_0 = \mathbf{A}$. Here and further on we use the $^{+-}$ superscript when referring to objects similar to those with a $^+$, the difference being that in the $^{+-}$ case we are dealing with the problem on $\mathbb{R} \setminus \{0\}$ introduced in Remark A.1.5.

Let Y_λ denote the characteristic function of $(-\infty, \lambda]$. If we formally set $f = Y_\lambda$ in (A.3.15) we get

$$\text{“}\mathrm{Tr}\,(Y_\lambda(\mathbf{A}^{+-}) - Y_\lambda(\mathbf{A}))\text{”} = \text{“}\mathrm{Tr}\,(\mathbf{E}_\lambda^{+-} - \mathbf{E}_\lambda)\text{”} = -\xi(\lambda). \tag{A.3.17}$$

Here we put quotation marks to stress the fact that this is just a formal calculation, since the operators involved are not, generally speaking, of trace class; see the discussion after Definition A.3.5.

By symmetry (see Remark A.1.5)

$$\mathbf{f}^{+-}(z) := \mathrm{Tr}(\mathbf{R}_z^{+-} - \mathbf{R}_z) = 2\,\mathrm{Tr}(\mathbf{R}_z^{+} - \theta\mathbf{R}_z\theta^*) = 2\mathbf{f}^+(z)\,, \tag{A.3.18}$$

$$\mathrm{Tr}((\mathbf{E}_\nu^{+-} - \mathbf{E}_\nu)\chi(\delta x)) = 2\,\mathrm{Tr}((\mathbf{E}_\nu^{+} - \mathbf{E}_\nu)\chi(\delta x))\,. \tag{A.3.19}$$

Applying Lemma A.3.6 and formula (A.3.19) we obtain

$$\begin{aligned}\text{shift}^{+-}(\nu) &:= \lim_{\delta\to+0} \operatorname{Tr}((\mathbf{E}_\nu^{+-} - \mathbf{E}_\nu)\chi(\delta x)) \\ &= 2\lim_{\delta\to+0} \operatorname{Tr}((\mathbf{E}_\nu^{+} - \mathbf{E}_\nu)\chi(\delta x)) = 2\,\text{shift}^{+}(\nu)\,.\end{aligned} \tag{A.3.20}$$

"Formula" (A.3.17) and formula (A.3.20) suggest

PROPOSITION A.3.13. *We have*

$$\text{shift}^{+-}(\nu) = -\xi(\nu) \quad \text{a. a.} \quad \nu \in \mathbb{R}.$$

PROOF OF PROPOSITION A.3.13. From Proposition A.3.7 it easily follows that

$$\text{shift}^{+-}(\nu) = -\frac{1}{2\pi i}\oint_{\gamma_\nu} \operatorname{Tr}(\mathbf{R}_z^{+-} - \mathbf{R}_z)\,dz = -\lim_{\theta\to+0}\frac{1}{2\pi i}\int_{\gamma_\nu^\theta} \operatorname{Tr}(\mathbf{R}_z^{+-} - \mathbf{R}_z)\,dz\,.$$

It follows from (A.3.15) with $f(\lambda) = (\lambda - z)^{-1}$ that

$$\operatorname{Tr}(\mathbf{R}_z^{+-} - \mathbf{R}_z) = -\int_{-\infty}^{\infty} (\lambda - z)^{-2}\,\xi(\lambda)\,d\lambda\,. \tag{A.3.21}$$

For fixed $\theta > 0$ this is a uniformly bounded continuous function of $z \in \gamma_\nu^\theta$. Moreover, it is no restriction to assume that γ_ν is given by $z(\varphi) = \nu + i\varphi$ for small φ. We conclude using the Fubini–Tonelli theorem that

$$\begin{aligned}\text{shift}^{+-}(\nu) &= \lim_{\theta\to+0}\frac{1}{2\pi i}\int_{-\infty}^{\infty}\left(\int_{\gamma_\nu^\theta}(\lambda - z)^{-2}\,dz\right)\xi(\lambda)\,d\lambda \\ &= -\lim_{\theta\to+0}\frac{1}{\pi}\int_{-\infty}^{\infty}\frac{\theta}{(\nu-\lambda)^2+\theta^2}\,\xi(\lambda)\,d\lambda\,.\end{aligned}$$

It follows from well-known properties of the Poisson kernel that the right-hand side equals $-\xi(\nu)$ for almost all $\nu \in \mathbb{R}$. □

We are now prepared to proceed to the

PROOF OF PROPOSITION A.3.8. We let $\eta(t)$ denote the spectral shift function for the pair of operators $-\mathbf{R}_{-L}^{+-}$, $-\mathbf{R}_{-L}$.

It follows from Theorem 8.2.1 in [**Ya**] that

$$-s_+ \leqslant \eta(t) \leqslant s_-\,. \tag{A.3.22}$$

On the other hand, if $f \in C_0^\infty(-L, +\infty)$ and $g(t) = f\left(-L - \frac{1}{t}\right)$, then, applying formula (A.3.15) twice, we get

$$\begin{aligned}\int f'(s)\,\xi(s)\,ds &= \operatorname{Tr}(f(\mathbf{A}^{+-}) - f(\mathbf{A})) = \operatorname{Tr}(g(-\mathbf{R}_{-L}^{+-}) - g(-\mathbf{R}_{-L})) \\ &= \int g'(t)\,\eta(t)\,dt = \int f'(s)\,\eta\left(-\frac{1}{s+L}\right)ds\,.\end{aligned}$$

In view of the facts that $\xi(s) = 0$ when $s < \min\sigma(\mathbf{A}^+)$ and that $\eta(t) = 0$ when $t < \min\sigma(-\mathbf{R}_{-L}^{+-})$, it follows by Proposition A.3.13 that

$$\text{shift}^{+-}(\nu) = -\xi(\nu) = -\eta\left(-\frac{1}{\nu+L}\right)$$

for almost all $\nu \in \mathbb{R}$. The required result (A.3.11) now follows from the above formula and (A.3.22), (A.3.20). $\square$

REMARK A.3.14. One has the following expression for $\eta(t)$:

$$\eta(t) = \lim_{\varepsilon \to +0} \arg D(t + i\varepsilon)$$

where

$$D(w) = \det(I + K_L^{+-}(-\mathbf{R}_{-L} - w)^{-1}), \quad \operatorname{Im} w > 0,$$

and the argument is chosen in such a way that $\lim_{y\to\infty} \arg D(iy) = 0$. This gives

$$\xi(s) = \lim_{\varepsilon \to +0} \arg E(s + i\varepsilon),$$

where

$$E(z) = \det(I + (z + L)K_L^{+-}(I + (L + z)\mathbf{R}_z)), \quad \operatorname{Im} z > 0,$$

and the argument is chosen in such manner that $\xi(0) = 0$. If we choose an orthonormal basis $f_1, \ldots, f_{s_+ + s_-}$ for the range of K_L^{+-}, this becomes the determinant of the matrix with elements

$$\delta_{jk} + (z + L)(K_L^{+-} f_k, f_j) + (z + L)^2(K_L^{+-}\mathbf{R}_z f_k, f_j).$$

This is, however, less handy for practical calculations. Note that the presence of the operator $\mathbf{R}_z$ in the last term couples the two half-axes.

A consequence of Proposition A.3.13 is the following

COROLLARY A.3.15. *If* $\mu \in \mathbb{C} \setminus [\min \sigma(\mathbf{A}^+), +\infty)$, *then*

$$\mathbf{f}^+(\mu) = \int_{-\infty}^{\infty} (\lambda - \mu)^{-2} \operatorname{shift}^+(\lambda)\, d\lambda\,.$$

PROOF OF COROLLARY A.3.15. In view of (A.3.18) and (A.3.20) this follows if we insert $\xi(\lambda) = -\operatorname{shift}^{+-}(\lambda)$ into (A.3.21). $\square$

Corollary A.3.15 gives us a method for determining the jumps of $\operatorname{shift}^+(\nu)$.

COROLLARY A.3.16. *For any* $\nu \in \mathbb{R}$

$$\operatorname{shift}^+(\nu + 0) - \operatorname{shift}^+(\nu) = \lim_{\substack{\mu \to \nu \\ \mu \in \mathbb{C} \setminus \sigma(\mathbf{A}^+)}} (\nu - \mu)\, \mathbf{f}^+(\mu)\,.$$

PROOF OF COROLLARY A.3.16. For $\nu < 0$ and $\nu \in \mathbb{R}_+ \setminus \Lambda$ the proof is obvious, so further on we assume $\nu \in \mathbb{R}_+ \cap \Lambda$.

As $\mathbf{f}^+(\overline{z}) = \overline{\mathbf{f}^+(z)}$ it is sufficient to prove

$$\operatorname{shift}^+(\nu + 0) - \operatorname{shift}^+(\nu) = \lim_{\substack{\mu \to \nu \\ \mu \in \mathbb{C} \setminus \sigma(\mathbf{A}^+) \\ \operatorname{Im} \mu \geqslant 0}} (\nu - \mu)\, \mathbf{f}^+(\mu)\,.$$

We know that $\mathbf{f}^+(z)$, $\operatorname{Im} z \geqslant 0$, has a Laurent expansion in fractional powers of $z - \nu$, so it is sufficient to prove the statement with $\mu = \nu + i\varepsilon$. From Corollaries A.3.15 and A.3.9 it follows that

$$|\mathbf{f}^+(\nu + i\varepsilon)| = \varepsilon \left| \int \frac{\operatorname{shift}^+(\nu + \varepsilon s)}{((\varepsilon s + \nu) - \nu - i\varepsilon)^2}\, ds \right| \leqslant \frac{m\pi}{\varepsilon}\,.$$

The Laurent-type behaviour of $\mathbf{f}^+(\nu + i\varepsilon)$ and the above formula imply that the limit $-i\lim_{\varepsilon\to+0} \varepsilon\, \mathbf{f}^+(\nu + i\varepsilon)$ exists.

Let $g_\nu(\lambda)$ be a bounded function on $\mathbb{R}$ which is continuous at ν and satisfies $g_\nu(\nu) = 0$. Clearly

$$\lim_{\varepsilon\to+0} \varepsilon \int_{-\infty}^{\infty} \frac{g_\nu(\lambda)}{(\lambda - \nu - i\varepsilon)^2}\, d\lambda = 0\,.$$

We have

$$\operatorname{shift}^+(\lambda) = h_\nu(\lambda) + g_\nu(\lambda)\,,$$

where

$$h_\nu(\lambda) = \begin{cases} \operatorname{shift}^+(\nu) & \text{for} \quad \lambda \leqslant \nu, \\ \operatorname{shift}^+(\nu + 0) & \text{for} \quad \lambda > \nu, \end{cases}$$

and g_ν has the required properties. Consequently

$$\lim_{\substack{\mu\to\nu \\ \mu\in\mathbb{C}\setminus\sigma(\mathbf{A}^+) \\ \operatorname{Im}\mu\geqslant 0}} (\nu - \mu)\, \mathbf{f}^+(\mu) \;=\; -i \lim_{\varepsilon\to+0} \varepsilon\, \mathbf{f}^+(\nu + i\varepsilon) \;=\; -i \lim_{\varepsilon\to+0} \varepsilon \int_{-\infty}^{\infty} \frac{h_\nu(\lambda)}{(\lambda - \nu - i\varepsilon)^2}\, d\lambda\,.$$

Direct evaluation of the integral on the right proves the corollary. □

LEMMA A.3.17. *On each connected component of* $\mathbb{R}_+ \setminus \Lambda$ *for each continuous branch of* $\arg\det(iR(\nu))$ *there exists an integer* M *such that*

$$2\pi \operatorname{shift}^+(\nu) \;=\; \arg\det(iR(\nu)) + \pi M\,.$$

PROOF OF LEMMA A.3.17. By Proposition A.2.18 the scattering matrix for the pair $\mathbf{A}^{+-}$, $\mathbf{A}$ is the matrix from Lemma A.2.10, and consequently

$$\det S(\nu) = (-1)^m \det{}^2 R(\nu) = \det{}^2(iR(\nu))\,.$$

In view of formulae (A.3.16), (A.3.20) and Proposition A.3.13 we obtain

$$\det{}^2(iR(\nu)) \;=\; e^{4\pi i \operatorname{shift}^+(\nu)}$$

for almost all ν. Since $R(\nu)$ and $\operatorname{shift}^+(\nu)$ are smooth functions on $\mathbb{R}_+ \setminus \Lambda$, the required statement follows from this. □

The above result and the fact that $\operatorname{shift}^+(\nu) = N^+(\nu)$ when $\nu \leqslant 0$ makes the following proposition plausible.

PROPOSITION A.3.18. *On each connected component of* $\mathbb{R} \setminus \Lambda_{00}$ *there exists a continuous branch* $\arg_0 \det(iR(\nu))$ *of the argument* $\arg\det(iR(\nu))$ *such that*

$$\operatorname{shift}^+(\nu) = N^+(\nu) + \frac{1}{2\pi} \arg_0 \det(iR(\nu))\,, \qquad \nu \in \mathbb{R} \setminus \Lambda_{00}. \tag{A.3.23}$$

Here $\det(iR(\nu)) := 1$ *for* $\nu < 0$.

Recall that Λ_{00} denotes the set of $\nu \in \Lambda_0$ such that $A(\xi) = \nu$ has multiple real ξ-roots. Recall also that at points $\nu \in \Lambda \setminus \Lambda_{00}$ we define the reflection matrix $R(\nu)$ by continuity (it is, in fact, analytic at such points). Finally, recall that we choose all our functions to be left-continuous. Clearly, under such a convention formula (A.3.23) extends to all real ν.

For the proof of Proposition A.3.18 we shall need several auxiliary results.

DEFINITION A.3.19. We call a point $\nu \in \Lambda_{00}$ a *normal* threshold if the algebraic equation $A(\xi) = \nu$ has only one multiple real root $\zeta = \zeta^*$ and $A''(\zeta^*) \neq 0$.

DEFINITION A.3.20. We call the normal threshold ν *soft* if there is a generalized eigenfunction corresponding to ν of the form

$$v(x) = e^{ix\zeta^*} + w(x), \tag{A.3.24}$$

where $w(x)$ is exponentially decreasing in x, and *rigid* if it has no solution of this form.

LEMMA A.3.21. *Let ν be a normal threshold such that there is a corresponding generalized eigenfunction of the form*

$$v_1(x) = xe^{ix\zeta^*} + ae^{ix\zeta^*} + \sum_{\zeta_k^+ \neq \zeta^*} c_k e^{ix\zeta_k^+(\nu)} + w_1(x), \tag{A.3.25}$$

where w_1 is exponentially decaying. Then ν is rigid.

PROOF OF LEMMA A.3.21. Assume that ν is soft, and let v be the corresponding generalized eigenfunction of the form (A.3.24). Since both v and v_1 satisfy the boundary conditions corresponding to $\mathbf{A}^+$ we have $\mathcal{A}^+(v, v_1) = 0$. But on the other hand, it follows from Corollary A.1.3 that

$$\mathcal{A}^+(v, v_1) = \frac{i}{2} A''(\zeta^*) \neq 0\,,$$

which is a contradiction. □

LEMMA A.3.22. *Let $\nu_0 \in \Lambda_{00}$ be a normal threshold. Then*

$$\det(iR(\nu_0 + 0)) = \pm i \det(iR(\nu_0)),$$

where the plus sign applies if ν_0 is soft, and the minus sign if ν_0 is rigid. In particular, if zero is a simple threshold we have $\det(iR(+0)) = \pm i$.

PROOF OF LEMMA A.3.22. Let us first consider the case $q(\nu_0 + 0) > q(\nu_0)$. Then $\zeta_s^+(\nu_0 + 0) = \zeta_s^-(\nu_0 + 0)$ is our double real root (for some $1 \leqslant s \leqslant q(\nu_0 + 0)$).

Denote for brevity $q = q(\nu_0 + 0)$. Then, of course, $q(\nu_0) = q - 1$.

For simplicity we shall assume that ζ^* is the only multiple ξ-root of the equation $A(\xi) = \nu_0$. Indeed, if we have other multiple ξ-roots they are necessarily nonreal (by the definition of a normal threshold), and multiple nonreal roots can be easily dealt with by a choice of basis described in the proof of Lemma A.1.11.

In what follows we assume that ν and z lie in a small neighbourhood of ν_0, with $\nu \in \mathbb{R}$ and $z \in \mathbb{C}$, $\operatorname{Im} z \geqslant 0$.

For $\nu > \nu_0$ the reflection matrix $R(\nu)$ is a $q \times q$ matrix function given by formula (A.2.3). Denote by $\mathcal{R}(z)$ the analytic continuation of $R(\nu)$, $\nu > \nu_0$, into the upper complex half-plane. Then $R(\nu)$, $\nu < \nu_0$, can be identified with a submatrix of $\mathcal{R}(\nu)$ of size $(q-1) \times (q-1)$.

Naturally, the elements of $\mathcal{R}(z)$ admit power series expansions in fractional powers of $z - \nu_0$. Since $R(\nu)$, $\nu > \nu_0$, is unitary, these expansions cannot contain negative powers. Hence $R(\nu_0)$ is a unitary "submatrix" of the unitary matrix $R(\nu_0 + 0)$, and this in turn implies that $R_{ks}(\nu_0 + 0) = R_{sk}(\nu_0 + 0) = 0$ for $k \neq s$, and that $|R_{ss}(\nu_0 + 0)| = 1$. We conclude that

$$\det(iR(\nu_0 + 0)) = iR_{ss}(\nu_0 + 0) \det(iR(\nu_0)).$$

Consequently, in order to prove the lemma we must prove that $R_{ss}(\nu_0+0)=1$ if the threshold is soft, and that $R_{ss}(\nu_0+0)=-1$ if the threshold is rigid.

To determine $R_{ss}(\nu_0+0)$, we shall study the behaviour of the generalized eigenfunction

$$(A.3.26)\quad \begin{aligned} u_s(\nu,x) &\equiv (\Gamma^+_{\mathrm{in}}(\nu)^* e^{(s)})(x) \\ &= \frac{e^{ix\zeta_s^-(\nu)}}{(-2\pi A'(\zeta_s^-(\nu)))^{1/2}} + \sum_{k=1}^{q} R_{ks}(\nu)\,\frac{e^{ix\zeta_k^+(\nu)}}{(2\pi A'(\zeta_k^+(\nu)))^{1/2}} \\ &\quad + \sum_{k=q+1}^{m} D_{ks}(\nu)\, e^{ix\xi_k^+(\nu)}, \qquad \nu>\nu_0 \end{aligned}$$

(see also Proposition A.2.1), under analytic continuation counterclockwise along a small circle around ν_0. Here $e^{(s)}\in\mathbb{C}^q$ is a vector with entry 1 in the sth position, and zeros everywhere else.

In (A.3.26) we have

$$(A.3.27)\quad D_{ks}(\nu) = -\sum_{r=1}^{m}(Q^+(\nu+i0))^{-1}_{kr}\,\frac{B_r(\zeta_s^-(\nu))}{(-A'(\zeta_s^-(\nu)))^{1/2}}, \qquad q+1\leqslant k\leqslant m.$$

Before proceeding further note that

$$(A.3.28)\quad R_{ks}(\nu) = O(|\nu-\nu_0|^{1/4}), \qquad k\neq s,$$

$$(A.3.29)\quad D_{ks}(\nu) = d_{ks}(\nu-\nu_0)^{-1/4} + O(|\nu-\nu_0|^{1/4}),$$

for some constants d_{ks} as $\nu\to\nu_0$. Indeed, all these functions must have expansions of the form $(\nu-\nu_0)^{-1/4}\sum_{p=-M}^{\infty} c_p(\nu-\nu_0)^{p/2}$, where the factor $(\nu-\nu_0)^{-1/4}$ comes from the square roots in (A.2.3) and (A.3.27). Since we know that the $R_{ks}(\nu)$ are bounded and that $R_{ks}(\nu_0+0)=0$, the estimate (A.3.28) is obvious. To obtain (A.3.29) we note that if some of the $D_{ks}(\nu)$ grew faster than $(\nu-\nu_0)^{-1/4}$ as $\nu\to\nu_0+0$, then $(\nu-\nu_0)^{t/4}u_s(\nu)$, for some $t>1$, would converge to an eigenfunction of $\mathbf{A}^+$ as $\nu\to\nu_0+0$. The latter is impossible since $(\,u_s(\nu),\,u_0\,)_+=0$ (an eigenfunction is orthogonal to all generalized eigenfunctions).

When we continue $u_s(\nu)$ analytically along a small circle around ν_0, $\zeta_s^+(\nu)$ turns into $\zeta_s^-(\nu)$ and vice versa, but all the other roots are single-valued in a neighbourhood of $z=\nu_0$. Our solution is transformed into

$$\begin{aligned} (u_s(\nu,x))_{\mathrm{anal}} &= \frac{e^{ix\zeta_s^+(\nu)}}{i\,(2\pi A'(\zeta_s^+(\nu)))^{1/2}} + (R_{ss}(\nu))_{\mathrm{anal}}\,\frac{e^{ix\zeta_s^-(\nu)}}{i(-2\pi A'(\zeta_s^-(\nu)))^{1/2}} \\ &\quad + \sum_{k\neq s}(R_{ks}(\nu))_{\mathrm{anal}}\,\frac{e^{ix\zeta_k^+(\nu)}}{i\,(2\pi A'(\zeta_k^+(\nu)))^{1/2}} + \sum_{k=q+1}^{m}(D_{ks}(\nu))_{\mathrm{anal}}\,e^{ix\xi_k^+(\nu)}. \end{aligned}$$

This is also a generalized eigenfunction, and furthermore

$$(A.3.30)\quad i(R_{ss}(\nu))_{\mathrm{anal}}\,u_s(\nu,x) + (\,u_s(\nu,x)\,)_{\mathrm{anal}} \equiv 0\,.$$

Indeed, the left-hand side contains no incoming terms and hence the statement follows from Corollary A.1.4 and the fact that $\nu>\nu_0$ is not an eigenvalue. Collecting the coefficients at $e^{ix\zeta_s^+(\nu)}$ and at $e^{ix\xi_k^+(\nu)}$ in (A.3.30), and letting $\nu \to \nu_0 + 0$, we conclude that $(R_{ss}(\nu_0+0))^2 = 1$ and that

$$R_{ss}(\nu_0+0)\, d_{ks} = d_{ks}. \tag{A.3.31}$$

It remains to note that if $R_{ss}(\nu_0+0) = 1$, then $(\nu-\nu_0)^{1/4}u_s$ converges to a generalized eigenfunction of the form (A.3.24) as $\nu \to \nu_0+0$, while if $R_{ss}(\nu_0+0) = -1$, then (A.3.31) implies that $d_{ks} = 0$. In the latter case $(\nu-\nu_0)^{-1/4}u_s(\nu)$ converges to a generalized eigenfunction of the form (A.3.25) as $\nu \to \nu_0+0$. Thus, $R_{ss}(\nu_0+0) = 1$ if the threshold is soft, and $R_{ss}(\nu_0+0) = -1$ if the threshold is rigid. This proves the lemma in the case $q(\nu_0+0) > q(\nu_0)$.

Let us now consider the case $q(\nu_0+0) < q(\nu_0)$. Then $\zeta_{s-1}^+(\nu_0) = \zeta_s^-(\nu_0)$ is our double real root (for some $2 \leqslant s \leqslant q(\nu_0)$).

An argument similar to the one at the beginning of the proof now shows that

$$\det(iR(\nu_0)) = -iR_{s-1\,s}(\nu_0)\det(iR(\nu_0+0)),$$

where the minus sign comes from the rule for the expansion of a determinant along its sth column. Thus the leading oscillatory terms in u_s involving our roots are

$$\frac{e^{ix\zeta_s^-(\nu)}}{(-2\pi A'(\zeta_s^-(\nu)))^{1/2}} + R_{s-1\,s}(\nu)\frac{e^{ix\zeta_{s-1}^+(\nu)}}{(2\pi A'(\zeta_{s-1}^+(\nu)))^{1/2}}.$$

Performing a procedure similar to the one described above we conclude that $R_{s-1\,s}(\nu_0) = 1$ if the threshold is soft, and that $R_{s-1\,s}(\nu_0) = -1$ if the threshold is rigid, which yields the required result. □

LEMMA A.3.23. *Let $\nu_0 \in \mathbb{R} \setminus \Lambda_{00}$. Then*

$$\operatorname{shift}^+(\nu_0+0) - \operatorname{shift}^+(\nu_0) = N^+(\nu_0+0) - N^+(\nu_0). \tag{A.3.32}$$

On the other hand, if $\nu_0 \in \Lambda_{00}$ is a normal threshold, then

$$\operatorname{shift}^+(\nu_0+0) - \operatorname{shift}^+(\nu_0) = N^+(\nu_0+0) - N^+(\nu_0) \pm \frac{1}{4}, \tag{A.3.33}$$

where the plus or minus sign is chosen according to whether the threshold ν is soft or rigid, respectively.

PROOF OF LEMMA A.3.23. We prove (A.3.33) first. We start by discussing the case $q := q(\nu_0+0) > q(\nu_0)$. Then our double real root is $\zeta^* = \zeta_s^+(\nu_0+0) = \zeta_s^-(\nu_0+0)$ for some $1 \leqslant s \leqslant q$. For simplicity we shall assume that the algebraic equation $A(\xi) = \nu_0$ has no multiple nonreal roots. (If there are such roots they can easily be dealt with using a different basis for the exponentially decaying solutions as in the proof of Lemma A.1.11.)

It follows from Corollary A.3.16 and the fact that $\mathbf{f}^+(z)$ has an expansion in fractional powers of $z-\nu_0$ that it suffices to prove that $\lim_{\nu\to\nu_0+0}(\nu_0-\nu)\mathbf{f}^+(\nu+i0)$ equals the right-hand side of (A.3.33). Here

$$\mathbf{f}^+(\nu+i0) = \lim_{\delta\to+0}\int_0^\infty (\mathbf{r}_{\nu+i0}^+(x,x) - \mathbf{r}_{\nu+i0}(x,x))\,\chi(\delta x)\,dx \tag{A.3.34}$$

by Lemma A.3.2.

In what follows we assume that $\nu > \nu_0$ lies in a small neighbourhood of ν_0.

Let $p_0(x,y)$ denote the integral kernel of the orthogonal projection onto the space spanned by the eigenfunctions corresponding to ν_0. We shall prove that, for $x, y \in \mathbb{R}_+$,

$$\mathbf{r}^+_{\nu+i0}(x,y) - \mathbf{r}_{\nu+i0}(x,y) = (\nu_0 - \nu)^{-1} p_0(x,y) + \sum_{j,k=1}^{2} \gamma_{jk}(\nu + i0, x, y), \tag{A.3.35}$$

where the γ_{jk} are given by

$$\gamma_{11}(\nu+i0,x,y) = i \sum_{k,r=1}^{q} R_{kr}(\nu) \frac{e^{i(x\zeta_k^+(\nu) - y\zeta_l^-(\nu))}}{(-A'(\zeta_k^+(\nu))\, A'(\zeta_l^-(\nu)))^{1/2}},$$

$$\gamma_{21}(\nu+i0,x,y) = i\,(2\pi)^{1/2} \sum_{r=1}^{q} \sum_{k=m-q+1}^{m} D_{kr}(\nu) \frac{e^{i(x\xi_k^+(\nu) - y\zeta_r^-(\nu))}}{(-A'(\zeta_l^-(\nu)))^{1/2}},$$

$$\gamma_{12}(\nu+i0,x,y) = i\,(2\pi)^{1/2} \sum_{k=1}^{q} \sum_{p=m-q+1}^{m} \sum_{l=1}^{q} R_{lk}(\nu)\, \overline{D_{pk}(\nu) \frac{e^{i(x\zeta_l^+(\nu) - y\xi_k^-(\nu)))}}{(A'(\zeta_l^+(\nu)))^{1/2}}},$$

$$\gamma_{22}(\nu+i0,x,y) = \sum_{k,l=m-q+1}^{m} H_{kl}(\nu)\, e^{i(x\xi_k^+(\nu) - y\xi_l^-(\nu))}.$$

Here $R(\nu)$ and $D(\nu)$ are as in thc proof of Lemma A.3.22, and $H_{kl}(\nu) = O((\nu-\nu_0)^{-1/2})$ when $\nu \to \nu_0 + 0$.

Clearly we can split the right-hand side in formula (A.3.1), for $z = \nu + i0$, into four parts, the first of which contains terms that are oscillating in x and y, the second exponentially decaying in x and oscillating in y, etc. To identify these pieces we can observe that for $0 < x < y$ and fixed y the kernel $\mathbf{r}^+_z(x,y)$ looks like a generalized eigenfunction for $\mathbf{A}^+$. In particular, the incoming terms in $\mathbf{r}^+_{\nu+i0}(x,y)$ come from $\mathbf{r}_{\nu+i0}(x,y)$ and the part which is oscillating in y is

$$\gamma_{11}(\nu+i0,x,y) + \gamma_{21}(\nu+i0,x,y) = \Gamma^+_{\text{in}}(\nu)^*(a^-(y)),$$

where $a_l^-(y) = i(2\pi)^{1/2} \frac{e^{-iy\zeta_l^-(\nu)}}{(-2\pi A'(\zeta_l^-(\nu)))^{1/2}}$, $1 \leqslant l \leqslant q$ (cf. A.1.4). In view of (A.3.26) this gives γ_{11} and γ_{21}. To derive the formula for γ_{12} we use that $\overline{\mathbf{r}_{\nu+i0}(x,y)} = \mathbf{r}_{\nu-i0}(y,x)$ and, by an argument similar to the one above, we have

$$\begin{aligned}\gamma_{11}(\nu-i0,x,y) &+ \gamma_{21}(\nu-i0,x,y) \\ &= \Gamma^+_{\text{out}}(\nu)^*(a^+(y)) = \Gamma^+_{\text{in}}(\nu)^*(R(\nu))^*(a^+(y)), \qquad x<y,\end{aligned}$$

where $a_l^+(y) = -i(2\pi)^{1/2} \frac{e^{-iy\zeta_l^+(\nu)}}{(2\pi A'(\zeta_l^+(\nu)))^{1/2}}$. Finally, γ_{22} is what remains. It is easy to see that the $H_{kl}(\nu)$ admit Laurent-type expansions in half-integer powers of $\nu - \nu_0$. Suppose that there exists an integer $j \geqslant 2$ such that $H_{kl}(\nu) = O((\nu-\nu_0)^{-j/2})$ for all indices k, l, and $\lim_{\nu\to\nu_0+0}(\nu-\nu_0)^{j/2} H_{kl}(\nu) \neq 0$ for some k, l. Then

$$g_y(x) = \lim_{\nu\to\nu_0+0} (\nu-\nu_0)^{j/2} (\mathbf{r}^+_{\nu+i0}(x,y) - (\nu_0-\nu)^{-1} p_0(x,y)) \tag{A.3.36}$$

would be an eigenfunction corresponding to the eigenvalue ν_0 (here we use that $\lim_{\nu\to\nu_0+0}(\nu-\nu_0)^{j/2}\mathbf{r}_{\nu+i0}(x,y)=0$). From (A.3.36) we get

$$\begin{aligned} 0 &\neq \int_0^{+\infty} |g_y(x)|^2\,dx \\ &= \lim_{\nu\to\nu_0+0}(\nu-\nu_0)^{j/2}\int_0^{+\infty}(\mathbf{r}^+_{\nu+i0}(x,y)-(\nu_0-\nu)^{-1}p_0(x,y))\,\overline{g_y(x)}\,dx\,. \end{aligned} \tag{A.3.37}$$

But the spectral theorem implies

$$\int_0^{+\infty}(\mathbf{r}^+_z(x,y)-(\nu_0-z)^{-1}p_0(x,y))\,\overline{g_y(x)}\,dx \;=\; 0$$

for $z\in\mathbb{C}\setminus\mathbb{R}$, so

$$\int_0^{+\infty}(\mathbf{r}^+_{\nu+i0}(x,y)-(\nu_0-\nu)^{-1}p_0(x,y))\,\overline{g_y(x)}\,dx \;=\; 0\,.$$

Substituting the latter into (A.3.37) we obtain a contradiction.

It follows from formulae (A.3.34), (A.3.35) and our estimates for the γ_{jk} that

$$\begin{aligned} \mathbf{f}^+(\nu+i0) &= (\nu_0-\nu)^{-1}(N^+(\nu_0+0)-N^+(\nu_0)) \\ &\quad - R_{ss}(\nu)\,\frac{1}{(-A'(\zeta_s^+(\nu))A'(\zeta_s^-(\nu)))^{1/2}}\cdot\frac{1}{\zeta_s^+(\nu)-\zeta_s^-(\nu)} \\ &\quad + O((\nu-\nu_0)^{-3/4})\,. \end{aligned} \tag{A.3.38}$$

Consequently,

$$\begin{aligned} &\lim_{\nu\to\nu_0+0}(\nu_0-\nu)\,\mathbf{f}^+(\nu+i0) \\ &\quad = N^+(\nu_0+0)-N^+(\nu_0) \\ &\qquad - R_{ss}(\nu_0+0)\lim_{\nu\to\nu_0+0}\frac{(\nu_0-\nu)}{(-A'(\zeta_s^+(\nu))A'(\zeta_s^-(\nu)))^{1/2}}\cdot\frac{1}{\zeta_s^+(\nu)-\zeta_s^-(\nu)}\,. \end{aligned} \tag{A.3.39}$$

If we now use the fact that

$$\zeta_s^\pm(\nu) \;=\; \zeta^*\pm\left(\frac{2(\nu-\nu_0)}{A''(\zeta^*)}\right)^{1/2} + O(\nu-\nu_0) \qquad \text{as} \qquad \nu\to\nu_0+0,$$

we obtain

$$\begin{aligned} &(-A'(\zeta_s^+(\nu))A'(\zeta_s^-(\nu)))^{1/2}\cdot(\zeta_s^+(\nu)-\zeta_s^-(\nu)) \\ &\quad = \left(A''(\zeta^*)^2\left|\frac{2(\nu-\nu_0)}{A''(\zeta^*)}\right|\right)^{1/2}\cdot 2\left|\frac{2(\nu-\nu_0)}{A''(\zeta^*)}\right|^{1/2} + O((\nu-\nu_0)^{3/2}) \\ &\quad = 4(\nu-\nu_0)\,+\,O((\nu-\nu_0)^{3/2})\,. \end{aligned}$$

This implies

$$\lim_{\nu\to\nu_0+0}(\nu_0-\nu)\,\mathbf{f}^+(\nu+i0) \;=\; N^+(\nu_0+0)-N^+(\nu_0)\,+\,\frac14\,R_{ss}(\nu_0+0),$$

where $R_{ss}(\nu_0+0)=1$ if ν_0 is soft, and $R_{ss}(\nu_0+0)=-1$ otherwise (see the proof of Lemma A.3.22).

The proof of (A.3.33) is now complete in the case $q(\nu_0+0)>q(\nu_0)$. In the case $q(\nu_0)>q(\nu_0+0)$ the double real root is $\zeta^*=\zeta_{s-1}^+(\nu_0)=\zeta_s^-(\nu_0)$ for some

$2 \leqslant s \leqslant q(\nu_0)$. A calculation similar to the one above shows that the jump of shift$^+$ is

$$N^+(\nu_0+0) - N^+(\nu_0) \\ - R_{s-1\,s}(\nu_0) \lim_{\nu\to\nu_0-0} \frac{(\nu_0-\nu)}{(-A'(\zeta^+_{s-1}(\nu))A'(\zeta^-_s(\nu)))^{1/2}} \cdot \frac{1}{\zeta^+_{s-1}(\nu)-\zeta^-_s(\nu)}$$

(cf. the proof of Lemma A.3.22) . Comparing the expression under the limit sign with the one in the second term on the right in (A.3.39), we see that both the numerator and the second factor in the denominator have changed signs, whereas the first factor in the denominator is still positive. Carrying out the calculations we see that the statement of the lemma remains valid in this case.

Finally, let us prove (A.3.32). Now $\nu_0 \notin \Lambda_{00}$ and we can repeat our argument along the lines of (A.3.34), (A.3.35) in a simplified form (there are no double real roots). This gives

$$\mathbf{f}^+(\nu+i0) = (\nu_0-\nu)^{-1}(N^+(\nu_0+0)-N^+(\nu_0)) + O(1)$$

as $\nu \to \nu_0$ (cf. (A.3.38)), and we easily obtain the required statement. □

PROOF OF PROPOSITION A.3.18. It follows from Lemma A.3.17 that the function

$$U(\nu) \,:\, \mathbb{R}\setminus\Lambda \ni \nu \mapsto \det(iR(\nu))e^{-2\pi i\,\mathrm{shift}^+(\nu)}$$

is locally constant and may take only values 1 or -1. We know also that $U(\nu)=1$ when $\nu \leqslant 0$. Moreover, combining Lemmas A.3.22 and A.3.23 we see that $U(t)\equiv 1$ if Λ_{00} consists only of normal thresholds. It remains to prove that $U(\nu_0+0) = U(\nu_0-0)$ when ν_0 is not a normal threshold.

Let t be a small real parameter. It is easy to see that one can find continuous families of polynomials $A_t(\xi)$, $B_{j,t}(\xi)$ such that the corresponding operators $\mathbf{A}^+_t$ are self-adjoint, $\mathbf{A}^+_0 = \mathbf{A}^+$, and each $\mathbf{A}^+_t$, $t\neq 0$, has only normal thresholds in a neighbourhood of ν_0. Let us choose a $\delta>0$ such that $[\nu_0-\delta,\nu_0+\delta]\cap\Lambda=\{\nu_0\}$. Then, for sufficiently small t the points $\nu_0-\delta$, $\nu_0+\delta$ are outside the set Λ_t. Clearly, the reflection matrices $R_t(\nu_0+\delta)$ and $R_t(\nu_0-\delta)$ are continuous in t at the point $t=0$ (since the right-hand side in formula (A.2.3) depends continuously on the polynomials A_t, $B_{j,t}$). Similarly, the quantities $\mathrm{shift}^+_t(\nu_0\pm\delta)$ are continuous in t at the point $t=0$; see the proof of Lemma A.4.1 below. Since every $\mathbf{A}^+_t$, $t\neq 0$, has only normal thresholds in the interval $[\nu_0-\delta,\nu_0+\delta]$, it follows that $U(\nu_0+\delta) = \lim_{t\to 0} U_t(\nu_0+\delta) = \lim_{t\to 0} U_t(\nu_0-\delta) = U(\nu_0-\delta)$. □

A.4. Dependence on parameters

The purpose of this section is to derive some estimates needed in Section 5.4 in the main text of this book. We shall assume that we have a family $\mathbf{A}^+_s$, $s\in\Omega\subset\mathbb{R}^n$, of operators of the form discussed in this appendix which depend smoothly on s in the sense that the polynomials $A_s(\xi)$, $B_{s,j}(\xi)$ are smooth functions of s; here Ω is some small (open) neighbourhood of the origin. In contrast to the previous sections we shall assume that $\inf_{\xi\in\mathbb{R}} A_s(\xi)\geqslant 0$, and that, moreover, $\sigma(\mathbf{A}^+_s)\subset\mathbb{R}_+$ for all $s\in\Omega$. Clearly, the spectrum $\sigma(\mathbf{A}^+_s)$ depends continuously on s.

LEMMA A.4.1. *Let $\nu_0 \in \mathbb{R}_+$ be such that the equation $A_0(\xi) = \nu_0$ has no multiple real ξ-roots (i.e. ν_0 is not a threshold) and ν_0 is not an eigenvalue of $\mathbf{A}_0^+$. Then there is an (open) neighbourhood $O \subset \Omega$ of the origin and an $\varepsilon > 0$ such that $\mathrm{shift}_s^+(\nu)$ (the spectral shift function corresponding to $\mathbf{A}_s^+$) is a smooth function of (s, ν) on $\mathcal{O} = O \times (\nu_0 - \varepsilon, \nu_0 + \varepsilon)$.*

PROOF OF LEMMA A.4.1. It follows from Proposition A.3.7 that

$$\mathrm{shift}_s^+(\nu) = -\frac{1}{2\pi i} \oint_{\gamma_\nu} \mathbf{f}_s^+(z)\, dz\,, \tag{A.4.1}$$

where $\mathbf{f}_s^+(z) = \mathrm{Tr}(\mathbf{R}_z^+(s) - \mathbf{R}_z(s))$ and γ_ν satisfies the conditions stated before Lemma A.3.4 and depends smoothly on ν. If $A_0(\xi) = \nu_0$ has no multiple ξ-roots, we can choose γ_ν in such a way that $\gamma_\nu \cap \Lambda_s = \varnothing$ when $(s, \nu) \in \mathcal{O}$ and $\mathcal{O}$ is sufficiently small. It then follows easily from (A.3.2) and (A.1.10) that $\mathbf{f}_s^+(z)$ is a smooth function of (s, z) near our curves. In view of (A.4.1) this gives the required result. If $A_0(\xi) = \nu_0$ has multiple nonreal ξ-roots, it is still true that $f_s^+(z)$ is a smooth function of (s, z). This can be seen by deriving an analogue of (A.3.2) using a different basis for the exponentially decaying solutions of $A_s(\xi) = z$, as in the proof of Lemma A.1.11. □

We shall denote a pair $(s, z) \in \Omega \times \mathbb{C}$, $z \notin \sigma(\mathbf{A}_s^+)$, by w. Set $T(w; x, y) = \mathbf{r}_z^+(s, x, y) - \mathbf{r}_z(s, x, y)$. Following the discussion in A.1.4 we have $T(w; x, y) = \sum_{k=1}^m a_k(w, y)\, f_{k,w}(x)$, where

$$f_{k,w}(x) = \int_{-\infty}^{\infty} e^{ix\xi} \frac{\xi^{k-1}}{A_s(\xi) - z}\, d\xi\,.$$

Moreover, by symmetry we have $T(w; x, y) = \sum_{l=1}^m b_l(w, x)\, f_{l,w}(-y)$.

We conclude that

$$T(w; x, y) = \sum_{k,l=1}^m F_{kl}(w)\, f_{k,w}(x)\, f_{l,w}(-y)\,.$$

It is easy to see that $f_{k,w}(x)$, $k = 1, \ldots, m$, are linearly independent on the positive and on the negative semi-axes when $z \notin \sigma(\mathbf{A}_s)$, and that they are uniformly exponentially decaying as $x \to \pm\infty$ when w belongs to a compact set such that the z-component does not intersect $\sigma(\mathbf{A}_s)$.

Throughout this section we shall denote by χ a C^∞ function on $\mathbb{R}_+$ which vanishes identically in a neighbourhood of the origin and which is constant for sufficiently large x. We shall denote the resolvents of $\mathbf{A}_s^+$ and $\mathbf{A}_s$ (the corresponding "free" operator) by $\mathbf{R}_z^+(s)$ and $\mathbf{R}_z(s)$, respectively. The integral kernels of the latter will be denoted by $\mathbf{r}_z^+(s, x, y)$ and $\mathbf{r}_z(s, x, y)$.

LEMMA A.4.2. *Let $z_0 \notin \sigma(\mathbf{A}_0^+)$. Then there is an (open) neighbourhood $W \subset \Omega \times \mathbb{C}$ of the point $(0, z_0)$ such that*

$$\int_0^{+\infty} \chi(x/X)\, (\mathbf{r}_z^+(s, x, x) - \mathbf{r}_z(s, x, x))\, dx = O(X^{-\infty})$$

uniformly on W as $X \to +\infty$.

PROOF OF LEMMA A.4.2. In view of the uniform exponential decay of the functions $f_{k,w}(x)\, f_{l,w}(-x)$ it only remains to verify that $F_{kl}(w)$, $1 \leqslant k, l \leqslant m$, are uniformly bounded when w is sufficiently close to $(0, z_0)$. For $1 \leqslant q, r \leqslant m$ we have

$$\text{(A.4.2)}\qquad \sum_{k,l=1}^{m} G_{qk}^{+}(w)\, F_{kl}(w)\, G_{lr}^{-}(w) \;=\; \int_0^\infty \int_0^\infty \overline{f_{q,w}(x)}\, T(w; x, y)\, \overline{f_{r,w}(-y)}\, dx\, dy,$$

where $G_{qk}^{+}(w) = \int_0^\infty \overline{f_{q,w}(x)}\, f_{k,w}(x)\, dx$, $G_{lr}^{-}(w) = \int_0^\infty f_{l,w}(-y)\, \overline{f_{r,w}(-y)}\, dy$, and the modulus of the right-hand side is bounded by

$$\|\, f_{q,w}\, \|_{+}\, \|\, \mathbf{R}_z^{+}(s) - \mathbf{R}_z(s)\, \|\, \|\, f_{r,w}\, \|_{-}\,,$$

which, in turn, is uniformly bounded when w is sufficiently close to $(0, z_0)$. Since the matrix functions $G^{\pm}(w)$ are continuous and positive, we now conclude from (A.4.2) that the components of $F(w)$ are uniformly bounded in a sufficiently small neighbourhood of $(0, z_0)$. □

For small $\delta > 0$ we set $L_\delta = \{z = e^{i\varphi} : \delta \leqslant \varphi \leqslant 2\pi - \delta\}$.

LEMMA A.4.3. *Let $p > 1/2$. Then there is an (open) neighbourhood $O \subset \Omega$ of the origin such that*

$$\text{(A.4.3)}\qquad \int_0^{+\infty} \chi(x/X)\, (\mathbf{r}_z^{+}(s, x, x) - \mathbf{r}_z(s, x, x))\, dx \;=\; O(\delta^{+\infty})$$

uniformly in $z \in L_\delta$, $s \in O$ and $X \geqslant \delta^{-p}$, as $\delta \to +0$.

PROOF OF LEMMA A.4.3. We may restrict our attention to $z \in L_\delta \setminus L_{\delta_0}$ for some fixed small $\delta_0 > 0$. Indeed, we can split L_{δ_0} into a finite number of parts on which we may apply Lemma A.4.2.

Set

$$f_{k,w}^{\pm}(x) = \sum_{j=1}^{m} ((G^{\pm}(w))^{-1/2})_{kj}\, f_{j,w}(\pm x), \qquad 1 \leqslant k \leqslant m, \qquad x > 0\,.$$

Then $f_{k,w}^{+}(x)$, $1 \leqslant k \leqslant m$, and $f_{k,w}^{-}(x)$, $1 \leqslant k \leqslant m$, form orthonormal systems in $L_2(\mathbb{R}_+)$, which depend continuously on w, and satisfy the differential equation $(A_s(\pm D) - z)u = 0$ on $\mathbb{R}_+$. The solutions of the initial-value problems

$$\begin{aligned} (A_s(\pm D) - z)u &= 0 \quad \text{on } \mathbb{R}_+,\\ (D^{r-1}u)(+0) &= a_r^{\pm}\,, \qquad 1 \leqslant r \leqslant 2m\,, \end{aligned}$$

are linear in $a^{\pm}$ and continuous in z and (small) s. Since

$$\int_0^1 |f_{k,w}^{\pm}(x)|^2\, dx \;<\; \|\, f_{k,w}^{\pm}\, \|_{+}^{2} \;=\; 1\,,$$

it is therefore easy to see that the Cauchy data $a_{kr}^{\pm}(w) = (D^{r-1} f_{k,w}^{\pm})(+0)$, $1 \leqslant r \leqslant 2m$, are uniformly bounded (independently of δ) when $w \in O \times L_\delta$, if O is sufficiently small. We conclude that

$$f_{k,w}^{\pm}(x) \;=\; \int_{-\infty}^{\infty} e^{\pm i x \xi}\, \frac{p_{k,w}^{\pm}(\xi)}{A_s(\xi) - z}\, d\xi\,,$$

where $p^\pm_{k,w}(\xi) = \sum_{r=1}^{2m} a^\pm_{kr}(w)\xi^{r-1}$ have uniformly bounded coefficients, and the integrands have no poles in $\{\pm \operatorname{Im} z \leqslant 0\}$. Moreover,

$$T(w;x,y) = \sum_{k,l=1}^{m} \widetilde{F}_{kl}(w)\, f^+_{k,w}(x)\, f^-_{l,w}(y)\,,$$

where $\widetilde{F}_{kl}(w) = ((\mathbf{R}^+_z(s) - \mathbf{R}_s(s))\,\overline{f^-_{l,w}},\, f^+_{k,w})_+$. Clearly,

$$|\,\widetilde{F}_{kl}(w)\,| \leqslant \|\, \mathbf{R}^+_z(s) - \mathbf{R}_s(s)\,\| = O(\delta^{-1})\,.$$

Consequently it only remains to verify that

$$\int_0^\infty \chi(x/X)\, f^+_{k,w}(x)\, f^-_{l,w}(x)\, dx = O(\delta^\infty)\,, \tag{A.4.4}$$

uniformly over $O \times (L_\delta \setminus L_{\delta_0})$, $X \geqslant \delta^{-p}$, and $k,l = 1,\dots,m$.

We have

$$\begin{aligned}
&\int_0^\infty \chi(x/X)\, f^+_{k,w}(x)\, f^-_{l,w}(x)\, dx \\
&\quad = \int_0^\infty \chi(x/X) \left(\oint_{\gamma_+(w)} \oint_{-\gamma_-(w)} e^{ix(\xi-\eta)} \frac{p^+_{k,w}(\xi)\, p^-_{l,w}(\eta)}{(A_s(\xi)-z)\,(A_s(\eta)-z)}\, d\eta\, d\xi \right) dx \\
&\quad = \oint_{\gamma_+(w)} \oint_{-\gamma_-(w)} \left(\int_0^\infty \chi(x/X)\, e^{ix(\xi-\eta)}\, dx \right) \frac{p^+_{k,w}(\xi)\, p^-_{l,w}(\eta)}{(A_s(\xi)-z)\,(A_s(\eta)-z)}\, d\eta\, d\xi
\end{aligned}$$

for any simple, closed and positively oriented curves $\gamma_\pm(w) \subset \{\pm \operatorname{Im} z > 0\}$ which encircle all the ξ-roots of $A_s(\xi) = z$ in $\{\pm \operatorname{Im} \xi > 0\}$, respectively. We shall denote these roots by $\xi^\pm_k(w)$, $k = 1,\dots,m$.

We shall prove below that there is a constant $C_1 > 0$ such that

$$\inf_{s\in O} \min_{k,l} |\, \xi^+_k(s,z) - \xi^-_l(s,z)\,| \geqslant C_1\, |\operatorname{Im} z|^{1/2}\,, \qquad |z| = 1\,, \tag{A.4.5}$$

and it is obvious that there is a constant $C_2 > 0$ such that $|\operatorname{Im} \xi^\pm_k(s,z)| \geqslant C_2 |\operatorname{Im} z|$. This will imply that we can choose the curves $\gamma_\pm(w)$ in such a way that their lengths are bounded and for all $\xi \in \gamma_+(w)$, $\eta \in \gamma_-(w)$ and $k = 1,\dots,m$ we have

$$|\,\xi - \eta\,| \geqslant C_3\, \delta^{1/2}\,, \qquad |\,\xi - \xi^+_k(w)\,| \geqslant C_4\, \delta\,, \qquad |\,\eta - \xi^-_k(w)\,| \geqslant C_4\, \delta$$

with some positive constants C_3, C_4. The latter inequalities and the fact that $\xi^\pm_k(w)$ are uniformly bounded will imply the uniform estimates

$$\left| \frac{p^+_{k,w}(\xi)\, p^-_{l,w}(\eta)}{(A_s(\xi)-z)\,(A_s(\eta)-z)} \right| \leqslant C_5\, \delta^{-2m}\,, \tag{A.4.6}$$

$\xi \in \gamma_+(w)$, $\eta \in \gamma_-(w)$, $k,l = 1,\dots,m$, with some positive constant C_5. Integrating by parts we will conclude that

$$\int_0^\infty \chi(x/X)\, e^{ix(\xi-\eta)}\, dx = O\left(\left(\frac{1}{X(\xi-\eta)} \right)^{l_1} \right) = O(\, \delta^{l_1(p-\frac{1}{2})}\,)$$

for any $l_1 \in \mathbb{N}$, uniformly over $\xi \in \gamma_+(w)$, $\eta \in \gamma_-(w)$. This together with (A.4.6) and the uniform bounds for $\xi^\pm_k(w)$ will imply (A.4.4).

In order to derive (A.4.5) we consider two roots $\xi^+(w)$ and $\xi^-(w)$ of $A_s(\xi) = z$ such that $\operatorname{Im}\xi^+(w) > 0 > \operatorname{Im}\xi^-(w)$. We let $\xi^*(w) \in \mathbb{R}$ denote the point where the line segment $[\xi^-(w), \xi^+(w)]$ intersects the real axis. Clearly

$$(A.4.7) \qquad |\operatorname{Im} z| \leqslant |A_s(\xi^*) - z| = \prod_{1 \leqslant k \leqslant 2m} |\xi^* - \xi_k(w)|$$

where $\xi_k(w)$, $1 \leqslant k \leqslant 2m$, are all the roots of $A_s(\xi) = z$. Obviously, $|\xi_k(w)| \leqslant C_6$, so the inequality (A.4.7) implies that

$$(2C_6)^{2m-2} |\xi^+(w) - \xi^*| |\xi^-(w) - \xi^*| \geqslant |\operatorname{Im} z|.$$

Since

$$|\xi^+(w) - \xi^-(w)| = |\xi^+(w) - \xi^*(w)| + |\xi^*(w) - \xi^-(w)|,$$

we obtain (A.4.5). □

Appendix B

Fourier Tauberian Theorems

Michael Levitin

The objective of this appendix is to formulate and prove Fourier Tauberian theorems as theorems of **classical analysis** without any reference to partial differential equations, spectral theory, etc.

The notion of a *Tauberian theorem* covers a wide range of different mathematical results; see the history of the subject in [**Ga**]. These results have the following in common. Suppose that we have some mathematical object with highly irregular behaviour (say, a discontinuous function or a divergent series) and suppose that we apply some averaging procedure which makes our object substantially more regular (say, a transformation which turns our discontinuous function into an infinitely smooth one or makes our divergent series absolutely convergent). A Tauberian theorem in our understanding is a mathematical result which recovers properties of the original irregular object from the properties of the averaged object.

Tauberian theorems described in this appendix are associated mainly with the Fourier transform. We give four main results — Theorems B.2.1, B.3.1, B.4.1 and B.5.1. Theorem B.2.1 essentially repeats the original Fourier Tauberian Theorem of B. M. Levitan; see also [**Hö3,** vol. 3, Lemma 17.5.6]. Theorem B.3.1 provides a useful rough estimate. Theorem B.4.1 is a refined (with improved remainder estimate) theorem of the type introduced by Safarov [**Sa2**]–[**Sa6**] for studying general (i.e., not necessarily polynomial) spectral asymptotics. Finally, Theorem B.5.1 is a special version of Theorem B.4.1 specifically designed for studying polynomial two-term spectral asymptotics; a theorem of this sort was implicitly used by Duistermaat and Guillemin in their pioneering paper [**DuiGui**].

Hereinafter in this appendix we use the "hat" to denote the Fourier transform of a function. By a prime we denote the derivative.

B.1. Introductory remarks

It is well known that, under certain conditions, the asymptotic behaviour of a function at infinity is determined by the singularities of its Fourier transform. We illustrate this fact by the following elementary example.

EXAMPLE B.1.1. Let $\hat f(t)$ be a complex-valued function on $\mathbb{R}$ which is infinitely smooth outside the point $t=0$, at which the function $\hat f(t)$ together with all its derivatives has finite left and right limits. We denote

$$\delta_k = \hat f^{(k)}(+0) - \hat f^{(k)}(-0), \qquad k = 0,1,\dots .$$

Supported by a Royal Society grant.

If $\hat{f}(t)$ and all its derivatives vanish faster than any given negative power of $|t|$ as $t \to \infty$, then the inverse Fourier transform $f(\lambda) = \mathcal{F}^{-1}_{t\to\lambda}[\hat{f}(t)]$ admits the following asymptotic expansion:

$$f(\lambda) \sim \frac{1}{2\pi} \sum_{k=0}^{\infty} \frac{\delta_k}{(-i\lambda)^{k+1}}, \qquad \lambda \to \infty. \tag{B.1.1}$$

We shall consider a more complicated situation. First, we allow $\hat{f}$ to be a distribution. Second, we allow $\hat{f}$ to have singularities not only at the origin. Third, we assume that information on $\hat{f}(t)$ is given only on a finite time interval rather than for all $t \in \mathbb{R}$. Under these assumptions it is impossible to construct a full asymptotic expansion of the type (B.1.1). Nevertheless, it turns out that for a monotone function f one can still relate the behaviour of $f(\lambda)$ at infinity to the singularities of its Fourier transform. This relation is the subject of the Fourier Tauberian theorems formulated in this appendix.

We shall denote by F_+ the class of real-valued monotone nondecreasing functions N on $\mathbb{R}$ such that $N(\lambda) = 0$ for $l \leqslant 0$, and $\lambda^{-p} N(\lambda) \to 0$ as $\lambda \to +\infty$, where p is some positive number depending on the particular function N. The latter condition can be rewritten as

$$N(\lambda) = o(\lambda^p), \qquad \lambda \to +\infty. \tag{B.1.2}$$

Let us fix a real-valued function ρ on $\mathbb{R}$ satisfying the following five conditions:
1. $\rho \in \mathcal{S}(\mathbb{R})$.
2. $\rho(\lambda) > 0$ for all $\lambda \in \mathbb{R}$.
3. $\hat{\rho}(0) = \int \rho(\lambda)\, d\lambda = 1$.
4. $\operatorname{supp} \hat{\rho}$ is compact.
5. The function $\rho(\lambda)$ is even.

Such a function ρ exists (see, for instance, [**Hö3**, Vol. 3, Sect. 17.5]). Note that under conditions (1)–(5) the Fourier transform $\hat{\rho}$ is a real even function.

We shall denote $\rho_T(\lambda) := T\rho(T\lambda)$, $\hat{\rho}_T(t) := \hat{\rho}(t/T)$, where T is a positive parameter. Obviously, if the function ρ satisfies the five conditions stated above, then ρ_T satisfies these conditions as well.

Throughout this appendix ν will denote some real number, not necessarily positive. All subsequent asymptotics in this appendix are written with respect to $\lambda \to +\infty$. Of course, asymptotics with respect to $\lambda \to -\infty$ are of no interest because for all $N \in F_+$ we have $N(\lambda) \equiv 0$ and $(N * \rho)(\lambda) = O(|\lambda|^{-\infty})$ as $\lambda \to -\infty$.

B.2. Basic theorem

In this section we will be using the notation C for various positive constants which may depend only on the choice of the function ρ and on the number ν but are independent of the function N.

THEOREM B.2.1. *If $N \in F_+$, and*

$$(N' * \rho)(\lambda) \leqslant \lambda^{\nu}, \qquad \forall \lambda \geqslant 1, \tag{B.2.1}$$

then

$$|N(\lambda) - (N * \rho)(\lambda)| \leqslant C\lambda^{\nu}, \qquad \forall \lambda \geqslant 1. \tag{B.2.2}$$

By an elementary rescaling argument Theorem B.2.1 immediately implies

COROLLARY B.2.2. *If* $N \in F_+$ *and* $(N' * \rho)(\lambda) = O(\lambda^\nu)$, *then*

$$N(\lambda) \;=\; (N * \rho)(\lambda) \;+\; O(\lambda^\nu)\,. \tag{B.2.3}$$

Moreover, if the estimate $(N' * \rho)(\lambda) = O(\lambda^\nu)$ *holds uniformly on some subset of* F_+ *, then* (B.2.3) *is also uniform.*

Clearly, formula (B.2.3) can be rewritten as

$$N(\lambda) \;=\; \mathcal{F}^{-1}_{t\to\lambda}[\hat\rho(t)\hat N(t)] + O(\lambda^\nu)\,,$$

which means that, as promised, we have established the relation between the behaviour of $N(\lambda)$ at infinity and the behaviour of $\hat N(t)$ in the neighbourhood of the origin.

The proof of Theorem B.2.1 is based on the following technical lemma.

LEMMA B.2.3. *Under the conditions of Theorem* B.2.1

$$|N(\lambda+s) - N(\lambda)| \leqslant C(1+|s|)^{1+|\nu|}\lambda^\nu\,, \qquad \forall\lambda \geqslant 1\,, \tag{B.2.4}$$

uniformly over all $s \in \mathbb{R}$.

PROOF. It is sufficient to proof (B.2.4) in the case when λ and s are integers. Indeed, the general case is reduced to this one by perturbing λ and s to one of their two nearest integers and using the monotonicity of N . Note also that the case $s = 0$ is trivial. Thus, in order to prove the lemma it is sufficient to establish the following two estimates:

$$N(\lambda+s) - N(\lambda) \leqslant Cs^{1+|\nu|}\lambda^\nu\,, \qquad \forall\lambda, s \in \mathbb{N}\,, \tag{B.2.5}$$

$$N(\lambda) - N(\lambda-s) \leqslant Cs^{1+|\nu|}\lambda^\nu\,, \qquad \forall\lambda, s \in \mathbb{N}\,. \tag{B.2.6}$$

We have

$$\begin{aligned}(N' * \rho)(\lambda) \;&=\; \int \rho(\lambda-\mu)\,dN(\mu) \;\geqslant\; C\int_{\lambda-1}^{\lambda+1} dN(\mu)\\ &=\; C\big((N(\lambda+1) - N(\lambda)) + (N(\lambda) - N(\lambda-1))\big)\end{aligned} \tag{B.2.7}$$

(here we have used the fact that $\rho(\Lambda) \geqslant C > 0$ for $\Lambda \in [-1,1]$). Formulae (B.2.1), (B.2.7) imply

$$N(\lambda+1) - N(\lambda) \leqslant C\,\lambda^\nu\,, \qquad \forall\lambda \in \mathbb{N}\,, \tag{B.2.8}$$

$$N(\lambda) - N(\lambda-1) \leqslant C\,\lambda^\nu\,, \qquad \forall\lambda \in \mathbb{N}\,. \tag{B.2.9}$$

Adding up the inequalities (B.2.8) we get

$$\begin{gathered}N(\lambda+s) - N(\lambda) = \sum_{k=1}^{s}(N(\lambda+k) - N(\lambda+k-1))\\ \leqslant C\sum_{k=1}^{s}(\lambda+k-1)^\nu \leqslant \begin{cases} Cs(\lambda+s-1)^\nu & \text{if } \nu \geqslant 0\,,\\ Cs\lambda^\nu & \text{if } \nu < 0\,.\end{cases}\end{gathered}$$

Since in the case $\nu \geqslant 0$ we have $(\lambda+s-1)^\nu \leqslant s^\nu\lambda^\nu$, the above formula implies (B.2.5).

We now prove (B.2.6). It is sufficient to prove (B.2.6) for $s \leqslant \lambda$ because for $s > \lambda$ we have $N(\lambda) - N(\lambda - s) = N(\lambda) - N(\lambda - \lambda)$. So further on $s \leqslant \lambda$. Adding up the inequalities (B.2.9) we get

$$N(\lambda) - N(\lambda - s) = \sum_{k=1}^{s} (N(\lambda - k + 1) - N(\lambda - k))$$

$$\leqslant C \sum_{k=1}^{s} (\lambda - k + 1)^{\nu} \leqslant \begin{cases} Cs\lambda^{\nu} & \text{if } \nu \geqslant 0, \\ Cs(\lambda - s + 1)^{\nu} & \text{if } \nu < 0. \end{cases}$$

Since in the case $\nu < 0$ we have $(\lambda - s + 1)^{\nu} \leqslant s^{|\nu|}\lambda^{\nu}$, the above formula implies (B.2.6). □

PROOF OF THEOREM B.2.1. Using (B.2.4) we obtain

$$|(N * \rho)(\lambda) - N(\lambda)| = \left| \int (N(\lambda - \mu) - N(\lambda))\, \rho(\mu)\, d\mu \right|$$

$$\leqslant \int |N(\lambda - \mu) - N(\lambda)|\, \rho(\mu)\, d\mu \leqslant C\, \lambda^{\nu} \int (1 + |\mu|)^{1+|\nu|}\, \rho(\mu)\, d\mu = C\, \lambda^{\nu}$$

for all $\lambda \geqslant 1$. □

The following is a weighted version of Theorem B.2.1.

THEOREM B.2.4. *Let* $N \in F_+$ *and*

(B.2.10) $$(N' * \rho_T)(\lambda) \leqslant d\, \lambda^{\nu}, \qquad \forall \lambda \geqslant T^{-1},$$

where $d > 0$ *and* $T > 0$ *are parameters. Then*

(B.2.11) $$|N(\lambda) - (N * \rho_T)(\lambda)| \leqslant C\, d T^{-1} \lambda^{\nu}, \qquad \forall \lambda \geqslant T^{-1}.$$

Here the constant C *is the same as in Theorem* B.2.1.

PROOF. Set $\widetilde{\lambda} := T\lambda$, $\widetilde{N}(\widetilde{\lambda}) := d^{-1} T^{1+\nu} N(\widetilde{\lambda}/T)$. Then (B.2.10) is equivalent to

$$(\widetilde{N}' * \rho)(\widetilde{\lambda}) \leqslant \widetilde{\lambda}^{\nu}, \qquad \forall \widetilde{\lambda} \geqslant 1.$$

Therefore by Theorem B.2.1

$$|\widetilde{N}(\widetilde{\lambda}) - (\widetilde{N} * \rho)(\widetilde{\lambda})| \leqslant C\, \widetilde{\lambda}^{\nu}, \qquad \forall \widetilde{\lambda} \geqslant 1.$$

The latter is equivalent to (B.2.11). □

B.3. Rough estimate for the nonzero singularities

In this section we give a simple result which shows that the singularities of the distribution $\mathcal{F}_{\lambda \to t}[N'(\lambda)]$ at different t are not totally independent. Namely, we show that the singularities at $t \neq 0$ cannot be stronger than the singularity at $t = 0$.

THEOREM B.3.1. *If* $N \in F_+$ *and*

$$(N' * \rho)(\lambda) = O(\lambda^\nu), \tag{B.3.1}$$

then for any function γ *such that* $\hat\gamma \in C_0^\infty(\mathbb{R})$ *we have*

$$(N' * \gamma)(\lambda) = O(\lambda^\nu), \tag{B.3.2}$$

and moreover

$$\limsup_{\lambda\to+\infty} \frac{|(N' * \gamma)(\lambda)|}{\lambda^\nu} \leqslant C_{\rho,\gamma} \limsup_{\lambda\to+\infty} \frac{(N' * \rho)(\lambda)}{\lambda^\nu}, \tag{B.3.3}$$

where $C_{\rho,\gamma} > 0$ *is a constant independent of the choice of the function* N *and of the number* ν.

PROOF. Let us choose a number $\delta > 0$ such that $\hat\rho(t) \neq 0$ on $[-\delta, \delta]$. Further on we assume without loss of generality that $\operatorname{diam}(\operatorname{supp}\hat\gamma) \leqslant 2\delta$. Indeed, this can always be achieved by representing the original function γ as a finite sum of functions possessing this property.

Let us choose a $\tau \in \mathbb{R}$ such that $\operatorname{supp}\hat\gamma \in [\tau - \delta, \tau + \delta]$ and denote

$$\alpha(\lambda) = e^{i\tau\lambda}\rho(\lambda), \quad \beta(\lambda) = \mathcal{F}^{-1}_{t\to\lambda}\left[\frac{\hat\gamma(t)}{\hat\alpha(t)}\right].$$

Then

$$(N' * \gamma)(\lambda) = ((N' * \alpha) * \beta)(\lambda). \tag{B.3.4}$$

As ρ is a nonnegative function and dN is a nonnegative measure we have

$$|(N' * \alpha)(\lambda)| \leqslant (N' * \rho)(\lambda). \tag{B.3.5}$$

Formulae (B.3.4), (B.3.5), (B.3.1) imply (B.3.3) with $C_{\rho,\gamma} = \int |\beta(\mu)|\, d\mu$. □

B.4. General refined theorem

In this section we denote by C positive constants (perhaps, different) which may depend only on the functions N_j, ρ and on the number ν, but not on the parameters s, T and ε. If a constant depends on some of the parameters s, T or ε this will be indicated by respective subscripts.

THEOREM B.4.1. *Let* $N_j \in F_+$, $(N_j' * \rho)(\lambda) = O(\lambda^\nu)$, $j = 1, 2$,

$$(N_2 * \rho)(\lambda) = (N_1 * \rho)(\lambda) + o(\lambda^\nu) \tag{B.4.1}$$

and

$$(N_2' * \gamma)(\lambda) = (N_1' * \gamma)(\lambda) + o(\lambda^\nu) \tag{B.4.2}$$

for any function γ *such that* $\hat\gamma \in C_0^\infty(\mathbb{R})$, $\operatorname{supp}\hat\gamma \subset (0, +\infty)$. *Then*

$$N_1(\lambda - \varepsilon) - o(\lambda^\nu) \leqslant N_2(\lambda) \leqslant N_1(\lambda + \varepsilon) + o(\lambda^\nu), \qquad \forall \varepsilon > 0,$$

or equivalently

$$N_1(\lambda - o(1)) - o(\lambda^\nu) \leqslant N_2(\lambda) \leqslant N_1(\lambda + o(1)) + o(\lambda^\nu). \tag{B.4.3}$$

Formula (B.4.3) means that there exists a (positive) function f such that $f(\lambda) \to 0$ as $\lambda \to +\infty$, and

$$N_1(\lambda - f(\lambda)) \;-\; \lambda^\nu f(\lambda) \;\leqslant\; N_2(\lambda) \;\leqslant\; N_1(\lambda + f(\lambda)) \;+\; \lambda^\nu f(\lambda)\,.$$

Formula (B.4.3) differs from standard asymptotic formulae (like (B.2.3) above or (B.5.2) below) in the sense that we are comparing the graphs of the functions $N_1(\lambda)$ and $N_2(\lambda)$ not only in the "vertical" direction, but in the "horizontal" direction as well. This is natural when one compares the graphs of two counting functions: trying to increase accuracy we inevitably have to start comparing the graphs in both directions due to the possible presence of discontinuities. This phenomenon is well known in probability theory, where graphs of monotone functions are compared using the so-called *Lévy metric*, [**GnKo,** Chapter 2, Section 9]. The basic idea in the definition of the classical Lévy metric is to measure the distance between the graphs in the direction forming an angle of $\frac{3\pi}{4}$ with the positive λ-semiaxis. Theorem B.4.1 can be reformulated in terms of a weighted Lévy metric, in which the angle is a function of λ.

We shall prove Theorem B.4.1 in several steps. First, we prove

LEMMA B.4.2. *Let* $N \in F_+$ *and* $(N' * \rho)(\lambda) = O(\lambda^\nu)$. *Then for all* $s \geqslant 0$, $T \geqslant 1$, $\lambda \geqslant 1$ *we have the uniform estimate*

$$(N*\rho_T)(\lambda-s)-C(1+sT)^{-1}\lambda^\nu \leqslant N(\lambda) \leqslant (N*\rho_T)(\lambda+s)+C(1+sT)^{-1}\lambda^\nu\,. \tag{B.4.4}$$

PROOF. For brevity, we shall prove only the left inequality (B.4.4); the right one is proved in a similar way.

We have

$$\begin{aligned}(N * \rho_T)(\lambda - s) - N(\lambda) &= \int \big(N(\lambda - s - T^{-1}\tau) - N(\lambda)\big)\,\rho(\tau)\,d\tau \\ &\leqslant \int_{-\infty}^{-sT} \big(N(\lambda - s - T^{-1}\tau) - N(\lambda)\big)\,\rho(\tau)\,d\tau\end{aligned}$$

(here we used the monotonicity of N). Estimating the integrand by Lemma B.2.3 we obtain

$$(N * \rho_T)(\lambda - s) - N(\lambda) \;\leqslant\; C\,\lambda^\nu \int_{sT}^{+\infty} (1 - s + T^{-1}\tau)^{1+|\nu|}\rho(\tau)\,d\tau \tag{B.4.5}$$

for $\lambda \geqslant 1$. But

$$(1 - s + T^{-1}\tau)^{1+|\nu|} \;\leqslant\; (1+\tau)^{1+|\nu|}\,, \qquad \forall \tau \geqslant sT$$

(here we used the inequalities $s \geqslant 0$, $T \geqslant 1$), and

$$\rho(\tau) \;\leqslant\; C\,(1+\tau)^{-3-|\nu|}\,, \qquad \forall \tau \geqslant 0\,,$$

so

$$\int_{sT}^{+\infty} (1 - s + T^{-1}\tau)^{1+|\nu|}\rho(\tau)\,d\tau \;=\; C\int_{sT}^{+\infty} (1+\tau)^{-2}\,d\tau \;=\; C(1+sT)^{-1}\,.$$

Substituting the latter into (B.4.5) we arrive at (B.4.4). □

REMARK B.4.3. Obviously, if the estimate $(N' * \rho)(\lambda) = O(\lambda^\nu)$ holds uniformly on some subset of F_+, then the constant C in (B.4.4) is independent of the particular function N from this subset (i.e., (B.4.4) is also uniform).

Lemma B.4.2 implies

LEMMA B.4.4. *Let* $N_j \in F_+$, $(N_j' * \rho)(\lambda) = O(\lambda^\nu)$, $j = 1, 2$, *and*

$$(N_2 * \rho_T)(\lambda) = (N_1 * \rho_T)(\lambda) + o(\lambda^\nu), \qquad \forall T \geqslant 1. \tag{B.4.6}$$

Then (B.4.3) holds.

PROOF. Note that both functions N_2 and N_1 satisfy the conditions of Lemma B.4.2, and therefore estimates (B.4.4) hold. With account of (B.4.6) the inequalities (B.4.4) can be rewritten as

$$\begin{aligned}&(N_k * \rho_T)(\lambda - s) - C(1 + sT)^{-1}\lambda^\nu - o((\lambda - s)^\nu)\\ &\quad \leqslant N_j(\lambda) \leqslant (N_k * \rho_T)(\lambda + s) + C(1 + sT)^{-1}\lambda^\nu + o((\lambda + s)^\nu),\end{aligned} \tag{B.4.7}$$

where $j, k = 1, 2$, $j \neq k$. Formulae (B.4.7) imply

$$\begin{aligned}&N_1(\lambda - 2s) - C(1 + sT)^{-1}(\lambda^\nu + (\lambda - s)^\nu) - o((\lambda - s)^\nu)\\ &\quad \leqslant N_2(\lambda)\\ &\quad \leqslant N_1(\lambda + 2s) + C(1 + sT)^{-1}(\lambda^\nu + (\lambda + s)^\nu) + o((\lambda + s)^\nu).\end{aligned} \tag{B.4.8}$$

Formula (B.4.8) holds for any fixed $s \geqslant 0$ and $T \geqslant 1$. So we can set $s = T^{-1/2}$. Then (B.4.8) takes the form

$$\begin{aligned}&N_1(\lambda - 2T^{-1/2}) - CT^{-1/2}(\lambda^\nu + (\lambda - T^{-1/2})^\nu) - o((\lambda - T^{-1/2})^\nu)\\ &\quad \leqslant N_2(\lambda)\\ &\quad \leqslant N_1(\lambda + 2T^{-1/2}) + CT^{-1/2}(\lambda^\nu + (\lambda + T^{-1/2})^\nu)\\ &\qquad + o((\lambda + T^{-1/2})^\nu).\end{aligned} \tag{B.4.9}$$

It remains only to set $T = T(\lambda)$, where $T(\lambda)$ is an increasing function which tends to $+\infty$ as $\lambda \to +\infty$. The function $T = T(\lambda)$ can be chosen to increase so slowly that for both o-terms appearing in (B.4.9) we have $o\left((\lambda \mp T^{-1/2}(\lambda))^\nu\right) = o(\lambda^\nu)$; this remark is necessary because our o-terms (originating from (B.4.6)) might depend on T.

Formula (B.4.9) with the substitution $T = T(\lambda)$ implies (B.4.3). □

PROOF OF THEOREM B.4.1. In view of Lemma B.4.4, it is sufficient to show that (B.4.1) and (B.4.2) imply (B.4.6). Let us split $\hat{\rho}_T$ into the sum of functions $\hat{\rho}_{T,1}, \hat{\rho}_{T,2} \in C_0^\infty(\mathbb{R})$ such that $\hat{\rho}(t) \geqslant C > 0$ for $t \in \operatorname{supp} \hat{\rho}_{T,1}$, and $\hat{\rho}_{T,2}(t)$ vanishes in a neighbourhood of $t = 0$. Then (B.4.2) implies (B.4.6) for $\rho_{T,2}$ because

$$(N_j * \rho_{T,2})(\lambda) = \mathcal{F}^{-1}_{t \to \lambda}\left[(it)\hat{N}_j(t)\,(it)^{-1}\hat{\rho}_{T,2}(t)\right] = (N_j' * \widetilde{\rho}_{T,2})(\lambda)$$

where $j = 1, 2$, and $\widetilde{\rho}_{T,2}(\lambda) = \mathcal{F}^{-1}_{t \to \lambda}\left[(it)^{-1}\hat{\rho}_{T,2}(t)\right]$. Obviously,

$$N_j * \rho_{T,1} = N_j * \rho * \beta, \qquad \beta(\lambda) = \mathcal{F}^{-1}_{t \to \lambda}\left[\frac{\hat{\rho}_{T,1}(t)}{\hat{\rho}(t)}\right], \quad j = 1, 2. \tag{B.4.10}$$

Since the function β is rapidly decreasing, (B.4.1) and (B.4.10) imply (B.4.6) for $\rho_{T,1}$. □

B.5. Special version of the general refined theorem

We formulate below a special version of the general refined theorem specifically oriented towards the case when N has a polynomial two-term asymptotics, i.e., when the nonzero singularities of the Fourier transform of N' are weaker than the singularity at the origin.

THEOREM B.5.1. *Let* $N \in F_+$, $(N' * \rho)(\lambda) = O(\lambda^\nu)$, *and*

$$(N' * \gamma)(\lambda) = o(\lambda^\nu), \tag{B.5.1}$$

for any function γ *such that* $\hat\gamma \in C_0^\infty(\mathbb{R})$, $\operatorname{supp}\hat\gamma \subset (0, +\infty)$. *Then*

$$N(\lambda) = (N * \rho)(\lambda) + o(\lambda^\nu). \tag{B.5.2}$$

REMARK B.5.2. As will be clear from the proof, if the conditions of Theorem B.5.1 are fulfilled uniformly on some subset of F_+, then (B.5.2) also holds uniformly.

Theorem B.5.1 has a simpler formulation than Theorem B.4.1, and it can be viewed as a natural extension of Corollary B.2.2. However, Theorem B.5.1 requires more restrictive conditions on the functions involved than Theorem B.4.1. Basically, the conditions of Theorem B.5.1 ensure that the (discontinuous) monotone function N does not have very big "jumps", and this is why we are able to compare the graphs in the "vertical" direction only.

PROOF OF THEOREM B.5.1. According to the Lagrange formula we have for some $\theta_\mp(\lambda) \in (0,1)$

$$\begin{aligned} &(N * \rho)(\lambda \mp o(1)) - (N * \rho)(\lambda) \\ &\quad = o(1)\,(N' * \rho)\,(\lambda \mp \theta_\mp(\lambda)\,o(1)) = o(1)\,O(\lambda^\nu) = o(\lambda^\nu). \end{aligned} \tag{B.5.3}$$

In view of Lemma B.4.2 and (B.5.3), it is sufficient to prove that

$$(N * \rho)(\lambda) - (N * \rho_T)(\lambda) = (N * (\rho - \rho_T))\,(\lambda) = o(\lambda^\nu), \qquad \forall T \geqslant 1. \tag{B.5.4}$$

Indeed, then (B.5.2) is obtained from (B.4.4) by substitution $s = f_1(\lambda)$, $T = f_2(\lambda)$, where f_1 and f_2 are arbitrary functions such that

$$f_1(\lambda) \to 0, \quad f_2(\lambda) \to +\infty, \quad f_1(\lambda)\,f_2(\lambda) \to +\infty, \qquad \lambda \to +\infty.$$

Since $\hat\rho(0) = 1$ and $\hat\rho'(0) = 0$, the function $\hat\rho(t) - \hat\rho_T(t)$ has a second order zero at $t = 0$. Therefore (B.5.4) is a consequence of the following lemma, which completes the proof of Theorem B.5.1.

LEMMA B.5.3. *Let* $N \in F_+$ *satisfy the conditions of Theorem B.5.1. Assume that* $\hat\rho_0 \in C_0^\infty(\mathbb{R})$ *and* $\hat\rho_0(t) = t^2\hat\alpha(t)$, *where* $\hat\alpha \in C_0^\infty(\mathbb{R})$. *Then*

$$(N * \rho_0)(\lambda) = o(\lambda^\nu). \tag{B.5.5}$$

PROOF. We have

$$(N * \rho_0)(\lambda) \;=\; \mathcal{F}^{-1}_{t\to\lambda}[\hat{N}(t)\, t^2 \hat{\alpha}(t)] \;=\; (N' * \beta)(\lambda)\,, \tag{B.5.6}$$

where $\beta = \mathcal{F}^{-1}_{t\to\lambda}[-it\, \hat{\alpha}(t)]$.

Let $\hat{f} \in C_0^\infty(\mathbb{R})$ and $\hat{f}(t) = 1$ in a neighbourhood of the origin. Denote $\hat{f}_\delta(t) = \hat{f}(t/\delta)$ and $\widetilde{f}_\delta(\lambda) = \mathcal{F}^{-1}_{t\to\lambda}[-it\, \hat{f}_\delta(t)]$. Then $\widetilde{f}_\delta(\lambda) = \delta^2\, \mathcal{F}^{-1}_{t\to\delta\lambda}[-it\, \hat{f}(t)]$ and

$$\int |\widetilde{f}_\delta(\lambda)|\, d\lambda \;=\; \delta\, C_f\,, \tag{B.5.7}$$

where $C_f = \int |\mathcal{F}^{-1}_{t\to\lambda}[-it\, \hat{f}(t)]\,|\, d\lambda$.

Clearly,

$$\beta(\lambda) \;=\; \mathcal{F}^{-1}_{t\to\lambda}[-it\,(1 - \hat{f}_\delta(t))\, \hat{\alpha}(t)] \;+\; \mathcal{F}^{-1}_{t\to\lambda}[-it\, \hat{f}_\delta(t)\, \hat{\alpha}(t)]\,. \tag{B.5.8}$$

In view of (B.5.1), the contribution to (B.5.6) of the first term on the right-hand side of (B.5.8) is $o(\lambda^n)$. The contribution to (B.5.6) of the second term on the right-hand side of (B.5.8) is $(N' * \alpha * f_\delta)$. By Theorem B.3.1, $(N' * \alpha)(\lambda)$ is estimated by $C\,(\lambda^\nu + 1)$, so

$$|\,(N' * \alpha * f_\delta)(\lambda)\,| \;\leqslant\; \delta\, C\, C_f\, (\lambda^\nu + 1)\,,$$

Thus,

$$|\,(N * \rho_0)(\lambda)\,| \;\leqslant\; \delta\, C\, C_f\, (\lambda^\nu + 1) \;+\; o(\lambda^n)\,.$$

Since δ can be chosen arbitrarily small, this implies (B.5.5). □

Appendix C

Stationary Phase Formula

1. The main theorem. First of all we recall the classical stationary phase formula (see, for example, [**Hö3,** Vol. 1, Sect. 7.7]).

THEOREM C.1. *Let* $v \in C_0^\infty(\mathbb{R}^N)$, *and let* f *be a real* C^∞*-function defined in a neighbourhood of* $\operatorname{supp} v$ *such that for some point* $z_0 \in \operatorname{supp} v$ *we have* $f_z(z_0) = 0$, $\det f_{zz}(z_0) \neq 0$, *and* $f_z \neq 0$ *if* $z \neq z_0$. *Then*

$$\begin{aligned} &\Big| \int v(z)\, e^{i\lambda f(z)}\, dz \\ &\qquad - e^{i\lambda f(z_0)}\, e^{i\pi\sigma/4}\, \big|\det\big(\lambda f_{zz}(z_0)/2\pi\big)\big|^{-1/2} \sum_{j<p} \lambda^{-j}\, L_j v(z_0) \Big| \\ &\leqslant \text{const}\ \lambda^{-p} \sum_{|\alpha|\leqslant 2p} \sup_z |\partial_z^\alpha v(z)|\,, \qquad \forall \lambda > 0\,, \end{aligned} \tag{C.1}$$

where σ *is the signature of the matrix* $f_{zz}(z_0)$ *and* $L_j = L(z, D_z)$ *are differential operators of order* $2j$ *defined as follows. Let*

$$f_0(z) \;=\; f(z) \;-\; f(z_0) \;-\; \langle\, f_{zz}(z_0)\,(z - z_0)\,,\, z - z_0 \,\rangle\,/2 \tag{C.2}$$

(*this function has a third order zero at the point* z_0). *Then*

$$L_j v(z) \;=\; i^{-j} \sum_{\substack{\mu,\nu\,:\,\nu-\mu=j,\\ 3\mu\leqslant 2\nu}} (\mu!\,\nu!)^{-1}\, 2^{-\nu}\, \langle\, \big(f_{zz}(z_0)\big)^{-1} D_z\,,\, D_z \,\rangle^\nu\, (f_0^\mu\, v)(z)\,. \tag{C.3}$$

If the functions v, f *and the stationary point* z_0 *smoothly depend on some parameters* (*one of which may be* λ) *and all the conditions are fulfilled uniformly with respect to these parameters, then the estimate* (C.1) *is also uniform.*

The following simple result supplements Theorem C.1.

PROPOSITION C.2. *Assume that under the conditions of Theorem* C.1 *the phase* f *has no stationary points* (*i.e.,* $f_z \neq 0$ *on* $\operatorname{supp} v$). *Then for any integer* p

$$\Big| \int v(z)\, e^{i\lambda f(z)}\, dz \Big| \;\leqslant\; \text{const}\ \lambda^{-p} \sum_{|\alpha|\leqslant p} \sup_z |\partial_z^\alpha v(z)|\,, \qquad \forall \lambda > 0\,. \tag{C.1$'$}$$

PROOF. We can replace $e^{i\lambda f}$ in the integral by $|f_z|^{-2} \sum_k f_{z_k} D_{z_k} e^{i\lambda f}$ and integrate by parts. Repeating this procedure we obtain the required estimate. □

Note that the remainder estimate in Proposition C.2 is slightly better than in Theorem C.1. But it is not essential for our aims, and we will consider (C.1′) as a particular case of (C.1) (with $L_j \equiv 0$).

2. Stationary phase formula for oscillatory integrals. We apply the stationary phase formula to oscillatory integrals. This requires some explanation because such integrals do not converge in the usual sense. In this subsection we prove a simple lemma which enables us to justify the stationary phase formula for an abstract oscillatory integral.

Let $z' \in \Omega' \subset \mathbb{R}^N$, where Ω' is a bounded open set. We consider a local oscillatory integral smoothly depending on some parameters z''

$$I_{\varphi,a}(z',z'') = \int e^{i\varphi(z',z'',\zeta)}\, a(z',z'',\zeta)\, đ\zeta\,, \qquad \zeta \in \mathbb{R}^n\,,\ a \in S^l\,,$$

such that the inequality $|\varphi_{z'}| \geqslant \mathrm{const}\,|\zeta|$ holds on $\overline{\mathrm{cone\,hull}\,(\mathrm{supp}\,a)}$ with some positive constant.

LEMMA C.3. *Let $g(z',z'')$ be a smooth real function such that $|g_{z'}| \geqslant \mathrm{const} > 0$. Denote*

$$\Omega_{\varphi,g} = \{(z',z'',\zeta) : |g_{z'}(z',z'')|/3 \leqslant |\varphi_{z'}(z',z'',\zeta)| \leqslant 3\,|g_{z'}(z',z'')|\}\,.$$

Let $v(z',z'',\zeta)$ be an arbitrary smooth cut-off function, bounded with all its derivatives, such that $v(z',z'',\zeta) = 1$ in a neighbourhood of the set $C_\varphi \cap \mathrm{cone\,supp}\,a \cap \Omega_{\varphi,g}$. Then for any positive integer p

$$\begin{aligned} &\Big| \int e^{i\lambda g(z',z'')}\, e^{i\varphi(z',z'',\zeta)}\, a(z',z'',\zeta)\, đ\zeta\, dz' \\ &\qquad - \int e^{i\lambda g(z',z'')}\, e^{i\varphi(z',z'',\zeta)}\, v(z',z'',\zeta/|\lambda|)\, a(z',z'',\zeta)\, đ\zeta\, dz' \Big| \\ &\leqslant \mathrm{const}\,(1+|\lambda|)^{-p} \end{aligned} \tag{C.4}$$

uniformly with respect to z'' and $\lambda \in \mathbb{R}$.

PROOF. Clearly, if $a \in S^{-\infty}$, then both integrals on the left-hand side of (C.4) are rapidly decreasing. Indeed, if

$$L(z',z'',D_{z'}) = |g_{z'}|^{-2} \sum_k g_{z'_k}\, D_{z'_k}\,,$$

then integrating by parts we obtain

$$\begin{aligned} \int e^{i\lambda g(z',z'')}\, \mathcal{I}_{\varphi,\tilde a}\, dz' &= \lambda^{-p} \int \mathcal{I}_{\varphi,\tilde a}\, L^p e^{i\lambda g(z',z'')}\, dz' \\ &= \lambda^{-p} \int e^{i\lambda g(z',z'')}\, (L^T)^p \mathcal{I}_{\varphi,\tilde a}\, dz'\,, \end{aligned}$$

and similarly for the second integral. Therefore we can assume that all the amplitudes we are dealing with vanish in a neighbourhood of $\{\zeta = 0\}$ and satisfy $\overline{\mathrm{cone\,hull}\,(\mathrm{supp}\,a)} = \mathrm{cone\,supp}\,a$.

Moreover, in order to apply the above procedure we only need the functions defined by our oscillatory integrals to be smooth. Therefore we can also assume that $\mathrm{cone\,supp}\,a$ lies in an arbitrarily small conic neighbourhood of C_φ. Let us choose this neighbourhood to be so small that $v = 1$ in a neighbourhood of the set $\mathrm{cone\,supp}\,a \cap \Omega_{\varphi,g}$.

Let us consider the integral

$$\text{(C.5)} \qquad \int e^{i\lambda g(z',z'')+i\varphi(z',z'',\zeta)} \left(1 - v(z',z'',\zeta/|\lambda|)\right) a(z',z'',\zeta)\, đ\zeta\, dz'\,.$$

On the support of the function $\left(1 - v(z',z'',\zeta/|\lambda|)\right) a(z',z'',\zeta)$ we have the estimates

$$\begin{aligned} \left| \lambda g_{z'}(z',z'') - \varphi_{z'}(z',z'',\zeta) \right| &\geqslant |\lambda| \left| |g_{z'}(z',z'')| - |\varphi_{z'}(z',z'',\zeta/|\lambda|)| \right| \\ &\geqslant \frac{|\lambda|}{2} \left(|g_{z'}(z',z'')| + |\varphi_{z'}(z',z'',\zeta/|\lambda|)| \right) \geqslant \text{const}\,(|\lambda| + |\zeta| + 1) \end{aligned}$$

(in the last inequality we also used the fact that $|\zeta| \geqslant \text{const} > 0$ on $\operatorname{supp} a$). Now we replace $e^{i\lambda g+i\varphi}$ in (C.5) by

$$|\lambda g_{z'} + \varphi_{z'}|^{-2} \sum_k (\lambda g_{z'_k} + \varphi_{z'_k})\, D_{z'_k}\, e^{i\lambda g+i\varphi}$$

and then integrate by parts with respect to z'. Repeating this procedure we can transform (C.5) into an oscillatory integral with the same phase function and an amplitude which is estimated with all its derivatives by $\text{const}\,(|\lambda|+|\zeta|+1)^{-p}$ for an arbitrary positive integer p. Obviously, such an integral defines a smooth function rapidly decreasing with all its derivatives as $|\lambda| \to \infty$. □

Lemma C.3 enables us to use the stationary phase formula in the following situation. Suppose that the phase function φ is real and we are looking for asymptotics of the integral

$$\text{(C.6)} \qquad \int e^{i\lambda g(z',z'')}\, \mathcal{I}_{\varphi,a}\, dz' = \int e^{i\lambda g(z',z'')}\, e^{i\varphi(z',z'',\zeta)}\, a(z',z'',\zeta)\, đ\zeta\, dz'$$

as $|\lambda| \to \infty$. By Lemma C.3 we may insert in this integral a cut-off function v satisfying the conditions of Proposition C.2. Obviously, we can choose $\operatorname{supp} v$ to be compact and separated from the set $\{\zeta = 0\}$. Then we obtain the integral

$$\int e^{i\lambda g(z',z'')}\, e^{i\varphi(z',z'',\zeta)}\, v(z',z'',\zeta/|\lambda|)\, a(z',z'',\zeta)\, đ\zeta\, dz'$$

over a compact set which has the same asymptotic behaviour as (C.6). After the change of variables $\zeta \to |\lambda|\zeta$ this integral takes the form

$$\text{(C.7)} \qquad |\lambda|^n \int e^{i|\lambda|\left((\operatorname{sgn}\lambda) g(z',z'')+\varphi(z',z'',\zeta)\right)}\, v(z',z'',\zeta)\, a(z',z'',|\lambda|\zeta)\, đ\zeta\, dz'\,.$$

Since $a \in S^l$, we have

$$\begin{aligned} \left| \partial^\alpha_{z',z''} \partial^\beta_\zeta \left(a(z',z'',|\lambda|\zeta)\right) \right| &= |\lambda|^{|\beta|} \left| (\partial^\alpha_{z',z''} \partial^\beta_\zeta a)(z',z'',|\lambda|\zeta) \right| \\ &\leqslant \text{const}\, |\lambda|^{|\beta|} (|\lambda\zeta| + 1)^{l-|\beta|} \leqslant \text{const}\, (|\lambda| + 1)^l \end{aligned}$$

on $\operatorname{supp} v$. So the amplitude in (C.7) is bounded with all its derivatives by $\text{const}\,(|\lambda|+1)^l$ uniformly with respect to $\lambda \in \mathbb{R}$.

When the function $(\operatorname{sgn}\lambda)g + \varphi$ satisfies the conditions of Theorem C.1 (or of Proposition C.2) we can apply to the integral in (C.7) the stationary phase formula. As a result we obtain an expansion of the form (C.1) with remainder estimate $O(|\lambda|^{l-p})$ (of course, the factor $|\lambda|^n$ has to be accounted for separately). The terms in this expansion depend only on the values of the phase function, the amplitude and their derivatives at the stationary point. But in the neighbourhood of this point $\upsilon = 1$, so we would have obtain the same terms had we formally applied the stationary phase formula to the integral without the cut-off υ. Thus, we have proved

COROLLARY C.4. *Let φ be a real phase function. Let us consider the integral*

$$
\begin{aligned}
&\int e^{i\lambda g(z',z'')}\,\mathcal{I}_{\varphi,a}\,dz' \\
\text{(C.8)}\qquad &= |\lambda|^n \times \begin{cases} \int e^{i\lambda\left(g(z',z'')+\varphi(z',z'',\zeta)\right)}\,a(z',z'',\lambda\zeta)\,đ\zeta\,dz', & \lambda>0, \\ \int e^{i|\lambda|\left(\varphi(z',z'',\zeta)-g(z',z'')\right)}\,a(z',z'',|\lambda|\zeta)\,đ\zeta\,dz', & \lambda<0. \end{cases}
\end{aligned}
$$

If in a neighbourhood of $C_\varphi \cap \operatorname{cone\,supp} a$ the function $f = g \pm \varphi$ satisfies the conditions of Theorem C.1 *(or Proposition* C.2*) with $z = (z',\zeta)$ uniformly with respect to z'', then for the corresponding integral on the right-hand side of* (C.8) *the stationary phase formula* (C.1) *holds with the remainder $O(|\lambda|^{l-p})$.*

REMARK C.5. Suppose that we change variables $\zeta \to (\zeta'(\zeta),\zeta''(\zeta))$ in the integral (C.8) so that $(\zeta'(\lambda\zeta),\zeta''(\lambda\zeta)) = (\lambda\zeta'(\zeta),\zeta''(\zeta))$, $\forall\lambda>0$. Then we can regard ζ'' as parameters. If the conditions of Corollary C.4 are fulfilled for $z = (z',\zeta')$ uniformly with respect to (z'',ζ''), then (C.1) holds for the integral with respect to (z',ζ'').

3. Standard oscillatory integrals. In this subsection we prove Proposition 2.7.7.

Suppose that Proposition 2.7.7 is false. Then there exists a standard oscillatory integral $\mathcal{I}_{\varphi,\mathfrak{a}}$ of the form (2.7.11) which defines a smooth half-density, and such that $\mathfrak{a} \in S^l$ and $\mathfrak{a}_l \not\equiv 0$ on $\mathfrak{D}_i$. Let us choose a point $(t_0;y_0,\eta_0) \in \mathfrak{D}_i$ such that $\mathfrak{a}_l(t_0;y_0,\eta_0) \neq 0$ and $x_0 := x^*(t_0;y_0,\eta_0) \notin \partial M$. Set $\xi_0 := \xi^*(t_0;y_0,\eta_0)$.

Let us take a test half-density of the form $v_\lambda(x) = \chi(x)e^{i\lambda\psi(x)}$ where the function ψ is such that $\operatorname{Im}\psi \geqslant 0$, $\psi_x(x_0) = \xi_0$, $\lambda>0$ is a parameter, and $\chi \in C_0^\infty(\overset{\circ}{M})$ is such that $\chi(x_0) = 1$ and $\operatorname{supp}\chi$ lies in a small neighbourhood of the point x_0. As we assumed $\mathcal{I}_{\varphi,\mathfrak{a}}$ to be smooth we have

$$
\text{(C.9)}\qquad \int_M \mathcal{I}_{\varphi,\mathfrak{a}}(t_0,x,y_0)\,\overline{v_\lambda(x)}\,dx = O(\lambda^{-\infty}), \qquad \lambda\to+\infty.
$$

Substituting (2.7.11) in (C.9) we obtain an integral with respect to $dx\,đ\eta$. It is easy to see that η which do not lie in a small conic neighbourhood of η_0 *a priori* give an $O(\lambda^{-\infty})$ contribution to the integral (C.9). Indeed, the presence of the two cut-offs $\varsigma(t_0,x;y_0,\eta)$ and $\chi(x)$ means that we are integrating over η such that $x^*(t_0;y_0,\eta)$ is close to x_0. This implies that if η is not conically close to η_0, then $\xi^*(t_0;y_0,\eta)$ is not conically close to ξ_0 (here we use the nondegeneracy of our canonical transformation). Thus, for η lying outside a small conic neighbourhood of η_0 we have the uniform estimate $|\varphi_x - \lambda\psi_x| \geqslant \text{const}\,(\lambda+|\eta|)$, $\forall\lambda>0$, with some const >0, on $\operatorname{supp}(\varsigma(t_0,\cdot\,;y_0,\cdot)\chi(\cdot))$, which allows us to integrate by parts with respect to x. So further on we assume without loss of generality that $\operatorname{supp}\mathfrak{a}(t_0;y_0,\cdot)$ lies in a small conic neighbourhood of the point η_0.

Let us choose some local coordinates x in which $\det\xi^*_\eta(t_0;y_0,\eta_0) \neq 0$; this is possible by Lemma 2.3.2. Without loss of generality we may assume that our phase function is linear in x in a conic neighbourhood of the point $(t_0,x_0;y_0,\eta_0)$. Indeed, a locally linear phase function exists by Lemma 2.4.13 and we can transform our original oscillatory integral to an oscillatory integral with a linear phase function by the argument from the proof of Theorem 2.7.10 (note that this procedure does not change the principal symbol). It may seem that here we are making a logical loop because the proof of Theorem 2.7.10 involves the use of the operator $\mathfrak{S}$ which was defined as the operator mapping the amplitude into **the** symbol, and we have not yet established the uniqueness of the symbol. However, our operator $\mathfrak{S}$ was constructed explicitly (see 2.7.3) and there is nothing wrong in using it as long as we (temporarily) view it as an operator mapping the amplitude into **a** symbol.

Obviously, we may also take ψ to be linear in our chosen coordinates.

We can now rewrite the integral in (C.9) as

$$\int e^{i(\langle x-x^*(t_0;y_0,\eta),\xi^*(t_0;y_0,\eta)\rangle-\lambda\langle x,\xi_0\rangle)}\mathfrak{a}(t_0;y_0,\eta)\,d_\varphi(t_0,x;y_0,\eta)\,\overline{\chi(x)}\,dx\,đ\eta\,.$$

Applying to this integral Corollary C.4 and taking the leading term of the asymptotic expansion we conclude that $\mathfrak{a}_l(t_0;y_0,\eta_0)=0$, which contradicts our assumption.

4. Boundary oscillatory integrals. The proof of Proposition 2.9.6 is the same as that of Proposition 2.7.10, one only has to replace the coordinates x by (x',t). Note that in order to perform a reduction to the case of linear phases (see above) it may be necessary to make a change of coordinates (x',t) which mixes the spatial coordinates and time.

5. Boundary layer oscillatory integrals. In this subsection we prove Proposition 2.8.7.

Suppose that Proposition 2.8.7 is false. Then there exists a boundary layer oscillatory integral $\mathcal{I}_{\varphi,\mathfrak{a}}$ of the form (2.7.11) which defines a smooth half-density, and such that $\mathfrak{a}\in S^l$ and

$$\partial_t^k\mathfrak{a}_l\big|_{t=t_i^*}\not\equiv 0\quad\text{on } O \tag{C.10}$$

for some nonnegative integer k.

The smoothness of $\mathcal{I}_{\varphi,\mathfrak{a}}$ implies the smoothness of $\mathcal{I}_{\varphi,\mathfrak{a}}|_{\partial M_x}$. By Lemma 2.9.11 the latter is a boundary Lagrangian distribution with principal symbol proportional by a nonzero factor to $\mathfrak{a}_l|_{t=t_i^*}$. As we have already proved Proposition 2.9.6 this implies

$$\mathfrak{a}_l\big|_{t=t_i^*}\equiv 0\quad\text{on } O\,. \tag{C.11}$$

Let us now consider the oscillatory integral $\left((D_t^{2m}-A(x,D_x)\right)\mathcal{I}_{\varphi,\mathfrak{a}}$. Excluding the variable x from the amplitude (see Section 3.3 and 3.4.3) and using (C.11) we see that $\left((D_t^{2m}-A(x,D_x)\right)\mathcal{I}_{\varphi,\mathfrak{a}}=\mathcal{I}_{\varphi,\mathfrak{a}^{(1)}}$ where $\mathfrak{a}^{(1)}\in S^{l+2m-1}$ is an amplitude independent of x and $\mathfrak{a}^{(1)}_{l+2m-1}\big|_{t=t_i^*}$ is proportional by a nonzero factor to $\partial_t\mathfrak{a}_l|_{t=t_i^*}$. But we already know that the smoothness of $\mathcal{I}_{\varphi,\mathfrak{a}^{(1)}}$ implies $\mathfrak{a}^{(1)}_{l+2m-1}\big|_{t=t_i^*}\equiv 0$ on O, so $\partial_t\mathfrak{a}_l|_{t=t_i^*}\equiv 0$ on O.

Now we can consider the oscillatory integral $\left((D_t^{2m}-A(x,D_x)\right)^2\mathcal{I}_{\varphi,\mathfrak{a}}$, and so on. Iterating our procedure we obtain a contradiction with (C.10).

Appendix D

Hamiltonian Billiards: Proofs

In the first two sections it will be convenient to deal with the extended manifold $\widehat{M}$; see 1.1.2. In this way we will avoid (until special notice) the necessity of considering reflections, at least in explicit form. For the sake of simplicity we will omit the "wide hats" in the notation for points and local coordinates on the extended manifold. We will retain the "wide hat" only in the notation for the manifold, in order to distinguish the original manifold M from its extension $\widehat{M}$.

Points from $S^*\widehat{M}$ will be denoted by $(x, \widetilde{\xi})$. Whenever we have a function $f(x, \xi)$ defined of $T'\widehat{M}$, we will write its restriction to $S^*\widehat{M}$ as $f(x, \widetilde{\xi})$. For example, we shall denote Hamiltonian trajectories on $S^*\widehat{M}$ by $(x^*(t; y, \widetilde{\eta}), \widetilde{\xi}^*(t; y, \widetilde{\eta}))$, $(y, \widetilde{\eta}) \in S^*\widehat{M}$ being the starting point.

Points from $S^*M|_{\partial M}$ will be denoted by $(x', \widetilde{\xi})$. For $(x', \widetilde{\xi}) \in S^*M|_{\partial M}$ we shall denote by $(\widetilde{\xi})'$ the first $n-1$ components of the corresponding covector $\xi \in T'M$. The notation $(\widetilde{\xi})'$ should not be confused with $\widetilde{\xi}'$: the latter will be used to denote points on the cosphere $S^*_{x'}\partial M$.

By $\operatorname{meas}(\,\cdot\,)$, $\operatorname{meas}_{\partial M}(\,\cdot\,)$ we shall denote natural measures on $S^*\widehat{M}$, $S^*M|_{\partial M}$ with elements $dx\, d\widetilde{\xi}$, $\left|h_{\xi_n}(x', 0, \widetilde{\xi})\right| dx'\, d\widetilde{\xi}$, respectively; see 1.1.10 for the definition of $d\widetilde{\xi}$.

D.1. Measure of "awkward" starting points

In our construction of the wave group we have excluded certain "awkward" reflections; see conditions (2), (3) of Definition 1.3.25. The aim of this section is to demonstrate that the measure of starting points of such billiard trajectories is zero.

Below, we introduce the operators Φ^t, Ψ^T, $\Theta^\pm$, Ξ mapping subsets of $S^*\widehat{M}$, $S^*M|_{\partial M}$, $S^*M|_{\partial M}$, $S^*M|_{\partial M}$ to subsets of $S^*\widehat{M}$, $S^*\widehat{M}$, $S^*M|_{\partial M}$, $S^*M|_{\partial M}$, respectively.

Definition D.1.1. For any $P \subset S^*\widehat{M}$ and $t \in \mathbb{R}$ we define

$$\Phi^t(P) = \bigcup_{(y,\widetilde{\eta}) \in P} (x^*(t; y, \widetilde{\eta}), \widetilde{\xi}^*(t; y, \widetilde{\eta})).$$

Definition D.1.2. For any $Q \subset S^*M|_{\partial M}$ and $T \in \mathbb{R}$ we define

$$\Psi^T(Q) = \bigcup_{t \in I_T} \Phi^t(Q), \quad \text{where } I_T = \begin{cases} [0,T] & \text{if } T > 0, \\ \{0\} & \text{if } T = 0, \\ [T,0] & \text{if } T < 0. \end{cases}$$

Further on we will often have to deal with the set

$$G := \{(x',\widetilde{\xi}) \in S^*M|_{\partial M} : h_{\xi_n}(x',0,\widetilde{\xi}) = 0\}\,. \tag{D.1.1}$$

Obviously, this set is closed and

$$\mathrm{meas}_{\partial M}\, G = 0\,. \tag{D.1.2}$$

Given a point $(y',\widetilde{\eta}) \in S^*M|_{\partial M} \setminus G$ we denote

$$\begin{aligned} T_+(y',\widetilde{\eta}) &= \min\{t:\ t>0 \text{ and } x^*(t;y',0,\widetilde{\eta}) \in \partial M\}\,,\\ T_-(y',\widetilde{\eta}) &= \max\{t:\ t<0 \text{ and } x^*(t;y',0,\widetilde{\eta}) \in \partial M\}\,. \end{aligned}$$

If $x^*(t;y',0,\widetilde{\eta}) \notin \partial M$, $\forall t>0$, or $x^*(t;y',0,\widetilde{\eta}) \notin \partial M$, $\forall t<0$, we set $T_+(y',\widetilde{\eta}) = +\infty$ or $T_-(y',\widetilde{\eta}) = -\infty$, respectively.

Denote

$$\begin{aligned} D(\Theta^\pm) = \{(y',\widetilde{\eta}) \in S^*M|_{\partial M} :\ & (y',\widetilde{\eta}) \notin G,\ T_\pm(y',\widetilde{\eta}) \neq \pm\infty,\\ & (x^{*\prime}(T_\pm(y',\widetilde{\eta});y',0,\widetilde{\eta}), \widetilde{\xi}^*(T_\pm(y',\widetilde{\eta});y',0,\widetilde{\eta})) \notin G\}\,. \end{aligned}$$

The sets $D(\Theta^\pm)$ are open in $S^*M|_{\partial M}$, and consequently measurable.

DEFINITION D.1.3. For any $Q \subset S^*M|_{\partial M}$ we define

$$\Theta^\pm(Q) = \bigcup_{(y',\widetilde{\eta})\in Q\cap D(\Theta^\pm)} (x^{*\prime}(T_\pm(y',\widetilde{\eta});y',0,\widetilde{\eta}), \widetilde{\xi}^*(T_\pm(y',\widetilde{\eta});y',0,\widetilde{\eta}))\,.$$

DEFINITION D.1.4. For any $Q \subset S^*M|_{\partial M}$ we define

$$\begin{aligned} \Xi(Q) = \{(x',\widetilde{\xi}) \in S^*M|_{\partial M} :\ & x' = y',\ (\widetilde{\xi})' = (\widetilde{\eta})',\\ & h_{\xi_n}(x',0,\widetilde{\xi})\, h_{\xi_n}(y',0,\widetilde{\eta}) < 0 \text{ for some } (y',\widetilde{\eta}) \in Q\}\,. \end{aligned}$$

An obvious property of the operators $F = \Phi^t, \Psi^T, \Theta^\pm, \Xi$ is

$$F(P_1 \cup P_2) = F(P_1) \cup F(P_2)\,, \qquad \forall P_1, P_2\,.$$

The following Lemmas D.1.5, D.1.7, D.1.8, D.1.10 describe the measure properties of the operators Φ^t, Ψ^T, $\Theta^\pm$, Ξ.

LEMMA D.1.5. *If the set $P \subset S^*\widehat{M}$ is measurable, then* $\mathrm{meas}\,\Phi^t(P) = \mathrm{meas}\,P$.

PROOF. Proposition 2.3.5 implies that Φ^t is a restriction to $S^*\widehat{M}$ of the canonical transformation (2.3.10); note that here we do not deal with reflections, extending (2.3.10) to $T'\widehat{M}$. Therefore the lemma follows from the fact that (2.3.10) preserves the symplectic volume (see the beginning of 2.3.1) and the Hamiltonian (see the beginning of 1.3.1)

LEMMA D.1.6. *For any $T \in \mathbb{R}$ we have* $\mathrm{meas}\,\Psi^T(G) = 0$.

PROOF OF LEMMA D.1.6. As the rays originating from G are tangent to ∂M at $t=0$, for sufficiently small $|\tau|$ we have

$$\Psi^\tau(G) \subset S^*\widehat{M} \cap \{|x_n| \leqslant c\tau^2\} := P^\tau ,$$

where $c > 0$ is some constant. Obviously, for any $k \in \mathbb{N}$

$$\begin{aligned}\Psi^{k\tau}(G) &= \Psi^\tau(G) \cup \Phi^\tau(\Psi^\tau(G)) \cup \Phi^{2\tau}(\Psi^\tau(G)) \cup \cdots \cup \Phi^{(k-1)\tau}(\Psi^\tau(G)) \\ &\subset P^\tau \cup \Phi^\tau(P^\tau) \cup \Phi^{2\tau}(P^\tau) \cup \cdots \cup \Phi^{(k-1)\tau}(P^\tau) ,\end{aligned}$$

so, with account of Lemma D.1.5,

$$\mathrm{meas}(\Psi^{k\tau}(G)) \leqslant k \,\mathrm{meas}(P^\tau) \leqslant Ck\tau^2 ,$$

where $C > 0$ is some constant. Setting $\tau = T/k$ and letting $k \to \infty$ we obtain $\mathrm{meas}\,\Psi^T(G) = 0$. $\sqcap$

LEMMA D.1.7. *If the set $Q \subset S^*M|_{\partial M}$ is measurable, then for any $T \in \mathbb{R}$ the set $\Psi^T(Q) \in S^*\widehat{M}$ is also measurable and*

$$\mathrm{meas}\,\Psi^T(Q) \leqslant |T| \,\mathrm{meas}_{\partial M}\, Q . \tag{D.1.3}$$

PROOF OF LEMMA D.1.7. In view of Lemma D.1.6 and formula (D.1.2) it is sufficient to prove Lemma D.1.7 under the assumption that

$$\overline{Q} \cap G = \varnothing . \tag{D.1.4}$$

Indeed, as G is closed we can choose a sequence of closed sets $Q_1 \subset Q_2 \subset \cdots$ separated from G and such that $\bigcup_j Q_j = Q \setminus G$; if we establish (D.1.3) for Q_j, then the result for Q will follow by going to the limit as $j \to \infty$.

Under the condition (D.1.4) for sufficiently small $|\tau|$ there is a bijection between $I_\tau \times \overline{Q}$ and $\Psi^\tau(\overline{Q})$. Moreover, in a neighbourhood of $\Psi^\tau(\overline{Q})$ we can change our original local coordinates $(x,\widetilde{\xi})$ to coordinates $(y', t, \widetilde{\eta})$ defined by

$$(x^*(t; y', 0, \widetilde{\eta}), \widetilde{\xi}^*(t; y', 0, \widetilde{\eta})) = (x, \widetilde{\xi}) .$$

We have

$$\left| \det \frac{\partial(x,\widetilde{\xi})}{\partial(y', t, \widetilde{\eta})} \right|_{t=0} = |h_{\xi_n}(y', 0, \widetilde{\eta})| ,$$

which implies

$$\mathrm{meas}\,\Psi^\tau(Q) = |\tau| \,\mathrm{meas}_{\partial M}\, Q + O(\tau^2) , \qquad \tau \to 0 . \tag{D.1.5}$$

As in the proof of Lemma D.1.6 we have

$$\mathrm{meas}(\Psi^{k\tau}(Q)) \leqslant k \,\mathrm{meas}\,\Psi^\tau(Q) . \tag{D.1.6}$$

Setting $\tau = T/k$ and letting $k \to \infty$ we obtain from (D.1.5), (D.1.6) the required estimate (D.1.3). $\square$

LEMMA D.1.8. *If the set* $Q \subset S^*M|_{\partial M}$ *is measurable, then so are* $\Theta^\pm(Q) \subset S^*M|_{\partial M}$ *and* $\operatorname{meas}_{\partial M} \Theta^\pm(Q) = \operatorname{meas}_{\partial M}(Q \cap D(\Theta^\pm)) \leqslant \operatorname{meas}_{\partial M} Q$.

PROOF OF LEMMA D.1.8. As $D(\Theta^\pm)$ are open in $S^*M|_{\partial M}$ it is sufficient to prove Lemma D.1.8 under the assumption $\overline{Q} \subset D(\Theta^\pm)$ (cf. (D.1.4)). Under the latter condition for sufficiently small $\tau > 0$ there is a bijection between $[0,\tau] \times \overline{Q}$ and $\Psi^\tau(\overline{Q})$, as well as between $[0,\tau] \times \Theta^\pm(\overline{Q})$ and $\Psi^\tau(\Theta^\pm(\overline{Q}))$, so

$$\begin{aligned} \operatorname{meas} \Psi^\tau(Q) &= \tau \operatorname{meas}_{\partial M} Q + O(\tau^2), \qquad & \tau \to 0, \\ \operatorname{meas} \Psi^\tau(\Theta^\pm(Q)) &= \tau \operatorname{meas}_{\partial M} \Theta^\pm(Q) + O(\tau^2), \qquad & \tau \to 0 \end{aligned}$$

(cf. (D.1.5)). But Definition D.1.3 and Lemma D.1.6 imply

$$\operatorname{meas} \Psi^\tau(Q) = \operatorname{meas} \Psi^\tau(\Theta^\pm(Q)).$$

The latter three formulae yield $\operatorname{meas}_{\partial M} Q = \operatorname{meas}_{\partial M} \Theta^\pm(Q)$. □

For a given point $(x', \widetilde{\xi'}) \in S^*\partial M$ let us denote

$$\mathbf{S}(x', \widetilde{\xi'}) = \{(y', \widetilde{\eta}) \in S^*M|_{\partial M} : y' = x', \ (\widetilde{\eta})' = \lambda \widetilde{\xi'} \ \text{ for some } \lambda \geqslant 0\}.$$

Obviously, $\mathbf{S}(x', \widetilde{\xi'})$ is a smooth connected one-dimensional manifold with boundary. The manifold $\mathbf{S}(x', \widetilde{\xi'})$ has the topology of a line segment and its boundary consists of two points. We also have $S^*M|_{\partial M} = \bigcup_{(x', \widetilde{\xi'}) \in S^*\partial M} \mathbf{S}(x', \widetilde{\xi'})$.

LEMMA D.1.9. *Let* $Q \subset S^*M|_{\partial M}$ *be such that for any* $(x', \widetilde{\xi'}) \in S^*\partial M$ *the set* $Q \cap \mathbf{S}(x', \widetilde{\xi'})$ *contains a finite number of points. Then* $\operatorname{meas}_{\partial M} Q = 0$.

PROOF OF LEMMA D.1.9. Let $0 \leqslant \theta \leqslant \pi$ be a coordinate parametrizing $\mathbf{S}(x', \widetilde{\xi'})$; we use here the notation θ to highlight the analogy with standard spherical coordinates. Then we can use $(x', \widetilde{\xi'}, \theta)$ as coordinates on $S^*M|_{\partial M}$. We have $\left|h_{\xi_n}(x', 0, \widetilde{\xi})\right| dx'\, d\widetilde{\xi} = dx'\, d\widetilde{\xi'}\, d\mu_\theta$, where $d\mu_\theta$ is the element of a continuous measure on $\mathbf{S}(x', \widetilde{\xi'})$. The result follows by Tonelli's theorem. □

LEMMA D.1.10. *If the set* $Q \subset S^*M|_{\partial M}$ *is measurable, then so is* $\Xi(Q) \subset S^*M|_{\partial M}$ *and* $\operatorname{meas}_{\partial M} \Xi(Q) \leqslant m \operatorname{meas}_{\partial M} Q$.

PROOF OF LEMMA D.1.10. Denote

$$K := \{(x', \widetilde{\xi}) \in S^*M|_{\partial M} : x' = y', \ (\widetilde{\xi})' = (\widetilde{\eta})' \text{ for some } (y', \widetilde{\eta}) \in G\}. \tag{D.1.7}$$

Clearly, $G \subset K$. The set K is closed and satisfies the conditions of Lemma D.1.9 because for any $(x', \widetilde{\xi'}) \in S^*\partial M$ the set $K \cap \mathbf{S}(x', \widetilde{\xi'})$ contains at most $(2m-1)^2$ points. Consequently

$$\operatorname{meas}_{\partial M} K = 0. \tag{D.1.8}$$

We also have

$$\Xi(K) \subset K. \tag{D.1.9}$$

In view of (D.1.8), (D.1.9) it is sufficient to prove Lemma D.1.10 under the assumption

$$Q \subset S^*M|_{\partial M} \setminus K\,.$$

Note that $\Xi\left(S^*M|_{\partial M} \setminus K\right) = S^*M|_{\partial M} \setminus K$.

For $k = 1, 2, \ldots, m$ we define H_k as the set of all $(x', \widetilde{\xi}) \in S^*M|_{\partial M} \setminus K$ such that there are exactly $2k$ distinct points $(y', \widetilde{\eta}) \in S^*M|_{\partial M} \setminus K$ satisfying the conditions $x' = y'$, $(\widetilde{\xi})' = (\widetilde{\eta})'$. It is easy to see that H_k are nonoverlapping open subsets of $S^*M|_{\partial M}$ such that $\Xi(H_k) = H_k$. Some of them may be empty, but always $H_1 \neq \varnothing$. We have

$$S^*M|_{\partial M} \setminus K = H_1 \cup H_2 \cup \cdots \cup H_m\,.$$

Denote $H_k^\pm := \{(x', \widetilde{\xi}) \in H_k \,:\, \pm h_{\xi_n}(x', 0, \widetilde{\xi}) > 0\}$. The sets $H_k^\pm$ are open in $S^*M|_{\partial M}$ and satisfy $H_k^+ \cup H_k^- = H_k$, $H_k^+ \cap H_k^- = \varnothing$, $\Xi(H_k^\pm) = H_k^\mp$. Therefore it is sufficient to prove that

$$\text{(D.1.10)} \qquad \overline{Q} \subset H_k^\pm$$

(cf. (D.1.4)) implies

$$\text{(D.1.11)} \qquad \operatorname{meas}_{\partial M} \Xi(Q) \leqslant k \operatorname{meas}_{\partial M} Q\,.$$

For $(x', \widetilde{\xi}) \in H_k^\pm$ let us denote by $\widetilde{\eta}^{(j,\mp)}(x', \widetilde{\xi}) \in C^\infty(H_k^\pm)$, $j = 1, 2, \ldots, k$, the distinct $\widetilde{\eta}$-solutions of $(\widetilde{\xi})' = (\widetilde{\eta})'$, $(x', \widetilde{\eta}) \in H_k^\mp$. Consider the mappings $\Xi_{k,j}^\mp$ defined by $H_k^\pm \ni (x', \widetilde{\xi}) \mapsto \left(x', \widetilde{\eta}^{(j,\mp)}(x', \widetilde{\xi})\right) \in H_k^\mp$, and denote $H_{k,l}^\pm := \Xi_{k,l}^\pm(H_k^\mp)$. The sets $H_{k,l}^\pm$ are open in $S^*M|_{\partial M}$ and satisfy $H_{k,1}^\pm \cup H_{k,2}^\pm \cup \cdots \cup H_{k,k}^\pm = H_k^\pm$, $H_{k,l}^\pm \cap H_{k,l'}^\pm = \varnothing$ for $l \neq l'$, $\Xi_{k,j}^\mp(H_{k,l}^\pm) = H_{k,j}^\mp$. Obviously, in order to prove (D.1.11) under the assumption (D.1.10) it is sufficient to show that

$$\text{(D.1.12)} \qquad \operatorname{meas}_{\partial M} \Xi_{k,j}^\mp(Q) = \operatorname{meas}_{\partial M} Q \qquad \text{if} \quad \overline{Q} \subset H_{k,l}^\pm\,.$$

Formula (D.1.12) is proved by the same argument as in the proof of Lemma D.1.8. Namely, we have

$$\begin{aligned} \operatorname{meas} \Psi^\tau(Q) &= \tau \operatorname{meas}_{\partial M} Q + O(\tau^2)\,, & \tau \to +0\,,\\ \operatorname{meas} \Psi^\tau(\Xi_{k,j}^\mp(Q)) &= \tau \operatorname{meas}_{\partial M} \Xi_{k,j}^\mp(Q) + O(\tau^2)\,, & \tau \to +0\,,\\ \operatorname{meas} \Psi^\tau(Q) &= \operatorname{meas} \Psi^\tau(\Xi_{k,j}^\mp(Q))\,, & \end{aligned}$$

which yields (D.1.12). □

Further on we denote

$$\Psi(Q) = \bigcup_{t \in \mathbb{R}} \Phi^t(Q) \subset S^*\widehat{M}\,.$$

Lemma D.1.7 implies

COROLLARY D.1.11. *If* $\operatorname{meas}_{\partial M} Q = 0$, *then* $\operatorname{meas} \Psi(Q) = 0$.

Let $\Lambda_{\pm 1}(\cdot) := \Xi(\Theta^{\pm}(\cdot))$, and then inductively $\Lambda_{\pm(p+1)}(\cdot) := \Lambda_{\pm 1}(\Lambda_{\pm p}(\cdot))$, $p = 1, 2, \dots$; these operators map subsets of $S^*M|_{\partial M}$ to subsets of $S^*M|_{\partial M}$. Let Λ_0 be the identity map. Lemmas D.1.8, D.1.10 imply

COROLLARY D.1.12. *If* $\operatorname{meas}_{\partial M} Q = 0$, *then* $\operatorname{meas}_{\partial M} \Lambda_k(Q) = 0$, $\forall k \in \mathbb{Z}$.

From now on we start considering reflections from the boundary. For any $Q \subset S^*M|_{\partial M}$ we define $\Sigma(Q) \subset S^*\overset{\circ}{M}$ as the set of points $(y, \widetilde{\eta})$ for which there exists a $\tau \in \mathbb{R} \setminus \{0\}$ and a billiard trajectory $(x^*(t; y, \widetilde{\eta}), \widetilde{\xi}^*(t; y, \widetilde{\eta}))$ experiencing a finite number of transversal reflections on the open time interval between 0 and τ, and such that $(x^*(\tau \mp 0; y, \widetilde{\eta}), \widetilde{\xi}^*(\tau \mp 0; y, \widetilde{\eta})) \in Q$ for $\pm\tau > 0$.

The main result of this section is

THEOREM D.1.13. *If* $\operatorname{meas}_{\partial M} Q = 0$, *then* $\operatorname{meas} \Sigma(Q) = 0$.

PROOF OF THEOREM D.1.13. We have $\Sigma(Q) \subset \bigcup_{k \in \mathbb{Z}} \Psi(\Lambda_k(Q))$, so the required result follows from Corollaries D.1.11, D.1.12. □

It remains now to choose the set $Q \subset S^*M|_{\partial M}$ covering the awkward situations we want to exclude (conditions (2), (3) of Definition 1.3.25).

DEFINITION D.1.14. Put $\mathcal{Q} := \mathcal{Q}_1 \cup \mathcal{Q}_2$, where $\mathcal{Q}_1$ and $\mathcal{Q}_2$ are sets of points $(x', \widetilde{\xi}) \in S^*M|_{\partial M}$ satisfying the following conditions, respectively:

1. The algebraic equation $A_{2m}\left(x', 0, (\widetilde{\xi})', \zeta\right) = 1$ has a multiple ζ-root (not necessarily real).
2. The number 1 is an eigenvalue of the auxiliary one-dimensional spectral problem (1.1.7), (1.1.8) with $\xi' = (\widetilde{\xi})'$.

Note that $K \subset \mathcal{Q}_1$, where K is the set defined by (D.1.7).

LEMMA D.1.15. $\operatorname{meas}_{\partial M} \mathcal{Q} = 0$.

PROOF OF LEMMA D.1.15. Let $(x', \widetilde{\xi}')$ be an arbitrary point from $S^*\partial M$. The set $Q_1 \cap \mathbf{S}(x', \widetilde{\xi}')$ contains at most $(2m-1)^2$ points and, by Proposition A.1.13, the set $Q_2 \cap \mathbf{S}(x', \widetilde{\xi}')$ contains at most

$$(m_1 + m_2 + \cdots + m_m)\frac{(2m-1)!}{(m!)^2}$$

points. Therefore the required result follows from Lemma D.1.9. □

Obviously, Lemma D.1.15 implies Lemmas 1.3.21 and 1.3.28.

D.2. Dead-end trajectories

1. Proof of Lemma 1.3.29. Let $(y, \eta) \in T'\overset{\circ}{M}$ be a T_+-admissible point. Denote by $\mathfrak{N}(y, \eta)$ the set of all types of billiard trajectories originating from (y, η); the elements of this set are some multiindices $\mathfrak{n} = \mathfrak{n}_1|\mathfrak{n}_2|\dots|\mathfrak{n}_i|\dots|\mathfrak{n}_r$ of various (but finite!) length; see the end of 1.3.3. Clearly, $1 \leqslant \mathfrak{n}_i \leqslant m$.

Suppose that the set $\mathfrak{N}(y, \eta)$ has infinitely many elements. Denote by $\mathfrak{N}_l(y, \eta)$ the subset of $\mathfrak{N}(y, \eta)$ which consists of all multiindices from $\mathfrak{N}(y, \eta)$ having the number l in the first position. We have $\mathfrak{N}(y,\eta) = \bigcup_{l=1}^{m} \mathfrak{N}_l(y,\eta)$. Hence for at least one value of l the set $\mathfrak{N}_l(y,\eta)$ has infinitely many elements. Let us fix such an l and denote it by $\mathfrak{m}_1$.

We can apply the same procedure to the set $\mathfrak{N}_{\mathfrak{m}_1}(y,\eta)$ and find a natural number $\mathfrak{m}_2$ such that there are infinitely many multiindices from $\mathfrak{N}_{\mathfrak{m}_1}(y,\eta)$ having the number $\mathfrak{m}_2$ in the second position. Repeating our procedure infinitely many times we obtain a multiindex $\mathfrak{m} = \mathfrak{m}_1|\mathfrak{m}_2|\mathfrak{m}_3|\dots$ of infinite length. The corresponding billiard trajectory experiences an infinite number of reflections on the time interval $(0\,,T_+)$, which contradicts Definitions 1.3.25, 1.3.26. □

Clearly, Lemma 1.3.29 implies Lemma 1.3.27.

2. Proof of Lemma 1.3.31. The proof given below is similar to that of Lemma 2 in [**CorFomSin,** Chap. 6, Sect. 1]. The following result (cf. Lemma D.1.10) plays a crucial role in our arguments.

LEMMA D.2.1. *Let the simple reflection condition be fulfilled. Then, if the set* $Q \subset S^*M|_{\partial M}$ *is measurable, so is* $\Xi(Q)$ *and* $\operatorname{meas}_{\partial M}\Xi(Q) = \operatorname{meas}_{\partial M} Q$.

PROOF OF LEMMA D.2.1. The proof of Lemma D.2.1 repeats that of Lemma D.1.10, but with obvious simplifications: for $k=1$ (D.1.11) turns into an equality, whereas for $k>1$ we get $H_k = \varnothing$. □

Further on in this subsection we assume simple reflection.

Let us view Ξ as an operator mapping points of $D(\Xi) := S^*M|_{\partial M} \setminus K = H_1$ (see the proof of Lemma D.1.10) to points of $D(\Xi)$. This is possible because of simple reflection (recall that in Definition D.1.4 we viewed Ξ as an operator mapping sets to sets). The mapping $\Xi : D(\Xi) \to D(\Xi)$ is a smooth bijection with a smooth inverse $\Xi^{-1} = \Xi$. By (D.1.8), $D(\Xi) \subset S^*M|_{\partial M}$ is measurable, and by Lemma D.2.1, Ξ preserves $\operatorname{meas}_{\partial M}(\,\cdot\,)$. Thus, Ξ is an *endomorphism* of the space $D(\Xi)$; see Definition 2 in [**CorFomSin,** Chap. 1, Sect. 1].

Let us denote by U^+ (U^-) the set of points $(y',\widetilde{\eta}) \in D(\Xi)$ such that the billiard trajectory originating from $\Xi(y',\widetilde{\eta})$ experiences an infinite number of transversal reflections in a finite positive (negative) time. Obviously, a dead-end trajectory goes through $\Xi(U^+)$ or $\Xi(U^-)$ at the moment it hits the boundary for the first time; more precisely, by "first time" we understand $t=\tau-0$ and $t=\tau+0$ in the case of positive and negative moment of reflection τ, respectively. So by Corollary D.1.11 in order to prove Lemma 1.3.31 it is sufficient to show that

$$\operatorname{meas}_{\partial M} U^{\pm} = 0\,. \tag{D.2.1}$$

From now on we shall also view the operators $\Theta^{\pm}$ as operators mapping points to points (cf. Definition D.1.3); here, in order to avoid introducing new notation, we still allow the trajectories to go onto the extended manifold. Thus, we have smooth bijections $\Theta^{\pm} : D(\Theta^{\pm}) \to D(\Theta^{\mp})$. By Lemma D.1.8 the operators $\Theta^{\pm}$ preserve $\operatorname{meas}_{\partial M}(\,\cdot\,)$; however, they are not necessarily endomorphisms because in the general case $D(\Theta^{\pm}) \ne D(\Theta^{\mp})$.

Denote $\widetilde{D}(\Theta^{\pm}) = \Theta^{\mp}(D(\Theta^{\mp}) \setminus K) \setminus K$. Of course, $\widetilde{D}(\Theta^{\pm}) \subset D(\Theta^{\pm})$ and by (D.1.8) and Lemma D.1.8, $\operatorname{meas}_{\partial M}(D(\Theta^{\pm}) \setminus \widetilde{D}(\Theta^{\pm})) = 0$. It will be convenient for us to view the operators $\Theta^{\pm}$ as operators acting from $\widetilde{D}(\Theta^{\pm})$ to $\widetilde{D}(\Theta^{\mp})$ (this is still a smooth bijection). The advantage is that $\widetilde{D}(\Theta^{\pm}) \subset D(\Xi)$.

Consider the operators

$$\Lambda_{\pm 1}(\,\cdot\,) := \Xi(\Theta^{\pm}(\,\cdot\,)) : D(\Lambda_{\pm 1}) \to \Xi(D(\Lambda_{\mp 1}))\,, \qquad D(\Lambda_{\pm 1}) = \widetilde{D}(\Theta^{\pm})\,,$$

and then inductively $\Lambda_{\pm(p+1)}(\,\cdot\,) := \Lambda_{\pm 1}(\Lambda_{\pm p}(\,\cdot\,))$, $p = 1, 2, \dots$. We have

$$\Lambda_{\pm(p+1)} : D(\Lambda_{\pm(p+1)}) \to \Xi(D(\Lambda_{\mp(p+1)}))\,,$$

where $D(\Lambda_{\pm(p+1)}) = \Xi(\Lambda_{\mp p}(D(\Lambda_{\mp p}) \cap \Xi(D(\Lambda_{\pm 1}))))$.

By Λ_0 we shall denote the identity mapping in $D(\Xi)$. The operators Λ_k, $k \in \mathbb{Z}$, are bijective and preserve $\mathrm{meas}_{\partial M}(\,\cdot\,)$. Moreover, $\Lambda_k^{-1}(\,\cdot\,) = \Xi(\Lambda_{-k}(\Xi(\,\cdot\,)))$.

Denote $V^{\pm} = \bigcap_{p\in\mathbb{N}} D(\Lambda_{\pm p})$. The set V^+ (V^-) is the set of points $(y', \widetilde{\eta}) \in D(\Xi)$ such that the billiard trajectory originating from $\Xi(y', \widetilde{\eta})$ experiences an infinite number of transversal reflections in positive (negative) time. Clearly, $U^{\pm} \subset V^{\pm}$, where the points from $U^{\pm}$ are singled out by the condition that the infinite number of reflections occurs in a finite time. It is also clear from the inductive construction given above that the sets $V^{\pm} \subset S^*M|_{\partial M}$ are measurable.

We have

$$V^{\pm} \supset \Lambda_{\pm 1}(V^{\pm}) \supset \Lambda_{\pm 2}(V^{\pm}) \supset \cdots\,. \tag{D.2.2}$$

Denote

$$V^{\pm}_{\infty} = \bigcap_{p\in\mathbb{N}} \Lambda_{\pm p}(V^{\pm})\,. \tag{D.2.3}$$

The sets $V^{\pm}_{\infty}$ are the sets of points $(y', \widetilde{\eta}) \in D(\Xi)$ such that the billiard trajectory originating from $\Xi(y', \widetilde{\eta})$ experiences an infinite number of transversal reflections *both* in positive and negative time. The sets V^+_{∞} and V^-_{∞} are related by the identity $V^{\pm}_{\infty} = \Xi(V^{\mp}_{\infty})$.

Formulae (D.2.2), (D.2.3) and the fact that the operators $\Lambda_{\pm p}$ preserve $\mathrm{meas}_{\partial M}(\,\cdot\,)$ imply that

$$\mathrm{meas}_{\partial M}(V^{\pm}) = \mathrm{meas}_{\partial M}(V^{\pm}_{\infty})\,;$$

consequently, proving (D.2.1) is equivalent to proving

$$\mathrm{meas}_{\partial M}\, W^{\pm} = 0\,, \tag{D.2.4}$$

where $W^{\pm} = U^{\pm} \cap V^{\pm}_{\infty}$.

The mappings $\Lambda_{\pm 1}$ are endomorphisms of the spaces $V^{\pm}_{\infty}$, so we can apply standard results from [**CorFomSin**]. Set $f_{\pm}(\,\cdot\,) = \pm T_{\pm}(\Xi(\,\cdot\,)) : V^{\pm}_{\infty} \to (0, +\infty)$, where $T_{\pm}$ are the functions defined at the beginning of subsection 1 (the time of next reflection in the $^+$ case and the time of previous reflection in the $^-$ case). The functions $f_{\pm}$ are positive measurable functions, so by Lemma 1 of [**CorFomSin,** Chap. 1, Sect. 1] for almost all points $(y', \widetilde{\eta}) \in V^{\pm}_{\infty}$ we have

$$\sum_{p=1}^{\infty} f_{\pm}(\Lambda_{\pm p}(y', \widetilde{\eta})) = \infty \tag{D.2.5}$$

(this abstract result is a consequence of the Poincaré Recurrence Theorem). But condition (D.2.5) is equivalent to $(y', \widetilde{\eta}) \notin W^{\pm}$, which proves (D.2.4). □

Other sets of sufficient conditions which guarantee nonblocking without requiring simple reflection are given in [**Va6**], [**Va7**].

D.3. Convexity and concavity

Throughout this section the billiard system is assumed to satisfy the strong simple reflection condition (but is not necessarily a geodesic one).

1. Proof of Lemma 1.3.17. Assuming that the manifold M is strongly convex let us prove that there are no grazing billiard trajectories. Suppose that a grazing trajectory exists, i.e., there is a $(y,\eta)\in T'\overset{\circ}{M}$ and a $\tau\in\mathbb{R}\setminus\{0\}$ such that the trajectory $(x^*(t;y,\eta),\xi^*(t;y,\eta))$ experiences a finite number of transversal reflections on the open time interval between 0 and τ, and $x_n^*(\tau;y,\eta)=\dot{x}_n^*(\tau;y,\eta)=0$. For definiteness we shall deal with the case $\tau>0$. By Taylor's formula

$$x_n^*(t;y,\eta) = -\frac{\mathbf{k}(x^{*\prime}(\tau;y,\eta),\xi^{*\prime}(\tau;y,\eta))}{2}(t-\tau)^2 + O((t-\tau)^3)$$

as $t\to\tau-0$, so $x_n^*(t;y,\eta)<0$ for $t\in(0,\tau)$ sufficiently close to τ. This is a contradiction because our billiard trajectory is not allowed to go outside the manifold.

Still assuming the manifold M to be strongly convex let us prove that there are no dead-end billiard trajectories. For definiteness we shall deal with the case when there is an infinite number of reflections in a finite positive time. Let t_k^*, t_{k+1}^*, and t_{k+2}^* be three consecutive moments of reflection of a given billiard trajectory. Elementary analysis of the Taylor expansions for the trajectory in powers of $t-t_k^*$ on the interval (t_k^*,t_{k+1}^*), and in powers of $t-t_{k+1}^*$ on the interval (t_{k+1}^*,t_{k+2}^*) gives the estimate

$$\text{(D.3.1)}\qquad t_{k+2}^*-t_{k+1}^* = t_{k+1}^*-t_k^* + O((t_{k+1}^*-t_k^*)^2)$$

as $t_{k+1}^*-t_k^*\to 0$. The estimate (D.3.1) is uniform over all possible reflections of all possible trajectories. Now consider a dead-end trajectory and denote for brevity $f_k=t_{k+1}^*-t_k^*$. Then $f_k\to+0$ as $k\to\infty$, and we obtain from (D.3.1)

$$\text{(D.3.2)}\qquad f_{k+1}\geqslant f_k - C\,f_k\,f_{k+1}\,,\qquad k\geqslant k_0\,,$$

where k_0 is a sufficiently large natural number and C is a positive constant independent of k. Let $\mathbf{f}_k$ be the solution of the recurrent equation

$$\text{(D.3.3)}\qquad \mathbf{f}_{k+1} = \mathbf{f}_k - C\,\mathbf{f}_k\,\mathbf{f}_{k+1}\,,\qquad k\geqslant k_0\,,$$

subject to the initial condition $\mathbf{f}_{k_0}=f_{k_0}$. Comparing (D.3.2) and (D.3.3) it is easy to see (by induction) that

$$\text{(D.3.4)}\qquad f_k\geqslant\mathbf{f}_k>0\,,\qquad k\geqslant k_0\,.$$

The equation (D.3.3) is solved explicitly

$$\text{(D.3.5)}\qquad \mathbf{f}_k = \frac{f_{k_0}}{1+Cf_{k_0}(k-k_0)}\,.$$

Formulae (D.3.4), (D.3.5) imply that $\sum_{k=k_0}^{\infty}f_k=\infty$, which contradicts the fact that the billiard trajectory is a dead-end one.

Finally, assuming that the manifold M is strongly concave let us prove that there are no dead-end billiard trajectories. This follows immediately from the fact that the time $t_{k+1}^*-t_k^*$ between two successive reflections is bounded from below by a positive constant uniformly over all possible reflections of all possible trajectories. □

2. Sufficient conditions for nonperiodicity. This subsection is devoted to the proof of Lemma 1.3.34. First we will prove the following auxiliary

LEMMA D.3.1. *In the analytic case, if the manifold M is convex then there are no grazing billiard trajectories.*

The statement of Lemma D.3.1 is similar to that of the first part of Lemma 1.3.17, but now the manifold M is not required to be strongly convex. Recall also that convexity implies the fulfilment of the strong simple reflection condition.

PROOF OF LEMMA D.3.1. Suppose that a grazing trajectory exists, i.e., there is a $(y,\eta)\in T'\overset{\circ}{M}$ and a $\tau\in\mathbb{R}\setminus\{0\}$ such that the trajectory $(x^*(t;y,\eta),\xi^*(t;y,\eta))$ experiences a finite number of transversal reflections on the open time interval between 0 and τ, and $x^*_n(\tau;y,\eta)=\dot{x}^*_n(\tau;y,\eta)=0$. For definiteness we shall deal with the case $\tau>0$. Further on we assume that $t\in(0,\tau)$ is sufficiently close to τ, and that (x,ξ) is sufficiently close to $(x^*(\tau-0;y,\eta),\xi^*(\tau-0;y,\eta))$.

We have

$$x^*_n(t) \;=\; c\,(\tau-t)^p \;+\; O((\tau-t)^{p+1}) \tag{D.3.6}$$

as $t\to\tau-0$, where $p\geqslant 2$ is some natural number and

$$c>0 \tag{D.3.7}$$

is some constant. Note that here we have used analyticity, because in the nonanalytic case $x^*_n(t)$ might have had an infinite order zero. Differentiating (D.3.6) we get

$$\dot{x}^*_n(t) \;=\; -\,p\,c\,(\tau-t)^{p-1} \;+\; O((\tau-t)^p)\,, \tag{D.3.8}$$

$$\ddot{x}^*_n(t) \;=\; p\,(p-1)\,c\,(\tau-t)^{p-2} \;+\; O((\tau-t)^{p-1})\,. \tag{D.3.9}$$

In view of the strong simple reflection condition the equation $\dot{x}^*_n=h_{\xi_n}(x^*,\xi^*)$ can be resolved with respect to ξ_n. Substituting the resulting $\xi_n=\xi_n(x^*,\xi^{*\prime},\dot{x}^*_n)$ into the equation $\ddot{x}^*_n=-\{h_{\xi_n},h\}(x^*,\xi^*)$ we get

$$\ddot{x}^*_n=-k(x^*,\xi^{*\prime},\dot{x}^*_n) \tag{D.3.10}$$

with some analytic function k. Of course,

$$k(x',0,\xi',0)=\mathbf{k}(x',\xi')\geqslant 0\,, \tag{D.3.11}$$

where $\mathbf{k}$ is the Hamiltonian curvature of ∂M. Substituting (D.3.6), (D.3.8), (D.3.9) into (D.3.10) we get

$$p\,(p-1)\,c\,(\tau-t)^{p-2} \;=\; -\,\mathbf{k}(x^{*\prime}(t),\xi^{*\prime}(t)) \;+\; O((\tau-t)^{p-1})\,. \tag{D.3.12}$$

Formulae (D.3.12), (D.3.11) imply $c\leqslant 0$, which contradicts (D.3.7). □

Further on in this subsection we always assume analyticity and convexity.

DEFINITION D.3.2. Let T be a positive real number, and l a nonnegative integer. We shall say that the point $(y,\eta) \in T'\overset{\circ}{M}$ is (T,l)-periodic if the billiard trajectory originating from (y,η) is periodic with period T after l reflections.

Note that in the above definition the reflections are transversal by Lemma D.3.1.

DEFINITION D.3.3. We shall say that the point $(y_0,\eta_0) \in T'\overset{\circ}{M}$ is absolutely (T,l)-periodic if it is (T,l)-periodic and all the points (y,η) from a (small) neighbourhood of (y_0,η_0) are (T,l)-periodic with the same T and l.

By analyticity the above definition is equivalent to

DEFINITION D.3.3′. We shall say that the point $(y_0,\eta_0) \in T'\overset{\circ}{M}$ is absolutely (T,l)-periodic if the function

$$|x^*(T;y,\eta) - y|^2 \ + \ |\xi^*(T;y,\eta) - \eta|^2$$

of the variables (y,η) has an infinite order zero at (y_0,η_0).

Of course, Definition D.3.3′ is an analogue of Definition 1.3.2.

LEMMA D.3.4. *If there exists a point $(y^{(0)},\eta^{(0)}) \in T'\overset{\circ}{M}$ which is absolutely (T,l)-periodic, then all the points $(y,\eta) \in T'\overset{\circ}{M}$ are absolutely (T,l)-periodic with the same T and l.*

PROOF OF LEMMA D.3.4. Consider an arbitrary point $(y^{(1)},\eta^{(1)}) \in T'\overset{\circ}{M}$. As we assume our manifold to be connected, the points $(y^{(0)},\eta^{(0)})$ and $(y^{(1)},\eta^{(1)})$ can be connected by a continuous curve $(y^{(s)},\eta^{(s)}) \in T'\overset{\circ}{M}$, $s \in [0,1]$. Denote by S the set of $s \in [0,1]$ such that $(y^{(s)},\eta^{(s)})$ is absolutely (T,l)-periodic with given T and l. Definition D.3.3 implies that the set S is open in $[0,1]$, whereas Definition D.3.3′ implies that the set S is closed in $[0,1]$. The only two subsets of $[0,1]$ which are simultaneously open and closed in $[0,1]$ are the empty set and the set $[0,1]$ itself. Our set S is nonempty because $0 \in S$, so $S = [0,1]$ and, consequently, the point $(y^{(1)},\eta^{(1)})$ is absolutely (T,l)-periodic. □

We are now prepared to proceed to the

PROOF OF LEMMA 1.3.34. Suppose that under the conditions of Lemma 1.3.34 the nonperiodicity condition is not fulfilled. Then by Lemmas 1.3.31 and 1.3.24 the set of absolutely periodic points has nonzero measure. Consequently by Lemma D.3.4 all the points $(y,\eta) \in T'\overset{\circ}{M}$ are absolutely (T,l)-periodic with the same T and l. As $\partial M \neq \varnothing$, $\mathbf{k} \geqslant 0$ and $\mathbf{k} \not\equiv 0$, we can choose a point $(y_0',\eta_0') \in T'\partial M$ at which $\mathbf{k}(y_0',\eta_0') > 0$. Let δ be a small positive parameter. Set $y(\delta) = (y_0',\delta)$ and take $\eta(\delta)$ such that $\dot{x}_n^*(0;y(\delta),\eta(\delta)) = 0$ (that is, we send the ray parallel to ∂M). Denote by $t_j^*(\delta) > 0$ the moment of the jth reflection of the billiard trajectory originating from $(y(\delta),\eta(\delta))$. Elementary analysis of the formula (D.3.10)) gives

$$t_j^*(\delta) \ = \ (2j-1)\sqrt{\frac{2\delta}{\mathbf{k}(y_0',\eta_0')}} + O(\delta), \qquad j = 1,2,\dots, \tag{D.3.13}$$

as $\delta \to 0$ (here the remainder term depends, in general, on j). Formula (D.3.13) implies that for sufficiently small δ we have $t_{l+1}^*(\delta) \leqslant T$, which contradicts the fact that the point $(y(\delta),\eta(\delta))$ is absolutely (T,l)-periodic. □

D.4. Measurability of sets and functions

1. Proof of Lemmas 1.3.23 and 1.8.2. Let T_+ be an arbitrary positive number. Assuming that the nonblocking condition is fulfilled, we shall prove that the set

$$\Pi(T_+) := \bigcup_{T\in(0,T_+]} \Pi_T \subset S^*\overset{\circ}{M}$$

is measurable, which by countable additivity would imply that Π is measurable.

Let O_{T_+} be the set of all T_+-admissible points from $S^*\overset{\circ}{M}$. In view of Lemma 1.3.27 the set O_{T_+} is open in $S^*\overset{\circ}{M}$, and consequently measurable. Moreover, by Lemma 1.3.28 $\operatorname{meas}(S^*\overset{\circ}{M} \setminus O_{T_+}) = 0$.

Let us fix an arbitrary point $(y_0,\widetilde{\eta}_0) \in O_{T_+}$. Then, in order to prove the measurability of $\Pi(T_+)$ it is sufficient to prove the measurability of $\Pi(T_+) \cap O_{T_+}(y_0,\widetilde{\eta}_0)$, where $O_{T_+}(y_0,\widetilde{\eta}_0) \subset O_{T_+}$ is some small neighbourhood of $(y_0,\widetilde{\eta}_0)$. Indeed, this would imply that for any compact set $K \subset O_{T_+}$ the set $\Pi(T_+)\cap K$ is measurable, and then one has only to choose a sequence of compact sets $K_1 \subset K_2 \subset \cdots$ such that $\bigcup_j K_j = O_{T_+}$.

By Lemma 1.3.29 the number of types of billiard trajectories originating from $O_{T_+}(y_0,\widetilde{\eta}_0)$ is finite if $O_{T_+}(y_0,\widetilde{\eta}_0)$ is sufficiently small. We can now enumerate all legs of all types of billiard trajectories by the index j which runs through a finite number of values. Let us denote by $\Pi_j(T_+)$ the set of $(y,\widetilde{\eta}) \in O_{T_+}(y_0,\widetilde{\eta}_0)$ such that the billiard trajectory originating from $(y,\widetilde{\eta})$ returns to $(y,\widetilde{\eta})$ on the jth leg at some moment of time $t\in(0,T_+]$. Clearly, in order to prove the measurability of $\Pi(T_+)\cap O_{T_+}(y_0,\widetilde{\eta}_0)$ it is sufficient to prove the measurability of $\Pi_j(T_+)\cap O_{T_+}(y_0,\widetilde{\eta}_0)$ for each j.

Denote by $P_j(T_+)$ the set of $(t;y,\widetilde{\eta}) \in (0,T_+]\times O_{T_+}(y_0,\widetilde{\eta}_0)$ such that the billiard trajectory originating from $(y,\widetilde{\eta})$ returns to $(y,\widetilde{\eta})$ on the jth leg at the time t. Clearly, the set $P_j(T_+)$ is closed in $(0,T_+]\times O_{T_+}(y_0,\widetilde{\eta}_0)$. Moreover, it is easy to see that it is closed in $[0,T_+]\times O_{T_+}(y_0,\widetilde{\eta}_0)$. The set $\Pi_j(T_+)$ is a projection of $P_j(T_+)$, so $\Pi_j(T_+)$ is closed in $O_{T_+}(y_0,\widetilde{\eta}_0)$, and consequently measurable. Lemma 1.3.24 is proved.

Lemma 1.8.2 is proved in a similar way. First we note that the set $O_{y,T_+} \subset S^*_yM$ of admissible "directions" is open in S^*_yM. Since y is a regular point, the complement of this set in S^*_yM is of measure zero. Then we take a sufficiently small neighbourhood $O_{y,T_+}(\widetilde{\eta}_0) \subset O_{y,T_+}$ of a given $\widetilde{\eta}_0$ and introduce the sets $\Pi_{y,j}(T_+)$ which correspond to different legs of billiard trajectories originating from $O_{y,T_+}(\widetilde{\eta}_0)$. The sets $\Pi_{y,j}(T_+)$ are closed in $O_{y,T_+}(\widetilde{\eta}_0)$. This implies that the sets $\Pi_y(T_+)\cap O_{y,T_+}(\widetilde{\eta}_0)$, and consequently $\Pi_y(T_+) = \bigcup_{T\in(0,T_+]}\Pi_{y,T}$, are measurable.

2. Proof of Lemmas 1.3.24 and 1.8.3. As in the previous subsection we conclude that in order to prove Lemma 1.3.24 it is sufficient to show that

$$\operatorname{meas}\Big(\bigcup_{0<T\leqslant T_+}(\Pi_{T,j}\setminus\Pi^a_{T,j})\cap O_{T_+}(y_0,\widetilde{\eta}_0)\Big) = 0,$$

where T_+ is an arbitrary positive number, $(y_0,\widetilde{\eta}_0)$ is an arbitrary point from O_{T_+}, $O_{T_+}(y_0,\widetilde{\eta}_0)\subset O_{T_+}$ is a small neighbourhood of $(y_0,\widetilde{\eta}_0)$, and $\Pi_{T,j}$ and $\Pi^a_{T,j}$ are the sets of T-periodic and absolutely T-periodic points corresponding to the jth legs of the billiard trajectories. But this immediately follows from Corollary 4.1.9.

Similarly, Lemma 1.8.3 follows from Corollary 4.1.15.

3. Proof of Lemma 1.8.9. By similar arguments we conclude that it is sufficient to prove Lemma 1.8.9 assuming that

$$\Omega \subset \{\widetilde{\eta} \in O_{y,T_+}(\widetilde{\eta}_0) \,:\, \mathbf{T}_y(\widetilde{\eta}) \leqslant T_+\},$$

where $O_{y,T_+}(\widetilde{\eta}_0) \subset O_{y,T_+}$ is a sufficiently small neighbourhood of $\widetilde{\eta}_0$. But for such Ω the required result has already been established in Lemma 4.3.3.

4. Proof of Lemmas 1.7.5 and 1.8.8. In order to prove Lemma 1.7.5 it is sufficient to show that the restrictions of the functions $\mathbf{T}$, $\mathbf{q}$ to the set

$$\bigcup_{0<T\leqslant T_+} \Pi_T^a \cap O_{T_+}(y_0, \widetilde{\eta}_0)$$

are measurable. Assuming the neighbourhood $O_{T_+}(y_0,\widetilde{\eta}_0)$ to be sufficiently small, we choose a covering of $[0, T_+]$ by small intervals (t_j, t'_j), $j = 0, 1, \ldots, p$, such that

1. For any j the quantity $2(t'_j - t_j)$ is less than $\inf \mathbf{T}$ (see Lemma 1.7.4) and less than the minimal time which is needed for trajectories originating at $O_{T_+}(y_0,\widetilde{\eta}_0)$ to reach the boundary.
2. $t_{j-1} < t_j$, $t'_{j-1} < t'_j$, and $[t_{j-1}, t'_{j-1}] \cap [t_{j+1}, t'_{j+1}] = \varnothing$.
3. If $x^*(t; y, \eta) \notin \partial M$ for all $t \in [t_j, t'_j]$ and $(y,\eta) \in O_{T_+}(y_0,\widetilde{\eta}_0)$, then there exists an associated phase function φ_j which is real for all $(y,\eta) \in O_{T_+}(y_0,\widetilde{\eta}_0)$, $t \in [t_j, t'_j]$, and x close to $x^*(t; y, \eta)$.

Clearly, if $x^*(t; y, \eta) \in \partial M$ for some $t \in [t_j, t'_j]$ and $(y,\eta) \in O_{T_+}(y_0,\widetilde{\eta}_0)$, then $(x^*(t; y, \eta), \xi^*(t; y, \eta)) \neq (y, \eta)$ for all $t \in [t_j, t'_j]$ (so the time interval $[t_j, t'_j]$ does not contain any periods of the trajectories originated from $O_{T_+}(y_0,\widetilde{\eta}_0)$). Therefore we need to consider only the j which satisfy (3).

Let $\widetilde{t}_\star^{(j)}(y,\eta)$ be the function defined by $\varphi_j(\widetilde{t}_\star^{(j)}, y; y, \eta) = 0$. By Lemma 4.1.8,

$$\Pi^{(j)} = \bigcup_{t_j \leqslant T \leqslant t'_j} \Pi_T^a \cap O_{T_+}(y_0, \widetilde{\eta}_0)$$

coincides with the set of points at which $\nabla \widetilde{t}_\star^{(j)}$ has infinite order zeros, and

$$\widetilde{t}_\star^{(j)}(y,\eta) = k\,\mathbf{T}(y,\eta), \qquad \forall (y,\eta) \in \bigcup_{t_j \leqslant T \leqslant t'_j} \Pi_T^a \cap O_{T_+}(y_0, \widetilde{\eta}_0). \tag{D.4.1}$$

Here $k \in \mathbb{N}$ may depend on (y,η). However, k is uniquely determined by j and (y,η) due to condition (1).

Let $\Pi_k^{(j)}$ be the subset of $\bigcup_{0<T\leqslant T} \Pi_T^a \cap O_{T_+}(y_0,\widetilde{\eta}_0)$ where (D.4.1) holds for these particular values of j and k. We have $\Pi^{(0)} = \varnothing$, $\Pi^{(1)} = \Pi_1^{(1)}$ and, in view of conditions (1) and (2),

$$\Pi_1^{(j)} = \Pi^{(j)} \setminus \bigcup_{j' < j-1} \Pi^{(j')}.$$

Since the sets $\Pi^{(j)}$ are closed in $O_{T_+}(y_0,\widetilde{\eta}_0)$ (and consequently measurable), the sets $\Pi_1^{(j)}$ are measurable. Obviously,

$$\{(y,\eta) \in \Pi^a \cap O_{T_+}(y_0,\widetilde{\eta}_0) : t_j \leqslant \mathbf{T}(y,\eta) \leqslant t'_j\} = \Pi_1^{(j)}$$

and the restriction of $\mathbf{T}$ to this set coincides with the restriction of the smooth function $\widetilde{t}_\star^{(j)}$. This implies that the restriction of $\mathbf{T}$ to $\bigcup_{0<T\leqslant T_+}\Pi_T^a\cap O_{T_+}(y_0,\widetilde{\eta}_0)$ is measurable. The function $\mathbf{q}$ is measurable as a composition of a smooth function $\mathfrak{f}$ and a measurable function $\mathbf{T}$ (here we use the fact that $\mathfrak{f}_c$ is a restriction of a smooth function to the set $\{(y,\eta):x_\eta^*(y,\eta)=0\}$; see 1.5.3).

Similarly, Lemma 1.8.8 follows from Lemma 4.1.14 if we assume that $2(t_j'-t_j)$ is less than $\inf\mathbf{T}_y$ (see Lemma 1.8.10).

D.5. Lengths of loops and periodic trajectories

All the results in this section were formulated in Sections 1.7, 1.8 under the simple reflection condition. However, it will be clear from the proofs that they remain valid in the general case.

1. Estimates for lengths of loops. First, we give the

PROOF OF LEMMA 1.8.10. Sufficiently short loops originating from $S^*M|_K$ do not experience reflections. Therefore the required result follows from the fact that the smooth vector function $\dot{x}^*(t;y,\widetilde{\eta})$ does not vanish and, consequently, for small $t\neq 0$ we have $x^*(t;y,\widetilde{\eta})\neq y$ uniformly over all $(y,\widetilde{\eta})\in S^*M|_K$. □

In the case of a manifold without boundary we can take $K=M$, which gives $\inf_{y\in M,\,\widetilde{\eta}\in\Pi_y}\mathbf{T}_y(\widetilde{\eta})>0$. The following lemma shows that in the case of a manifold with boundary the condition $K\subset\overset{\circ}{M}$ is essential.

LEMMA D.5.1. *If* $\partial M\neq\varnothing$, *then* $\inf_{y\in M,\,\widetilde{\eta}\in\Pi_y}\mathbf{T}_y(\widetilde{\eta})=0$.

PROOF. Take an arbitrary point $z=(z',0)\in\partial M$. Set $\zeta=(0,\dots,0,\zeta_n)$, where ζ_n is the negative number defined by the condition $h(z,\zeta)=1$. For small $\tau>0$ set $(y,\widetilde{\eta})=(x^*(-\tau;z,\zeta),\widetilde{\xi}^*(-\tau;z,\zeta))$. Then the billiard trajectory originating from $(y,\widetilde{\eta})$ forms a loop of length 2τ with one reflection (here we use the symmetry of our Hamiltonian, $h(x,\xi)=h(x,-\xi)$). As τ can be taken arbitrarily small this proves the lemma. □

2. Estimate for periods of periodic trajectories. Let $(y,\widetilde{\eta})\in S^*\overset{\circ}{M}$ and $T_\Gamma>0$. Consider a billiard trajectory

$$\Gamma=\left(x^*(t;y,\widetilde{\eta}),\,\widetilde{\xi}^*(t;y,\widetilde{\eta})\right)\subset S^*M,\qquad 0\leqslant t\leqslant T_\Gamma,$$

which experiences a finite number p_Γ of transversal reflections on the time interval $(0,T_\Gamma)$, and satisfies the conditions

$$\begin{aligned}(x^*(t;y,\widetilde{\eta}),\widetilde{\xi}^*(t;y,\widetilde{\eta}))&\neq(y,\widetilde{\eta}),\qquad\forall t\in(0,T_\Gamma),\\(x^*(T_\Gamma;y,\widetilde{\eta}),\widetilde{\xi}^*(T_\Gamma;y,\widetilde{\eta}))&=(y,\widetilde{\eta}).\end{aligned}$$

Such a billiard trajectory will be called a periodic trajectory of period T_Γ with p_Γ reflections. The set of all periodic trajectories will be denoted by $\mathfrak{P}$.

Any periodic trajectory is either a loop itself or contains a loop, so $T_\Gamma\geqslant\mathbf{T}_y(\widetilde{\eta})$. Therefore in the case $\partial M=\varnothing$ the required statement follows from Lemma 1.8.10. However, the case $\partial M\neq\varnothing$ cannot be handled in such a simple manner in view of Lemma D.5.1.

Let $\partial M \neq \varnothing$, and let us prove that $\inf_{\Gamma\in\mathfrak{P}} T_\Gamma > 0$.

Suppose that $\inf_{\Gamma\in\mathfrak{P}} T_\Gamma = 0$. Then there exists a sequence of periodic trajectories

$$\Gamma_j = (x^*(t;y^{(j)},\widetilde{\eta}^{(j)}),\widetilde{\xi}^*(t;y^{(j)},\widetilde{\eta}^{(j)})), \qquad j=1,2,\dots,$$

of period T_{Γ_j} with p_{Γ_j} reflections such that $T_{\Gamma_j}\to 0$ as $j\to\infty$.

We have

$$x_n^*(t;y^{(j)},\widetilde{\eta}^{(j)}) \to 0 \quad \text{as} \quad j\to\infty \tag{D.5.1}$$

uniformly over $t\in[0,T_{\Gamma_j}]$. Otherwise we could extract a subsequence of periodic trajectories Γ_{j_k}, $k=1,2,\dots$, separated from the boundary and apply Lemma 1.8.10.

We also have

$$\dot{x}_n^*(t;y^{(j)},\widetilde{\eta}^{(j)}) \to 0 \quad \text{as} \quad j\to\infty \tag{D.5.2}$$

uniformly over $t\in[0,T_{\Gamma_j}]$ (here at the moments of reflection we consider both one-sided derivatives). Indeed, suppose that (D.5.2) does not hold. Then there exists an $\varepsilon>0$, a subsequence of periodic trajectories Γ_{j_k}, $k=1,2,\dots$, and a sequence of numbers $\tau_k\in[0,T_{\Gamma_{j_k}}]$ such that $\dot{x}_n^*(\tau_k;y^{(j_k)},\widetilde{\eta}^{(j_k)})>\varepsilon$, or a similar subsequence with $\dot{x}_n^*(\tau_k;y^{(j_k)},\widetilde{\eta}^{(j_k)})<-\varepsilon$. For definiteness let us deal with the first case. We can choose a $\delta>0$ such that for all $k=1,2,\dots$ the trajectory originating from the point $(x^*(\tau_k;y^{(j_k)},\widetilde{\eta}^{(j_k)}),\widetilde{\xi}^*(\tau_k;y^{(j_k)},\widetilde{\eta}^{(j_k)}))$ does not reach the boundary on the time interval $(0,\delta]$. Using the periodicity of Γ_{j_k} and applying our usual arguments for loops without reflection we conclude that $T_{\Gamma_{j_k}}\geqslant \text{const}>0$, which is a contradiction.

Since S^*M is compact, we may assume without loss of generality that

$$(y^{(j)},\widetilde{\eta}^{(j)}) \to (z,\widetilde{\zeta}) \quad \text{as} \quad j\to\infty. \tag{D.5.3}$$

In view of (D.5.1), $z=(z',0)\in\partial M$.

In local coordinates, we have

$$(\widetilde{\xi}^*(t;y^{(j)},\widetilde{\eta}^{(j)}))' = (\widetilde{\eta}^{(j)})' - \int_0^t h_{x'}(x^*(\tau;y^{(j)},\widetilde{\eta}^{(j)},\widetilde{\xi}^*(\tau;y^{(j)},\widetilde{\eta}^{(j)}))\,d\tau$$

(note that a similar formula does not hold for ξ_n^* because ξ_n^* is discontinuous at reflections). This formula and (D.5.3) imply

$$(\widetilde{\xi}^*(t;y^{(j)},\widetilde{\eta}^{(j)}))' \to (\widetilde{\zeta})' \quad \text{as} \quad j\to\infty \tag{D.5.4}$$

uniformly over $t\in[0,T_{\Gamma_j}]$.

In our local coordinates $\widetilde{\zeta}\in S_z^*M$ is identified with a vector $\zeta\in\mathbb{R}^n$ such that $h(z,\zeta)=1$. From (D.5.2), (D.5.4) and Euler's identity we obtain

$$\langle \zeta\,,\,\dot{x}^*(t;y^{(j)},\widetilde{\eta}^{(j)})\rangle \to 1 \quad \text{as} \quad j\to\infty \tag{D.5.5}$$

uniformly over $t\in[0,T_{\Gamma_j}]$. Note that the vector function $\dot{x}^*(t;y^{(j)},\widetilde{\eta}^{(j)})$ itself is, generally speaking, discontinuous, and may not have a limit as $j\to\infty$.

Integrating (D.5.7) with respect to t we get

$$\frac{\langle \zeta\,,\,x^*(T_{\Gamma_j};y^{(j)},\widetilde{\eta}^{(j)})-y^{(j)}\rangle}{T_{\Gamma_j}} \to 1 \quad \text{as} \quad j\to\infty,$$

which contradicts the periodicity of Γ_j. □

3. Proof of Lemmas 1.7.8 and 1.8.15. For $(y,\widetilde{\eta})\in\Pi^a$ let $\mathbf{T}^a(y,\widetilde{\eta})=\inf\{T:(y,\widetilde{\eta})\in\Pi^a_T\}$. Let $\Omega\subset\Pi^a$ be an open connected subset of Π^a. In the analytic case, if some point $(y_0,\widetilde{\eta}_0)\in\Omega$ is absolutely T-periodic, then all the points $(y,\widetilde{\eta})\in\Omega$ are absolutely T-periodic. This implies that $\mathbf{T}^a(y,\widetilde{\eta})\leqslant\mathbf{T}^a(y_0,\widetilde{\eta}_0)$ for all $(y,\widetilde{\eta})\in\Omega$ and $(y_0,\widetilde{\eta}_0)\in\Omega$. Therefore the function $\mathbf{T}^a$ is constant on Ω. Now Lemma 1.7.8 follows from (1.3.7).

Lemma 1.8.15 is proved in the same manner (of course, in this case one has to apply Lemma 1.8.3 instead of (1.3.7)).

D.6. Maslov index

1. Argument of the determinant of symmetric matrices. Let $d\in\mathbb{N}$. On the set of complex symmetric nonsingular $d\times d$-matrices C with $\operatorname{Re}C\geqslant 0$ one can single out a branch of $\arg\det C$ such that

1. $\arg\det C=0$ if $\operatorname{Im}C=0$ (and $\operatorname{Re}C>0$).
2. $\arg\det C$ depends continuously on C

(see, for example, [**Hö3,** Vol. 1, Sect. 3.4]). We denote this specified branch by $\arg\det_+C$.

It is easy to see that $\arg\det_+(S^T\cdot C\cdot S)=\arg\det_+C$ for all real nonsingular $d\times d$-matrices S. One can prove [**Hö3,** Vol. 1, Sect. 3.4] that for a purely imaginary matrix C

$$\arg\det_+C=\frac{\pi}{2}\operatorname{sgn}(C/i)\,,\tag{D.6.1}$$

where $\operatorname{sgn}(\cdot)$ denotes the signature of a real symmetric matrix (i.e., the difference between the numbers of its strictly positive and strictly negative eigenvalues).

LEMMA D.6.1. *Let C be a complex symmetric $n\times n$-matrix with $\operatorname{Re}C\geqslant 0$ and $\operatorname{rank}C=d$. Then there exists a real nonsingular $n\times n$-matrix S such that*

$$S^T\cdot C\cdot S=\begin{pmatrix}C_d&0\\0&0\end{pmatrix}\tag{D.6.2}$$

where C_d is a nonsingular $d\times d$-matrix.

PROOF. For any complex vector $\mathbf{z}$ we have

$$\operatorname{Re}\langle C\mathbf{z},\bar{\mathbf{z}}\rangle=\langle(\operatorname{Re}C)\mathbf{z},\bar{\mathbf{z}}\rangle.$$

Since $\operatorname{Re}C\geqslant 0$, from this equality it follows that for all $\mathbf{z}\in\ker C$ we have $(\operatorname{Re}C)\mathbf{z}=0$, and consequently $(\operatorname{Im}C)\mathbf{z}=0$. Therefore

$$\ker C=\ker(\operatorname{Re}C)\cap\ker(\operatorname{Im}C)$$

and the orthogonal projection Π_C on $\ker C$ is a real matrix. Clearly, the equality (D.6.2) holds for a real matrix S if and only if

$$S\begin{pmatrix}0\\\mathbf{z}'\end{pmatrix}\in\Pi_C\,\mathbb{R}^n\,,\qquad\forall\mathbf{z}'\in\mathbb{R}^d\,.\tag{D.6.3}$$

Moreover, if S is nonsingular then C_d is also nonsingular. □

We now extend the domain of definition of the function $\arg\det_+ C$ to the set of all (maybe, singular) symmetric complex $n\times n$-matrices C with $\operatorname{Re} C \geqslant 0$ in the following way. Given a matrix C we choose a real nonsingular matrix S satisfying (D.6.2) and put

$$\arg\det_+ C \;:=\; \arg\det_+ C_d\,.$$

Certainly, the matrix C_d is not determined by C uniquely. However, if

$$S^T\cdot C\cdot S \;=\; \begin{pmatrix} C_d & 0\\ 0 & 0\end{pmatrix}, \qquad (S')^T\cdot C\cdot S' \;=\; \begin{pmatrix} C'_d & 0\\ 0 & 0\end{pmatrix}$$

for real nonsingular $n\times n$-matrices S and S', then, in view of (D.6.3), $C'_d = Q_d^T\cdot C_d\cdot Q_d$ for some real nonsingular $d\times d$-matrix Q_d (one can take for Q_d the upper left $d\times d$-block of $S^{-1}\cdot S'$). Therefore, $\arg\det_+ C$ does not depend on the choice of S.

Now $\arg\det_+ C$ is not continuous with respect to C on the whole class of symmetric complex matrices with $\operatorname{Re} C \geqslant 0$ (it may have jumps when $\operatorname{rank} C$ changes). But $\arg\det_+ C$ is continuous on any subclass of matrices C having the same kernel. Note also that (D.6.1) remains valid for singular purely imaginary matrices C.

REMARK D.6.2. A complex symmetric $n\times n$-matrix C can be identified with a complex-valued quadratic form on the space $\mathbb{R}^n$ with a given basis. When we change the basis, C transforms into $S^T\cdot C\cdot S$ where S is the (real) transformation matrix. The identity $\arg\det_+ C = \arg\det_+ (S^T\cdot C\cdot S)$ means that $\arg\det_+ C$ is an invariant characteristic of the corresponding quadratic form.

REMARK D.6.3. The function $\arg\det_+ C$ appears in the stationary phase formula for a complex phase function [**Hö3,** Vol. 1, Sect. 7.7].

2. One more definition. In this subsection we prove Lemma 1.5.3 and Proposition 1.5.2. Besides, we introduce the third definition of the Maslov index and prove that all three definitions are equivalent.

PROOF OF LEMMA 1.5.3. Since the matrix x^*_η behaves as a tensor under changes of coordinates x, it is sufficient to prove Lemma 1.5.3 in a one coordinate system. We choose a coordinate system with specified "normal" coordinate x_n; see 1.1.2.

At the moments of reflection t^*_k we have

$$x^*_\eta(t^*_k-0;y,\eta) \;+\; \dot x^*(t^*_k-0;y,\eta)\,(t^*_k)_\eta(y,\eta) \;=\; \frac{d}{d\eta}\,x^*(t^*_k-0;y,\eta)$$
$$=\; \frac{d}{d\eta}\,x^*(t^*_k+0;y,\eta) \;=\; x^*_\eta(t^*_k+0;y,\eta) \;+\; \dot x^*(t^*_k+0;y,\eta)\,(t^*_k)_\eta(y,\eta)$$

and

$$0 \;=\; \frac{d}{d\eta}\,x^*_n(t^*_k-0;y,\eta) \;=\; (x^*_n)_\eta(t^*_k-0;y,\eta) \;+\; \dot x^*_n(t^*_k-0;y,\eta)\,(t^*_k)_\eta(y,\eta)\,,$$
$$0 \;=\; \frac{d}{d\eta}\,x^*_n(t^*_k+0;y,\eta) \;=\; (x^*_n)_\eta(t^*_k+0;y,\eta) \;+\; \dot x^*_n(t^*_k+0;y,\eta)\,(t^*_k)_\eta(y,\eta)\,.$$

Since $\dot{x}_n^*(t_k^* - 0; y, \eta) \neq 0$ and $\dot{x}_n^*(t_k^* + 0; y, \eta) \neq 0$, these equalities imply that the matrices $x_\eta^*(t_k^* - 0; y, \eta)$ and $x_\eta^*(t_k^* + 0; y, \eta)$ have the same kernel. Therefore they have the same rank. □

Let $\varphi_k \in \mathfrak{F}_k$, $k = 0, 1, \ldots, \mathbf{r}$, be a matching set of phase functions corresponding to the family of billiard trajectories $(x^*(t; y, \eta), \xi^*(t; y, \eta))$ (see Definition 2.6.2). We will use the notation $\Phi_{xx}^{(k)}$, $\Phi_{x\eta}^{(k)}$ and $\Phi_{\eta\eta}^{(k)}$ for the matrix functions (2.4.7) defined on the sets $\mathfrak{D}_k$ (2.4.1).

When we change coordinates y, the matrix $\Phi_{\eta\eta}^{(k)}$ is multiplied by the Jacobian matrix and its transposed, i.e., it behaves as a quadratic form. Since $\operatorname{Im}\varphi_k \geqslant 0$, the imaginary part of $\Phi_{\eta\eta}^{(k)}$ is nonnegative. So, in view of Remark D.6.2, $\arg\det_+ (\Phi_{\eta\eta}^{(k)}/i)$ is well defined and independent of the choice of coordinates.

By (2.4.9)

$$\Phi_{x\eta}^{(k)} = \xi_\eta^* - \Phi_{xx}^{(k)} \cdot x_\eta^*, \tag{D.6.4}$$

$$\Phi_{\eta\eta}^{(k)} = -(x_\eta^*)^T \cdot (\xi_\eta^* - \Phi_{xx}^{(k)} \cdot x_\eta^*) = -(\xi_\eta^* - \Phi_{xx}^{(k)} \cdot x_\eta^*)^T \cdot x_\eta^*. \tag{D.6.5}$$

Since the phase function φ_k is nonsingular, (D.6.5) implies $\ker \Phi_{\eta\eta}^{(k)} = \ker x_\eta^*$. Therefore, $\arg\det_+ (\Phi_{\eta\eta}^{(k)}/i)$ is continuous with respect to $\varphi_k \in \mathfrak{F}_k$. If $\Phi_{\eta\eta}^{(k)}$ is real, then by (D.6.1)

$$\arg\det_+ (\Phi_{\eta\eta}^{(k)}/i) = -\frac{\pi}{2} \operatorname{sgn} \Phi_{\eta\eta}^{(k)}. \tag{D.6.6}$$

LEMMA D.6.4. *For all $(t; y, \eta) \in \mathfrak{D}_k$ we have*

$$\arg(\det{}^2 \Phi_{x\eta}^{(k)}) = 2 \arg\det_+ (\Phi_{\eta\eta}^{(k)}/i) + \pi \mathcal{R} \pmod{2\pi}. \tag{D.6.7}$$

PROOF. Let Π and $\widetilde{\Pi}$ be the orthogonal projections on $\ker x_\eta^*$ and $\ker (x_\eta^*)^T$, respectively, and $\Pi' := I - \Pi$, $\widetilde{\Pi}' := I - \widetilde{\Pi}$. We have

$$\widetilde{\Pi}' \cdot \xi_\eta^* \cdot \Pi = 0. \tag{D.6.8}$$

Indeed, proving (D.6.8) is equivalent to proving that for any column $u \in \mathbb{R}^n$

$$(\widetilde{\Pi}' u)^T \cdot \xi_\eta^* \cdot \Pi = 0. \tag{D.6.9}$$

But $\widetilde{\Pi}' u = x_\eta^* v$ for some column $v \in \mathbb{R}^n$, and (D.6.9) follows from (2.3.6).

Formulae (D.6.4), (D.6.5), (D.6.8) imply

$$\Phi_{x\eta}^{(k)} = \widetilde{\Pi} \cdot \xi_\eta^* \cdot \Pi + \widetilde{\Pi}' \cdot \Phi_{x\eta}^{(k)} \cdot \Pi' + \widetilde{\Pi} \cdot \Phi_{x\eta}^{(k)} \cdot \Pi', \tag{D.6.10}$$

$$\Phi_{\eta\eta}^{(k)} = -\Pi' \cdot (x_\eta^*)^T \cdot \widetilde{\Pi}' \cdot \Phi_{x\eta}^{(k)} \cdot \Pi'. \tag{D.6.11}$$

By (D.6.10), in special bases associated with the decompositions $\mathbb{R}^n = \Pi'\mathbb{R}^n \oplus \Pi\mathbb{R}^n$ and $\mathbb{R}^n = \widetilde{\Pi}'\mathbb{R}^n \oplus \widetilde{\Pi}\mathbb{R}^n$ the matrix $\Phi_{x\eta}^{(k)}$ is triangular:

$$\Phi_{x\eta}^{(k)} = \begin{pmatrix} \mathcal{A} & 0 \\ \mathcal{B} & \mathcal{C} \end{pmatrix},$$

where $\mathcal{A}$ is the upper left $\mathcal{R} \times \mathcal{R}$-block of $\widetilde{\Pi}' \cdot \Phi^{(k)}_{x\eta} \cdot \Pi'$, and $\mathcal{C}$ is the lower right $(n-\mathcal{R}) \times (n-\mathcal{R})$-block of $\widetilde{\Pi} \cdot \xi^*_\eta \cdot \Pi$. Thus, $\det \Phi^{(k)}_{x\eta}$ is equal to the determinant of $\mathcal{A}$ modulo a real factor.

By formula (D.6.11), in the special basis associated with the decomposition $\mathbb{R}^n = \Pi'\mathbb{R}^n \oplus \Pi\mathbb{R}^n$ the matrix $\Phi^{(k)}_{\eta\eta}/i$ takes the form (D.6.2), and the determinant of the corresponding $\mathcal{R} \times \mathcal{R}$-matrix $C_{\mathcal{R}}$ coincides with $i^{-\mathcal{R}} \det \mathcal{A}$ modulo a real factor. This implies (D.6.7). □

Obviously, $\arg(\det^2 \Phi^{(k)}_{x\eta})$ coincides with $\arg f$ modulo 2π, where f is the function defined by (1.5.10). Let us choose the branches of the arguments such that $\arg(\det^2 \Phi^{(k)}_{x\eta}) = \arg_0 f$ on $\mathfrak{D}_k$, $k = 0, \dots, \mathbf{r}$, and introduce the functions

$$\Theta_k \ := \ (2\pi)^{-1} \arg(\det^2 \Phi^{(k)}_{x\eta}) \ - \ \pi^{-1} \arg \det_+ (\Phi^{(k)}_{\eta\eta}/i) \ + \ \mathcal{R}/2$$

defined on $\mathfrak{D}_k$. By Lemma D.6.4 the Θ_k take only integer values.

As the phase functions φ_k are matching, $\arg_0 f$ is continuous, and since we have already proved Lemma 1.5.3, we have

$$\Theta_{k-1}(t^*_k; y, \eta) \ = \ \Theta_k(t^*_k; y, \eta), \qquad k = 1, \dots, \mathbf{r}.$$

So we can consider the function $\Theta(t; y, \eta)$ which is defined on the set $[0, T] \times O$ and coincides with $\Theta_k(t; y, \eta)$ for $t^*_k(y, \eta) \leqslant t \leqslant t^*_{k+1}(y, \eta)$.

PROPOSITION D.6.5. *The integer-valued function Θ is independent of the choice of the matching set of phase functions $\varphi_k \in \mathfrak{F}_k$.*

PROOF. Since Θ_k depends continuously on $\varphi_k \in \mathfrak{F}_k$, it does not change under a continuous variation of φ_k. Therefore the proposition follows from the fact that any two matching sets of phase functions $\varphi_k \in \mathfrak{F}_k$ can be continuously transformed one into another (Lemma 2.6.4). □

REMARK D.6.6. Note that the branch of $\arg(\det^2 \Phi^{(k)}_{x\eta})$ is determined by the full history of the trajectory, whereas $\arg \det_+ (\Phi^{(k)}_{\eta\eta}/i)$ and $\operatorname{rank} x^*_\eta$ are purely local objects. Therefore the function Θ_k still contains information about the whole trajectory.

PROOF OF PROPOSITION 1.5.2. Without loss of generality we assume Ω_j and $\check{\mathfrak{D}}_j \subset \mathfrak{D}_k$ to be sufficiently small. Let us choose a phase function φ_k which coincides with $\langle x - x^*, \xi^* \rangle$ on the set $\Omega_j \times \check{\mathfrak{D}}_j$ (see Lemma 2.4.13). Then $\Phi^{(k)}_{\eta\eta} = -(\xi^*_\eta)^T \cdot x^*_\eta$, and by (D.6.6) we obtain

$$\text{(D.6.12)} \qquad r^-_j \ = \ \frac{1}{2} \Big(\mathcal{R} \ - \ \operatorname{sgn}\big((\xi^*_\eta)^T \cdot x^*_\eta\big) \Big) \ = \ \Theta_k \ - \ (2\pi)^{-1} \arg(\det^2 \Phi^{(k)}_{x\eta}).$$

The function φ_k is real on $\Omega_j \times \check{\mathfrak{D}}_j$, and therefore

$$\text{(D.6.13)} \qquad \arg(\det^2 \Phi^{(k)}_{x\eta})\big|_{\check{\mathfrak{D}}_j} \ = \ \text{const}\,.$$

Now Proposition D.6.5 implies that the function $r^-_j(t; y, \eta)$ is independent of the choice of the coordinates x modulo a constant function. □

DEFINITION D.6.7. The integer $-\Theta(T; y_0, \eta_0)$ is called the Maslov index of the trajectory $(1.5.1)_0$.

THEOREM D.6.8. *Under the condition* (1.5.2), *Definitions* 1.5.1, 1.5.9, *and* D.6.7 *are equivalent.*

PROOF. In view of (1.5.2) we have

$$\Theta(T; y_0, \eta_0) = (2\pi)^{-1} \arg(\det{}^2 \Phi^{(r)}_{x\eta})(T; y_0, \eta_0) = (2\pi)^{-1} \arg_0 f(T; y_0, \eta_0).$$

This implies the equivalence of Definitions D.6.7 and 1.5.9. On the other hand, choosing a phase function of the form $\langle x - x^*, \xi^* \rangle$ we obtain by (D.6.12) and (D.6.13)

$$\Theta(t_{j+1}; y_0, \eta_0) - \Theta(t_j; y_0, \eta_0) = r_j^-(t_{j+1}; y_0, \eta_0) - r_j^-(t_j; y_0, \eta_0).$$

Therefore

$$\begin{aligned} \Theta(T; y_0, \eta_0) &= \sum_{j=0}^{N-1} \left(\Theta(t_{j+1}; y_0, \eta_0) - \Theta(t_j; y_0, \eta_0)\right) \\ &= \sum_{j=0}^{N-1} \left(r_j^-(t_{j+1}; y_0, \eta_0) - r_j^-(t_j; y_0, \eta_0)\right), \end{aligned}$$

which implies the equivalence of Definitions D.6.7 and 1.5.1. □

REMARK D.6.9. When (1.5.2) is not fulfilled, Definitions 1.5.1 and 1.5.9 do not make sense. Indeed, in this case $r_N^-(T; y_0, \eta_0)$ may depend on the choice of the coordinates x on Ω_N, and $\arg_0 f(T; y_0, \eta_0)$ may depend on the choice of φ_r (moreover, $(2\pi)^{-1} \arg_0 f(T; y_0, \eta_0)$ is not necessarily an integer). However, Definition D.6.7 makes sense even in this case.

3. Proof of Lemmas 1.5.5 and 1.5.6. By Taylor's formula

$$\begin{aligned} x^*_\eta(t+s; y, \eta) &= x^*_\eta(t; y, \eta) + s \frac{d}{dt} x^*_\eta(t; y, \eta) + O(s^2) \\ &= x^*_\eta(t; y, \eta) + s \frac{d}{d\eta} (h_\xi(x^*, \xi^*)) + O(s^2) \\ &= x^*_\eta(t; y, \eta) + s\, \partial_{\xi x} h(x^*, \xi^*) \cdot x^*_\eta + s\, \partial_{\xi\xi} h(x^*, \xi^*) \cdot \xi^*_\eta + O(s^2), \\ \xi^*_\eta(t+s; y, \eta) &= \xi^*_\eta(t; y, \eta) + s \frac{d}{dt} \xi^*_\eta(t; y, \eta) + O(s^2) \\ &= \xi^*_\eta(t; y, \eta) - s \frac{d}{d\eta} (h_x(x^*, \xi^*)) + O(s^2) \\ &= x^*_\eta(t; y, \eta) - s\, \partial_{xx} h(x^*, \xi^*) \cdot x^*_\eta - s\, \partial_{x\xi} h(x^*, \xi^*) \cdot \xi^*_\eta + O(s^2) \end{aligned}$$

as $s \to 0$, so

$$\begin{aligned} C_j(t+s; y, \eta) &= \left(\xi^*_\eta(t+s; y, \eta)\right)^T \cdot x^*_\eta(t+s; y, \eta) \\ &= (\xi^*_\eta)^T \cdot x^*_\eta - s\, (x^*_\eta)^T \cdot \partial_{xx} h(x^*, \xi^*) \cdot x^*_\eta \\ &\quad + s\, (\xi^*_\eta)^T \cdot \partial_{\xi\xi} h(x^*, \xi^*) \cdot \xi^*_\eta + O(s^2). \end{aligned} \tag{D.6.14}$$

Here $x^* = x^*(t;y,\eta)$, $\xi^* = \xi^*(t;y,\eta)$, $x^*_\eta = x^*_\eta(t;y,\eta)$, $\xi^*_\eta = \xi^*_\eta(t;y,\eta)$.

By Euler's identity $x^*_\eta\,\eta \equiv 0$ and $\partial_{\xi\xi}h(x^*,\xi^*)\cdot\xi^*_\eta\,\eta \equiv \partial_{\xi\xi}h(x^*,\xi^*)\,\xi^* \equiv 0$. Therefore for all s the kernels of all the matrices in (D.6.14) (including the matrix $O(s^2)$) contain the column η. Moreover, since $\operatorname{rank}\partial_{\xi\xi}h(x^*,\xi^*) = n-1$, the kernel of $(\xi^*_\eta)^T\cdot\partial_{\xi\xi}h(x^*,\xi^*)\cdot\xi^*_\eta$ coincides with the corresponding one-dimensional subspace.

When $x^*_\eta = 0$ we obtain from (D.6.14)

$$C_j(t+s;y,\eta) \;=\; s\,(\xi^*_\eta)^T\cdot\partial_{\xi\xi}h(x^*,\xi^*)\cdot\xi^*_\eta \;+\; O(s^2)\,,$$

which immediately implies Lemma 1.5.5.

Now let $r_+(x^*,\xi^*) = n-1$. Denote by Π_η the orthogonal projection on the one-dimensional linear subspace generated by η. Then the symmetric matrix

$$\mathcal{C}_j \;=\; \mathcal{C}_j(t;y,\eta) \;:=\; (\xi^*_\eta)^T\cdot\partial_{\xi\xi}h(x^*,\xi^*)\cdot\xi^*_\eta + \Pi_\eta$$

is positive, and

$$\mathcal{C}_j^{-1/2}\cdot(\xi^*_\eta)^T\cdot\partial_{\xi\xi}h(x^*,\xi^*)\cdot\xi^*_\eta\cdot\mathcal{C}_j^{-1/2} \;=\; I-\Pi_\eta\,.$$

Therefore we can rewrite (D.6.14) as

$$\begin{aligned}
&\mathcal{C}_j^{-1/2}\cdot C_j(t+s;y,\eta)\cdot\mathcal{C}_j^{-1/2}\\
&\quad= s\,(I-\Pi_\eta)\\
&\qquad+\mathcal{C}_j^{-1/2}\cdot\big((\xi^*_\eta)^T - s\,(x^*_\eta)^T\cdot\partial_{xx}h(x^*,\xi^*)\big)\cdot x^*_\eta\cdot\mathcal{C}_j^{-1/2} + O(s^2);
\end{aligned}\tag{D.6.15}$$

note that $\eta = \mathcal{C}_j^{1/2}\eta$ and the column η lies in the kernels of all the three terms on the right-hand side of (D.6.15). Since $\det\xi^*_\eta \ne 0$, the eigenvalues of the symmetric matrix

$$\mathcal{C}_j^{-1/2}\cdot\big((\xi^*_\eta)^T \;-\; s\,(x^*_\eta)^T\cdot\partial_{xx}h(x^*,\xi^*)\big)\cdot x^*_\eta\cdot\mathcal{C}_j^{-1/2}$$

are either identically equal to zero or uniformly bounded away from zero for small s. Therefore the rank of the matrix (D.6.15) is equal to $n-1$ for small $s\ne 0$. Moreover, formula (D.6.15) implies that for small s the difference

$$r_j^-(t+s;y,\eta) - r_j^-(t;y,\eta)$$

coincides with the number of negative eigenvalues of the matrix $s\,\widehat{\Pi}\cdot(I-\Pi_\eta)\cdot\widehat{\Pi}$, where $\widehat{\Pi}$ is the orthogonal projection on

$$\ker\big(\mathcal{C}_j^{-1/2}\cdot\big((\xi^*_\eta)^T \;-\; s\,(x^*_\eta)^T\cdot\partial_{xx}h(x^*,\xi^*)\big)\cdot x^*_\eta\cdot\mathcal{C}_j^{-1/2}\big) \;=\; \ker(x^*_\eta\cdot\mathcal{C}_j^{-1/2})\,.$$

But the matrix $\widehat{\Pi}\cdot(I-\Pi_\eta)\cdot\widehat{\Pi}$ is nonnegative and has precisely $n-1-\mathcal{R}(t;y,y)$ positive eigenvalues (the corresponding eigenvectors lie in $\ker(x^*_\eta\cdot\mathcal{C}_j^{-1/2})$ and are orthogonal to η). This proves Lemma 1.5.6.

Appendix E

Factorization of Smooth Functions and Taylor-type Formulae

In this appendix we describe one technical construction which plays an essential role in the procedure of reduction of the amplitude of an oscillatory integral; see sections 2.7–2.9. The construction described below is in fact a simplified version of the well-known Malgrange preparation theorem [**Hö3,** Vol. 1, Theorem 7.5.7]. This simplified version will be sufficient for our purposes due to the special properties of our classes of phase functions $\mathfrak{F}_i$, $\mathfrak{F}_i^{\mathrm{bl}}$, $\mathfrak{F}_i'$.

The construction given in this subsection has its own advantages.

1. It gives explicit formulae which can be used in the computation of the symbol.
2. It possesses certain invariance properties, which is important when working on a manifold M.
3. It allows us to control the support of the functions involved, which is necessary when $\partial M \neq \varnothing$.

1. Basic problem. Consider the Euclidean space $\mathbb{R}^n$ equipped with Cartesian coordinates $z = (z_1, \ldots, z_n)$. Let $\mathcal{W} \subset \mathbb{R}^n$ be a neighbourhood of the origin, and let $a(z) \in C^\infty(\mathcal{W})$ be a complex-valued scalar function. Our first objective will be to find an n-component complex-valued column function $g(z) = (g_1(z), \ldots, g_n(z)) \in C^\infty(\mathcal{W})$ such that

$$a(z) = a(0) + \langle z, g(z) \rangle . \tag{E.1}$$

Here $\langle z, g(z)\rangle := \sum_{j=1}^n z_j\, g_j(z)$.

Let us introduce the differential operators

$$Z_0 := 1, \qquad Z_k := \sum_{|\alpha|=k} \frac{z^\alpha}{\alpha!}\,\partial_z^\alpha, \qquad k = 1, 2, \ldots, \tag{E.2}$$

where $\alpha = (\alpha_1, \ldots, \alpha_n)$ are multiindices. It is easy to see that

$$(k+1) Z_{k+1} = Z_1 Z_k - k Z_k = \sum_{j=1}^n z_j\, Z_k\, \partial_{z_j}, \qquad k = 0, 1, 2, \ldots . \tag{E.3}$$

Let $a_q(z) \in C^\infty(\mathbb{R}^n \setminus \{0\})$ be positively homogeneous of degree $q \in \mathbb{R}$. Then for $k = 1, 2, \ldots$ we have

$$Z_k\, a_q = \frac{q(q-1)(q-2)\cdots(q-k+1)}{k!}\, a_q \tag{E.4}$$

(when $k = 1$ this is the classical Euler identity). Formula (E.4) is established by induction on k with the help of the left equality (E.3). In the special case $q \in \mathbb{N}$ formula (E.4) can be rewritten as

$$Z_k \, a_q \;=\; \begin{cases} \frac{q!}{k!\,(q-k)!}\, a_q & \text{if} \quad k \leqslant q\,, \\ 0 & \text{if} \quad k > q\,. \end{cases} \tag{E.5}$$

Let us introduce the sequence of real numbers $c_1, c_2, \dots$, defined as the solution of the following recursive system:

$$\sum_{k=1}^{q} \frac{q!}{k!\,(q-k)!}\, c_k \;=\; 1\,, \qquad q = 1, 2, \dots\,. \tag{E.6}$$

Solving (E.6) we get

$$c_k \;=\; (-1)^{k-1}\,, \qquad k = 1, 2, \dots\,. \tag{E.7}$$

Set

$$a_q(z) \;:=\; \sum_{|\alpha|=q} (\partial_z^\alpha a)|_{z=0}\, \frac{z^\alpha}{\alpha!}\,.$$

According to Taylor's formula we have

$$a(z) \;\sim\; a(0) + \sum_{q=1}^{\infty} a_q(z)\,, \qquad z \to 0\,. \tag{E.8}$$

The $a_q(z)$ are polynomials homogeneous in z of degree q, so from (E.5), (E.6) we obtain

$$a_q(z) \;= \sum_{k=1}^{\infty} c_k\, Z_k\, a_q\,, \qquad q = 1, 2, \dots\,. \tag{E.9}$$

Substituting (E.9) in (E.8) we get

$$a(z) \;\sim\; a(0) + \sum_{k,q=1}^{\infty} c_k\, Z_k\, a_q \;\sim\; a(0) + \sum_{k=1}^{\infty} c_k\, Z_k\, a\,. \tag{E.10}$$

Here the interchange of order of summation is justified because the double sum contains only a finite number of nonzero terms of a given degree q (degree in powers of z). Using (E.3) we can rewrite formula (E.10) as

$$a(z) \;\sim\; a(0) + \sum_{j=1}^{n} z_j \left(\sum_{k=1}^{\infty} c_k\, k^{-1}\, Z_{k-1}\, \partial_{z_j} a \right). \tag{E.11}$$

Let $\chi_k(z) \in C_0^\infty(\mathbb{R}^n)$, $0 \leqslant \chi_k(z) \leqslant 1$, $k = 1, 2, \dots$, be cut-off functions which are identically equal to 1 in some neighbourhoods of the origin. Set

$$g' \;=\; \left(\sum_{k=1}^{\infty} c_k\, k^{-1}\, \chi_k\, Z_{k-1} \right) \partial_z a\,, \tag{E.12}$$

where ∂_z is the column operator of first order partial differentiations in z. We shall assume that the sequence of cut-offs $\chi_1(z)$, $\chi_2(z)$, ... is chosen in such a way that the series (E.12) converge uniformly over $\mathcal{W}$, as well as all their partial derivatives with respect to z. This can always be achieved by setting, for example, $\chi_k(z) =$

$\chi(d_k|z|)$, where χ is an arbitrary function from $C_0^\infty(\mathbb{R})$ which is identically equal to 1 in some neighbourhood of zero, and d_1, d_2, ..., is a sequence of positive numbers tending to ∞; the necessary convergence properties hold if the sequence d_1, d_2, ..., tends to ∞ sufficiently fast (cf. Lemma 2.1.2).

Formulae (E.11), (E.12) imply

$$a(z) = a(0) + \langle z, g'(z)\rangle + O(|z|^\infty), \qquad z \to 0, \tag{E.13}$$

where $O(|z|^\infty)$ denotes a function which has an infinite order zero at $z=0$. Set

$$g = g' + g'', \tag{E.14}$$

$$g'' = \frac{a - a|_{z=0} - \langle z, g'\rangle}{z^T B\, z}\, B\, z, \tag{E.15}$$

where $z = (z_1, \dots, z_n)$ is understood as a column, and $B = B(z) \in C^\infty(\mathcal{W})$ is an arbitrary positive Hermitian $n \times n$-matrix function. Formulae (E.15), (E.13) imply that the column function $g'' = g''(z)$ is infinitely smooth, and moreover,

$$g'' = O(|z|^\infty). \tag{E.16}$$

Combining (E.13)–(E.15) we obtain (E.1).

2. Generalization of the basic problem. Now let us generalize our original problem (E.1). Let $f(z) = \{f_1(z), \dots, f_n(z)\} \in C^\infty(\mathcal{W})$ be a given smooth complex-valued n-component column function such that

$$f = 0 \text{ if and only if } z = 0, \tag{E.17}$$

$$\det J|_{z=0} \neq 0, \tag{E.18}$$

where

$$J = J(z) := \partial_z f^T. \tag{E.19}$$

We want to find a column function $g(z) = (g_1(z), \dots, g_n(z)) \in C^\infty(\mathcal{W})$ such that

$$a(z) = a(0) + \langle f(z), g(z)\rangle. \tag{E.20}$$

Obviously, this problem coincides with (E.1) in the special case $f(z) \equiv z$.

Set

$$\partial_{f_j} := e_j^T J^{-1} \partial_z, \qquad j = 1, 2, \dots, n, \tag{E.21}$$

where e_j is the jth basis column (column with 1 in the jth row and zeros elsewhere). By $\partial_f := J^{-1}\partial_z$ we shall denote the column of operators (E.21).

Let us prove that the operators (E.21) commute. In view of (E.21) we have $\partial_{f_j}\partial_{f_k} - \partial_{f_k}\partial_{f_j} = b\,\partial_z$, where $b = b(z)$ is a row function defined by the formula

$$b(z) = \left(e_j^T J^{-1} \partial_z e_k^T - e_k^T J^{-1} \partial_z e_j^T\right) J^{-1}.$$

Let $\mathbf{z}$ be an arbitrary fixed point from $\mathcal{W}$, and $\mathbf{J} := J(\mathbf{z})$ (constant matrix). Using the formula for the derivative of the inverse matrix and formula (E.19) we get

$$\begin{aligned}
b(\mathbf{z}) &= \left\{\left(e_k^T \mathbf{J}^{-1} \partial_z e_j^T - e_j^T \mathbf{J}^{-1} \partial_z e_k^T\right) \mathbf{J}^{-1} J\right\}\big|_{z=\mathbf{z}} \mathbf{J}^{-1} \\
&= \left\{\left(e_k^T \mathbf{J}^{-1} \partial_z e_j^T - e_j^T \mathbf{J}^{-1} \partial_z e_k^T\right) \mathbf{J}^{-1} \partial_z f^T\right\}\big|_{z=\mathbf{z}} \mathbf{J}^{-1} \\
&= \left\{\left((e_k^T \mathbf{J}^{-1} \partial_z)(e_j^T \mathbf{J}^{-1} \partial_z) - (e_j^T \mathbf{J}^{-1} \partial_z)(e_k^T \mathbf{J}^{-1} \partial_z)\right) f^T\right\}\big|_{z=\mathbf{z}} \mathbf{J}^{-1}.
\end{aligned}$$

But $e_k^T \mathbf{J}^{-1} \partial_z$ and $e_j^T \mathbf{J}^{-1} \partial_z$ are scalar differential operators with constant coefficients, so they commute. This implies $b(\mathbf{z}) = 0$.

The fact that the operators ∂_{f_j} and ∂_{f_k} commute is not really surprising because formally they can be interpreted as derivatives with respect to the new independent variables f_j and f_k. However, in the general case, when $f(z)$ is complex-valued and nonanalytic, we cannot justify the change of variables $z \to f$, and for this reason we have given the detailed arguments above.

As a generalization of (E.2), (E.12), (E.15) set

$$\text{(E.22)} \qquad F_0 := 1\,, \qquad F_k := \sum_{|\alpha|=k} \frac{f^\alpha}{\alpha!}\,\partial_f^\alpha\,, \qquad k = 1, 2, \ldots,$$

$$\text{(E.23)} \qquad g' = \left(\sum_{k=1}^{\infty} c_k\, k^{-1}\, \chi_k\, F_{k-1} \right) \partial_f a\,,$$

$$\text{(E.24)} \qquad g'' = \frac{a - a\big|_{z=0} - \langle f, g' \rangle}{f^T B \overline{f}}\, B\overline{f}\,.$$

Here we have the right to use the notation ∂_f^α because we know that the operators (E.21) commute.

The $g(z)$ constructed in accordance with formulae (E.14), (E.19), (E.21)–(E.24) satisfies the required equality (E.20). Let us substantiate this claim. As in subsection 1, it is sufficient to establish that

$$\text{(E.25)} \qquad a(z) = a(0) + \langle f(z), g'(z) \rangle + O(|z|^\infty)\,,$$

cf. (E.13). Moreover, it is sufficient to establish (E.25) under the assumption that $a(z)$ and $f(z)$ are polynomials, because any $a(z), f(z) \in C^\infty(\mathcal{W})$ can be approximated by polynomials in z with arbitrary accuracy. But in the case when $a(z)$, $f(z)$ are polynomials the construction (E.19), (E.21)–(E.23) is reduced to (E.2), (E.12) by a change of independent variables $z \to f$.

Note that in the above argument the change $z \to f$ leads, generally speaking, to complex independent variables f. The necessity of dealing with complex variables forced us to consider polynomials as an intermediate step.

3. Further generalization with "trivial" parameters u. Let us generalize the problem (E.20) further by introducing additional real parameters $u = (u_1, \ldots, u_m)$. Thus, we study the function $a(z; u)$, where $z \in \mathbb{R}^n$, $u \in \mathbb{R}^m$. The function a is defined in some neighbourhood of the set $\{z = 0\}$. Here, as well as in the next subsection, we do not specify more precisely the domain of definition of the function a in all its variables because in our applications (sections 2.7–2.9) it is clear from the context. We also have a given complex-valued n-component column function $f(z; u) = \{f_1(z; u), \ldots, f_n(z; u)\}$ with the same domain of definition as $a(z; u)$, and such that (E.17), (E.18) hold. We want to find a $g(z; u) = (g_1(z; u), \ldots, g_n(z; u))$ such that

$$\text{(E.26)} \qquad a(z; u) = a(0; u) + \langle f(z; u), g(z; u) \rangle\,.$$

It is easy to see that our previous construction (E.14), (E.19), (E.21)–(E.24) gives the required $g(z; u)$. The only difference is that now everywhere we have the dependence on the additional parameters u. In particular, the cut-offs $\chi_k = \chi_k(z; u)$ appearing in (E.23) depend on the additional parameters u; as usual,

$0 \leqslant \chi_k \leqslant 1$ and $\chi_k = 1$ in some neighbourhoods of the set $\{z = 0\}$. These cut-offs should (and can) be chosen in such a way that the series (E.23) converge uniformly over any compact set in the $(n+m)$-dimensional domain of definition of the function $a(z;u)$, as well as all their partial derivatives with respect to z and u.

4. Final generalization with "nontrivial" parameter v. Finally, let us generalize the problem (E.26) further by introducing one more real parameter v. Thus, we study the function $a(z;u;v)$, where $z \in \mathbb{R}^n$, $u \in \mathbb{R}^m$, $v \in \mathbb{R}$. The function a is defined in some neighbourhood of the set $\{z = 0\}$. We also have a given complex-valued n-component column function $f(z;u;v) = \{f_1(z;u;v), \dots, f_n(z;u;v)\}$ with the same domain of definition as $a(z;u;v)$, and such that

$$f = 0 \text{ if and only if } z = 0,\ v = 0, \tag{E.27}$$

$$\det J|_{z=0,v=0} \neq 0, \tag{E.28}$$

$$\operatorname{Im}\,(f_v^T J^{-1})\big|_{z=0,v=0} \neq 0, \tag{E.29}$$

where $f_v := \partial_v f$. We want to find a $g(z;u;v) = (g_1(z;u;v), \dots, g_n(z;u;v))$ such that

$$a(z;u;v) = a_0(u;v) + \langle f(z;u;v), g(z;u;v)\rangle, \tag{E.30}$$

where a_0 is independent of z. The latter of course implies

$$a_0(u;v) := a(0;u;v) - \langle f(0;u;v), g(0;u;v)\rangle. \tag{E.31}$$

Note that (E.27) is substantially different from (E.17) in that (E.27) has the additional condition $v = 0$. This means that v is a "nontrivial" parameter compared to the "trivial" parameters u.

Note also that the condition (E.29) guarantees that for small z and v

$$|f(z;u;v)| \geqslant c(u)\,(|z| + |v|) \tag{E.32}$$

with some $c(u) > 0$.

Set

$$g'' = \frac{(a - \langle f, g'\rangle) - (a - \langle f, g'\rangle)|_{z=0}}{f^T B \overline{f}}\, B\overline{f} \tag{E.33}$$

(cf. (E.24)). Then the construction (E.14), (E.19), (E.21)–(E.23), (E.33) gives the required $g(z;u;v)$. Only now everywhere we have the dependence on the additional parameters u, v. In particular, the cut-offs $\chi_k = \chi_k(z;u;v)$ appearing in (E.23) depend on the additional parameters u, v. The cut-offs satisfy the condition $0 \leqslant \chi_k \leqslant 1$ and are identically equal to 1 in some neighbourhoods of the set $\{z = 0, v = 0\}$. These cut-offs should (and can) be chosen in such a way that the series (E.23) converge uniformly over any compact set in the $(n+m+1)$-dimensional domain of definition of the function $a(z;u)$, as well as all their partial derivatives with respect to z, u, and v.

Our claim that the construction (E.14), (E.19), (E.21)–(E.23), (E.33) solves the problem (E.30) can be substantiated in the following way, similar to that in subsection 2.

It is sufficient to establish that

$$a(z;u;v) = a_0'(u;v) + \langle f(z;u;v), g'(z;u;v)\rangle + O((|z| + |v|)^\infty) \tag{E.34}$$

(cf. (E.25)), where

$$a_0'(u;v) := a(0;u;v) - \langle f(0;u;v), g'(0;u;v)\rangle \tag{E.35}$$

(cf. (E.31)). Then formulae (E.32)–(E.34) would imply that the column function $g'' = g''(z;u;v)$ is infinitely smooth, and moreover,

$$g'' = O((|z| + |v|)^\infty) \tag{E.36}$$

(cf. (E.16)). Combining (E.14), (E.33)–(E.35) we obtain (E.30), (E.31).

Let us prove (E.34). It is sufficient to prove (E.34) under the assumption that $a(z;u;v)$ and $f(z;u;v)$ are polynomials in z. In this case we can change the independent variable $z \to f$ and carry out our standard construction (E.2), (E.12) in this variable. Returning to our original variable z we get

$$a(z;u;v) = a(\widetilde{z}(u;v);u;v) + \langle f(z;u;v), \widetilde{g}'(z;u;v)\rangle + O(|z - \widetilde{z}(u;v)|^\infty), \tag{E.37}$$

$$\widetilde{g}' = \left(\sum_{k=1}^{\infty} c_k\, k^{-1}\, \widetilde{\chi}_k\, F_{k-1}\right) \partial_f a \tag{E.38}$$

(cf. (E.25), (E.23)), where $z = \widetilde{z}(u;v)$ is the (complex) z-solution of the equation $f(z;u;v) = 0$, and $\widetilde{\chi}_k = \widetilde{\chi}_k(z;u;v)$ are some cut-offs which are identically equal to 1 in some neighbourhoods of the (complex) set $\{z = \widetilde{z}(u;v)\}$. But the remainder $O(|z - \widetilde{z}(u;v)|^\infty)$ can be estimated as $O((|z| + |v)|)^\infty)$. As

$$\widetilde{\chi}_k - \chi_k = O((|z| + |v)|)^\infty),$$

we have

$$\widetilde{g}' - g' = O((|z| + |v)|)^\infty),$$

and this in turn implies that replacing $\widetilde{g}'$ by g' we change the right-hand side of (E.38) by $O((|z| + |v)|)^\infty)$. Thus, we can rewrite (E.37) as

$$a(z;u;v) = a(\widetilde{z}(u;v);u;v) + \langle f(z;u;v), g'(z;u;v)\rangle + O((|z| + |v)|)^\infty). \tag{E.39}$$

Setting $z = 0$ in (E.39) and comparing with (E.35) we get

$$a(\widetilde{z}(u;v);u;v) = a_0'(u;v) + O(|v|^\infty). \tag{E.40}$$

Formulae (E.39), (E.40) imply (E.34).

5. Invariance properties. The operators Z_k are invariant under linear changes of coordinates z, that is, under changes of the type $\widehat{z} = A\,z$ where A is a constant nonsingular $n \times n$-matrix. This fact is easily checked for $k = 0, 1$, and for $k \geqslant 2$ it is established by induction on k with the help of the left equality (E.3).

An immediate consequence is that the operators F_k are invariant under linear changes of the column function f, namely, under changes of the type $\widehat{f} = A\,f$ where A is a nonsingular $n \times n$-matrix function independent of z. This implies (see formula (E.23)) that under a linear change $\widehat{f} = A\,f$ the column function g' also changes linearly: $\widehat{g}' = (A^{-1})^T g'$.

Suppose now that the matrix function B appearing in the definition of g'' (see (E.15), (E.24) or (E.33)) changes according to the law $\widehat{B} = (A^{-1})^T \cdot B \cdot A^{-1}$. Then the whole column function $g = g' + g''$ changes according to the law $\widehat{g} = (A^{-1})^T g$.

In particular, let $m \geqslant n$, and let $(u_1, \ldots, u_n)$ be local coordinates on some manifold. If f is a vector with respect to these local coordinates, a and the χ_k are functions, and B is a covariant tensor, then our g will be a covector.

Moreover, suppose that z are also local coordinates on some manifold, and that f, a, χ_k, B behave as functions with respect to these local coordinates. Then our g behaves as a function with respect to z. This fact follows from the invariance of the operator ∂_f under changes of coordinates z.

6. Nonuniqueness of g. The n-component column function g in formulae (E.1), (E.20), (E.26), (E.30) is defined nonuniquely. One reason for this is that on the right-hand side of (E.3) we could have performed the factorization with respect to z in different ways, and another reason is that we can always add to g a nontrivial column function which is orthogonal to f and has an infinite order zero at $z = 0$ (in the case of formulae (E.1), (E.20), (E.26)) or at $z = 0$, $v = 0$ (in the case of formula (E.30)). However, $g|_{z=0}$ (in the case of formulae (E.1), (E.20), (E.26)) and $g|_{z=0,v=0}$ (in the case of formula (E.30)) are uniquely defined.

References

[Agm1] S. Agmon, *On kernels, eigenvalues, and eigenfunctions of operators related to elliptic problems*, Comm. Pure Appl. Math. **18** (1965), 627–663.

[Agm2] ———, *Asymptotic formulas with remainder estimates for eigenvalues of elliptic operators*, Arch. Rat. Mech. Anal. **28** (1967/68), no. 3, 165–183.

[AgmDoNir] S. Agmon, A. Douglis, and L. Nirenberg, *Estimates near the boundary for solutions of elliptic partial differential equations satisfying general boundary conditions*, I, II, Comm. Pure Appl. Math. **12** (1959), no. 4, 623–727; **17** (1964), no. 1, 35–92.

[AgrVi] M. S. Agranovich and M. I. Vishik, *Elliptic problems with a parameter and parabolic problems of general type*, Uspekhi Mat. Nauk **19** (1964), no. 3 (117), 53–161; English transl., Russian Math. Surveys **19** (1964), no. 3, 53–157.

[Ar1] V. I. Arnol′d, *On a characteristic class entering into conditions of quantization*, Funktsional. Anal. i Prilozhen. **1** (1967), no. 1, 1–14; English transl., Functional Anal. Appl. **1** (1967), 1–13.

[Ar2] ———, *Mathematical methods of classical mechanics*, "Nauka", Moscow, 1974; English transl., Springer-Verlag, New York, 1989.

[AsLid] A. G. Aslanyan and V. B. Lidskiĭ, *The distribution of eigenfrequencies of thin elastic shells*, "Nauka", Moscow, 1974. (Russian)

[AsLidVa] A. G. Aslanyan, V. B. Lidskiĭ, and D. Vassiliev, *Frequencies of free vibrations of a thin shell interacting with a liquid*, Funktsional. Anal. i Prilozhen. **15** (1981), no. 3, 1–9; English transl., Functional Anal. Appl. **15** (1981), 157–164.

[BaHilf] H. P. Baltes and E. R. Hilf, *Spectra of finite systems: a review of Weyl's problem*, Bibliographisches Institut, Zurich, 1976.

[Be] A. Besse, *Manifolds all of whose geodesics are closed*, Springer-Verlag, New York, 1978.

[Big] N. L. Biggs, *Discrete mathematics*, Clarendon Press, Oxford, 1990.

[BirSo] M. Birman and M. Solomyak, *Spectral theory of self-adjoint operators in Hilbert space*, Leningrad Univ., Leningrad, 1980; English transl., Reidel, Dordrecht, 1987.

[BdM] L. Boutet de Monvel, *Boundary problems for pseudodifferential operators*, Acta Math. **126** (1971), 11–51.

[Br] J. Brüning, *Zur Abschätzung der Spektralfunktion elliptischer Operatoren*, Math. Z. **137** (1974), 75–85.

[CodLson] E. A. Coddington and N. Levinson, *Theory of ordinary differential equations*, Krieger, Malabar, FL, 1984.

[CorFomSin] I. P. Cornfeld, S. V. Fomin, and Ya. G. Sinai, *Ergodic theory*, "Nauka", Moscow, 1980; English transl., Springer-Verlag, New York, 1982.

[Cou] R. Courant, *Über die Schwingungen eigenspannter Platten*, Math. Z. **15** (1922), 195–200.

[CouHilb] R. Courant and D. Hilbert, *Methods of mathematical physics*, vol. 1, Wiley, New York, 1989.

[De] P. Debye, *Zur Theorie der spezifischen Wärmen*, Ann. Phys. **39,** H. 4 (1912), no. 14, 789–839.

[DuiGui] J. J. Duistermaat and V. W. Guillemin, *The spectrum of positive elliptic operators and periodic bicharacteristics*, Invent. Math. **25** (1975), 39–79.

[DuiGuiHö] J. J. Duistermaat, V. W. Guillemin, and L. Hörmander, *Fourier integral operators: selected classical articles* (J. Brüning and V. W. Guillemin, eds.), Springer-Verlag, New York, 1994.

[DupMazOn] M. Dupuis, R. Mazo, and L. Onsager, *Surface specific heat of an isotropic solid at low temperatures*, J. Chem. Phys. **33** (1960), no. 5, 1452–1461.

[FlLtinVa1] J. Fleckinger, M. Levitin, and D. Vassiliev, *Heat equation on the triadic von Koch snowflake: asymptotic and numerical analysis*, Proc. London Math. Soc. **71** (1995), 372–396.

[FlLtinVa2] ———, *Heat content of the triadic von Koch snowflake*, Intern. J. Appl. Science and Computations **2** (1995), no. 2, 289–305.

[FlVa1] J. Fleckinger and D. Vassiliev, *Tambour fractal: exemple d'une formule asymptotique a deux termes pour la "fonction de comptage"*, C. R. Acad. Sci. Paris Série I **311** (1990), 867–872.

[FlVa2] ———, *An example of a two-term asymptotic formula for the "counting function" of a fractal drum*, Trans. Amer. Math. Soc. **337** (1993), no. 1, 99–116.

[Ga] T. H. Ganelius, *Tauberian remainder theorems*, Lecture Notes in Math., vol. 232, Springer-Verlag, Berlin and New York, 1971.

[GnKo] B. V. Gnedenko and A. N. Kolmogorov, *Limit distributions for sums of independent random variables*, "Nauka", Moscow, 1949; English transl., Addison–Wesley, Reading, MA, 1968.

[GolLidTo] A. L. Gol′denveizer, V. B. Lidskiĭ, and P. E. Tovstik, *Free vibrations of thin elastic shells*, "Nauka", Moscow, 1979. (Russian)

[GolVa] A. L. Gol′denveizer and D. Vassiliev, *Distribution of free vibration frequencies in two- and three-dimensional elastic bodies*, Mekhanika i nauchno-tekhnicheskii progress, vol. 3, Mekhanika deformiruemogo tverdogo tela, "Nauka", Moscow, 1988, pp. 223–236; English transl., Mechanical Engineering and Applied Mechanics, vol. 3, Mechanics of Deformable Solids (N. Kh. Arutiunian, I. F. Obraztsov, and V. Z. Parton, eds.), Hemisphere Publ., New York, 1991, pp. 227–242.

[Gon] V. S. Gontkevich, *Natural oscillations of plates and shells*, "Naukova Dumka", Kiev, 1964; German transl., VEB Fachbuchverlag, Leipzig, 1967.

[Gui] V. W. Guillemin, *The Radon Transform on Zoll Surfaces*, Adv. Math. **22** (1976), 85–119.

[GuiSt] V. W. Guillemin and S. Sternberg, *Geometric asymptotics*, Amer. Math. Soc., Providence, RI, 1977.

[Gur] T. E. Gureev, *Asymptotics of the spectrum of a biharmonic operator on a hemisphere*, Vestnik Leningrad. Univ. Math. Mech. Astronom. (Ser. 1) **1988**, no. 3, 94–95; English transl., Vestnik Leningrad. Univ. Math. **21** (1988), no. 3, 59–61.

[Hö1] L. Hörmander, *The spectral function of an elliptic operator*, Acta Math. **121** (1968), 193–218.

[Hö2] ———, *Fourier integral operators* I, Acta Math. **127** (1971), 79–183.

[Hö3] ———, *The analysis of linear partial differential operators*, Vol. 1–4, Springer-Verlag, Berlin, 1983, 1985.

[Iv1] V. Ya. Ivrii, *On the second term of the spectral asymptotics for the Laplace-Beltrami operator on manifolds with boundary*, Funktsional. Anal. i Prilozhen. **14** (1980), no. 2, 25–34; English transl., Functional Anal. Appl. **14** (1980), 98–106.

[Iv2] ———, *Precise spectral asymptotics for elliptic operators acting in fiberings over manifolds with boundary*, Lecture Notes in Math., vol. 1100, Springer-Verlag, New York, 1984.

[Iv3] ———, *Semiclassical microlocal analysis and precise spectral asymptotics*, 9 preprints, Centre de Mathématiques, École Polytechnique, Palaiseau, 1990–1992.

[Iv4] ———, *Semiclassical microlocal analysis and spectral asymptotics*, Springer-Verlag, New York (to appear).

[Ja] M. Jammer, *The conceptual development of quantum mechanics*, The History of Modern Physics 1800–1950, vol. 12, American Institute of Physics, New York, 1989.

[Je] J. H. Jeans, *On the partition of energy between matter and ether*, Philosophical Magazine, 6th series **10** (1905), 91–98.

[LanLif] L. D. Landau and E. M. Lifshits, *Theory of elasticity*, Course of theoretical physics, vol. 7, "Nauka", Moscow, 1965; English transl., Pergamon Press, Oxford, 1986.

[LapSaVa] A. Laptev, Yu. Safarov, and D. Vassiliev, *On global representation of Lagrangian distributions and solutions of hyperbolic equations*, Comm. Pure Appl. Math. **47** (1994), no. 11, 1411–1456.

[Ltan] B. M. Levitan, *On the asymptotic behavior of the spectral function of a self-adjoint differential second order equation*, Izv. Akad. Nauk SSSR Ser. Mat. **16** (1952), 325–352.

[LtanSar] B. M. Levitan and I. S. Sargsjan, *Introduction to spectral theory: selfadjoint ordinary differential operators*, "Nauka", Moscow, 1970; English transl., Amer. Math. Soc., Providence, RI, 1991.

[LtinSu] M. Levitin and A. G. Sudakov, *Eigenvalues of a problem on the semi-axis with constant coefficients*, Mathematical Methods of Control and Data-Processing, Moscow Institue of Physics and Technology Publishing House, Moscow, 1985, pp. 48–53. (Russian)

[LtinVa1] M. Levitin and D. Vassiliev, *Some examples of two-term spectral asymptotics for sets with fractal boundary*, Partial differential operators and mathematical physics, Proc. Conf. on Partial Differential Equations (Holzhau, July 3–9, 1994) (M. Demuth and B.-W. Schulze, eds.), Operator Theory: Advances and Applications, vol. 78, Birkhäuser, Basel, Boston, and Berlin, 1995, pp. 227–233.

[LtinVa2] ———, *Spectral asymptotics, renewal theorem, and the Berry conjecture for a class of fractals*, Proc. London Math. Soc. **72** (1996), 188–214.

[LtinVa3] ———, *Vibrations of shells contacting fluid: asymptotic analysis*, Acoustic Interaction with Submerged Elastic Structures. Part I: Acoustic Scattering and Resonances, World Scientific, Singapore (to appear).

[LioMag] J.-L. Lions and E. Magenes, *Non-homogeneous boundary value problems and their applications*, vol. 1, Springer-Verlag, Berlin, Heidelberg, and New York, 1972.

[Me] R. Melrose, *Weyl's conjecture for manifolds with concave boundary*, Proc. Sympos. Pure Math., vol. 36, Amer. Math. Soc., Providence, RI, 1980, pp. 257–274.

[Nio] F. I. Niordson, *The spectrum of free vibrations of a thin elastic spherical shell*, Intern. J. Solids and Structures **24** (1988), no. 9, 947–961.

[Pl] Å. Pleijel, *On a theorem by P. Malliavin*, Israel J. Math. **1** (1963), no. 3, 166–168.

[Ra] Lord Rayleigh, *The dynamical theory of gases and of radiation*, Nature **72** (1905), 54–55, 243–244.

[ReSim] M. Reed and B. Simon, *Methods of modern mathematical physics*, vol. 1, Academic Press, New York, 1972; vol. 2, 1975; vol. 3, 1979; vol. 4, 1978.

[RoShSo] G. V. Rozenblum, M. A. Shubin, and M. Z. Solomyak, *Spectral theory of differential operators*, Current problems in mathematics: fundamental directions, Partial Differential Equations VII, vol. 64, VINITI, Moscow, 1989; English transl., Encyclopaedia of Mathematical Sciences, vol. 64, Springer-Verlag, New York, 1994.

[Sa1] Yu. Safarov, *On the asymptotics of the eigenvalues of diffraction problems*, Dokl. AN SSSR **281** (1985), no. 5, 1058–1061; English transl., Soviet Math. Dokl. **31** (1985), no. 2, 392–395.

[Sa2] ———, *Asymptotics of the spectrum of a pseudodifferential operator with periodic bicharacteristics*, Zap. Nauchn. Sem. Leningrad. Otd. Mat. Inst. Steklov. (LOMI) **152** (1986), 94–104; English transl. in J. Soviet Math. **40** (1988), no. 5.

[Sa3] ———, *On the second term of the spectral asymptotics of the transmission problem*, Acta Appl. Math. **10** (1987), no. 2, 101–130.

[Sa4] ———, *Asymptotics of the spectrum of a boundary value problem with periodic billiard trajectories*, Funktsional. Anal. i Prilozhen. **21** (1987), no. 3, 88–90; English transl., Functional Anal. Appl. **21** (1987), 337–339.

[Sa5] ———, *Exact asymptotics of the spectrum of a boundary value problem, and periodic billiards*, Izv. Akad. Nauk. SSSR Ser. Mat. **52** (1988), 1230–1251; English transl., Math. USSR-Izv. **33** (1989), 553–573.

[Sa6] ———, *Asymptotics of the spectral function of a positive elliptic operator without the non-trapping condition*, Funktsional. Anal. i Prilozhen. **22** (1988), no. 3, 53–65; English transl., Functional Anal. Appl. **22** (1988), 213–223.

[Sa7] ———, *Nonclassical two-term spectral asymptotics*, Symposium "Partial Differential Equations" (Holzhau, 1988), Teubner-Texte Math., vol. 112, Teubner, Leipzig, 1989, pp. 250–258.

[Sa8] ———, *Precise spectral asymptotics and inverse problems*, Integral Equation and Inverse Problems (V. Petkov and R. Lazarov, eds.), Longman, New York, 1991, pp. 239–240.

[Sa9] ———, *Pseudodifferential operators and linear connections*, Proc. London Math. Soc. (to appear).

[SaVa1] Yu. Safarov and D. Vassiliev, *Branching Hamiltonian billiards*, Dokl. AN SSSR **301** (1988), 271–275; English transl., Soviet Math. Dokl. **38** (1989), 64–68.

[SaVa2] ———, *The asymptotic distribution of eigenvalues of differential operators*, Spectral theory of operators (S. Gindikin, ed.), Amer. Math. Soc. Transl., Ser. 2, vol. 150, Amer. Math. Soc., Providence, RI, 1992, pp. 55–111.

[S.-HuS.-Pa] J. Sanchez-Hubert and E. Sanchez-Palencia, *Vibration and coupling of continuous systems: asymptotic methods*, Springer-Verlag, Berlin, 1989.

[S.-PaVa] E. Sanchez-Palencia and D. Vassiliev, *Remarks on the vibration of thin elastic shells and their numerical computation*, C. R. Ac. Sci. Paris Série II **314** (1992), 445–452.

[Se1] R. Seeley, *A sharp asymptotic remainder estimate for the eigenvalues of the Laplacian in a domain of* $\mathbb{R}^3$, Adv. Math. **29** (1978), 244–269.

[Se2] ———, *An estimate near the boundary for the spectral function of the Laplace operator*, Amer. J. Math. **102** (1980), 869–902.

[Sh] M. A. Shubin, *Pseudodifferential operators and spectral theory*, "Nauka", Moscow, 1978; English transl., Springer-Verlag, Berlin, 1986.

[Sz.-NaFoi] B. Sz.-Nagy and C. Foias, *Analyse harmonique des opérateurs de l'espace de Hilbert*, Akadémiai Kiadó, Budapest, 1967.

[Tr] F. Trèves, *Introduction to pseudodifferential and Fourier integral operators*, Plenum Press, New York, 1982.

[Va1] D. Vassiliev, *Asymptotics of the distribution function of the spectrum of pseudodifferential operators with parameters*, Funktsional. Anal. i Prilozhen. **14** (1980), no. 3, 65–66; English transl., Functional Anal. Appl. **14** (1980), 217–219.

[Va2] ———, *The distribution of eigenfrequencies of a thin elastic shell interacting with fluid*, Ph.D. Thesis, Moscow Institute of Physics and Technology, 1981. (Russian)

[Va3] ———, *Binomial asymptotics of the spectrum of a boundary value problem*, Funktsional. Anal. i Prilozhen. **17** (1983), no. 4, 79–81; English transl., Functional Anal. Appl. **17** (1983), 309–311.

[Va4] ———, *Two-term asymptotics of the spectrum of a boundary value problem under an interior reflection of general form*, Funktsional. Anal. i Prilozhen. **18** (1984), no. 4, 1–13; English transl., Functional Anal. Appl. **18** (1984), 267–277.

[Va5] ———, *Asymptotics of the spectrum of pseudodifferential operators with small parameters*, Mat. Sb. **121** (1983), no. 1, 60–71; English transl., Math. USSR-Sb. **49** (1984), no. 1, 61–72.

[Va6] ———, *Two-term asymptotics of the spectrum of a boundary value problem in the case of a piecewise smooth boundary*, Dokl. AN SSSR **286** (1986), no. 5, 1043–1046; English transl., Soviet Math. Dokl. **33** (1986), no. 1, 227–230.

[Va7] ———, *Asymptotics of the spectrum of a boundary value problem*, Trudy Moskov. Mat. Obshch. **49** (1986), 167–237; English transl., Trans. Moscow Math. Soc. **1987**, 173–245.

[Va8] ———, *Resonance phenomena in elasticity and hydroelasticity*, Doctor of Sciences Thesis, Institute for Problems in Mechanics, USSR Academy of Sciences, Moscow, 1988. (Russian)

[Va9] ———, *Two-term asymptotics of the spectrum of natural frequencies of a thin elastic shell*, Dokl. AN SSSR **310** (1990), no. 4, 777–780; English transl., Soviet Math. Dokl. **41** (1990), 108–112.

[Va10] ———, *One can hear the dimension of a connected fractal in* $\mathbb{R}^2$, Integral Equations and Inverse Problems (V. Petkov and R. Lazarov, eds.), Longman, New York, 1991, pp. 270–273.

[Va11] ———, *The distribution of eigenvalues of partial differential operators*, Équations aux Dérivées Partielles, Séminaire 1991–1992, École Polytechnique, Palaiseau, 1992, pp. XVII-1–XVII-17.

[Va12] ———, *Characteristic properties of distributions associated with the wave group*, Journees Équations aux Dérivées Partielles, Saint-Jean-De-Monts, École Polytechnique, Palaiseau (France), 1992, pp. XV-1–XV-14.

[Va13] ______, *Construction of the wave group for higher order partial differential operators*, Équations aux Dérivées Partielles, Séminaire 1994–1995, École Polytechnique, Palaiseau (France), 1995, pp. IX-1–IX-14.

[Va14] ______, *The symbol of a Lagrangian distribution*, Journees Équations aux Dérivées Partielles, Saint-Jean-De-Monts, École Polytechnique, Palaiseau (France), 1995, pp. V.1–V.12.

[We1] H. Weyl, *Über die Abhängigkeit der Eigenschwingungen einer Membran von der Begrenzung*, J. Reine Angew. Math. **141** (1912), 1–11.

[We2] ______, *Über die Randwertaufgabe der Strahlungstheorie und asymptotische Spektralgesetze*, J. Reine Angew. Math. **143** (1913), no. 3, 177–202.

[We3] ______, *Das asymptotische Verteilungsgesetz der Eigenschwingungen eines beliebig gestalteten elastischen Körpers*, Rend. Circ. Math. Palermo **39** (1915), 1–50.

[Wo] M. Wojtkowski, *Principles for the design of billiards with nonvanishing Lyapunov exponents*, Comm. Math. Phys. **105** (1986), no. 3, 391–414.

[Ya] D. R. Yafaev, *Mathematical scattering theory: general theory*, Amer. Math. Soc., Providence, RI, 1992.

[Zo] O. Zoll, *Über Flächen mit Scharen geschlossener geodätischer Linien*, Math. Ann. **57** (1903), 108–133.

Principal Notation

$\dot{}$ (dot)	denotes differentiation with respect to time t
$\hat{}$ (hat)	denotes the Fourier transform
$\widehat{}$ (wide hat)	stands for the extension from (of) our original manifold
$\simeq$	equality of formal Taylor expansions
$\lesssim$	less than or of the order of
$\{\tau\}$	$\in [0,1)$ fractional part of the real number τ
$\{\tau\}_{2\pi}$	$\in [-\pi,\pi)$ residue modulo 2π of the real number τ
$(\tau)_+$	$=\tau$ if $\tau \geqslant 0$, and $=0$ if $\tau < 0$
$\int$	absence of limits of integration implies integration over the whole space
$(M * N)(\lambda)$	$=\int M(\lambda-\mu)N(\mu)\,d\mu$ convolution
$\{f,g\}$	$=\langle f_\xi, g_x\rangle - \langle f_x, g_\xi\rangle$ Poisson brackets
∇	gradient
$\#$	the number of elements in a finite set

$\mathbf{A}$, $\mathbf{A}^+$	1.6.3
$\mathbf{A}(z')$, $\mathbf{A}^+(z')$	5.4.2
$\mathcal{A}$	self-adjoint operator generated by the boundary value problem, 1.1.7
$\overline{\mathcal{A}}$	$=(\mathcal{A}^*)^T$
C_φ	2.1.1, 2.1.2 and Section 2.2
$C^\infty_{\bar{B}}$, C^∞_B, $C^\infty_{B_x}$	3.1.1
D_{x_k}	$=-i\partial/\partial x_k$
$\mathcal{D}$	$=C_0^\infty$, Schwartz space of test functions
$\mathcal{D}'$	Schwartz space of distributions (dual of $\mathcal{D}$)
$\dot{\mathcal{D}}'$, $\dot{\mathcal{D}}'_{\bar{B}}$	3.1.1
E_λ	spectral projection of the operator $\mathcal{A}^{1/(2m)}$, 1.8.1
$\mathbf{E}_\nu$, $\mathbf{E}_\nu^+$	spectral projections of the operators $\mathbf{A}$, $\mathbf{A}^+$, 1.6.3
$\mathbf{E}_\nu(z')$, $\mathbf{E}_\nu^+(z')$	spectral projections of the operators $\mathbf{A}(z')$, $\mathbf{A}^+(z')$, 5.4.2
$\mathcal{E}$	$=C^\infty$
$\mathcal{E}'$	space of distributions with compact support (dual of $\mathcal{E}$)
F_+	class of monotone functions, Section B.1
$\mathcal{F}_{\lambda\to t}[f(\lambda)]$	$=\hat{f}(t)=\int e^{-it\lambda}f(\lambda)\,d\lambda$ one-dimensional Fourier transform
$\mathcal{F}^{-1}_{t\to\lambda}[\hat{f}(t)]$	$=f(\lambda)=(2\pi)^{-1}\int e^{it\lambda}\hat{f}(t)\,dt$ inverse one-dimensional Fourier transform
H^s	Sobolev space W_2^s
$\mathcal{I}_{\varphi,a}$	oscillatory integral with phase function φ and amplitude a, 2.1.1, 2.7.1
J_y	1.8.4

Meas	Riemannian $(n-1)$-dimensional volume of the boundary
$\overset{\circ}{M}$	interior of the manifold M
$N(\lambda)$	counting function of the boundary value problem, 1.2.1
$\mathbf{N}^+(\nu)$	counting function of the auxiliary one-dimensional problem, 1.6.3
$\mathbb{N}$	$=\{1,2,\dots\}$ the set of natural numbers
O	(often with indices) various open conic subsets of $T'\overset{\circ}{M}$
O_T	set of T-admissible points, 1.3.3
O_∞	set of admissible points, 1.3.3
$\mathcal{O}$	(often with indices) various open conic subsets of $(T_-,T_+)\times M\times O$
$\mathbf{O}$	conically compact conic subset of O, 2.7.1
$\mathbf{O}^\pm(d)$	2.10.2
P_1	principal symbol of a (pseudo)differential operator P of order 1
P_{sub}	subprincipal symbol of a (pseudo)differential operator P
$\mathcal{P}^{\mathrm{d}}$	1.3.2, 1.3.3
$\mathcal{P}^{\mathrm{g}}$	1.3.2, 1.3.3
$\mathbf{Q}(\lambda)$	1.7.2
$\mathbf{Q}(y,\lambda)$	1.8.4
$\mathbf{Q}_P(y,\lambda)$	1.8.4
$R(\nu)$	reflection matrix, Section 1.4
$\mathbf{R}_\mu$, $\mathbf{R}_\mu^+$	resolvents of the operators $\mathbf{A}$, $\mathbf{A}^+$, 1.6.3
$\mathbf{R}_\mu(z')$, $\mathbf{R}_\mu^+(z')$	resolvents of the operators $\mathbf{A}(z')$, $\mathbf{A}^+(z')$, 5.4.2
$\mathbb{R}^n$	n-dimensional Euclidean space
$\mathbb{R}^n_+$	$=\{x\in\mathbb{R}^n : x_n\geqslant 0\}$
$S(\nu)$	scattering matrix, 1.6.3
S^l	class of amplitudes, 2.1.1
$S^{-\infty}$	$=\bigcap_{l\in\mathbb{R}} S^l$, or $\bigcap_{l\in\mathbb{R}} S^l(1;1,\varepsilon_1)$ in 2.10.6
$S^l(0,0;1,\varepsilon_1)^\pm$	class of parameter-dependent amplitudes, 2.10.2
$S^{-\infty,\pm}$	$=\bigcap_{l\in\mathbb{R}} S^l(0,0;1,\varepsilon_1)^\pm$
$S^l(1;1,\varepsilon_1)$	class of parameter-dependent amplitudes, 2.10.6
S^*M	cosphere bundle, 1.1.10
$\mathcal{S}(\mathbb{R})$	Schwartz space of rapidly decreasing test functions
$\mathcal{S}'(\mathbb{R})$	Schwartz space of tempered distributions (dual of $\mathcal{S}(\mathbb{R})$)
$\mathbb{S}^k$	k-dimensional sphere (in $\mathbb{R}^{k+1}$)
$T^\pm_\pm(d)$	2.10.2
T_δ	3.4.1
$\mathbf{T}$	period function, 1.7.2
$\mathbf{T}_y$	1.8.4
$T'M$, $T'\overset{\circ}{M}$	cotangent bundles T^*M, $T^*\overset{\circ}{M}$ with the zero section $\{\xi=0\}$ excluded
$T'\partial M$	cotangent bundle $T^*\partial M$ with the zero section $\{\xi'=0\}$ excluded
Tr	trace of an operator
U_y, $U_{y,\lambda}$	1.8.4
$\mathbf{U}(t)$	wave group, Introduction to Chapter 3
$\mathbf{U}_P(t)$	$=\mathbf{U}(t)P$
$\mathbf{U}_{P,Q}(t)$	$=Q^*\mathbf{U}(t)P$
Vol	Riemannian n-dimensional volume
WF	wave front set, 2.1.3
$\arg_0$	a fixed branch of the argument

c_0, c_1	Weyl coefficients for the counting function
$c_0(y)$	first Weyl coefficient for $e(\lambda, y, y)$, 1.8.3
$c_1(x')$	5.4.1
$c_{0;P}(y)$	$= c_{0;P,P}(y)$
$c_{0;P,Q}(y)$	1.8.3
$c_{1;P}(y)$	1.8.3
cone hull	conic hull, Section 2.2
cone supp	conic support, 2.1.1
d	parameter characterizing distance to ∂M, Introduction to Section 2.10
d_φ	Section 2.2 and 2.7.1
d^∞	2.10.5–2.10.7
div	divergence
$đ\xi$	$= (2\pi)^{-n}\, d\xi$
$đ\xi'$	$= (2\pi)^{1-n}\, d\xi'$
$e(\lambda, x, y)$	spectral function (integral kernel of E_λ), 1.8.1
$e_{P,Q}(\lambda, x, y)$	integral kernel of $Q^* E_\lambda P$, 1.8.1
$\mathbf{e}$, $\mathbf{e}^+$	integral kernels of the operators $\mathbf{E}$, $\mathbf{E}^+$, 1.6.3
$\mathbf{f}^+(\mu)$	regularized trace of $\mathbf{R}_\mu^+$, 1.6.3
grad	gradient
$\mathbf{k}(x', \xi')$	Hamiltonian curvature, 1.3.2
λ_k	eigenvalues of the operator $\mathcal{A}^{1/(2m)}$
meas	canonical measure on the cosphere bundle S^*M, 1.1.10
meas_x	canonical measure on the unit cosphere S_x^*M, 1.1.10
$\text{meas}_{\partial M}$	measure on $S^*M\|_{\partial M}$, Introduction to Appendix D
$\mathbf{q}$	1.7.2
$\mathbf{q}_y$	1.8.4
$\mathbf{r}_\nu$, $\mathbf{r}_\nu^+$	integral kernels of the operators $\mathbf{R}$, $\mathbf{R}^+$, 1.6.3
shift^+	spectral shift of the auxiliary one-dimensional problem, 1.6.1
sing supp	singular support of a distribution, 2.1.1
$t_\star$	4.1.1
$\widetilde{t}_\star$	4.1.2
t_k^*	moments of reflection
tr	trace of a matrix
$\mathbf{u}(t, x, y)$	Schwartz kernel of the operator $\mathbf{U}(t)$
$\mathbf{u}_P(t, x, y)$	Schwartz kernel of the operator $\mathbf{U}_P(t)$
$\mathbf{u}_{P,Q}(t, x, y)$	Schwartz kernel of the operator $\mathbf{U}_{P,Q}(t)$
v_k	eigenfunctions (half-densities) of the operator $\mathcal{A}^{1/(2m)}$
vol	symplectic volume on T^*M, 1.1.10
vol′	symplectic volume on $T^*\partial M$, 1.1.10
(x^*, ξ^*)	Hamiltonian or billiard trajectory, 1.3.1–1.3.3
x_η^*	matrix with elements $(x_j^*)_{\eta_i}$ (j being the number of the row and i that of the column)
$\mathfrak{C}_i$	2.4.1
$\mathfrak{C}'_i$	2.4.1
$\mathfrak{C}^{(l)}$	3.4.3
$\mathfrak{F}^0$	class of phase functions corresponding to pseudodifferential operators, 2.7.5
$\mathfrak{F}_i$	class of standard phase functions, 2.4.1

$\mathfrak{F}'_i$	class of boundary phase functions, Section 2.5
$\mathfrak{F}_i^{\mathrm{bl}}$	class of boundary layer phase functions, 2.6.4
$\mathfrak{L}_{-r}$	2.7.3
$\mathfrak{N}(O)$	set of types of billiard trajectories originating from O, 3.4.2
$\mathfrak{D}_i$	2.4.1
$\mathfrak{D}^{(l)}$	3.4.3
$\hat{\mathfrak{D}}_j$	2.7.2
$\check{\mathfrak{D}}_j$	1.5.1
$\mathfrak{S}$, $\mathfrak{S}_{-r}$	2.7.3
$\mathfrak{S}'$, $\mathfrak{S}'_{-r}$	Section 2.9
$\mathfrak{T}_i(y,\eta)$	2.3.2
$\mathfrak{a}$, $\mathfrak{b}$, $\mathfrak{c}$	(full) symbols of Lagrangian distributions
$\mathfrak{a}_k$, $\mathfrak{b}_k$, $\mathfrak{c}_k$	homogeneous terms of (full) symbols
$\mathfrak{f}$	total phase shift, 1.7.2
$\mathfrak{f}_c$	phase shift generated by caustics, 1.7.2
$\mathfrak{f}_r$	phase shift generated by reflections, 1.7.2
$\mathfrak{f}_s$	phase shift generated by the subprincipal symbol, 1.7.2
$\mathfrak{m}$	$=\mathfrak{m}_1\vert\dots\vert\mathfrak{m}_{\mathbf{r}}$, 1.3.3
Δ	Laplace operator, Example 1.2.4
Π_T	1.3.1–1.3.3
Π	$=\bigcup_{T>0}\Pi_T$
Π_T^a	1.3.1–1.3.3
Π^a	$=\bigcup_{T>0}\Pi_T^a$
$\Pi_{y,T}$	1.8.2
Π_y	$=\bigcup_{T>0}\Pi_{y,T}$
$\Pi_{y,T}^a$	1.8.2
$\Pi_{y,T,i}$	4.1.2
Π_y^a	$=\bigcup_{T>0}\Pi_{y,T}^a$
$\Pi_{y,T,i}^a$	4.1.2
Φ_y	1.8.4
Φ_{xx}, $\Phi_{\eta\eta}$, $\Phi_{x\eta}$	2.4.1
Φ^t	Hamiltonian or billiard flow (see 1.3.1 and 1.3.2)
Ψ^l, Ψ_0^l	classes of pseudodifferential operators, 2.1.3
α_Γ	Maslov index of the trajectory Γ, 1.5.1
δ_l^k	Kronecker symbol
$\zeta_j^{\mp}$	Section 1.4
ν^{st}	threshold, Section 1.4
ξ_η^*	matrix with elements $(\xi_j^*)_{\eta_i}$ (j being the number of the row and i that of the column)
$\vert\xi\vert_x$	length of the covector $\xi\in T^*M$ (when M is a Riemannian manifold)
$\hat\rho$	function of the class $C_0^\infty(\mathbb{R})$, Introduction to Chapter 4
ρ_T, $\hat\rho_T$	Introductions to Section 4.2 and Chapter 5
ς	cut-off function in the oscillatory integral, Section 2.2
χ	(often with indices) various cut-off functions
$\chi_\pm$, χ_0	special cut-off functions on $\mathbb{R}_+$, 5.1.2
ω_n	$=\pi^{n/2}/\Gamma(1+n/2)$ the volume of the unit ball in $\mathbb{R}^n$

Index

Absolutely periodic point, 1.3.1
Admissible points and sets, 1.3.3
Amplitude, 2.1.1 and Section 2.2
  conic support of, 2.1.1

Billiard flow, 1.3.2
Billiard trajectories, 1.3.2
  absolutely periodic, 1.3.1–1.3.3
  admissible, 1.3.3
  branching of, 1.3.3
  dead-end, 1.3.2–1.3.3
  grazing, 1.3.2–1.3.3
  periodic, 1.3.1–1.3.3
  type of, 1.3.3
Boundary value problem
  ellipticity, 1.1.4
  positiveness, 1.1.6
  self-adjointness, 1.1.5

Canonical differential forms, 1.1.10
Canonical transformations, 2.3.1
Caustic set, 1.5.1
Clusters of eigenvalues, 1.7.1
Conic hull, Section 2.2
Conjugate point, 1.5.1
Convexity and concavity, 1.3.2
Coordinates
  boundary, 1.1.2
  "normal", 1.1.2
Cosphere bundle, 1.1.10
Counting function, 1.2.1

Density, 1.15

Eikonal equation, 2.4.2
Ergodicity, 1.3.4
Euclidean billiards, 1.3.5

Focal point, 1.8.2

Half-density, 1.1.5
Hamiltonian, 1.1.10
Hamiltonian curvature, 1.3.2
Hamiltonian flow, 1.3.1

Geodesic billiards, 1.3.2
Geodesic flow, 1.3.1

Lagrangian distribution
  boundary, Section 2.9
  boundary layer, Section 2.8
  standard, 2.7.2
Laplacian on half-densities, 1.2.1
Liouville's formula, 2.7.4

Maslov index, Section 1.5

Nonblocking condition, 1.3.3
Nonperiodicity condition, 1.3.1–1.3.3

Oscillatory integral
  boundary, Section 2.9
  boundary layer, Section 2.8
  global, Section 2.2
  local time-dependent, 2.1.2
  local time-independent, 2.1.1
  order of, 2.1.1
  standard, 2.7.1

Periodic point, 1.3.1
Phase functions, 2.1.1 and Section 2.2
  boundary, Section 2.5
  boundary layer, 2.6.4
  matching sets of, 2.6.2, 2.6.4
  nondegenerate, Section 2.2
  simple, Section 2.2
  standard, Section 2.5
Phase shift, Section 1.4 and 1.7.2
Pleijel's formula, 5.4.2
Polarization formula, 1.8.1
Positively homogeneous operator, 2.7.3
Principal symbol of a Lagrangian distribution
  boundary, Section 2.9
  boundary layer, Section 2.8
  standard, 2.7.4
Principal symbol of a (pseudo)differential operator, 1.1.3, 2.1.3
Pseudodifferential operator, 2.1.3
  conic support of, 2.1.3
  dual symbol of, 2.1.3
  order of, 2.1.3
  symbol of, 2.1.3

Pseudolocality, 2.1.3

Ray, 1.3.1
Reflection law, 1.3.2, 1.3.3
Reflection matrix, Section 1.4
Regular point, 1.8.2

Scattering matrix, 1.6.3
Simple reflection condition, 1.3.4 and Section 1.4
  strong, 1.3.4 and Section 1.4
Singular support, 2.1.1
Spectral function, 1.8.1
Spectral parameter, 1.1.1
Spectral projection, 1.8.1
Spectral shift, 1.6.3
Subprincipal symbol, 2.1.3
Symbol of a Lagrangian distribution
  boundary, Section 2.9
  boundary layer, Section 2.8
  standard 2.7.3
Symplectic volume, 1.1.10

Threshold, Section 1.4
  normal, Section 1.4
  rigid, 1.6.3
  soft, 1.6.3
Tubular neighbourhood, 1.1.2

Unitary dilation, 1.8.5

Wave front set, 2.1.4
Wave group, introduction to Chapter 3

Zoll surface, 1.3.1